BIOSTATISTICS:
Statistics in Biomedical, Public Health and Environmental Sciences

The Bernard G. Greenberg Volume

BIOSTATISTICS: Statistics in Biomedical, Public Health and Environmental Sciences

The Bernard G. Greenberg Volume

edited by

Pranab K. SEN

Department of Biostatistics
University of North Carolina
Chapel Hill, NC
U.S.A.

1985

NORTH-HOLLAND
AMSTERDAM · NEW YORK · OXFORD

ISBN: 0 444 87694 4

Published by:

ELSEVIER SCIENCE PUBLISHERS B.V.
P.O. Box 1991
1000 BZ Amsterdam
The Netherlands

Sole distributors for the U.S.A. and Canada:

ELSEVIER SCIENCE PUBLISHING COMPANY, INC.
52 Vanderbilt Avenue
New York, N.Y. 10017
U.S.A.

Library of Congress Cataloging in Publication Data

Main entry under title:

Biostatistics : statistics in biomedical, public health, and environmental sciences.

Commemorative monograph in honor of Bernard G. Greenberg's 65th birthday.
"Publications of B.G. Greenberg": p.
1. Medical statistics--Addresses, essays, lectures. 2. Biometry--Addresses, essays, lectures. 3. Greenberg, Bernard G.--Addresses, essays, lectures. I. Sen, Pranab Kumar, 1937- . II. Greenberg, Bernard G. [DNLM: 1. Biometry. WA 900 B616]
RA409.B49 1985 574'.072 84-28729
ISBN 0-444-87694-4 (U.S.)

Printed in The Netherlands

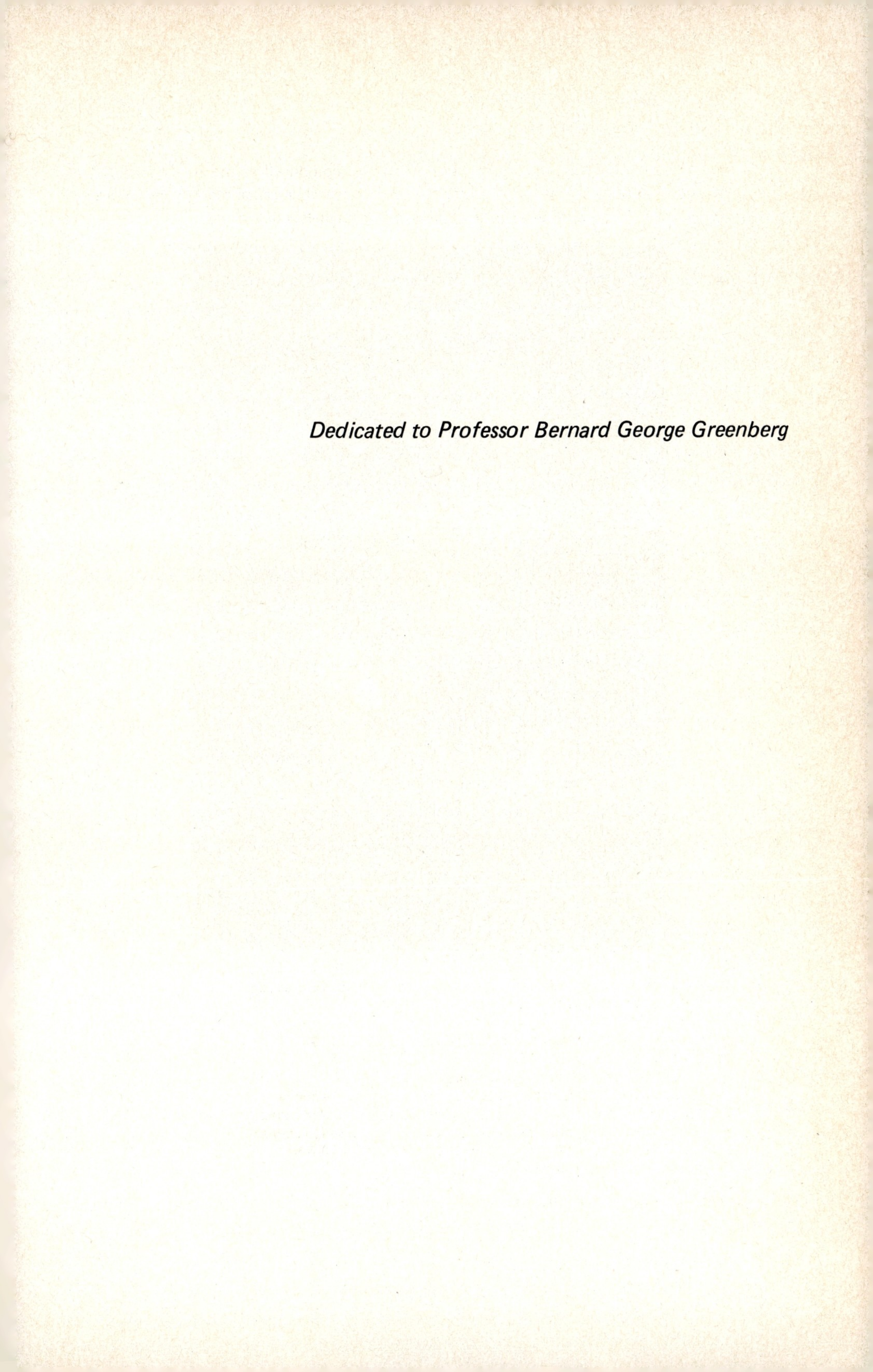

Dedicated to Professor Bernard George Greenberg

Professor Bernard G. Greenberg

PREFACE

This set of essays in a broad spectrum of Biostatistics is dedicated to our highly esteemed colleague, Bernard George Greenberg, Kenan Professor of Biostatistics and former Dean, School of Public Health, University of North Carolina, Chapel Hill. Professor Greenberg attained his sixty-fifth birthday on October 4, 1984, and this commemorative monograph is in honor of this occasion. It is a measure of the esteem and affection with which he is held, and reflects the tributes of his colleagues and students. For more than three decades, Dr. Greenberg has played a key role in the development of Biostatistics program in North Carolina (as well as in a national cum international setup), broadening the traditional corners of Biometry to include Public Health, Environmental and Biomedical Sciences. It is indeed by virtue of fruitful interaction of statistical methodology planning and the real world problems arising in Demography, Epidemiology, Clinical Trials, Occupational Health and Environmental Sciences, Health Administration and Education, Delivery of Health Services and a variety of Biomedical (including Neurological) Sciences that Biostatistics has emerged as a discipline on its own with significant academic as well as professional utility. The art of Biostatistics lies in the harmonious blending of the delicate art of understanding the salient features of the basic problems in experimental sciences, transmitting them to statistically meaningful and interpretable forms and drawing valid and reliable statistical inferences from them. The Department of Biostatistics, University of North Carolina, Chapel Hill represents an exemplary example of such interaction, and this is largely due to the outstanding leadership of Professor Greenberg and his contributions to both theoretical and applied Biostatistics.

There is hardly any area within the wide spectrum of Biostatistics that has not been enriched by the contributions of Professor Greenberg. An article in appreciation of his varied contributions, written by his younger son, Dr. Raymond S. Greenberg, included in this volume, depicts this picture clearly. A complete list of published work of Professor Greenberg is also included in this volume for a comprehensive account of his long career as a biostatistician academician. It is within this broad realm of Biostatistics, we have attempted to have contributions to this volume along with the hope that it would match his visions in spirit. It was very difficult to have an exhaustive set of contributions in this respect, and because of obvious limitations, we had to limit ourselves. We apologize if any areas has been left out inadvertently or any important contribution has been omitted.

I feel it a great honor and privilege to edit this volume. In the development of my professional career, Professor Greenberg has been the most influential person, and, on a personal level too, he has always been like my own elder brother. The task of editing this volume has been made easier by the warm and most friendly support I have received from most of my colleagues in this department as well as from the school. In particular, Dr. James E. Grizzle's consistent help and encouragement have always been very helpful. Most of the administrative work relating to this volume has been carreid out by Ms. Roberta Clark, and the whole monograph has been typed by Ms. Jennie Capparella, to both of whom I am deeply indebted for their able assistance. I would also like to record here the encouragement and support from Dr. Gerard Wanrooy of North Holland Publishing Company towards to the planning and preparation of this monograph. Finally, all the articles in this volume have been reviewed, mostly by other contributors and sometimes by other colleagues acknowledged at the end of some of the papers. To all of them, we express our sincere thanks for this whole hearted cooperation and professional work.

Professor Bernard G. Greenberg is planning for retirement by the end of June 1985. We all wish him a very happy and long life after retirement.

Chapel Hill, NC
November 13, 1984

P.K. Sen

CONTENTS

CONTRIBUTORS

[Numbers in parentheses relate to the initial page numbers of the articles contributed.]

M.G. Akritas, *Institute of Statistics, Texas A&M University, College Station, TX 77843 (221).*

Ingrid Amara, *Department of Biostatistics, University of North Carolina, Chapel Hill, NC 27514 (357).*

S.I. Bangdiwala, *Department of Biostatistics, University of North Carolina, Chapel Hill, NC 27514 (13).*

V.P. Bhapkar, *Department of Biostatistics, University of Michigan, Ann Arbor, MI 48109 (237).*

George E. Bonney, *Department of Biometry, Louisiana State University Medical Center, New Orleans, LA 70112 (437).*

N.E. Breslow, *Department of Biostatistics, University of Washington, Seattle, WA 98195 (55).*

Erica Brittain, *Department of Biostatistics, University of North Carolina, Chapel Hill, NC 27514 (31).*

Edward Bryant, *Ortho Pharmaceuticals, Raritan, NJ 08869 (251).*

Shoutir Kishore Chatterjee, *Department of Statistics, Calcutta University, Calcutta 700019, India (283).*

Prithwis Das Gupta, *Population Division, U.S. Bureau of the Census, Washington, DC 20233 (75).*

H.A. David, *Department of Statistics, Iowa State University, Ames, IA 50011 (305).*

C.E. Davis, *Department of Biostatistics, University of North Carolina, Chapel Hill, NC 27514 (31).*

Jeremy E. Dawson, *Department of Physiology, J.C.S.M.R., Australian National University, Canberra, A.C.T. 2601, Australia (345).*

N.E. Day, *University of Washington, Department of Biostatistics SC-32, Seattle, WA 98195 (55).*

Victor De Gruttola, *Department of Biostatistics, Harvard University School of Public Health, Boston, MA 02215 (421).*

Donald L. Doerfler, *Department of Biostatistics, University of North Carolina, Chapel Hill, NC 27514 (195).*

Neil Dubin, *Institute of Environmental Medicine, New York University Medical Center, 550 First Avenue, New York, NY 10016 (165).*

Regina C. Elandt-Johnson, *Department of Biostatistics, University of North Carolina, Chapel Hill, NC 27514 (109).*

Robert C. Elston, *Department of Biometry, Louisiana State University Medical Center, New Orleans, LA 70112 (437).*

K. Ruben Gabriel, *Department of Statistics and Division of Biostatistics, University of Rochester, Rochester, NY 14642 (315).*

Edmund A. Gehan, *Department of Biomathematics, University of Texas System Cancer Center, Houston, TX 77030 (39).*

Dennis Gillings, *Department of Biostatistics, University of North Carolina, Chapel Hill, NC 27514 (251).*

Raymond S. Greenberg, *Atlanta Cancer Surveillance Center, Emory University School of Medicine, Decatur, GA 30030 (1).*

Muhammad K. Habib, *Department of Biostatistics, University of North Carolina, Chapel Hill, NC 27514 (511).*

Frank Harrell, *Division of Biometry, Department of Community and Family Medicine, Duke University Medical Center, Durham, NC 27710 (333).*

Sherman A. James, *Department of Epidemiology, University of North Carolina, Chapel Hill, NC 27514 (179).*

Norman L. Johnson, *Department of Statistics, University of North Carolina, Chapel Hill, NC 27514 (345).*

Ellen B. Kaplan, *Department of Biostatistics, University of North Carolina, Chapel Hill, NC 27514 (455).*

Azza R. Karmous, *Department of Biostatistics, University of North Carolina, Chapel Hill, NC 27514 (121).*

David G. Kleinbaum, *Department of Biostatistics, University of North Carolina, Chapel Hill, NC 27514 (179).*

Gary G. Koch, *Department of Biostatistics, University of North Carolina, Chapel Hill, NC 27514 (357).*

Lawrence L. Kupper, *Department of Biostatistics, University of North Carolina, Chapel Hill, NC 27514 (179).*

Peter A. Lachenbruch, *Department of Preventive Medicine and Environmental Health, University of Iowa, Iowa City, IA 52240 (389).*

Kerry L. Lee, *Division of Biometry, Department of Community and Family Medicine, Duke University Medical Center, Durham, NC 27710 (333).*

Nripes Kuman Mandal, *Department of Statistics, Calcutta University, Calcutta 700019, India (283).*

Nathan Mantel, *Department of Mathematics, Statistics and Computer Science, The American University, Bethesda, MD 20814 (399).*

Kadambari K. Namboodiri, *Cancer Control Consortium of Ohio, 101A Hamilton Hall, 1645 Neil Avenue, Columbus, OH 43210 (455).*

El-Sayed Nour, *Department of Biostatistics, University of North Carolina, Chapel Hill, NC 27514 (143).*

Charles L. Odoroff, *Department of Statistics and Division of Biostatistics, University of Rochester, Rochester, NY 14642 (315).*

Bernard S. Pasternack, *Institute of Environmental Medicine, New York University Medical Center, 550 First Avenue, New York, NY 10016 (165).*

Ibrahim A. Salama, *Department of Biostatistics, University of North Carolina, Chapel Hill, NC 27514 (413).*

Sara H. Salama, *Department of Geography, University of North Carolina, Chapel Hill, NC 27514 (121).*

A.K.Md.E. Saleh, *Department of Mathematics & Statistics, Carleton University, Ottawa, Canada K1S 5B6 (221).*

A.E. Sarhan, *Institute of Statistical Research, Cairo University, Giza, Egypt (413).*

P.K. Sen, *Department of Biostatistics, University of North Carolina, Chapel Hill, NC 27514 (13, 221, 511).*

Julio M. Singer, *Departmento de Estatistica, Universidade de Sao Paulo, Brazil (357).*

C.M. Suchindran, *Department of Biostatistics, University of North Carolina, Chapel Hill, NC 27514 (143).*

Chung-Yi Suen, *Department of Biostatistics, University of North Carolina, Chapel Hill, NC 27514 (179).*

Michael Symons, *Department of Biostatistics, University of North Carolina, Chapel Hill, NC 27514 (195).*

K.W. Teoh, *Department of Statistics, University of Kentucky, Lexington, KY 40506 (237).*

James H. Ware, *Department of Biostatistics, Harvard University School of Public Health, Boston, MA 02215 (421).*

Robert F. Woolson, *Department of Preventive Medicine and Environmental Health, University of Iowa, Iowa City, IA 52242 (389).*

Yang C. Yuan, *Department of Biostatistics, University of North Carolina, Chapel Hill, NC 27514 (195).*

BIOSTATISTICS: Statistics in Biomedical, Public Health
and Environmental Sciences, edited by P.K. Sen

IN APPRECIATION OF BERNARD G. GREENBERG

Raymond S. Greenberg

Atlanta Cancer Surveillance Center
Emory University School of Medicine
Decatur, Georgia 30030

This book is dedicated to Bernard George Greenberg, Kenan Professor of Biostatistics at the University of North Carolina. The broad range of topics presented in this volume is a fitting tribute to Greenberg's contributions in both theoretical and applied biostatistics.

Bernard Greenberg was born on October 4, 1919 in New York City. He was educated in the public school system, followed by undergraduate studies at the City College of New York. At 19 years of age, Greenberg was elected to Phi Beta Kappa and graduated with a Bachelor of Science degree in mathematics.

After college, Greenberg was employed briefly as a Junior Statistician for the United States Bureau of the Census. In 1940, he accepted a position as an Assistant Statistician at the New York State Department of Health. This first exposure to the field of public health lasted only one year, but it played a central role in his subsequent choice of careers. Greenberg learned the importance of reliable data and field observations under the direction of a bright young physician, David Rutstein. The friendship that Greenberg and Rutstein developed in Albany remained intact throughout their long and distinguished careers in public health.

In 1941, Greenberg was inducted into the United States Army where his leadership skills were quickly recognized. He was selected as a candidate for the officer training program and upon successful completion of the training, Greenberg was appointed as an instructor. Later, he advanced to the rank of captain and commanded infantry troops in the European Theater.

Upon return to the United States, Greenberg married Ruth E. Marck, a Research Assistant in physiological chemistry at Yale University. Following discharge from military service, Greenberg returned temporarily to the New York State Department of Health, but decided to obtain further academic training in statistics. His fiancée arranged a meeting with Chester Bliss, a prominent biometrician on the staff of the Connecticut Agricultural Experiment Station and the faculty at Yale. Bliss suggested that the young couple should consider the newly formed Institute of Statistics in Raleigh, North Carolina (1). Acting upon this advice, the Greenbergs travelled to North Carolina in 1946 to attend a Special Summer Session.

The Institute of Statistics, under the direction of Gertrude Cox, was developing rapidly during the postwar years. Miss Cox, the first woman professor on the faculty of North Carolina State University, recruited Harold Hotelling, who chaired the Department of Mathematical Statistics in Chapel Hill, and William C. Cochran, who provided the statistical and intellectual leadership of the Department of Experimental Statistics in Raleigh. Research conducted at the Institute was expanded beyond the traditional role in biological and social sciences to include research in mathematical, physical and industrial statistics. As financial support increased, new faculty were hired, including Henry ´Curly´ Lucas, Francis McVay, Richard Anderson, Paul Peach and Jackson Rigney (2).

During the Special Summer Session of 1946, in addition to Bliss, courses were offered by R.A. Fisher, George Snedecor, Jacob Wolfowitz, William Cochran and Gertrude Cox. Greenberg took the courses taught by Fisher and Snedecor for credit and he audited the courses taught by Wolfowitz and Cochran. It was a stimulating academic environment, but one lecturer stood out from the rest. In Greenberg's own words (1): "Before the end of the first week, I knew that Cochran was the person and teacher that I most wanted to emulate and applied immediately to become a regular graduate student that fall."

Greenberg was assigned to work with Curly Lucas, whose background in animal nutrition was the most closely related to public health. As a mentor, Lucas was unusually adept in several aspects of statistical analysis. First and foremost, Lucas was interested in the mathematical modeling of biological phenomena. Also, he played a prominent role in the application of analysis of variance to experimental and nonexperimental data.

As already indicated, Greenberg chose Cochran as a special role model. In later years, Greenberg commented that: "Bill Cochran was the greatest influence in my statistical career... (he) was the perfect gentleman and scholar with a superb sense of humor... In my eyes, he was the world's greatest experimental statistician."(1)

The special relationship that Cochran maintained with the graduate students is best illustrated by his approach to Ph.D. written examinations. Instead of contributing questions, Cochran would ask the student to referee a manuscript submitted to him for consideration of publication in the Journal of the American Statistical Association. In Greenberg's case, the first assigned paper was written by Johannes Ipsen, on the topic of estimating the mean and standard deviation of a truncated normal distribution. Two previous referees had recommended disapproval of the paper on the basis of its unconventional mathematics. However, Cochran suspected that Ipsen's method was reasonable, and Greenberg was assigned the task of comparing the performance of the new approach with other procedures. Simulation studies indicated that Ipsen's method worked as well or better than the alternatives. Then, Greenberg and Cochran met together on almost a daily basis to reconstruct what the former thought should be the mathematics of the original paper. These private sessions offered a unique opportunity for Greenberg to observe Cochran's approach to solving a current research problem.

In 1949, Greenberg completed his doctoral dissertation in Experimental Statistics, as one of the first Ph.D. recipients in Statistics from North Carolina State College. On July 1 of that year, a Department of Biostatistics was established in the School of Public Health on the Chapel Hill campus. Greenberg was appointed as the chairman of this new department because of his background in public health and his strong ties to the Departments of Mathematical and Experimental Statistics. The role of the Department of Biostatistics was to provide basic instruction and consultation for students and faculty in public health and other allied disciplines. A training program for biostatisticians leading to the Master of Public Health degree was introduced in 1950. The curriculum was expanded to include courses in biological assay, quantitative epidemiology and the history and uses of vital statistics and demography.

C. Clark Cockerham was the second faculty member hired in Biostatistics and after a brief stay he was replaced by Harry Smith. At the time, Smith was a graduate student in Raleigh. Since the Department of Biostatistics did not have a doctoral program, a formal arrangement was established for doctoral students to specialize in Biostatistics while they fulfilled the degree requirements of one of the affiliated statistical departments. Under this arrangement, Smith was the first student to graduate, with Greenberg serving as the chair of his doctoral committee.

In 1953, the National Heart Institute awarded a grant to the University of North Carolina to facilitate the training of biostatisticians. This training grant provided funds for students, postdoctoral fellows, permanent and visiting faculty. Through this mechanism, the Department of Biostatistics was able to obtain the services of such prominent visiting professors as David R. Cox, J.O. Irwin, David B. Duncan, David J. Newell, Ahmed E. Sarhan, Herbert A. David, Robert C. Elston, K. Ruben Gabriel, Pranab K. Sen, and others, many of whom remained as tenured faculty.

Greenberg implemented a number of original programs during the 1950's. Master's degree students were encouraged to participate in field training under the supervision of advisors in various health agencies. These internships were intended to offer the students a practical experience in public health. The inspiration for this program may be traced to Greenberg's predoctoral work with the New York State Department of Health. The philosophy behind such field training can be found in an article written by Greenberg on this subject (3).

The Department of Biostatistics also contributed to the development of principles and methods for collaborative, multicenter, clinical trials. In 1955, the National Cancer Institute contracted with the Department to serve as the statistical coordinating center for the Southeastern Cooperative Cancer Chemotherapy Study Group. Through this collaborative group of about ten medical schools, research was conducted on the efficacy of new chemotherapeutic agents. A subsequent contract was signed with the Veterans Administration for the provision of statistical services for a multihospital study of the treatment of gastric ulcers. In 1959, Greenberg was the first investigator to publish a set of guidelines for the design and conduct of such collaborative medical trials (4).

With successful competition for federal research support, the Department of Biostatistics began to add new faculty at an unprecedented rate. The list of faculty associated with the Department is too long to be enumerated completely here. During Greenberg's tenure as chairman, he successfully recruited amongst others Harry Smith, Jr. (1953), Ahmed E. Sarhan and Edmund A. Gehan (1955), Thomas G. Donnelly (1956), Earl L. Diamond (1957), H. Bradley Wells and Roy R. Kuebler, Jr. (1958), Bernard S. Pasternack (1959), James E. Grizzle, Robert C. Elston and John Kosa (1960), Dana E.A. Quade and Khatab M. Hassanein (1962), Regina C. Elandt-Johnson, Herbert A. David and Jay J. Glasser (1964), James R. Abernathy, Elizabeth J. Coulter, Peter A. Lachenbruch and Pranab K. Sen (1965), Forrest E. Linder (1966), Gary G. Koch, Donna R. Brogan and Anthony F. Bartholomay (1967), Ronald W. Helms and Mindel C. Sheps (1968), Michael J. Symons (1969), O. Dale Williams, David G. Kleinbaum, Larry L. Kupper, Joan W. Lingner and J. Richard Stewart (1970), Dennis B. Gillings and Craig D. Turnbull (1971).

Under Greenberg's leadership, the Department of Biostatistics expanded its academic offerings. In 1965, a Ph.D. in Biostatistics was authorized by the Graduate School. That same year M.S.P.H. and Ph.D. programs in statistical aspects of demography and population studies were established. Also, a Ph.D. in statistical genetics was added in 1965. Later, masters and doctoral programs with emphases in mental health, environmental health, and health services research were established.

During two decades as chairman of the Department of Biostatistics, Greenberg continued to expand his research horizons. In collaboration with Ahmed E. Sarhan, he made fundamental contributions to order statistics (5) by means of a contract with the U.S. Army Office of Ordnance Research. With H. Bradley Wells and James R. Abernathy, Greenberg published seminal research on perinatal mortality (6,7). With A.E. Sarhan, he worked on approaches to matrix inversion (8). Later, he joined with several colleagues in advancing the theory and practice of the randomized response survey technique (9-11). Greenberg worked with Curtis G. Hames, a general practitioner in Evans County, Georgia and John C. Cassel, chairman of the Department of Epidemiology at Chapel Hill, to establish a major prospective cohort study of atherosclerotic cardiovascular disease among white and black participants.

Greenberg's contributions to statistics and public health have been recognized by many honors, including fellowship in the American Statistical Association, the American Public Health Association, the Institute of Mathematical Statistics and the American College of Epidemiology. Also, he was nominated and elected to membership in the International Statistical Institute, the American Epidemiological Society and the Institute of Medicine. He served as chairman of the Statistics Section of the American Public Health Association (1958-60), and the Training Section of the American Statistical Association (1959-60), as well as the president of the Biometric Society (ENAR, 1971-72) and vice president of the Association of Schools of Public Health (1973-75). Greenberg was selected for the editorial boards of the Journal of Chronic Diseases, the Review of the International Statistical Institute, the American Journal of Obstetrics and Gynecology, and the Journal of Statistical Planning and Inference.

Greenberg was appointed to seven study sections in the National Institutes of Health, including chairmanship of the Nursing Research Study Section, the Special Review Committee on Therapeutic Agents for Cardiovascular Disease, the Special Project Committee of the National Heart Institute, and the Epidemiological Studies Review Committee of the National Institute of Mental Health. He served as chairman of the Publications Advisory Board of the National Center for Health Services Research and Development from 1969 through 1972. He completed a two year term with the newly created Committee on National Statistics of the National Academy of Sciences and a three year term with the U.S. National Committee on Vital and Health Statistics. Greenberg has been an active consultant to the National Institutes of Health, the Children's Bureau, the National Center for Health Statistics, the Veterans Administration and the World Health Organization.

In 1966, Greenberg received the Bronfman Award from the American Public Health Association for his many contributions to research and education in the health field. Three years later, he was named Kenan Professor of Biostatistics at the University of North Carolina. In 1972, Greenberg was appointed as Dean of the School of Public Health in Chapel Hill. During the decade of his deanship, the School expanded its student enrollment and faculty, with concomitant increases in its research and service activities. Through his creative leadership, an undergraduate program was introduced leading to the B.S.P.H. degree, and alternative sources of funding were located to support academic programs that were threatened by reduced federal capitation. Also, he was honored by the Minority Student Caucus for his leadership of programs for black and native American students.

In 1983, Greenberg was inducted into the prestigious Order of the Golden Fleece for his service to the University of North Carolina. That same year he was the recipient of the O. Max Gardner Award for contributions to the welfare of the human race.

It is impossible to measure the full impact of Bernard Greenberg's career in biostatistics. He has been a visionary leader who inspired students and colleagues by his love of teaching, devoted service, hard work, and perseverance. Under his guidance, the Department of Biostatistics grew from a faculty of one in 1949 to its present size of over 30 regular full-time faculty. He bridged the gaps between statistics and other disciplines such as demography, epidemiology, mental health, survey research, environmental sciences, health services research, and clinical medicine. His broad range of interests and clarity of expression have made him a highly valued consultant. As an administrator, he was renowned for this sensitivity to the needs of both students and faculty, and his creativity in solving problems.

The author of this biographical sketch would be remiss not to note his own debt of gratitude to Bernard Greenberg. As the youngest of his three children, I have been priviliged to know Bernard Greenberg as a public figure and also as a devoted family man. Just as he treasures his private tutorials with Cochran, I fondly recall my private sessions on long evening walks with Bernard Greenberg. He succeeded in nurturing "a burning yearning for learning" in many individuals, including this writer.

REFERENCES

[1] Greenberg, B.G., personal communication, 1980.

[2] Greenberg, B.G., Cox, G.M., Mason, D.M., et al: Statistical Training and research: The University of North Carolina System. Intl. Stat. Rev. 46: 171-207, 1978.

[3] Greenberg, B.G.: Field training of biostatisticians. Am. Stat. 18: 19-22, 1964.

[4] Greenberg, B.G.: Conduct of cooperative field and clinical trials. Am. Stat. 13: 13-17, 1959.

[5] Sarhan, A.E. and Greenberg, B.G.: Contributions to Order Statistics. New York: John Wiley and Sons, Inc., 1962.

[6] Greenberg, B.G. and Wells, H.B.: Linear discriminant analysis in perinatal mortality. Am. J. Pub. Health. 53: 594-602, 1963.

[7] Abernathy, J.R., Greenberg, B.G., Wells, H.B., et al: Smoking as an independent variable in a multiple regression analysis upon birth weight and gestation. Am.J. Pub. Health. 56: 626-633, 1966.

[8] Sarhan, A.E. and Greenberg, B.G.: Matrix inversion, its interest and application in analysis of data. J. Am. Stat. Assn. 54: 755-766, 1959.

[9] Abul-Ela, A-L.A., Greenberg, B.G. and Horvitz, D.G.: A multi-proportions randomized response model. J. Am. Stat. Assn. 62: 990-1008, 1967.

[10] Greenberg, B.G., Abul-Ela, A-L.A., Simmons, W.R. et al: The unrelated question randomized response model: Theoretical framework. J. Am. Stat. Assn. 64: 520 -539, 1969.

[11 Greenberg, B.G., Kuebler, R.R., Abernathy, J.R., Horvitz, D.G.: Respondent hazards in the unrelated question randomized response model. J. Stat. Plan. Inf. 1: 53-60, 1977.

SELECTED ANNOTATED PUBLICATIONS OF BERNARD G. GREENBERG

[1] Greenberg, B.G., Wright, J.J., Sheps, C.G.: A technique for analyzing some factors affecting the incidence of syphilis. J. Am. Stat. Assn. 45: 373-399, 1950.

The authors considered the general case of a series of rates that are classified by age and period of time. They proposed a logarithmic model with age and period as main effects and the interaction term representing a cohort effect. This was the first attempt made to quantify the cohort effect. Age, period and cohort models are currently undergoing a popular revival along these lines. The authors were able to estimate the parameters in this example by placing restrictions on the age and cohort factors.

[2] Greenberg, B.G.: Why randomize? Biometrics. 7: 309-322, 1951.

The author discussed the advantages of randomization as contrasted with systematic schemes of allocating treatments. Greenberg illustrated the potential hazards of systematic designs, with an example from experimental parasitology. Also, the statistical efficiencies of various randomized designs were compared.

[3] Greenberg, B.G.: The use of analysis of covariance and balancing in analytical surveys. Am. J. Pub. Health. 43: 692-699, 1953.

In this paper, Greenberg discussed the concept which epidemiologists would later refer to as 'confounding'. He used a cross-sectional study of the growth of school children to illustrate an analysis of the efficiency of balancing (matching), which foreshadowed conclusions reached by others almost two decades later.

[4] Peterson, O.L., Andres, L.P., Spain, R.S. and Greenberg, B.G.: An analytical study of North Carolina general practice 1953-1954. J. Med. Educ. 31: 1-165, 1956.

This work was a landmark investigation in the fields of medical sociology and health services research. From a stratified random sample of general practitioners in North Carolina, the authors concluded that inadequate clinical training was responsible for the poor performance of many generalists. Physicians with better performances tended to come from the upper half of their medical school classes. However, the Medical College Aptitude Test was found to be a poor predictor of ultimate clinical performance. Older physicians were found to have less postgraduate training, with smaller patient loads, but a lower quality of work.

[5] Greenberg, B.G.: Conduct of cooperative field and clinical trials. Am. Stat. 13: 13-17, 28, 1959.

This paper represents the first published discussion of the design and conduct of multicenter experimental health research. Among the topics covered was the role of the statistical coordinating center. This model for collaborative research has been accepted as the standard approach to the organization of virtually all medical trials undertaken since 1960.

[6] Peacock, E.E. Jr., Greenberg, B.G. and Brawley, B.W.: The effect of snuff and tobacco on the production of oral carcinoma. Ann. Surg. 515: 542-550, 1960.

The authors summarized an experiment in which oral cancer could not be induced by the implantation of tobacco or snuff in the oral pouches of hamsters. A separate case-control study was used to demonstrate a significant association between oral cancer and use of snuff and tobacco among older persons of low socioeconomic status. The authors interpreted these results to imply that smokeless tobacco acted as a promoter but not an initiator of oral cancer. The case-control study had additional methodological importance since it represented one of the first attempts to use multiple comparison groups. A further contribution was the recognition and treatment of heterogeneity of effect, later to be referred to as 'effect modification'.

[7] White, K.L., Williams, T.F. and Greenberg, B.G.: The ecology of medical care. N. Eng. J. Med. 265: 885-892, 1961.

In this work, the authors used data from several population surveys to reach striking conclusions about the health-care behavior of adults. One out of every three adults with an illness consulted an physician for diagnosis and treatment. Of those persons who sought medical care, only one in 250 was referred to a university medical center. These statistics were interpreted to suggest that the clinical experience of medical students, and perhaps some of their mentors, in relevant to a limited proportion of illness in the general population. The appearance of this article set a precedent for publication of sociological work in traditional medical journals.

[8] Greenberg, B.G. and Wells, H.B.: Linear discriminant analysis in perinatal mortality. Am. J. Pub. Health. 53: 594-602, 1963.

Greenberg and Wells applied discriminant analysis to a case-control study of perinatal mortality. The authors found that the strongest predictors of perinatal death were gestational age, and complications of labor and delivery. After adjustment for these variables, traditional risk indicators, such as maternal age, race, education and parity were found to be unimportant in perinatal mortality. The best predictive linear model resulted in an appreciable number of misclassifications, a problem that has not been resolved in over 20 years of subsequent research.

[9] Greenberg, B.G.: Field training for biostatisticians. Am. Stat. 18: 19-22, 1964.

Greenberg presented in this review the principles and practice of structured field work for masters degree students in biostatistics. This plan was developed in the Department of Biostatistics at the University of North Carolina during the 1950's. The guidelines advocated in this article have been adapted by many other graduate programs in biostatistics.

[10] Abernathy, J.R., Greenberg, B.G., Wells, H.B. et al: Smoking as an independent variable in a multiple regression analysis upon birth weight and gestation. Am. J. Pub. Health. 56: 626-633, 1966.

The authors performed a multivariate linear regression analysis of data previously collected in Baltimore on an economically and racially homogeneous group of pregnant women. After adjustment for potential confounders, birth weight was found to be significantly associated with maternal smoking status. The lower birth weight in progeny of smokers was not attributable to shorter gestational periods.

[11] Abul-Ela, A-L.A., Greenberg, B.G. and Horvitz, D.G.: A multi-proportions randomized response model. J. Am. Stat. Assn. 62: 990-1008, 1967.

In this paper the authors extended the theory of the randomized response survey technique developed by Warner. This method was proposed for obtaining an unbiased estimate of the prevalence of a particular trait or characteristic which respondents may be reluctant to acknowledge to an interviewer. In Warner's original model, only binary responses with a single stigmatizing value were considered. The model presented in this paper allowed for multiple mutually exclusive stigmatizing responses. This extended model was illustrated for estimating the prevalances of mothers who were married prior to conception, mothers who were married after becoming pregnant and mothers who were unwed at the time of delivery.

[12] Greenberg, B.G., Abul-Ela, A-L.A., Simmons, W.R. et al: the unrelated question randomized response model: Theoretical framework. J. Am. Stat. Assn. 64: 520-539, 1969.

The authors developed the theoretical background for the randomized response method of Simmons. With this approach, a subject is asked to respond to one of two questions, only one of which is stigmatizing. The question posed is determined on a probability basis and the interviewer is blinded to which question is being answered. The statistical efficiency of this two question model was compared with that of the one question approach of Warner under situations of truthful and untruthful responses.

[13] Hulka, B.S., Grimson, R.C., Greenberg, B.G. et al: "Alternative" controls in a case-control study of endometrial cancer and exogenous estrogen. Am. J. Epideiol. 112: 376-387, 1980.

The authors addressed the issue of bias in estimation of effect attributable to the sources of comparison subjects in a case-control study of exogenous estrogen and endometrial cancer. Three separate comparison populations were chosen: hospitalized women who underwent dilatation and curretage (D and C), hospitalized women with other gynecologic problems, and a stratified random sample of women from the general population. History of estrogen use amongst women with endometrial cancer was compared separately against each of the control groups. A marked underestimate of effect was demonstrated for comparisons with the D and C control group. This paper is noteworthy for the rigor of its design and conduct, which helped to clarify sources of selection bias in case-control studies.

CLINICAL TRIALS

BIOSTATISTICS: Statistics in Biomedical, Public Health
and Environmental Sciences, edited by P.K. Sen

STATISTICAL MONITORING OF CLINICAL TRIALS WITH STAGGERED ENTRY

S.I. Bangdiwala and P.K. Sen

Department of Biostatistics
University of North Carolina at Chapel Hill

In the context of clinical trials and life testing situations, continuous statistical monitoring of the accumulating information is usually performed. Various repeated testing procedures are discussed and their applicability to potential censoring schemes is examined. A progressively censored linear rank statistic is proposed for the staggered entry situation and its utilization in the presence of concomitant variables is illustrated. A computer program for calculation of the statistic is included in the appendix.

KEY WORDS: Progressive censoring schemes, linear rank statistics, clinical trials monitoring, staggered entry, covariate adjustment

1. INTRODUCTION

Of ethical concern in clinical trials and life testing experimentation is the possibility of early termination for several reasons:

(i) initial results indicate a marked superiority of one treatment over the other (treatment efficacy),

(ii) toxic side effects of a treatment are found,

(iii) treatment differences are virtually non-existent (e.g. due to lack of adherence).

While (ii) and (iii) are mainly of ethical and financial concerns respectively, (i) is related to the hypothesis being tested. Therefore, the investigator would like to (and usually does) closely monitor the experiment on an ongoing basis. The multiple times the treatment difference test is performed need to be considered in the significance level of the statistical tests [Armitage, et al. (1969)]. Merely utilizing a Bonferroni procedure leads to a highly conservative test procedure because it does not account for the non-independence of the tests in a clinical trial setting with accumulating data. A variety of sequential testing procedures have been proposed for this problem. Pocock (1977) proposed using an adjusted critical value such that the overall level of significance α is achieved. O'Brien and Fleming (1979) consider a multiple testing procedure for a clinical trial comparing two independent treatments where the response is dichotomous and immediate. At each successive review of the experiment, with an "a priori" N number of looks planned, additional m_1 and m_2 subjects are given treatments 1 and 2, respectively, provided that the experiment is not terminated at that stage. At the Kth look at the data, the Pearson chi-squared statistic χ^2 is computed based on $K(m_1+m_2)$ subjects and $\frac{K}{N}\chi^2$ is compared to the single degree of freedom chi-squared critical value.

For most clinical trials, such as for cancer or cardiovascular disease, the measured response is not immediate. Seigel and Milton (1983) show that the O'Brien-Fleming procedure is also appropriate for such chronic disease models. However, the experimental design giving rise to the testing procedure is one that is not practical in the actual conduct of clinical trials, especially in ones involving human subjects. Usually most subjects are recruited during a given time period at the beginning of the study, and not throughout its conduct. The authors, based on simulation studies, propose that their stopping boundaries are still applicable as long as the proportion of subjects assigned to each of the two treatment groups remains constant at each testing point.

In the context of a proportional hazards assumption, Tsiatis (1981) modifies the testing procedure for the situation where the test statistic is calculated after every r number of failures, a more realistic situation. This is comparable to the O'Brien-Fleming procedure if the total number of failures is Nr. However, most clinical trial monitoring is conducted at specified time intervals, say every six months or so, not after certain number of failures accumulate. For equal time-span monitoring, one may believe that there would be approximately the same number of failures in each time interval. However, this is not the case. Even for exponential survival and thus a constant hazard, Davis (1978) shows that the number of expected failures in specified time periods is not constant or even close to it.

The O'Brien-Fleming procedure has been widely used in the context of multicenter clinical trials, such as the Lipid Research Clinics (1984) test of their Coronary Primary Prevention Trial treatment hypothesis. The LRC study modified the O'Brien-Fleming procedure to allow calculation after given time intervals, assuming an exponential failure distribution in both treatment groups.

Chatterjee and Sen (1973) propose a "progressively censored scheme" in which the test statistic is calculated at every failure time. Lan and DeMets (1983) propose a procedure for calculating discrete sequential boundaries, that is, the boundaries varying with time.

In Section 2, the situations under which the O'Brien-Fleming testing procedure is appropriate in the context of progressively censored schemes for clinical trials are examined. Section 3 examines the staggered entry case. A modified testing procedure based on linear rank statistics is suggested and its asymptotic properties are discussed. In Section 4, various procedures utilized for monitoring clinical trials are compared to the modified testing procedure presented in Section 3. The procedure is illustrated in section 5 on a hypothetical study based on data from the Lipid Research Clinics' Coronary Primary Prevention Trial.

2. PROGRESSIVE CENSORING SCHEMES

A "progressive censoring" scheme is what is actually performed in monitoring on-going clinical trials: at several time points during the course of the experiment, the investigator will examine the accumulated data in regard to the response of interest. At any given time point, subjects for whom the response has not occurred are considered "censored." There are two types of censoring of the data:

(i) Type I censoring (also called Truncation), where the looks at the data are at specified time points, and the number of failures occurring between these time points is a random variable.

(ii) Type II censoring, where the looks at the data are after predetermined numbers of failures occur, and the time between the looks is a random variable.

The O'Brien-Fleming procedure is appropriately used only under the Type II progressive censoring schemes. To accommodate Type I censoring situations, further assumptions about the distribution of failures between the specified time points are necessary. Under Type II censoring, the n_1 and n_2 subjects originally assigned to treatments 1 and 2 respectively can be thought of as giving rise to N recruiting batches of $m_1 = \frac{n_1}{N}$ and $m_2 = \frac{n_2}{N}$ subjects each. At the Kth stage of sampling the investigator would thus consider the $K(m_1+m_2) = \frac{K}{N}(n_1+n_2)$ subjects in the study. The N recruiting batches can be thought of as due to looking at the data after each of $r(>0)$ subjects fail, such that $(n_1+n_2) \geq rN$. Thus, at the Kth stage of the experiment, O'Brien and Fleming visualize Scheme 1.

Scheme 1	Start of Study			Kth Stage of Study			End of Study
Time	0	t_1	...	t_K	...	t_{N-1}	t_N
Stage of Study	0	1	...	K	...	N-1	N
Subjects in Study	$\frac{n_1}{N}+\frac{n_2}{N}$	$2(\frac{n_1}{N}+\frac{n_2}{N})$	...	$(K+1)(\frac{n_1}{N}+\frac{n_2}{N})$	...	n_1+n_2	n_1+n_2
# Failures	0	r		Kr	...	(N-1)r	Nr
# Alive	$\frac{n_1}{N}+\frac{n_2}{N}$	$2(\frac{n_1}{N}+\frac{N_2}{N})-r$	...	$(K+1)(\frac{n_1}{N}+\frac{n_2}{N})-Kr$	...	$(n_1+n_2)-(N-1)r$	$(n_1+n_2)-Nr$

However, the actual conduct of the study looks more like Scheme 2.

Scheme 2	Start of Study			Kth Stage of Study			End of Study
Time	0	t_1	...	t_K	...	t_{N-1}	t_N
Stage of Study	0	1	...	K	...	N-1	N
Subjects in Study	n_1+n_2	n_1+n_2	...	n_1+n_2	...	n_1+n_2	n_1+n_2
# Failures	0	r	...	Kr	...	(N-1)r	Nr
# Alive	(n_1+n_2)	$(n_1+n_2)-r$	...	$(n_1+n_2)-Kr$	...	$(n_1+n_2)-(N-1)r$	$(n_1+n_2)-Nr$

Thus, the O'Brien-Fleming procedure ignores the information of the extra people alive in the study. At the Kth stage of N predetermined stages, one constructs the following 2 x 2 table.

	Treatment 1	Treatment 2	
Alive	S_{1K}	S_{2K}	S_K
Failures	d_{1K}	d_{2K}	d_K
	Km_1	Km_2	$K(m_1+m_2)$

The O'Brien-Fleming test statistic is the logrank chi-squared statistic

$$\frac{K}{N} \frac{(m_1+m_2)(d_{1k}S_{2K} - d_{2K}S_{1K})^2}{Km_1m_2 \; d_K S_K} \tag{2.1}$$

which they show follows asymptotically the Brownian motion process $W(\frac{rK}{N})$ under the null hypothesis.

In order to show that the O'Brien-Fleming procedure is applicable under the Type II progressive censoring scheme, we show that it is equivalent to a linear rank statistic as proposed by Chatterjee and Sen (1973) that is appropriate for the situation. Therefore, define:

$N(m_1+m_2) = n$ total sample size

$X_1,\ldots,X_n$ independent r.v. from $F_1,\ldots,F_n$, absolutely continuous distribution functions

$Z_{n,1},\ldots,Z_{n,n}$ order statistics corresponding to the X_i
$R_{n1},\ldots,R_{nn}$ ranks of the X_i
$S_{n1},\ldots,S_{nn}$ antiranks
$a_n(i), i=1,\ldots,n$ scores
$c_i = \begin{cases} 1 \text{ for } i=1,\ldots,Nm_1 \\ 0 \text{ for } i=Nm_1+1,\ldots,n \end{cases}$ treatment indicator

At the sth observed failure, we define the following censored linear rank statistic:

$$T_{n,s} = \sum_{i=1}^{s} (c_{S_{ni}} - \bar{c}_n)\,[a_n(i) = a_n^*(s)], \text{ where } a_n^*(s) = \frac{1}{n-s}\sum_{i=s+1}^{n} a_n(i) \qquad (2.2)$$

and

$$\bar{c}_n = \frac{1}{n}\sum_{i=1}^{n} c_i = \frac{m_1}{m_1+m_2}\,.$$

Now, scores for the logrank statistic are Savage or "exponential scores," adjusted for their mean

$$a_n(i) = \frac{1}{n} + \frac{1}{n-1} + \ldots + \frac{1}{n-i+1} - 1 = \left(\sum_{j=1}^{i} \frac{1}{n-j+1}\right) - 1\,. \qquad (2.3)$$

Therefore, $a_n^*(s) = \frac{1}{n-s}\sum_{i=s+1}^{n}\left(\sum_{j=1}^{i}\frac{1}{n-j+1} - 1\right)$.

Now, consider the variance terms:

$$C_n^2 = \sum_{i=1}^{n}(c_i-\bar{c}_n)^2 = \sum_{i=1}^{Nm_1}\left(1 - \frac{Nm_1}{n}\right)^2 + \sum_{i=Nm_1+1}^{n}\left(-\frac{Nm_1}{n}\right)^2 = \frac{Nm_1m_2}{(m_1+m_2)}\,, \qquad (2.4)$$

and

$$A_{n,s}^2 = \frac{}{n-1}\left[\sum_{i=1}^{n}(a_n(i) - \bar{a}_n)^2 - \sum_{i=s+1}^{n}(a_n(i) - a_n^*(s))^2\right].$$

For exponential scores, $\bar{a}_n = 0$, and thus

$$A_{n,s}^2 = \frac{1}{n-1}\left[\sum_{i=1}^{s} a_n^2(i) + (n-s)a_n^{*2}(s)\right]. \qquad (2.5)$$

In order to show that the linear rank statistic with exponential scores is equivalent to the O'Brien-Fleming statistic, it suffices to show that they have the same asymptotic distribution. Under the null hypothesis of no group difference, the O'Brien-Fleming test statistic follows asymptotically a Gaussian process $W(\frac{rK}{N})$, based on the "time" points $\frac{s}{N}$. From Sen (1981), $E(T_{n,s}|H_o) = 0$ and $E(T_{n,s}^2|H_o) = C_n^2\,A_{n,s}^2$. Now, asymptotically, the exponential scores (2.3) are given by $\phi(u) = -\ln(1-u)-1$ since $a_n(i)$ are approximately $-\ln(1-\frac{i}{n+1}-1$. Thus, asymptotically $A_{n,s}^2 = \int_0^{s/N}\phi^2(u)du + \frac{1}{1-s/N}\left[\int_{s/N}^{1}\phi(u)du\right]^2 = s/N$. We thus see that both the linear rank statistic with exponential scores and the O'Brien-Fleming statistic have asymptotically the same Gaussian process approximation, once suitably scaled. In fact, the linear rank statistic setup is quite general and other scores suitable to particular situations can be utilized. Since $A_{n,s}^2$ will be known, $T_{n,s}$ can be suitably scaled by $C_n^{-1} A_{n,s}^{-1}$ and shown to be equivalent to the O'Brien-Fleming process to monitor the experiment.

3. STAGGERED ENTRY SITUATION

Staggered entry patterns, where the experimental units may not enter into the study all at the same point in time, are encountered in a variety of settings, especially in multicenter clinical trials. The experimental units may enter at different points in time, either at discrete time points or in batches. Therefore, at any given time of the experiment, s, there are n(s) subjects in the study and a few failures associated with them. Under a progressive censoring scheme, the experiment is monitored by the extension of the linear rank statistic to a two-dimensional time-parameter stochastic process (Majumdar and Sen (1978) and Sen (1976)).

In the context of a staggered entry situation, the use of the O'Brien-Fleming termination probabilities will lead to a very conservative testing procedure. Since the entry pattern is usually unknown, so are the sample sizes at the various specified time points at which the hypothesis is to be evaluated. Different subjects will have varying follow-up times and the procedure does not take this information into account. The use of a suitable upper bound has been suggested as a compromise for this situation, but this makes the procedure highly conservative.

Figure 1 schematically illustrates the number of subjects during the "recruitment" phase of the study at any point in time under two types of entry patterns. The recruitment time period is [0,T] and there is a total of n subjects to be enrolled in the study.

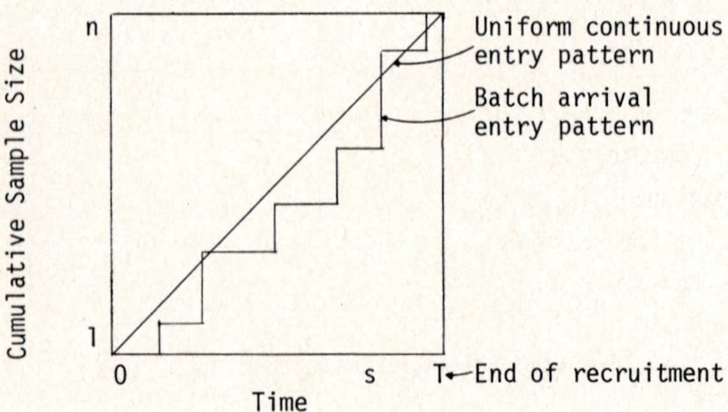

Figure 1. Examples of staggered entry patterns

Now, let us denote by q the number of failures and n(s) the number of subjects in the study, $1 \leq q \leq n(s) \leq n$. The possible failure patterns occupy the shaded lower triangular section of Figure 2. Now, a carefully constructed test statistic needs to take into account the entry pattern since the failure process is confounded within the staggered entry pattern. The proposed solution is to examine

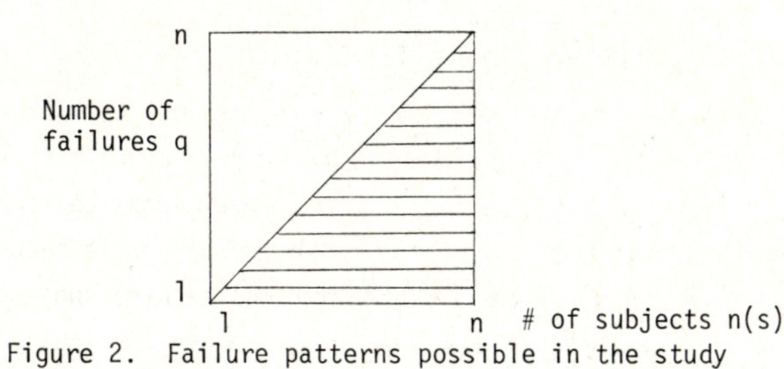

Figure 2. Failure patterns possible in the study

the process more carefully by defining homogeneous exposure groups based on their time of entry, essentially a cohort. The failures by time point s may correspond to different cohorts and this fact must be taken into account in the construction of the testing procedure. If the study is examined at some censoring time point T*, the subjects have exposure times $T^* - t_e$ (where t_e is their entry time) and actual failure times $t_f - t_e$ (where t_f is their time of failure). For each cohort which is determined by the pattern of entry times, one examines all the failures by time point T*. For some time point u, define

$N(T^*,u)$ = total number of subjects from all cohorts with exposures $\geq T^* - u$

$r(T^*,u)$ = total number of failures among these subjects with actual failure times $\leq u$

For all times points $u \in [0, T^* - t_1]$, one can compute the censored linear rank statistic (2.2) and thus obtain

$$T_{N(T^*,u),\ r(T^*,u)} \quad \text{for } 0 \leq u \leq T^* - t_1,\ T^* > t_1. \tag{3.1}$$

Careful examination of the process will reveal that $N(T^*,u)$ and $r(T^*,u)$ vary only either at an entry or actual failure point. Therefore, let T* be either a failure or entry time. Compute the following series of censored linear rank statistics at each time point u: $\{T_{N(T^*,u),q} : 0 \leq q \leq r(T^*,u)\}$. This set of statistics takes into account the entry pattern as well as the distribution of the failure points. If the censoring point T* = s, construct the following test statistic:

$$T = \max_{u \leq s} \{ \max_{0 \leq q \leq r(s,u)} \{T_{N(s,u),q}\}\}. \tag{3.2}$$

The testing procedure for T and simulated values of boundary crossing probabilities were presented by Majumdar and Sen (1978). The asymptotic distribution is approximated by the two-dimensional Bessel process (Sen (1981), Delong (1981)) over the restricted unit square, as determined by the entry pattern (see schematic diagram in Figure 3). Under the multidimensional process situation, the

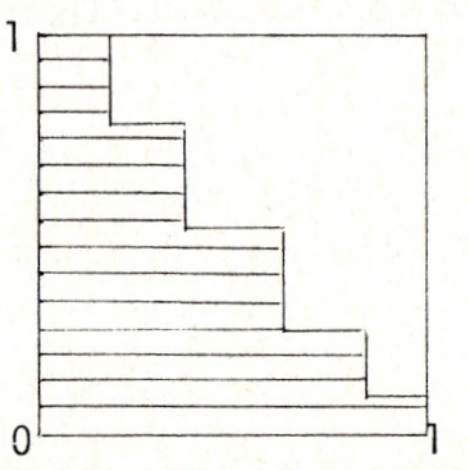

Figure 3. Asymptotic effective domain of the process

O'Brien-Fleming procedure becomes too complicated to be modified. On the other hand, the procedure presented here is quite general since it takes into account a variety of entry patterns. Sellke and Siegmund (1983) investigate the martingale structure of a process with a predetermined uniform entry pattern (diagonal line in Figure 1). However, such an entry pattern is rarely achieved, and departures from this pattern may affect their results. The procedure presented here does not violate the entry pattern and leads to valid results and although somewhat conservative, is not as conservative as modifications to the O'Brien-Fleming procedure for the staggered entry situation would be.

4. DISCUSSION

The utilization of the linear rank statistics approach to the survivorship problem may involve the utilization of complex calculations, but it has several advantages:

(i) applicable for the staggered entry problem
(ii) deals with stochastic covariables
(iii) no proportional hazards assumption required
(iv) rank procedures make use of covariables invariant to monotone transformation

Now, the procedure presented by Tsiatis (1981) assumes that the time of entry into the study is a bounded positive random variable in the interval [0,c] for some $c > 0$. Accrual is further assumed to be random during this interval and stochastically independent of failure time and the covariables of interest. These assumptions allow one to reduce the inherent two-dimensional staggered entry problem to one dimension by considering the following observable random variable at time t:

$$X(t) = \max\{\min(\text{failure time}, t\text{-entry time}), 0\}.$$

However, accrual is usually not at random within a specified recruitment period and departures from this assumption may affect the distribution of $X(t)$. With the introduction of stochastic covariables, the violation of the assumptions may have serious consequences that may be intractable to isolate. Furthermore, the proportional hazards assumption may provide a restriction on the applicability of the procedure.

In the case of non-staggered entry, Lan and DeMets (1983) propose a stochastic curtailment of the level of significance at failure points t such that the overall level of significance α desired for the experiment is achieved at the end (see Figure 4). This procedure thus forces the number of failures between the time points to be a random variable. This poses no problem in the simple case where there are no covariables since one can use an upper bound for an estimate of the expected number of failures in each interval in order to calculate the time points.

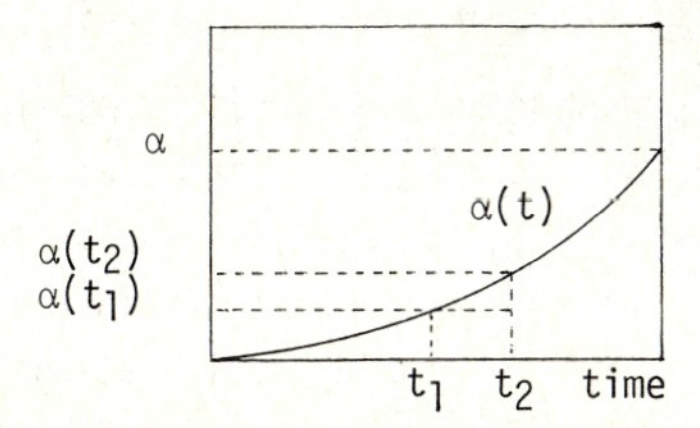

Figure 4. Lan and DeMets (1983) stochastic curtailment diagram

However, with the introduction of covariables, the risk sets are unknown in advance and there is no way of obtaining the time points for examining the accumulating data. In fact, at the kth stage of experimentation, one has observed

$Z_{n,1},\ldots,Z_{n,k-1}$	order statistics corresponding to the first k-1 failures
$S_{n,1},\ldots,S_{n,k-1}$	antiranks; that is, the number of the observation with the ith rank of failure time
$\underset{\sim}{X}_{S_{n,1}},\ldots,\underset{\sim}{X}_{S_{n,k-1}}$	corresponding vectors of covariables

However, since the failure times of the remaining n-k+1 observations are censored, $Z_{n,k},\ldots,Z_{n,n}$ are completely unknown, while $S_{n,k},\ldots,S_{n,n}$ are only partially known, since one would know which observations have not yet failed but one does not know which ordering they will have. At the same time, the vectors of stochastic covariables

$$\underset{\sim}{X}_{S_{n,k}},\ldots,\underset{\sim}{X}_{S_{n,n}}$$

are known, but their correspondence to the observations is unknown.

Lan and DeMets (1983) do not attempt to tackle the staggered entry problem. However, if one is able to justify the appropriateness of the proportional hazards assumption of Tsiatis (1981) in the case of stochastic covariables, one may try combining it with stochastic curtailment. However, in such a case one would still be subject to the criticism that the Tsiatis procedure would bring along.

Now, under the censored linear rank statistic methods proposed for the staggered entry situation in Section 3, a covariate adjusted rank statistic as proposed by Sen (1981, p. 365) can be readily adopted. To eliminate the effects of the covariates, one fits a linear regression of the primary variate censored linear rank statistic on the covariates and works with the residuals. The variances are suitably reduced. The variance $V_{nk} = C_n^2 A_{n,s}$ of $T_{n,s}$ is known beforehand, from (2.4) and (2.5), for all $1 \le k \le n$. The reduced variance due to the covariates $V^*_{n,k}$ is not known until the kth stage. However, since $\frac{V^*_{nk}}{V^*_{nr}} \ge \frac{V^*_{nk}}{V_{n,r}}$, at the kth stage one is protected from large fluctuations if a starting time $\varepsilon \ge 0$ such that $\frac{V^*_{n,k}}{V_{n,r}} \ge \varepsilon$ is utilized. This makes the procedure slightly conservative, but it is the only valid procedure that can be utilized in such situations. The next section illustrates how covariate adjustment is performed with data from the Lipid Research Clinics' Coronary Primary Prevention Trial.

5. EXAMPLE

Data from the Lipid Research Clinics' Coronary Primary Prevention Trial (CPPT) were utilized in a hypothetical study to illustrate the procedure for testing treatment efficacy in a continuously monitored clinical trial in the context of staggered entry and with the inclusion of covariables. The CPPT was a randomized double-blind multicenter study to test the efficacy of cholesterol lowering in reducing risk of coronary heart disease in 3806 asymptomatic middle-aged men with primary hypercholesterolemia (see LRC (1984)). The hypothetical study for this example is a non-randomized clinical trial to determine if there is a difference in the risk of cardiovascular events between two groups of clinical centers, comprising the 1900 men from the original CPPT's placebo treatment group. The groups of clinical centers for this study are presented in Table 1.

Table 1. "Groups" of Clinical Centers

Group 1	Group 2
George Washington Univ.	Baylor Univ.
Iowa Univ.	Johns Hopkins Univ.
Univ. of California (San Diego)	Stanford Univ.
Univ. of Minnesota	Univ. of Cincinnati
Univ. of Toronto	Univ. of Oklahoma
Washington Univ. (St. Louis	Univ. of Washington (Seattle)

Two factors related to cardiovascular disease will be considered covariables of interest: age at the beginning of the study (considered as the date of randomization into the CPPT) and quetelet index, a measure of obesity also measured at baseline. The cardiovascular end point is a conglomeration of various cardiovascular disease manifestations: coronary death, myocardial infarction, angina, intermittent claudication, stroke and congestive heart failure.

The CPPT was monitored during its course by a Safety and Data Monitoring Board (SDMB) twice every year. During the recruitment phase for the CPPT (1973-1976), the SDMB was primarily concerned with the accrual of participants and examined the study at regular time periods as well. For this example the actual dates at which the SDMB met during the recruitment phase of the CPPT will determine the seven dates at which the study results will be examined. The recruitment phase only is utilized since the target sample size (n=1900) has not been reached and the special features of the proposed monitoring procedure become more relevant. The maximum absolute value test statistic values obtained for each cumulative time period are presented in Table 2 for the test statistics utilizing Wilcoxon scores and Savage scores.

Table 2. Test Statistic (3.2) Values for Hypothetical Study

Date at which Study Hypothesis Is Tested	Current Sample Size	Cumulative Number of Failures (Events)	Statistic with Wilcoxon Scores	Statistic with Savage Scores
December 14, 1973	3	0	--	--
April 19, 1974	34	0	--	--
November 10, 1974	212	0	--	--
May 21, 1975	585	4	-2.0104	-2.0104
December 7, 1975	1331	22	-2.3351	-2.3352
May 9, 1976	1696	43	-1.8492	-1.8495
November 1, 1976	1900	83	-2.0231	-2.0254

The maximum in absolute value test statistic over all the times when the study hypothesis is tested is the test statistic compared with the critical values of Majumdar and Sen (1978). For this hypothetical study, the value is -2.335 for both types of scores, and thus, during the recruitment period, there would be sufficient evidence to warrant early stopping of the study at an $\alpha = .10$ due to different "treatment" efficacy between the two groups, but not for $\alpha = .05$. Since the sign of the test statistic with the largest absolute value was negative, Group 1 had a significantly ($p < .10$) smaller risk of a cardiovascular event than

Group 2, early on in the study, after adjustment for age and quetelet index at baseline.

The program utilized for calculation of the Wilcoxon two-covariate adjusted two-dimensional linear rank statistic is presented in the Appendix. The use of Savage scores or other scores requires merely changing one statement in the FORTRAN program. The program was specifically designed for a job classification with moderately inexpensive (or free) Central Processing Unit (CPU) time but relatively expensive Input/Output (I/O) cost. Run time in CPU minutes for a given sample size n is $O(n^2/2)$, but the I/O time is O(n).

ACKNOWLEDGEMENTS: The authors are thankful to Mr. T. Rowles for his programming assistance for the computation of the covariate adjusted two-dimensional PCS linear rank statistic. The work has been supported by the National Heart, Lung, and Blood Institute, Contract 1-HV-1-2243-L.

REFERENCES

[1] Armitage, P., McPherson, C.K., and Rowe, B.C., Repeated significance tests on accumulating data, J. Roy. Statist. Soc. A 132:235-244 (1969).

[2] Chatterjee, S.K. and Sen, P.K., Nonparametric testing under progressive censoring, Calcutta Statist. Assoc. Bull. 22:13-50 (1973).

[3] Davis, C.E., A two-sample Wilcoxon test for progressively censored data, Comm. in Statist. A7:389-398 (1978).

[4] DeLong, D.M., Crossing probabilities for a square root boundary by a Bessel process, Comm. in Statist. A10: 2197-2213 (1981).

[5] Lan, K.K.G. and DeMets, D.L., Discrete sequential boundaries for clinical trials, Biometrika 70:659-664 (1983).

[6] Lipid Research Clinics Program, The Lipid Research Clinics Coronary Primary Prevention Trial Results, I. Reduction in Incidence of Coronary Heart Disease, J. Amer. Med. Assoc. 251:351-364 (1984).

[7] Majumdar, H. and Sen, P.K., Nonparametric testing for single regression under progressive censoring with staggering entry and random withdrawal, Comm. in Statist. A7:349-371 (1978).

[8] O'Brien, P.C. and Fleming, T.R., A multiple testing procedure for clinical trials, Biometrics 35:549-556 (1979).

[9] Peto, R. and Peto, J., Asymptotically efficient rank invariant test procedures (with discussion), J. Roy. Statist. Soc. Ser. A 135:185-198 (1972).

[10] Pocock, S.J., Group sequential methods in the design and analysis of clinical trials, Biometrika 64:191-199 (1977).

[11] Seigel, D. and Milton, R.C., Further results on a multiple-testing procedure for clinical trials, Biometrics 39:921-928 (1983).

[12] Sellke, T. and Siegmund, D., Sequential analysis of the proportional hazards model, Biometrika 70:315-326 (1983).

[13] Sen, P.K., Sequential Nonparametrics, New York, John Wiley and Sons (1981).

[14] Sen, P.K., A two-dimensional functional permutational central limit theorem for linear rank statistics, Ann. of Probab. 4:13-26 (1976).

[15] Tsiatis, A., The asymptotic joint distribution of the efficient scores test for the proportional hazards model calculated over time, Biometrika 68:311-315 (1981).

APPENDIX:

```
+---------------------------------------------------------------+
|  FUNCTION   COMPUTES WILCOXON STATISTICS FOR                   |
|             PROGRESSIVE CENSORING WITH STAGGERING ENTRY,       |
|             TWO COVARIATES                                     |
+---------------------------------------------------------------+
 IMPLICIT REAL*8(A-H,O-Z)
 DIMENSION C(2000),X(2000),W0(2000),W00(2000),
1A0(2000),A1(2000),A2(2000),AA0(2000),AA1(2000),AA2(2000),
2XX(2000),C1(2000),W(2000),VK(6),VKINV(6),
3W1(2000),W2(2000),ITMPW(2000),W11(2000),W22(2000),
4XIKM(3),XXIM(3),IM(3),ICNTM(3),CMEANM(3),CINVSM(3),A0MNM(3),
5A1MNM(3),A2MNM(3),TK0M(3),TK1M(3),TK2M(3),VKM(3,6),VKINVM(3,6),
6TNKM(3)
 CALL UERSET(0,NULL)

 N=CURRENT SAMPLE SIZE
 READ(1,90) N
 IF(N.GT.2000) GO TO 5
 IF(N.LT.3) GO TO 6
 GO TO 7
5 PRINT 4,N
 STOP
6 PRINT 3,N
 STOP
7 CONTINUE
 DO 10 I=1,N

 C(I)=-1, IF SAMPLE 1; C(I)=+1, IF SAMPLE 2
 W(I)=FAILURE TIME, W(I) IS BLANK OR ZERO IF W(I)
 IS GREATER THAN CENSORING TIME
 W1(I)=FIRST COVARIATE
 W2(I)=SECOND COVARIATE
 READ(1,100) C(I),X(I),W(I),W1(I),W2(I)
 W0(I)=EFFECTIVE FAILURE TIME
 W0(I)=W(I)-X(I)
 IF(W0(I).LT.0.) W0(I)=0.
10 CONTINUE
 READ CUTPOINT
 READ(1,102) CTP
 PRINT 441,CTP
 WXX IS ARBITRARY POINT PAST CUTPOINT CTP
```

```
      WXX=CTP+5.
      DO 30 I=1,N
      IF(W(I).EQ.0.) W(I)=WXX
      IF(W0(I).EQ.0.) W0(I)=WXX
      XX(I) = EXPOSURE TIME (REGARDLESS OF FAILURE)
      XX(I)=CTP-X(I)
   30 CONTINUE

      CONSTRUCTION OF TWO-DIMENSIONAL STATISTICS;
      CUTPOINT IS SPECIFIED BY USER
      SAMPLE SIZE=N.  CENSORING POINT MAY BE THE N-TH ENTRY TIME OR
      ANY OF THE FAILURE TIMES BEFORE THE N+1-ST ENTRY
      DO 61 IL=1,3
   61 XIKM(IL)=0.
      LOOP THROUGH SUBJECTS
      DO 994 I=1,N
      EXIT WHEN CURRENT ENTRY TIME EXCEEDS CUTPOINT
      IF(X(I).GT.CTP) GO TO 995

      FIND MEAN OF C AND SUM OF SQUARES OF DIFFERENCES OF C FROM 1
      TO CURRENT CENSORING POINT I.
      CMEAN=0.
      DO 42 J=1,I
   42 CMEAN=CMEAN+C(J)
      CMEAN=CMEAN/FLOAT(I)
      CINVSQ=0.
      DO 43 J=1,I
   43 IF(C(J).NE.CMEAN) CINVSQ=CINVSQ+((C(J)-CMEAN)**2)
      CINVSQ=INVERSE OF SQUARE ROOT OF SUM OF SQUARES OF DIFFERENCES
      IF(CINVSQ.NE.0) CINVSQ=1./(DSQRT(CINVSQ))
      ICOUNT=0
      GET RANKS OF COVARIATES 1 AND 2
      DO 46 J=1,I
      W11(J)=W1(J)
      W22(J)=W2(J)
   46 ITMPW(J)=J
      CALL VSRTRD(W11,I,ITMPW)
      DO 51 J=1,I
      W11(J)=FLOAT(ITMPW(J))
   51 ITMPW(J)=J
      CALL VSRTRD(W11,I,ITMPW)
      DO 52 J=1,I
      W11(J)=FLOAT(ITMPW(J))
   52 ITMPW(J)=J
      CALL VSRTRD(W22,I,ITMPW)
      DO 53 J=1,I
      W22(J)=FLOAT(ITMPW(J))
   53 ITMPW(J)=J
      CALL VSRTRD(W22,I,ITMPW)
      DO 49 J=1,I
   49 W22(J)=FLOAT(ITMPW(J))

      TAKING STOCK OF NUMBER OF FAILURES WITH EFFECTIVE FAILURE
      TIMES LESS THAN OR EQUAL TO THE TOTAL CURRENT EXPOSURE
      TIME RELATING TO EACH CUMULATIVE SAMPLE SIZE
      DO 45 J=1,I
      PUT FAILURES INTO C1 ARRAY AND LOWER PART OF A1,A2 ARRAY
      DONT CONSIDER AS FAILURES IF FAILURE TIME EXCEEDS CUTPOINT
      IF(W0(J).GT.XX(I)) GO TO 44
      IF(W(J).GT.CTP) GO TO 44
```

```
      ICOUNT=ICCUNT+1
      W00(ICOUNT)=W0(J)
      A1(ICCUNT)=W11(J)
      A2(ICOUNT)=W22(J)
      C1(ICOUNT)=C(J)
   44 CONTINUE
   45 CONTINUE
      STATISTIC CANNCT EE CALCULATED IF NO FAILURES UNDER CURRENT
      CENSORING PCINT
      IF(ICOUNT.EQ.0) GC TO 994
      IC2=ICOUNT+1
      DO 245 J=1,I
      PUT NONFAILURES INTO UPPER PART OF A1,A2 ARRAY
      IF(W0(J).LE.XX(I)) GO TC 244
      IC2=IC2+1
      W00(IC2)=W0(J)
      A1(IC2)=W11(J)
      A2(IC2)=W22(J)
      C1(IC2)=C(J)
  244 CONTINUE
  245 CONTINUE
      SORT SUBSET OF FAILURES ON W00 (EFFECTIVE FAILURE TIME)
      DO 50 IJ=1,ICCUNT
      DO 50 IK=IJ,ICCUNT
      IF(IK.EQ.IJ) GO TO 50
      TMPW00=W00(IJ)
      TMPA1=A1(IJ)
      TMPA2=A2(IJ)
      TMPC1=C1(IJ)
      IF(W00(IK).LE.W00(IJ)) GO TO 60
      GO TO 50
   60 W00(IJ)=W00(IK)
      A1(IJ)=A1(IK)
      A2(IJ)=A2(IK)
      C1(IJ)=C1(IK)
      W00(IK)=TMPW00
      A1(IK)=TMPA1
      A2(IK)=TMPA2
      C1(IK)=TMPC1
   50 CONTINUE
      COMPUTATICNS FCR EACH SAMPLE SIZE I=1,N
      CALCULATE A0,A1,A2 AND THEIR MEANS
      A0MEAN=0.
      A1MEAN=0.
      A2MEAN=0.
      DO 70 J=1,I
      A0(J)=0.
      JJ=I-J+1
      DO 75 K=JJ,I
      WILCOXON STATISTIC
      ** NOTE:  TO CALCULATE USING WIICCXON, REPLACE THIS
      **        LINE ONLY.
      A0(J)=FLOAT(J)/FLCAT(I+1)
   75 CONTINUE
      A0MEAN=A0MEAN+A0(J)
      A1MEAN=A1MEAN+A1(J)
      A2MEAN=A2MEAN+A2(J)
   70 CONTINUE
      A0MEAN=A0MEAN/FLCAT(I)
      A1MEAN=A1MEAN/FLOAT(I)
```

```
      A2MEAN=A2MEAN/FLOAT(I)
      MAX=ICOUNT
      IF((I.EQ.ICOUNT).AND.(I.EQ.1)) GO TO 82
      IF(ICOUNT.LT.I) MAX=ICOUNT
      IF((ICOUNT.EQ.I).AND.(I.GE.2)) MAX=ICOUNT-1
      CALCULATE THE AVERAGE A0,A1,A2 ABOVE K FOR EACH K
      DO 80 K=1,MAX
      AA0(K)=0.
      AA1(K)=0.
      AA2(K)=0.
      KK=K+1
      DO 85 KL=KK,I
      AA0(K)=AA0(K)+A0(KL)
      AA1(K)=AA1(K)+A1(KL)
      AA2(K)=AA2(K)+A2(KL)
   85 CONTINUE
      AA0(K)=AA0(K)/FLOAT(I-K)
      AA1(K)=AA1(K)/FLOAT(I-K)
      AA2(K)=AA2(K)/FLOAT(I-K)
   80 CONTINUE
      GO TO 83
   82 AA0(1)=A0(1)
      AA1(1)=A1(1)
      AA2(1)=A2(1)
   83 CONTINUE
      K=MAX
      CALCULATE TK(0,1,2)
      TK0=0.
      TK1=0.
      TK2=0.
      DO 95 KK=1,K
      TK0=TK0+(C1(KK)-CMEAN)*(A0(KK)-AA0(K))
      TK1=TK1+(C1(KK)-CMEAN)*(A1(KK)-AA1(K))
      TK2=TK2+(C1(KK)-CMEAN)*(A2(KK)-AA2(K))
   95 CONTINUE
      TMP00=0.
      TMP10=0.
      TMP20=0.
      TMP11=0.
      TMP21=0.
      TMP22=0.
      CREATE COVARIANCE MATRIX
      DO 92 J=1,K
      TMP00=TMP00+(A0(J)*A0(J))
      TMP10=TMP10+(A1(J)*A0(J))
      TMP20=TMP20+(A2(J)*A0(J))
      TMP11=TMP11+(A1(J)*A1(J))
      TMP21=TMP21+(A2(J)*A1(J))
      TMP22=TMP22+(A2(J)*A2(J))
   92 CONTINUE
      VK00=((1./FLOAT(I))*(TMP00+((I-K)*AA0(K)*AA0(K))))-(A0MEAN*A0MEAN)
      VK10=((1./FLOAT(I))*(TMP10+((I-K)*AA1(K)*AA0(K))))-(A1MEAN*A0MEAN)
      VK20=((1./FLOAT(I))*(TMP20+((I-K)*AA2(K)*AA0(K))))-(A2MEAN*A0MEAN)
      VK11=((1./FLOAT(I))*(TMP11+((I-K)*AA1(K)*AA1(K))))-(A1MEAN*A1MEAN)
      VK21=((1./FLOAT(I))*(TMP21+((I-K)*AA2(K)*AA1(K))))-(A2MEAN*A1MEAN)
      VK22=((1./FLOAT(I))*(TMP22+((I-K)*AA2(K)*AA2(K))))-(A2MEAN*A2MEAN)
      PREPARE AND PASS TO IMSL INVERSION ROUTINE
      VK(1)=VK00
      VK(2)=VK10
      VK(3)=VK11
```

```
      VK(4)=VK20
      VK(5)=VK21
      VK(6)=VK22
      CALL LINV1P(VK,3,VKINV,0,D1,D2,ERROR)
      AN NON-ZERO ERROR INDICATES A (NEAR-)SINGULAR MATRIX
      IF(ERROR.GT.0) GO TO 93
      VK00=VKINV(1)
      VK10=VKINV(2)
      VK11=VKINV(3)
      VK20=VKINV(4)
      VK21=VKINV(5)
      VK22=VKINV(6)
      TNK IS THE SUM CF ((VK(0,J)/VK(0,0)*TK(J)).
      TNK=TK0+((VK10/VK00)*TK1)+((VK20/VK00)*TK2)
      XIK=CINVSQ*DSQRT(VK00)*TNK
      GO TO 91
   93 XIK=0.
      IF MATRIX IS (NEAR-)SINGULAR, VK MATRIX CANNOT BE INVERTED
      AND XI CANNOT BE CALCULATED.
   91 CONTINUE

      KEEP THREE LARGEST (ABSOLUTE VALUE) XIKS AND ASSOCIATED
      STATISTICS

      FIND WHERE TC INSERT NEW MEMBER OF THREE LARGEST XIKS
      DO 63 J=1,3
      IF(DABS(XIK).LE.DABS(XIKM(J))) GO TO 67
   63 CONTINUE
   67 J=J-1
      IF NOT ONE OF THREE LARGEST , DO NEXT CENSORING POINT
      IF(J.EQ.0) GO TO 66
      IF NEW ONE IS LEAST OF THE THREE, JUST REPLACE
      IF(J.EQ.1) GO TO 68
      OTHERWISE, MOVE LESSER ONES DOWN AND INSERT THIS ONE
      KK=J-1
      DO 69 JJ=1,KK
      XIKM(JJ)=XIKM(JJ+1)
      XXIM(JJ)=XXIM(JJ+1)
      IM(JJ)=IM(JJ+1)
      ICNTM(JJ)=ICNTM(JJ+1)
      CMEANM(JJ)=CMEANM(JJ+1)
      CINVSM(JJ)=CINVSM(JJ+1)
      A0MNM(JJ)=A0MNM(JJ+1)
      A1MNM(JJ)=A1MNM(JJ+1)
      A2MNM(JJ)=A2MNM(JJ+1)
      TK0M(JJ)=TK0M(JJ+1)
      TK1M(JJ)=TK1M(JJ+1)
      TK2M(JJ)=TK2M(JJ+1)
      DO 72 L=1,6
      VKM(JJ,L)=VKM(JJ+1,L)
   72 VKINVM(JJ,L)=VKINVM(JJ+1,L)
      TNKM(JJ)=TNKM(JJ+1)
   69 CONTINUE
      GO TO 71
   68 JJ=1
   71 CONTINUE
      XIKM(JJ)=XIK
      XXIM(JJ)=XX(I)
      IM(JJ)=I
      ICNTM(JJ)=ICOUNT
```

```
      CMEANM(JJ)=CMEAN
      CINVSM(JJ)=CINVSQ
      A0MNM(JJ)=A0MEAN
      A1MNM(JJ)=A1MEAN
      A2MNM(JJ)=A2MEAN
      TK0M(JJ)=TK0
      TK1M(JJ)=TK1
      TK2M(JJ)=TK2
      DO 64 L=1,6
      VKM(JJ,L)=VK(L)
   64 VKINVM(JJ,L)=VKINV(L)
      TNKM(JJ)=TNK
   66 CONTINUE
  994 CONTINUE
  995 CONTINUE
      DO 65 J=1,3
      PRINT 140,IM(J),XXIM(J)
      PRINT 980,CMEANM(J),CINVSM(J)
      PRINT 983,A0MNM(J),A1MNM(J),A2MNM(J)
      PRINT 984,TK0M(J),TK1M(J),TK2M(J)
      PRINT 985,VKM(J,1),VKM(J,2),VKM(J,3)
      PRINT 986,VKM(J,4),VKM(J,5),VKM(J,6)
      PRINT 989,VKINVM(J,1),VKINVM(J,2),VKINVM(J,3)
      PRINT 990,VKINVM(J,4),VKINVM(J,5),VKINVM(J,6)
      PRINT 987,TNKM(J)
      PRINT 988,XIKM(J)
   65 CONTINUE
   90 FORMAT(I4)
  100 FORMAT(F3.0,4F9.4)
  102 FORMAT(F9.4)
  441 FORMAT(1H0,'      ****** CURRENT CENSORING TIME=',F9.4,' ******',
     1//' *** THREE LARGEST (IN ABSCLUTE VALUE) XIKS ARE PRINTED WITH',
     2/' *** ACCOMPANYING STATISTICS IN ORDER.'///)
  140 FORMAT(//'**** SAMPLE SIZE=',I4,'  CURRENT EXPOSURE TIME=',F9.4)
  980 FORMAT('    C MEAN=',F20.9,'  CINVSQ=',F20.9)
  983 FORMAT('    MEAN(A0)=',F20.9,' MEAN(A1)=',F20.9,' MEAN(A2)=',F20.9)
  984 FORMAT('    TK0=',F20.9,' TK1=',F20.9,' TK2=',F20.9)
  985 FORMAT('    VK(0,0)=',F20.9,' VK(1,0)=',F20.9,' VK(1,1)=',F20.9)
  986 FORMAT('    VK(2,0)=',F20.9,' VK(2,1)=',F20.9,' VK(2,2)=',F20.9)
  989 FORMAT('    VK-1(0,0)=',F20.9,' VK-1(1,0)=',F20.9,' VK-1(1,1)=',
     1F20.9)
  990 FORMAT('    VK-1(2,0)=',F20.9,' VK-1(2,1)=',F20.9,' VK-1(2,2)=',
     1F20.9)
  987 FORMAT('    TNK=',F20.9)
  988 FORMAT(' *** XI(K)=',F20.9)
    3 FORMAT(1H0,'*** SAMPLE SIZE OF',I4,' MUST BE AT LEAST 3')
    4 FORMAT(1H0,'*** SAMPLE SIZE OF',I4,' OVER THE LIMIT OF 2000')
      STOP
      END
```

BIOSTATISTICS: Statistics in Biomedical, Public Health
and Environmental Sciences, edited by P.K. Sen

ROBUSTNESS OF PROGRESSIVELY CENSORED COMPARISON OF EXPONENTIAL SURVIVAL CURVES TO DEPARTURES FROM EXPONENTIAL DISTRIBUTION

C.E. Davis and Erica Brittain

Department of Biostatistics
University of North Carolina

Breslow and Haug (1972) proposed a statistic for comparing two exponential survival curves as the data are accumulating. Simulations indicate that, using boundaries of Chatterjee and Sen (1973), a sample of 200 or more is sufficient for the asymptotic results to be useable for the null distribution. Further simulations indicate that the distributional result is robust to the constant hazard assumption when the hazard is "nearly" constant, but can be very poor if the hazard is increasing or decreasing rapidly. Power comparisons with a rank statistic indicate little difference between the parametric and rank procedure.

KEY WORDS AND PHRASES: Early stopping in clinical trials, Wiener process, Wilcoxon test, Savage scores

INTRODUCTION

There is great interest among persons analyzing data from long term clinical trials in using statistical procedures which allow early termination of the trial if the data warrants. If one analyzes data as it is accumulating with a fixed level of significance, it is well known that the size of the test is no longer α. A number of methods have been proposed for providing a statistical test of accumulating data while controlling the size of the test.

Among these is a procedure proposed by Breslow and Haug (1972) which assumes that the failure distributions in the two treatment groups are exponential. Breslow and Haug show that the statistic proposed converges in distribution to a Wiener process and therefore the statistic has independent increments for large samples.

In practice, one rarely knows whether the failure distribution is exponential, and in many cases one would doubt this assumption *a priori*. For example, if the study is composed of both males and females and if the hazard for males differs from that for females, then the failure distribution of the combined sample no longer is exponential. A second possibility is if the hazard is increasing (or decreasing) with time as it might if the hazard is related to age. In either of these cases, the assumption of an exponential failure distribution would not hold, and the asymptotic distributional result might not be adequate for practical use. In this manuscript, the robustness of the Breslow-Haug statistic to departures from the exponential distribution will be investigated via simulation studies.

For comparison, results on power and early stopping for rank statistics will also be given.

DESCRIPTION OF PROCEDURE

Assume two treatment groups with n_i patients and distribution of failure times

$$F_i(x) = 1-e^{\lambda_i x}; \lambda_i > 0, x > 0; i=1, 2.$$

The hypothesis to be tested is $H_0 : \lambda_1/\lambda_2=1$ versus either a one or two sided alternative; i.e. $H_{a1}:\lambda_1/\lambda_2>1$ or $H_{a2}:\lambda_1/\lambda_2\neq1$. Let X_{ij} and Y_{ij} be the survival and entry times of the jth patient in the ith treatment group; i=1, 2; j=1, 2. . .n_i. Define $\delta(x)$ = 0 or 1 according as $x < 0$ or $x > 0$ and let

$$k_i(t) = \sum_{j=1}^{n_i} \delta(t-X_{ij}-Y_{ij})$$

and

$$S_i(t) = \sum_{j=i}^{n_i} \min[\max(0,t-Y_{ij}), X_{ij}]; i=1, 2.$$

For time t, $k_i(t)$ is the number of failures and $S_i(t)$ is the total exposure time in the ith group. The maximum likelihood estimate of the hazard based on the data available at time t is

$$\hat{\lambda}_i(t) = k_i(t)/S_i(t); i=1, 2.$$

Let t_k be the time of the kth failure and define

$$X_k = \begin{cases} k \log[\hat{\lambda}_1(t_k)/\hat{\lambda}_2(t_k)] & \text{if } \hat{\lambda}_1(t_k) > 0 \text{ and } \hat{\lambda}(t_k) > 0 \\ 0 & \text{otherwise.} \end{cases}$$

Note that $k=k_1(t_k) + k_2(t_k)$. Breslow and Haug (1972) show that X_k converges to a Wiener process with variance 4. In view of this result, any method for monitoring the accumulating data which assumes independent increments could be used for constructing boundaries. These include the methods of O'Brien and Fleming (1979), Pocock (1977) or Chatterjee and Sen (1973). In the results reported here, the Chatterjee and Sen boundary, which is a horizontal boundary, will be used in view of its simplicity. Results on Wiener processes derived by Chatterjee and Sen can be used to show that for sufficiently large n

$$p(\max_{1\leq k\leq n} |X_k| \geq 2x \sqrt{n}) = 1 - \sum_{m=-\infty}^{\infty}(-1)^m (\Phi[(2m+1)x]-\Phi[(2m-1)x])$$

$$\doteq 4[1-\Phi(x)] \qquad (2.1)$$

and

$$P(\max_{1 \leq k \leq n} X_k \geq 2x\sqrt{n}) = 2[1-\Phi(x)], \tag{2.2}$$

where $\Phi(x)$ is the c.d.f. of the standard normal distribution.

This suggests the following procedure. The statistic X_k is computed after each failure. If X_k exceeds the critical values for the two-sided or one-sided alternatives given by (2.1) and (2.2) respectively, the trial is stopped with rejection of the null hypotheses.

The formulation above assumes that every patient in the trial will be followed until failure. If the trial is designed to end with the (np)th failure ($0<p<1$), then one replaces n with np in (2.1) and (2.2) and the results remain valid [see Chatterjee and Sen (1973)].

The distributional results used to obtain the boundaries defined by (2.1) and (2.2) are large sample results. A natural question is: how large must the sample size be before these boundaries are useful? As the derivation of the small sample properties of the statistic $\max_k X_k$ is mathematically intractable, a small simulation study was conducted to answer this question for the two-sided alternative. All simulations described in this section and hereafter were conducted for both $\alpha = .05$ and $\alpha = .10$. Only the results for $\alpha = .05$ are presented, however, as those for $\alpha = .10$ follow the same pattern and consequently add no information. The results for the case when all survival times are observed ($p=1$), are in Figure 1. Each point in the figure is based on 1000 samples with the exception of that for $n = 400$ which is based on 2000. It is apparent from the figure that for small samples, the procedure is conservative and that the results are useful for samples of 200 or greater (100 patients in each group). Figure 2 gives results of the simulation experiment for $n = 50$, 100, 200, or 400 with four values of p. The procedure apparently rejects H_0 too often when $p < 1$ although it should be noted that all of the points in Figure 2 are within sampling variation of 0.05. It is interesting to note that with $n = 100$, the distributional approximation is closer with less data ($p=.5$ vs. $p=1$).

In summary, if all survival times are to be observed, the asymptotic distribution is adequate for samples of 200 or larger. If $p < 1$, some caution must be used in applying the results. The utility of the asymptotic results depend on both n and p and the current study is too small to draw general conclusions.

ROBUSTNESS AND POWER OF THE PROCEDURE

A second simulation study was conducted to determine if the method is robust against departures from the constant hazard assumption. Samples from a Weibull distribution with the same scale parameter and several shape parameters were generated. The results are in Figure 3. It is obvious that the method is not robust when all survival times are observed, rejecting the hypothesis too frequently for decreasing hazard and not often enough for increasing hazard. For p = 1/2 or 1/4, the method is reasonably robust, especially when the shape parameter is near 1.

The power and early stopping characteristics of the method were compared with the rank procedures based on Chatterjee and Sen's results as described by Koziol and Petkau (1978) and Davis (1978). Joe, Koziol and Petkau (1981) have shown that the Savage scores procedure is conservative and have proposed a modification which has been used in the following.

Table I contains the comparative powers when the underlying distributions are exponential. For the cases summarized, the parametric procedure and the Savage scores procedure have nearly identical results. The Wilcoxon score test has better early stopping properties at the cost of a loss in power.

In some trials, it is assumed that the hazard in the treatment group decreases with time while that in the control group is constant. For example, in a trial of a cholesterol lowering drug, one would hypothesize that the hazard would decrease the longer the participant took the drug. The power and stopping times of the methods have been compared when one sample is from an exponential distribution and the other is from a Weibull distribution with decreasing hazard. The results are summarized in Table II. As with exponential failures, the power of the parametric statistic is only slightly greater than the Savage score statistic, but is considerably greater than that for the Wilcoxon score statistic.

CONCLUSION

Since the parametric procedure is not robust to departures from the constant hazard assumption and the Savage score statistic has essentially the same power and stopping characteristics, one can conclude that the parametric procedure has little to offer over the rank procedure as a method of monitoring a clinical trial. The results here indicated that this is the case when all patients enter the study simultaneously. However, if the patients enter over an extended period of time, the rank procedures must be modified as described by Majumdar and Sen (1978). In this situation, the Savage scores procedure may not be as efficient. However, these results do not speak to this issue.

Table I. Power and Mean Stopping Time Achieved by the Three Procedures When the Underlying Distributions are Exponential 1000 Samples

	n=200, p=1					
	λ_1/λ_2=.5		λ_1/λ_2=.75		λ_1/λ_2=.90	
	Power	Stop	Power	Stop	Power	Stop
Parametric	1.00	90	.49	173	.11	195
Savage	1.00	90	.47	172	.11	194
Wilcoxon	.98	69	.33	166	.09	191
	n=200, p=1/2					
	λ_1/λ_2=.5		λ_1/λ_2=.75		λ_1/λ_2=.90	
	Power	Stop	Power	Stop	Power	Stop
Parametric	.92	63	.30	92	.09	98
Savage	.92	64	.30	92	.09	98
Wilcoxon	.89	60	.27	91	.09	97

Table II. Power and Mean Stopping Time Achieved by the Three Procedures When One Underlying Distribution is Exponential and the Other Weibull 500 Samples

	n=200, p=1					
	Shape Parameter=.50		Shape Parameter=.75		Shape Parameter=.90	
	Power	Stop	Power	Stop	Power	Stop
Parametric	1.00	47	1.00	77	.87	149
Savage	1.00	50	1.00	79	.80	150
Wilcoxon	1.00	39	1.00	64	.53	152
	n=200, p=1/2					
	Shape Parameter=.50		Shape Parameter=.75		Shape Parameter=.90	
	Power	Stop	Power	Stop	Power	Stop
Parametric	1.00	38	.99	60	.38	92
Savage	1.00	41	.98	61	.38	92
Wilcoxon	1.00	39	.97	59	.32	91

FIG. 1. Observed Significance Level as a Function of Total Sample Size When all Survival Times are Observed

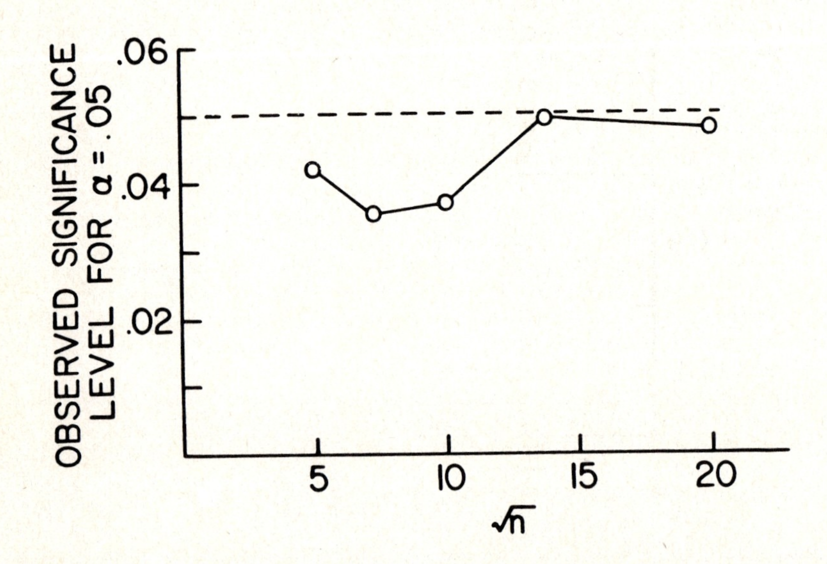

FIG. 2. Observed Significance Level as a Function of Proportion of Survival Times Observed

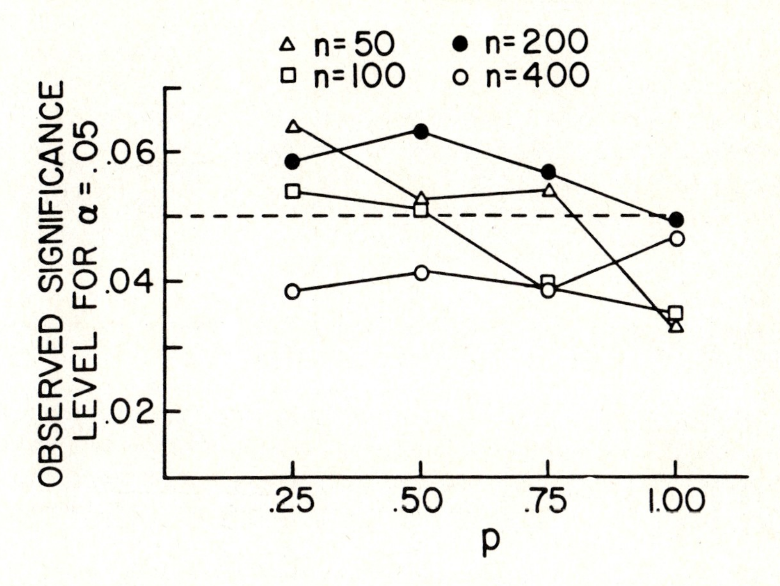

FIG. 3. Observed Significance Level as a Function of the Shape Parameter of the Weibull Distribution

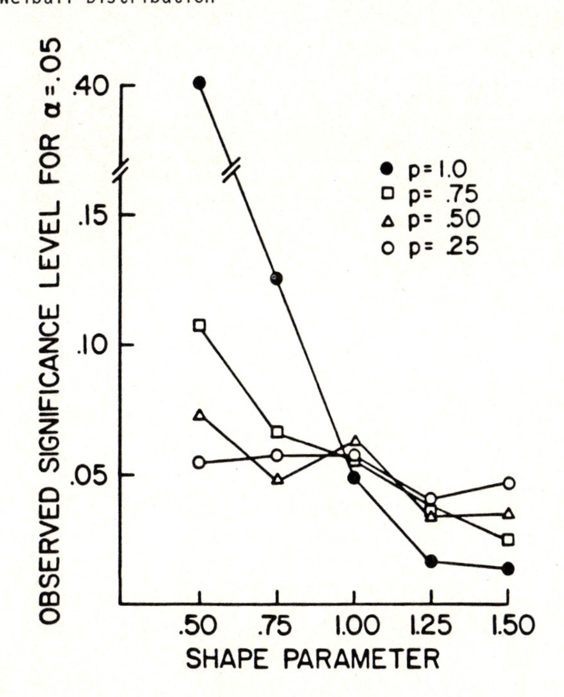

ACKNOWLEDGEMENTS: This work was supported by U.S. National Heart, Lung, and Blood Institute Contract NIH-NHLBI-71-2243 from the National Institutes of Health.

REFERENCES

[1] Breslow, N. and Haug, C., Sequential comparison of exponential survival curves, J. Amer. Statist. Assoc. 67 (339):691-697 (1972).

[2] Chatterjee, S.K. and Sen, P.K., Nonparametric testing under progressive censoring, Calcutta Statist. Assoc. Bull. 22 (1):13-50 (1973).

[3] Davis, C.E., A two sample Wilcoxon test for progressively consored data, Communications in Statistics - Theory and Methods A7 (4):389-398 (1978).

[4] Joe, H., Koziol, J.A. and Petkau, A.J., Comparison of procedures for testing the equality of survival distributions, Biometrics 37 (2): 327-40 (1981).

[5] Koziol, J.A. and Petkau, A.J., Sequential testing of the equality of two survival distributions using the modified savage statistic, Biometrika 65: 615-623 (1978).

[6] Majumdar, H. and Sen, P.K., Nonparametric testing for simple regression under progressive censoring with staggering entry and random withdrawal, Communications in Statistics - Theory and Methods A7:349-371 (1978).

[7] O'Brien, P.C. and Fleming, T.R., A multiple testing procedure for clinical trials, Biometrics 35:549-556 (1979).

[8] Pocock, S.J. and Simon, R., Sequential treatment assignment with balancing for prognostic factors in the controlled clinical trial, Biometrics 31 (1): 103-116 (1975).

[9] Pocock, S.J., Group sequential methods in the design and analysis of clinical trials, Biometrika 64:191-199 (1977).

BIOSTATISTICS: Statistics in Biomedical, Public Health
and Environmental Sciences, edited by P.K. Sen

THE USE OF PROGNOSTIC FACTORS IN THE PLANNING AND ANALYSIS OF CANCER CLINICAL TRIALS

Edmund A. Gehan

University of Texas System Cancer Center

Recent literature on the utilization of prognostic factors in clinical studies is reviewed with the objective of making recommendations that should prove useful in the conduct of cancer clinical trials. Since heterogeneity among patients with respect to prognostic factors often leads to more variability in results than is dependent upon the particular therapy administered, it is of prime importance to account properly for prognostic features of the patients. Valid inferences concerning the relative efficacies of treatments can be made only when there is proper adjustment for prognostic features. An example is given in acute leukemia which illustrates some of the main points and also outlines a strategy for the evaluation of new treatments in previously untreated patients.

KEY WORDS: Prognostic factors, planning and analysis of cancer clinical trials, acute leukemia

I. INTRODUCTION

In clinical trials comparing two or more treatments, the primary objective is to evaluate the relative effectiveness of the treatments. For many kinds of cancer, there is often more variability in outcomes following treatment depending upon the prognostic characteristics of the patients, rather than the particular therapeutic modality employed. Consequently, it is of great importance to take account of the prognostic features of the patients in both the planning and analysis of clinical trials so that correct inferences can be drawn concerning the relative efficacies of treatments.

In the last few years, several articles (Armitage and Gehan (1974), Zelen (1975), Brown (1980), Simon (1984), and Byar (1984)) have appeared considering various aspects of prognostic factors studies. The objective of this paper is to review the recent literature, especially those aspects of importance in the planning and analysis of cancer clinical trials and to make recommendations that should be useful in the conduct of cancer clinical trials. An example is given of a study in acute leukemia illustrating some of the main points.

II. PLANNING OF CLINICAL TRIALS

Knowledge of the prognostic characteristics of the patients can be used: to aid in the choice of treatments, to provide a basis for stratifying patients in a randomized clinical trial, to provide a basis for choosing a control group of patients in a non-randomized clinical trial, or in the development of statistical models used in the assignment of treatments.

Having knowledge of prognostic factors can help in deciding what kinds of treatments to propose in a clinical trial and in determining the eligibility of patients. For example, in successive studies of pediatric rhabdomyosarcoma, it was determined that the pathological characteristics of the tumor had a major influence on outcome. Consequently, a current study involves different comparisons of treatments depending upon the pathological type of tumor. Patients with more favorable prognosis will receive a treatment designed to be equally efficacious to a standard treatment with fewer complications, while patients with an unfavorable pathological type will receive multi-modality treatments that will be more intensive than those administered in previous studies. In general, being able to divide patients into sub-groups based upon expected prognosis permits planners of clinical trials to concentrate on making proposals for new treatments that are most appropriate for each sub-group. Without the capability of grouping patients by prognosis, it is likely that comparisons of therapies will be relatively inefficient from a statistical viewpoint (through not taking into account prognostic factors) and overall differences between therapies may not be evident within all sub-groups of patients.

In a continuing research program for a particular type of cancer, it is sometimes necessary to compromise between the goal of refining the definitions of patient sub-groups (and the treatment(s) to be employed within each) and that of obtaining definitive relative evaluations of treatments. Having too many sub-groups will make it difficult to accrue sufficient patients to obtain a meaningful comparison of patients within each sub-group. Having only a single group of patients makes it easier to accrue patients to a clinical trial and to achieve an overall comparison of two treatments, but it may be that sub-groups exist for which there are real treatments by sub-group interactions. If the sample size for the trial is determined without consideration of sub-groups, it is possible that an important treatment by sub-group interaction would not be detected.

Patients with a given type of cancer may differ in prognosis according to such characteristics as age, stage of disease, bone marrow status, and prior therapy. In non-Hodgkin's lymphoma, for example, histology and stage are important patient characteristics related to chance of response. In a randomized study designed to compare two treatments, patients could be stratified by these characteristics prior to allocation of treatment. Randomizing patients to treatments within each stratum would assure that patients on the two treatments were comparable in prognosis and number. Most statisticians favor a stratified random allocation of patients, especially since having comparable groups of patients is a sine qua non of a controlled clinical trial (Hill (1962)). Usually, there are two to eight strata and, in the analysis, treatments can be compared within strata and an overall comparison made by adding results over all strata. It is generally not practical to choose more than eight stratification categories, since ordinarily there will not be sufficient patients to distinguish prognosis among more than eight sub-groups. The extreme case of too many strata is when each patient constitutes a separate sub-group; in this circumstance, the advantages of stratification have completely disappeared and the design is equivalent to a simple randomized design.

Alternatively, a randomized clinical trial could be conducted by randomizing patients to treatments without regard to strata. Brown (1980) summarizes the arguments for and against stratification in clinical studies. Peto et al. (1976) have argued that in large clinical trials randomization without stratification will tend to have balanced treatment allocations within each stratification group. When there are N patients in a given cell, the amount of information achieved by complete balance (i.e., by assigning exactly one-half of the N patients to each treatment) can be achieved on the average by randomly assigning N+1 patients to the treatment arms without regard to balance. Peto et al. suggest that having different stratifications makes the design and conduct of a clinical trial more complex, possibly discouraging physicians from entering patients. Further, though prognostic factors may not be bases for stratification, they can be used for analysis and interpretation of the study. Brown (1980) finds the arguments against the use of prognostic factors in randomization of treatment assignments "logically valid but not persuasive." He argues that one should not randomize without stratification, a procedure expected to perform reasonably well on the average, when it is possible to achieve balance in every trial by stratification.

Adaptive stratification methods (Simon (1977)) allow more factors to be balanced for than do conventional procedures. Simon's adaptive procedure is designed to minimize the prognostic "distance" between treatment groups, distance being defined in terms of the important prognostic features. Taves (1974) replaces randomization by assignment of patients to achieve the best balance with respect to prognostic features. Finally, treatments have been allocated to individual patients within strata defined by mathematical models giving a prediction of probability of response. An example will be given of such a trial in acute leukemia.

In a non-randomized clinical trial, it is essential that the important patient characteristics related to prognosis be known so that there is an adequate basis for choosing a control group. Conducting a non-randomized clinical trial (Gehan and Freireich (1974)) can be very appealing to some planners of clinical trials, since substantially fewer patients are required and there are no ethical dilemmas for the clinical investigators either in having to decide whether to randomize patients or whether to stop the study early when results favor the new treatment. In the non-randomized study, a clinical investigator should always be administering the treatment program that he considers most likely to be successful. However, for a valid non-randomized study, it is always necessary to assume that factors associated with chronological time (e.g. patterns of referral for patients, factors related to selection of patients for study, diagnostic procedures, etc.) either do not change or are not of sufficient importance to explain observed treatment differences. If there are some patient characteristics related to prognosis that are not used in choosing a comparable control group, it must be assumed that these factors have the same distribution in each treatment group. In a randomized study, this assumption is valid by virtue of the randomization. Consequently, a non-randomized clinical trial should be undertaken only after having given careful consideration to the possible sources of bias and other uncertainties. A comprehensive knowledge of prognostic factors and methods of analysis taking appropriate account of them are essential prerequisites for carrying out an acceptable non-randomized study.

In some clinical trials applications, it may be possible to derive a regression model relating patient characteristics measured at the start of treatment to prognosis. Such a prediction could be of interest to a physician caring for patients or to the individual patients themselves (Gehan (1959)) or it could be used as a basis for stratifying and treating patients (Miettinen (1976)).

In a recent clinical trial in acute leukemia conducted at the University of Texas System Cancer Center (Keating et al. (1982)), a logistic regression model relating probability of complete remission to prognostic features (age, antecedent hematologic disorder, temperature, BUN value, hemoglobin value, and liver size) was used for the assignment of patients to treatment in the context of a strategy for developing new treatments for previously untreated patients. This trial is an interesting example of the utilization of prognostic factors in planning and analysis and further details are given in Section IV.

III. ANALYSIS OF CLINICAL TRIALS

Armitage and Gehan (1974) give three major reasons for utilizing prognostic factors in analyses. First, even in randomized clinical trials, there may be some differences between treatment groups in the distribution of prognostic variables. When these differences would cause a biased comparison between treatments, this should be corrected for in the analysis. In non-randomized clinical trials, it is of especial importance to take account of major prognostic characteristics in the comparison of treatment groups. Secondly, if prognostic variables have a strong relationship with outcome, much of the patient heterogeneity can be explained by these variables. Carrying out an analysis adjusting for patient characteristics will reduce the residual random variation and make comparisons between treatments more precise. Thirdly, in some circumstances, interactions between treatments and prognostic variables may be detected, i.e., the relative efficacies of the treatments differ according to the patient's prognosis.

In the analysis of a clinical trial, a common approach is to first compare the treatment groups with respect to possibly important prognostic variables. As pointed out by Simon (1984), this is "necessary but not sufficient," since lack of a statistically significant difference between treatment groups does not imply comparability. Dales and Ury (1978) demonstrate that it is possible to have relatively large imbalances in unimportant patient characteristics that do not bias a treatment comparison but relatively small differences in major prognostic determinants that do. Consequently, it is important to correct for this type if imbalance in the analysis.

There are two basic approaches to taking account of prognostic factors in analyses of clinical trials: methods based upon stratifications of the data and methods based on regression models. The approach based upon stratification is to divide the patients into disjoint sub-groups according to major prognostic features, calculate a measure of the difference between treatments within each sub-group, and form a weighted average of the differences over all sub-groups. This leads to a single measure of average treatment difference which is adjusted for the variability in patient outcome among strata. To compare response rates, Mantel developed a chi-square test (Mantel and Haenszel (1959)) which provides a way of evaluating the statistical significance of observed treatment differences over all strata. To compare survival experiences by treatment, Breslow (1984) describes logrank and generalized Wilcoxon type tests for stratified data and gives the circumstances for which each type of test may be most appropriate. The logrank test is most sensitive to differences between treatments arising late and is most powerful when the hazard functions are proportional among treatment groups. The Wilcoxon type tests are most sensitive to differences between treatments in the early part of the survival curves.

The second basic approach to taking account of prognostic factors in comparing treatments involves use of statistical regression models, logistic models for response (Cox, D.R. (1970)) or (most commonly) proportional hazards models (Cox,

D.R. (1972)) for survival data. A good exposition of the use of regression models in the analysis of clinical trials data has been given by Simon (1984), who gives procedures for fitting models to data, methods for examining the adequacy of fit, methods for selection and validation of models, and some uses of regression models in clinical oncology. He emphasizes the importance of calculating confidence intervals for treatment differences in interpreting results, rather than overuse of statistical significance tests. Having a confidence interval for a treatment difference, after adjustment for prognostic factors, indicates the precision of a statement that can be made concerning the relative merits of treatments. Further, it can indicate to what extent an increase in the number of patients might lead to a more precise statement of the difference between treatments.

A major question in the analysis of clinical trials is the extent to which sub-group analyses should be undertaken for the detection of treatment by covariate interactions. Of course, if the clinical trial was designed to detect treatment differences within major sub-groups of patients with a reasonable statistical significance level and power, then it is clearly appropriate to investigate these sub-groups for treatment by covariate interactions. A regression model with a separate treatment parameter for each sub-group can be compared with a model with a single treatment parameter by a likelihood ratio test. Evidence of significant interaction effect should be followed by further analyses to possibly explain the variation in relative treatment efficacy. However, if the clinical trial was planned only to have a sufficient number of patients for detecting a treatment difference over all sub-groups, then any significant treatment by sub-group interactions should be interpreted with extreme caution. This is especially true when numerous patient sub-groups were examined for possible interaction effects. Simon (1982) has recommended that, for a particular patient characteristic, statistical significance tests of treatment differences should not be performed separately by sub-group unless an overall statistical test for interaction was statistically significant.

Byar and Corle (1977) have indicated the importance of qualitative treatment co-variate interactions, especially because of their implications for the treatment of patients. If such interactions are present, there may be an optimal treatment or treatment of choice, i.e., some patients with a given disease should be treated in one way but other patients with the same disease should be treated differently. The choice of treatment depends upon the characteristics of the patients. This concept is illustrated by a randomized clinical trial of estrogen therapy vs. placebo for patients with Stage III or IV prostate cancer (Byar (1984)) where it was demonstrated that estrogen has a genuine effect on the disease for some patients, but has toxicity which outweighs its benefits for other categories of patients. An excess of cardiovascular deaths was noted among patients treated with estrogen whose optimal treatment was placebo, whereas an excess of deaths due to prostatic cancer was noted among patients treated with placebo whose optimal treatment was estrogen.

IV. EXAMPLE OF THE USE OF REGRESSION MODELS IN THE PLANNING, CONDUCT AND ANALYSIS OF A CLINICAL TRIAL IN ACUTE LEUKEMIA

An example of a non-randomized clinical trial in acute leukemia conducted at the University of Texas System Cancer Center (protocol DT7995) will be given to illustrate the ideas discussed. The overall objective of the study was to evaluate new strategies of treatment for previously untreated patients so that benefit/risk ratios would be maximized at defined time points during the patient's course of disease. Both at the beginning of remission induction and maintenance stages, statistical regression models were utilized to separate patients into prognostically favorable and unfavorable sub-groups. The strategy was to administer innovative treatments to patients with unfavorable prognosis (either for induction of remission or length of remission), while administering more conventional treatments to patients with favorable prognosis. The ethical argument was that a patient expected to have unfavorable prognosis after the administration of standard therapies, would not be compromised by the administration of innovative treatments. As promising treatments are identified in patients with unfavorable prognosis, they will be administered to patients with favorable prognosis.

Figure 1 gives the design and strategy of study for each newly diagnosed patient with adult acute leukemia. In the remission induction phase, patients with $<.4$ probability of complete remission received the new treatment regimen AMSA combined with OAP, while patients with predicted probabilities $\geq.4$ recieved Ad-AOP in a protected environment, if available. For patients achieving complete remission (CR) when the probability of CR was high, these patients were administered remission maintenance treatment with combination chemotherapy programs if the predicted probability of remaining in CR longer than 1 year was also high. For patients with a low probability of being in complete remission for longer than 1 year, patients received either bone marrow transplant or combined maintenance treatment with AMSA followed by POMP, depending upon whether an HLA compatible donor was available or not.

For patients achieving CR when the predicted probability was low, these patients received maintenance treatment with AMSA-OAP followed by POMP. In all cases, patients failing to respond or relapsing from complete remission were entered into a second induction treatment protocol.

For the assignment of patients in the remission induction phase, a logistic regression model was derived from a study of 325 previously untreated patients with acute leukemia, all of whom received a combination of Adriamycin and OAP treatment between 1973 and 1977. More than 20 possible prognostic factors were considered for their relationship to probability of complete remission, length of complete remission, and survival. The final form of the regression model was as follows:

$$\log\left\{\frac{p_i}{1-p_i}\right\} = .495 \quad -.744\,(\text{Age} - 2.54) - 1.680\,(\text{AHD} - .19)$$
$$-1.109\,(\text{Temp.} - 1.34) - 1.031\,(\text{BUN} - 1.12)$$
$$+1.039\,(\text{Hgb.} - 1.11) - 1.297\,(\text{Liver size} - .05)$$

where p_i = probability of complete remission for the ith patient

Age: 1 = less than 20 yrs, 2=20-49, 3=50-64, 4=≥65

Antecedent hematological disorder: 0=no, 1=yes

Temperature: 1=less than 101°F, 2=greater than or equal to 101°F.

BUN: 1=less than 12 gm%, 2=greater than or equal to 12 gm%

Liver: 0=not enlarged, 1=enlarged greater than 5 cm.

This logistic regression model was tested in two ways, first by fitting it to the data from which it was derived and, secondly, by fitting it to data from 107 patients studied prospectively. Observed complete remission rates were in close accord with the predicted probabilities of complete remission as obtained from the model.

As the clinical trial proceeded, initial results with AMSA-OAP were substantially better than expected and since an important objective was to evaluate this treatment in previously untreated patients, the criterion for assignment of patients in the remission induction phase was changed so that patients were classified as having favorable or unfavorable prognosis depending upon whether $p_i \geq .6$ or $p_i < .6$, respectively. A further complication was that relatively more patients in this study were treated in a protected environment than in the historical control group. Consequently, two general approaches were taken in the analysis of the data from the remission induction phase. First, a paired patient analysis was carried out, each patient in this study being paired with a pairmate from the historical series in accord with the way the study was conducted and some details of this analysis will be given. Secondly, several logistic regression analyses were done, including one variable for treatment (treatment in current study vs. treatment in historical control series) and various sets of prognostic factors.

One set of prognostic factors was that used for the assignment of treatment and separate analyses were carried out within sub-groups of patients with favorable or unfavorable prognosis. Two other logistic regression analyses were performed separately in patients with favorable and unfavorable prognosis, each analysis adjusting for prognostic factors relevant within the appropriate sub-group. Protected environment status was included as a variable in each regression equation. Both the paired data and logistic regression analyses lead to the same conclusions regarding the relative efficacies of the treatments. With a complex study design, performing multiple analyses leading to the same general conclusions gives more credence to the conclusions reached.

Patients from the total of 325 patients in the historical control group were selected as pairmates for each of the patients in the acute leukemia study (DT7995). The main basis for the pairing was the predicted probability of response derived from the logistic regression equation used in treatment assignment. When possible, a control patient was selected to have an equivalent predicted probability of response and protected environment (PE) status. When there were several pairmates meeting these criteria, the patient chosen was nearest to equivalent in age and diagnosis status. In the unfavorable prognosis patients, relatively more patients in the DT7995 study received treatment within a PE, so a stratified analysis was accomplished to examine patients matched and unmatched with respect to PE status. Because of the relatively large number of patients in the historical control series, the pairing process was successful in that there was zero average difference in predicted probability of response between patients in the DT7995 study and the control series, both for patients with favorable and unfavorable prognosis.

The conduct of the study resulted in two major sub-groups of patients which are shown in Table 1. The primary interest in the DT7995 study was in whether or not AMSA-OAP was equal to or superior in effectiveness to Ad-OAP in patients with unfavorable prognosis. However, since patients with favorable prognosis received Ad-OAP treatment both in the historical control and current study, it would be expected that the response rate would be about the same in both groups, unless there was some evidence of a real change in response rate with chronological time between 1973-1977 (historical control study) and the DT7995 study (1979-1983).

Table 2 shows that the 40% complete remission rate in AMSA-OAP patients that were matched for PE status was 10% higher than that for Ad-OAP patients in the historical control group. However, a paired data analysis indicated that the difference was not statistically significant (P=.31). When the AMSA-OAP treated patient was in a PE, Table 2 indicates that the 55% complete remission rate was 16% higher than that for an Ad-OAP patient not treated in a PE and the difference was not statistically significant (P=.36). Though there was an advantage in complete remission rate for patients receiving AMSA-OAP both for PE matched and unmatched patients, weighting the combined results by sample size did not demonstrate an overall advantage for AMSA-OAP treated patients (P=.10).

Table 1. Treatments Administered in Acute Leukemia (DT7995) Study and Historical Control Series

Prognosis	DT7995	Historical Controls
Unfavorable	AMSA-OAP ($p_i<.6$)*	Ad-OAP ($p_i<.6$)
Favorable	Ad-OAP ($p_i\geq.4$)	Ad-OAP ($p_i\geq.4$)

*p_i = probability of complete remission for the ith patient

Table 2. Complete Remissions (CR) in Paired Patients with Unfavorable Prognosis in the DT7995 Study and Historical Control Series According to Protected Environment (PE) Status

Matched for PE Status

		DT7995-AMSA-OAP CR	Fail	Total(%)
Historical Control	CR	8	9	17(30)
Ad-OAP	Fail	15	25	40(70)
	TOTAL(%)	23(40)	34(60)	57

Not Matched for PE Status

DT7995 Patient in PE, Control not in PE

		DT7995-AMSA-OAP CR	Fail	Total(%)
Historical Control	CR	6	7	13(39)
Ad-OAP	Fail	12	8	20(61)
	TOTAL(%)	18(55)	15(45)	33

The responses in the favorable prognosis patients are given in Table 3 which shows that there was essentially no difference in complete remission rate for paired patients receiving Ad-OAP treatment in 1973-1977 (complete remission rate of 71%) compared with that for patients receiving Ad-OAP on the DT7995 study (complete remission rate of 73%) (=p.88). Hence, this suggests that there has been no significant improvement in complete remission rate with chronological time over the period of these two studies, at least for the favorable prognosis patients, and suggests that it is reasonable to compare treatment programs in the unfavorable prognosis patients.

Table 3. Complete Remissions (CR) in Paired Patients with Favorable Prognosis in the DT7995 Study and Historical Series

		DT7995-Ad-OAP		
		CR	Fail	Total(%)
Historical Series	CR	56	21	77(71)
Ad-OAP	Fail	23	8	31(29)
	TOTAL(%)	79(73)	29(27)	108

Though no significant advances in remission induction treatment were demonstrated in the DT7995 study, a new treatment program (AMSA-OAP) was evaluated in previously untreated patients and was demonstrated to be at least as good and possibly better than the previous treatment program (Ad-OAP) in unfavorable prognosis patients. This study has continued so that all previously untreated patients received AMSA-OAP treatment without regard to predicted probability of complete remission. Hence, the DT7995 study illustrates the use of prognostic factors in both the design and analysis phases of the study. In clinical oncology, the study design provides an ethical way of evaluating new treatments in previously untreated patients with acute leukemia.

Figure 1
Design of Study and Strategy for Treatment
of Acute Leukemia Patients

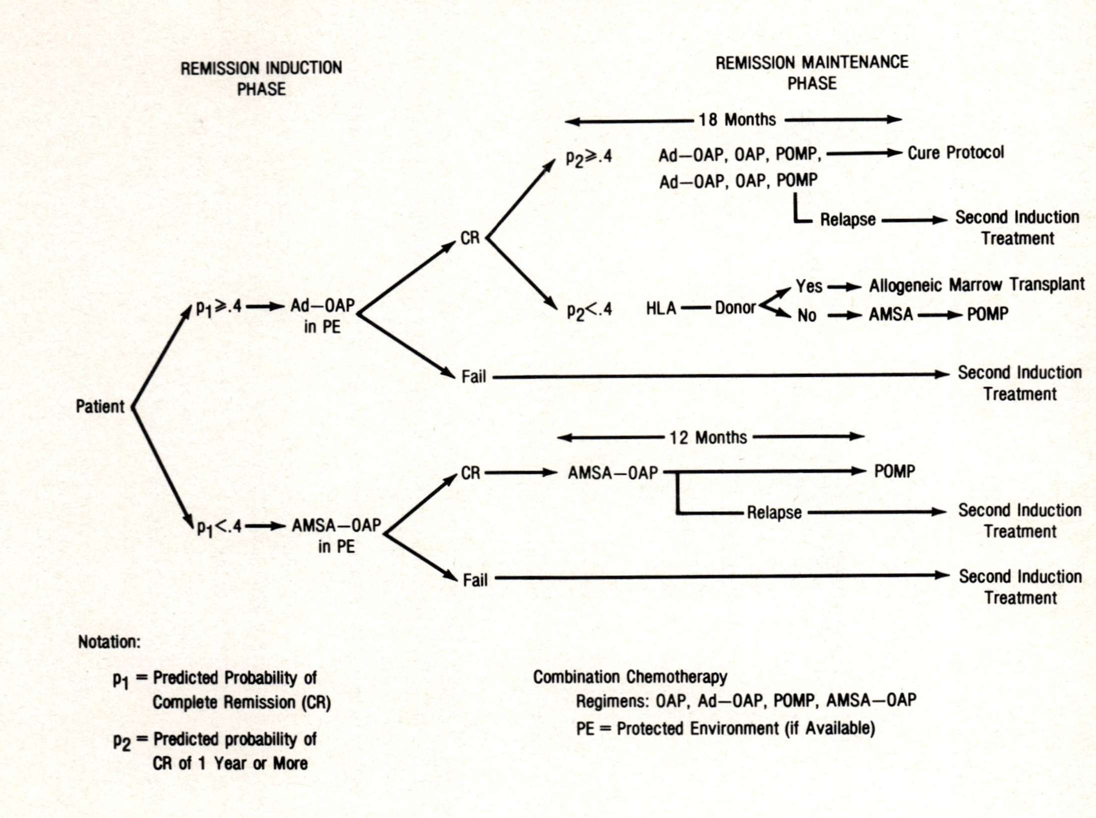

ACKNOWLEDGEMENT: I wish to acknowledge support from Grant CA-30138 in the preparation of this manuscript. Also, I'd like to thank Dr. Emil Freireich, Dr. Michael Keating and Mrs. Terry L. Smith for helpful discussions. Finally, I thank Mrs. Bettie Key for her timely assistance in typing the manuscript.

REFERENCES:

[1] Armitage, P. and Gehan, E.A., Statistical methods for the identification and use of prognostic factors, Int. J. Cancer 13:16-36 (1974).

[2] Breslow, N., Comparison of survival curves. In Cancer Clinical Trials, Methods and Practice, Buyse, M., Sylvester, R., and Staquet, M. (eds.), Chapter XXII, pp. 381-406 (1984).

[3] Brown, B.W., Designing for cancer clinical trials: selection of prognostic factors, Cancer Treat. Reports 64 (2-3):499-502 (1980).

[4] Byar, D., Identification of prognostic factors. In Cancer Clinical Trials, Methods and Practice, Buyse, M., Sylvester, R., and Staquet, M. (eds.), Chapter XXIV, pp. 423-443 (1984).

[5] Byar, D. and Corle, D.K., Selecting optimal treatment in clinical trials using covariate information, J. Chron. Dis. 30:445-59 (1977).

[6] Cox, D.R., The Analysis of Binary Data, Methuen, London (1970).

[7] Cox, D.R., Regression models and life tables (with discussion), J.R. Soc. Statist. Series B 34:187-220 (1972).

[8] Dales, L.G. and Ury, H.K., An improper use of statistical signficance testing in studying covariates, Int. J. Epidemiol. 7:373-375 (1978).

[9] Gehan, E.A., Use of medical measurements to predict the course of disease, Jour. of National Cancer Institute, Monograph No. 3, pp. 51-58 (1959).

[10] Gehan, E.A. and Freireich, E.J., Non-randomized controls in cancer clinical tirals, New Engl. J. Med. 290:198-203 (1974).

[11] Hill, A.B., Statistical Methods in Clinical and Preventive Medicine, Oxford University Press (1962).

[12] Keating, M.J., et al., A prognostic factor analysis for use in development of predictive models for response in adult acute leukemia, Cancer 50:457-465 (1982).

[13] Mantel, N. and Haenszel, W., Statistical aspects of the analysis of data from retrospective studies of disease, J. Nat. Cancer Inst. 22:719-48 (1959).

[14] Miettinen, O.S., Stratification by a multivariate confounder score, Am. J. Epidemiol. 104:609-620 (1976).

[15] Peto, R., et al., Design and analysis of randomized clinical trials requiring prolonged observation of each patient. I. Introduction and design, Br. J. Cancer 34:585-612 (1976).

[16] Simon, R., Adaptive treatment assignment methods and clinical trials, Biometrics 33:743-749 (1977).

[17] Simon, R., The design and conduct of clinical trials in oncology. In Principles and Practice of Oncology, DeVita, V.T., Hellman, S., and Rosenberg, S.A. (eds)., Philadelphia, Lippincott, pp. 198-225 (1982).

[18] Simon, R., Use of regression models: statistical aspects. In Cancer Clinical Trials, Methods and Practice, Buyse, M., Sylvester, R., and Staguet, M. (eds)., 444-66, Oxford University Press, Oxford (1984).

[19] Simon, R., Importance of prognostic factors in cancer clinical trials, Cancer Treat. Reports **68**:185-192 (1984).

[20] Smith T.L., et al., Prediction of remission in adult acute leukemia. Development and testing of predictive models, Cancer **50**:466-472 (1982).

[21] Taves, D.R., Minimization: a new method of assigning patients to treatment and control groups, Clin. Pharmac. Ther. **15**:443-453 (1974).

[22] Zelen, M., Importance of prognostic factors in planning therapeutic trials. In Cancer Therapy: Prognostic Factors and Criteria of Response, Staquet, M.J. (ed)., Raven Press, New York (1975).

DEMOGRAPHY AND FAMILY PLANNING

BIOSTATISTICS: Statistics in Biomedical, Public Health
and Environmental Sciences, edited by P.K. Sen

THE STANDARDIZED MORTALITY RATIO

N.E. Breslow
University of Washington, Seattle

N.E. Day
International Agency for Research on Cancer, Lyon

Statistical inference procedures for the standardized mortality ratio (SMR) are reviewed in the context of Poisson regression models for age specific death rates. Byar's approximate confidence limits perform well in comparison with those based on square root or logarithmic transforms. The non-comparability of SMRs for different exposure categories is due to statistical confounding and may be remedied by regression analysis of the SMR as a function of both exposure and confounding variables. Comparisons are made between SMR analyses that incorporate external background rates and analogous methods in which the background rates are estimated from the data. A proposal is made for estimation of the SMR as a continuous function of time since first employment in a given job or industry.

KEY WORDS: Cohort studies, indirect standardization, multiplicative models, non-parametric estimation, Poisson regression

1. INTRODUCTION

W.H. Farr introduced indirect age standardization as a method of comparing regional mortality rates in the late 1850's, presenting a detailed account of the calculations in an Appendix to the 20th annual report of the British Registrar General (Benjamin, 1968; Inskip et al., 1983). During the 1880's, similar comparisons of observed and expected deaths were used as an analytical statistical tool in occupational mortality studies by the Danish economist Westergaard (N. Keiding, personal communication). The indirectly standardized mortality ratio (SMR) has been in continuous use since 1931 in the Registrar General's decennial supplements on occupational mortality, supplanting the directly standardized cumulative mortality figure (CMF) that had been employed between 1881 and 1921 (Logan, 1982). Today, reports of industrial cohort studies that use the SMR for selected causes of death as the principal measure of the effects of occupation on health appear in nearly every issue of the Journal of Occupational Medicine and similar publications.

Considerable discussion has taken place in the statistical literature regarding the relative merits of the CMF and the SMR, the sensitivity of the former to sampling errors (e.g. Mosteller and Tukey, 1977) and the drawbacks of the latter when used to compare mortality between two occupations or other subgroups (Yule, 1934). The purpose of the present article is to review such properties of the SMR in the light of modern concepts of statistical inference. We evaluate several elementary techniques for hypothesis testing and interval estimation, we summarize the methods available for testing the equality of a series of SMRs against alternatives of heterogeneity and trend, and we show how the question of the comparability of two or more SMRs may be resolved in the context of regression analysis and statistical confounding.

2. DEFINITION AND NOTATION

Summary data from occupational cohort studies typically consist of a cross-classification of deaths d_{jk} and person-years denominators n_{jk} in J strata ($j=1,\dots,J$) and K comparison groups ($k=1,\dots,K$). For concreteness we may think of the strata as referring to age in five or ten year intervals and of the comparison groups as representing K categories of exposure. Quantitative dose levels x_k may be associated with the latter. When there is only a single SMR under consideration (K=1), the second index will be dropped. Often the data are collected in a cross-sectional survey rather than longitudinally, in which case the denominators n_{jk} may be obtained from a mid period census of the population under study rather than by active follow-up.

One object of the analysis is to compare the age-specific mortality rates in the study cohort, estimated by $\hat{\lambda}_j = d_j/n_j$, with a set of standard rates λ_j^* that are assumed known or are based on such large denominators that sampling errors are not at issue. The CMF is defined as the ratio of directly standardized rates for cohort vs. standard, namely

$$\mathrm{CMF} = \frac{\sum_{j=1}^{J} w_j d_j/n_j}{\sum_{j=1}^{J} w_j \lambda_j^*} , \qquad (1)$$

where the w_j are known weights. A number of papers have appeared in the demographic literature regarding the appropriate choice of weights in order that the corresponding standardized rates have the desired interpretation (e.g. Spiegelman and Marks, 1966; Keyfitz, 1966). A typical choice is for the w_j to be proportional to the age distribution of the standard population. Then the CMF may be interpreted as the ratio of the number of deaths that would be observed if age-specific cohort rates were applied to the standard population, divided by the observed number of standard deaths. The instability in the CMF stems from the fact that such weights take no account of the sampling variability of the age-specific rates, so that rates based on a small number of deaths may be weighted heavily in the summary measure. Alternatively, the CMF may be regarded as a weighted average of age-specific rate ratios $\hat{\lambda}_j/\lambda_j^*$ using weights $w_j\lambda_j^*$.

The SMR is defined to be the ratio of the total number of deaths observed in the cohort to the number expected by applying standard rates to its age-distribution

$$\mathrm{SMR} = \frac{\sum_{j=1}^{J} d_j}{\sum_{j=1}^{J} n_j \lambda_j^*} \tag{2}$$

One obvious advantage of this formula is that the individual age-specific numbers of deaths d_j are not required. It may be calculated from published data summaries in some cases where the CMF may not. The SMR is also a weighted average of the age-specific deaths in the jth group where the weights equal $n_j\lambda_j^*$, the expected number of deaths in the jth group. Thus the less precise ratios receive less weight.

As a statistical model for the data we assume that the numbers of deaths d_{jk} in each age × exposure cell have independent Poisson distributions with means $n_{jk}\lambda_{jk}$ where the λ_{jk} are unknown rates and the n_{jk} are constants (Armitage, 1966; Andersen, 1977; Gart, 1978; Gail, 1978). This widely used sampling model should be quite accurate provided that the numbers of deaths in each cell are small in comparison with the total number of individuals at risk. When the data are collected longitudinally, the same individuals may contribute to the rate denominator in more than one cell and the n_{jk} are more properly regarded as random variables. Nevertheless, the kernel of the Poisson likelihood is identical to that based on more elaborate failure time models which assume that the rates in each cell are constant, and this justifies the approach at least so far as large sample inferences are concerned (Holford, 1980; Laird and Olivier, 1981).

3. TESTS OF SIGNIFICANCE

The null hypothesis of interest is that the cohort and standard age-specific rates agree: $\lambda_j = \lambda_j^*$ for $1 \le j \le J$. This means that the observed number of deaths $D = \sum_j d_j$ is approximately Poisson distributed with mean $E = \sum_j n_j \lambda_j^*$ and that the SMR, D/E, equals unity in expectation. Conventional approaches (e.g. Armitage, 1971; Monson, 1980) to testing the departure of the observed to expected ratio from unity use the continuity corrected chi-square statistic defined by

$$\chi^2 = \frac{(|D-E|-0.5)^2}{E} \quad . \tag{3}$$

When the number of deaths is small the Poisson distribution is skew and the normal approximation implicit in the use of (3) may be inadequate. An "exact" test based on the tail probabilities of the Poisson or, equivalently, chi-square distributions could be substituted. However, these are tabulated only for a limited range of values of D and E (Pearson and Hartley, 1966). Byar (unpublished) suggested an approximation to this exact test based on the cubic Wilson-Hilferty formula used to approximate the chi-square distribution. This is obtained by calculating the equivalent normal deviate

$$\chi = 3\sqrt{\tilde{D}}\left(1 - \frac{1}{9\tilde{D}} - \left(\frac{E}{\tilde{D}}\right)^{1/3}\right), \tag{4}$$

where $\tilde{D} = D$ if $D > E$ and $\tilde{D} = D + 1$ otherwise (Rothman and Boice, 1979). Alternately, and what is somewhat easier to remember, we may use the variance stabilizing square root transform and calculate instead

$$\chi = 2(\sqrt{D} - \sqrt{E}). \tag{5}$$

As an example, suppose 15 deaths are observed in the study cohort as opposed to 8.33 expected on the basis of age-specific national rates. Formulas (4) and (5) yield $\chi = 1.98$ (p=0.048 two sided) and $\chi = 1.97$ (p=0.049), whereas (3) gives $\chi = 2.13$ (p=0.033) with the continuity correction and $\chi = 2.31$ (p=0.021) without. The exact Poisson p-value is 0.047. These and other similar calculations suggest that the simple χ^2 formula (3) should probably be avoided in small samples.

4. CONFIDENCE INTERVALS

"Exact" confidence intervals are obtained from the assumption that D is Poisson distributed with mean $\mu = \theta E$, where θ represents the unknown SMR. If μ_L and μ_U denote upper and lower limits for the Poisson mean, upper and lower limits for the SMR are given by $\theta_L = \mu_L/E$ and $\theta_U = \mu_U/E$, respectively. Haenszel and colleagues (1962) have tabulated the exact 95% limits for observed values of D ranging between 1 and 30. See also Table 1.

Approximate confidence limits based on Byar's approach are given by

$$\mu_L = D\left(1 - \frac{1}{9D} - \frac{Z_{\alpha/2}}{3\sqrt{D}}\right)^3 . \tag{6}$$

and

$$\mu_U = (D+1)\left(1 - \frac{1}{9(D+1)} + \frac{Z_{\alpha/2}}{3\sqrt{D+1}}\right)^3$$

where $Z_{\alpha/2}$ denotes the 100 $(1-\alpha/2)$ percentile of the standard normal distribution (Rothman and Boice, 1979). Less accurate but more easily remembered intervals are derived from extensions of the other test statistics (3) and (5) used to test the null hypothesis. Specifically, we solve $(|D-\theta E|-0.5)^2 = Z^2_{\alpha/2}\theta E$ for θ to find

$$\theta_L = \hat{\theta}(1 - \frac{1}{2D}) \left[1 + \frac{Z^2_{\alpha/2}}{(2D-1)} (1 - \sqrt{1 + (4D-2)/Z^2_{\alpha/2}})\right]$$

and (7)

$$\theta_U = \hat{\theta}(1 + \frac{1}{2D}) \left[1 + \frac{Z^2_{\alpha/2}}{(2D+1)} (1 + \sqrt{1 + (4D+2)/Z^2_{\alpha/2}})\right]$$

where $\hat{\theta} = D/E$ is the point estimate of the SMR. Alternatively, limits based on the square root transform are obtained from the equations $2(\sqrt{D} - \sqrt{\theta E}) = \pm Z_{\alpha/2}$ which yield

$$\theta_L = \hat{\theta}(1 - \frac{Z_{\alpha/2}}{2\sqrt{D}})$$

and (8)

$$\theta_U = \hat{\theta}(\frac{D+1}{D}) (1 + \frac{Z_{\alpha/2}}{2\sqrt{D+1}}) \ .$$

The replacement of D by D + 1 in the second equation is made on strictly empirical grounds in order to improve the approximation for small D. Analogous limits based on the log transform are

$$\theta_L = \hat{\theta}\exp(-Z_{\alpha/2}/\sqrt{D})$$

and (9)

$$\theta_U = \hat{\theta}\exp(+Z_{\alpha/2}/\sqrt{D}) \ .$$

The limits (6) - (9) may each be expressed in the form $\theta_L = \hat{\theta}M_L$ and $\theta_U = \hat{\theta}M_U$, i.e. as the point estimate of the SMR times upper and lower multipliers. Table 1 presents selected values of the multipliers for each approximate procedure and for that based on exact Poisson tail probabilities. Byar's approximation is seen to be accurate even for small numbers of deaths, and the square root transform performs reasonably well as soon as D exceeds ten or so. However, the approximations based on the simple chi-square statistic and log transform are not as satisfactory in small samples, especially not for the lower limit.

Returning to the previous example, we find $\hat{\theta} = 15/8.33 = 1.80$ and thus from Table 1 exact 95% confidence limits of $\theta_L = 1.80 \times 0.560 = 1.01$ and $\theta_U = 1.80 \times 1.650 = 2.97$. Those based on Byar's formula (6) are almost the same. The limits (8) based on the square root transform are slightly wider at (1.00, 2.98), those based on the chi-square test (7) equal (1.05, 3.05), and those from the log transform (9) are (1.09, 2.99).

5. COMPARABILITY OF SMR'S

In the Registrar General's latest decennial supplement (OPCS, 1978) SMRs for

Table 1. Exact and Approximate Multipliers for Computing Confidence Limits for the SMR[a]

No. of Deaths	Exact Limits		Byar's Approx. (Formula 6)		Chi-Square (Formula 7)		Square Root (Formula 8)		Log Transform (Formula 9)	
	Lower	Upper	Lower	Upper	Lower	Upper	Lower	Upper	Lower	Upper
95% Limits										
1	0.025	5.572	0.013	5.565	0.276	7.934	0.000	5.733	0.141	7.099
2	0.121	3.612	0.112	3.611	0.317	4.372	0.094	3.678	0.250	3.999
3	0.206	2.922	0.201	2.922	0.354	3.332	0.188	2.961	0.326	3.101
4	0.272	2.560	0.269	2.561	0.390	2.829	0.260	2.586	0.375	2.664
5	0.325	2.334	0.322	2.334	0.421	2.530	0.315	2.353	0.416	2.403
10	0.480	1.839	0.479	1.829	0.530	1.917	0.476	1.846	0.538	1.859
15	0.560	1.649	0.559	1.650	0.594	1.697	0.558	1.653	0.603	1.659
20	0.611	1.544	0.611	1.545	0.636	1.578	0.610	1.547	0.645	1.550
25	0.647	1.476	0.647	1.476	0.668	1.502	0.646	1.478	0.676	1.480
50	0.742	1.318	0.742	1.318	0.752	1.330	0.742	1.319	0.758	1.319
99% Limits										
1	0.005	7.430	0.000	7.471	0.266	12.154	0.000	7.301	0.076	13.144
2	0.052	4.637	0.038	4.656	0.277	6.185	0.008	4.561	0.162	6.181
3	0.113	3.659	0.102	3.671	0.296	4.500	0.066	3.604	0.226	4.425
4	0.168	3.149	0.160	3.157	0.321	3.704	0.127	3.105	0.276	3.625
5	0.216	2.830	0.209	2.836	0.346	3.236	0.180	2.794	0.316	3.165
10	0.372	2.140	0.369	2.142	0.448	2.302	0.351	2.120	0.443	2.258
15	0.460	1.878	0.458	1.879	0.513	1.975	0.445	1.864	0.514	1.945
20	0.518	1.733	0.517	1.734	0.559	1.802	0.507	1.723	0.562	1.779
25	0.560	1.640	0.559	1.641	0.594	1.693	0.551	1.632	0.597	1.674
50	0.673	1.425	0.673	1.426	0.691	1.449	0.669	1.421	0.695	1.440

[a] In order to obtain lower and upper limits for an SMR based on the indicated number of deaths, the point estimate of the SMR is multiplied by the values shown.

various causes of death are calculated for each occupation, for each social class and according to the levels of other explanatory variables. For example, the all causes SMR (in percent) for "miners and quarrymen" is 144 whereas that for "professionals" is 73. It could be tempting to deduce that the relative mortality for miners vs. professionals is on the order of 144/73 = 1.97, but this conclusion is not warranted without further analysis. Yule (1934) pointed out long ago that ratios of SMRs for two groups do not necessarily summarize the ratios of their component age-specific rates.

There is a precise analogy between the lack of comparability of SMRs and the arithmetic of statistical confounding in a set of 2 x 2 tables, a phenomenom that is also known as Simpson's (1951) paradox. Consider the data layout in Table 2. The ratios of SMRs for Cohort 1 vs. Cohort 2 within each age group equal the odds ratios calculated from the indicated 2 x 2 table and similarly the overall ratio of SMRs is just the odds ratio for the totals table. It follows (Simpson, 1951; Whittemore, 1978) that even if the two age-specific ratios are equal, they may differ from the overall SMR ratio if both (i) the SMRs for each cohort vary from one age group to another and (ii) the age distributions of expected deaths are different. Since the age-specific SMR ratios are nothing more than ratios of age-specific rates (assuming the same standard is used in their calculation), the ratio of two SMRs formed by pooling observed and expected numbers of deaths across age groups may lie entirely outside the range of age-specific rate ratios. Table 3, adapted from work of Kilpatrick (1963), provides a numerical illustration.

Table 2. Non-comparability of Two SMRs as an Example of Statistical Confounding in a Series of 2 x 2 Tables

	Age Group 1		Age Group 2		Total	
	Obs.	Exp.	Obs.	Exp.	Obs.	Exp.
Cohort 1	O_{11}	E_{12}	O_{12}	E_{12}	O_{1+}	E_{1+}
	O_{21}	E_{21}	O_{22}	E_{22}	O_{2+}	E_{2+}
$\frac{SMR_1}{SMR_2}$	$\frac{O_{11}E_{21}}{O_{21}E_{11}}$		$\frac{O_{12}E_{22}}{O_{12}E_{22}}$		$\frac{O_{1+}E_{2+}}{O_{2+}E_{1+}}$	

The CMF does not have this drawback since the ratio of two CMFs, being a ratio of directly standardized rates, can be expressed as a weighted average of the age-specific rate ratios. This has led some authors to conclude that the CMF is preferable to the SMR. In particular, Miettinen (1972) claims that SMRs are "internally standardized but not mutually comparable" and Kilpatrick (1963) notes that "the ratio of two CMFs is a CMF but the ratio of two SMRs is not an SMR." However, empirical investigations conducted by the Office of Population Censuses and Surveys (1978) show that the SMR and CMF tend in practice to give rather similar answers. When they do not, moreover, it is not clear that the CMF is necessarily the better statistic since the discrepancy may be due to the inordinate weighting given by the CMF to age groups having small numbers of deaths. Silcock (1959) determined analytically three conditions under which the SMR and CMF may yield substantially different answers: (i) the differences between the percentage age distributions of the standard population and the cohort are substantial; (ii) the ratios of age-specific rates (cohort to standard) vary considerably; and (iii) there is a strong correlation between the percentage differences and the rate ratios from one age group to the next.

The SMRs for two cohorts may be compared safely provided that the ratios of cohort to standard rates are approximately constant so that there is little or no confounding by age. When the age-specific ratios are not constant, at least not within the bounds of sampling error, the SMR has less appeal as a summary measure and probably should not be reported without additional qualification. Statistical tests of heterogeneity and trend in the age-specific SMRs help to validate the conditions required for comparability. Similar tests are used to compare the SMRs across exposure categories once the requisite stability with age has been demonstrated.

Table 3. Example of a Misleading Ratio of SMRs

	Age Group 1			Age Group 2			Total		
	0	E	SMR(%)	0	E	SMR(%)	0	E	SMR(%)
Cohort 1	100	200	50	1600	800	200	1700	1000	170
Cohort 2	80	120	67	180	60	300	250	180	144
SMR_1/SMR_2			75			67			118

6. TESTING FOR HETEROGENEITY AND TREND

The formal assumption which ensures that the SMR represents a good summary measure of the association between cohort and standard rates is that the latter are a constant multiple of the former, i.e.,

$$\lambda_j = \theta\lambda_j^*, \quad 1 \le j \le J \; . \tag{10}$$

Kilpatrick (1962) noted that the SMR is then the maximum likelihood estimate $\hat{\theta}$ of the multiplicative constant. He also suggested that the goodness of fit of this basic model, equivalent to a test of heterogeneity in the age-specific SMRs, could be evaluated via Pearson's chi-square statistic

$$\chi^2_{J-1} = \sum_{j=1}^{J} \frac{(d_j - \tilde{e}_j)^2}{\tilde{e}_j} , \tag{11}$$

where $\tilde{e}_j = \hat{\theta} n_j \lambda_j^*$ represents the expected number of deaths in the jth age group under the hypothesis. The corresponding single degree-of-freedom test for trend (Armitage, 1955) is

$$\chi^2_1 = \frac{\left\{\sum_{j=1}^{J} t_j(d_j - \tilde{e}_j)\right\}^2}{\sum_{j=1}^{J} t_j^2 \tilde{e}_j - \left(\sum_{j=1}^{J} t_j \tilde{e}_j\right)^2 / D} , \tag{12}$$

where t_j represents the midpoint of the jth age group and $D = \sum_j d_j$ represents the total deaths as defined earlier. The more general hypothesis that the age-specific rates for each of K comparison groups are proportional to those of the standard population may be written $\lambda_{jk} = \theta_k \lambda_j^*$. Gail (1978) suggests that this be tested by summing together K chi-square statistics of the form (11) to yield

$$\chi^2_{K(J-1)} = \sum_{j=1}^{K} \sum_{j=1}^{J} \frac{(d_{jk} - \tilde{e}_{jk})^2}{\tilde{e}_{jk}} , \tag{13}$$

where $\tilde{e}_{jk} = \hat{\theta}_k n_{jk} \lambda_j^*$ and $\hat{\theta}_k = SMR_k = D_k/E_k = \sum_j d_{jk} / \sum_j n_{jk} \lambda_j^*$.

Provided that the proportionality assumption holds, the hypothesis of equality among the SMRs in the K exposure categories ($\theta_1 = \theta_2 = \ldots = \theta_K$) is tested by

$$\chi^2_{K-1} = \sum_{k=1}^{K} \frac{(D_k - \tilde{E}_k)^2}{\tilde{E}_k} , \tag{14}$$

where now $\tilde{E}_k = (D+/E+)E_k$ are the expected numbers after multiplicative adjustment by the overall SMR $D+/E+ = \sum_k D_k / \sum_k E_k$ (Gail, 1978). Armitage (1966) calculates the exact mean and variance of this statistic under the null hypothesis, and on this basis suggests some modification so as to improve the applicability of the asymptotic chi-square distribution in finite samples. Assigning dose levels $x_1, \ldots, x_K$ to the K categories of exposure, he also derives the corresponding age-adjusted trend test

$$\chi_1^2 = \frac{\left\{\sum_{k=1}^{K} x_k(D_k - \tilde{E}_k)\right\}^2}{\sum_{k=1}^{K} x_k^2 \tilde{E}_k - \left(\sum_{k=1}^{K} x_k \tilde{E}_k\right)^2 / D_+} \tag{15}$$

following earlier work by Mantel (1963) and Birch (1964) for binomially distributed data.

Table 4 illustrates these calculations using summary data for workers employed before 1925 in a Montana copper smelter. A complete description of the study is given by Lee and Fraumeni (1969) and Lee-Feldstein (1983). Four exposure categories are defined in terms of duration of arsenic exposure. SMRs for respiratory cancer were calculated from national vital statistics using J=16 strata based on a 4 x 4 cross classification of age (40-49, 50-59, 60-69 and 70-79 years) and calendar year (1938-49, 1950-59, 1960-69, 1970-78). The test based on formula (14) provides strong evidence for change in the SMR with increasing years of exposure ($\chi_3^2 = 33.7$). The test (15) for linear trend yields $\chi_1^2 = 30.5$, where coded dose levels $x_1 = 1$, $x_2 = 2, \ldots, x_4 = 4$ are used for the quantitative exposure variable. However, there is also evidence that the SMRs vary according to calendar year and this raises some doubts about the validity of the comparison across exposure categories.

Table 4. Comparison of Respiratory Cancer SMRs by Duration of Arsenic Exposure for Workers Employed Before 1925 in a Montana Smelter

	Cumulative Years of Moderate/Heavy Arsenic[1]				
	0-0.9	1.0-4.9	5.0-14.9	15+	Total
Observed deaths	51	17	13	34	115
Person-Years	19,017	2,686	2,600	3,871	28,171
Standardized Rate[2] (per 1000 Py)	2.68	6.34	5.00	8.78	4.08
Expected Deaths[3]	21.47	2.95	2.76	4.44	31.62
CMR (%)	195.3	511.7	408.2	571.4	301.7
SMR (%)	237.5	577.1	471.7	765.0	353.7
Adjusted Expecteds ($\tilde{E}_k$)	78.10	10.71	10.02	16.17	115.00

[1] Exposure variable lagged two years.
[2] Directly standardized to the U.S. population in 1950.
[3] From U.S rates for white males by age and calendar year.

Quite a different approach to the problem of testing for heterogeneity of SMRs is taken by Kupper and Kleinbaum (1971). These authors make no assumption about the basic structure of the data akin to the multiplicative model (10). Instead, they define K unknown SMRs for k=1,...,K by $SMR_k = \sum_j n_{jk}\lambda_{jk}/\sum_j n_{jk}\lambda_j^*$ and treat the testing problem in terms of a non-linear hypothesis concerning the JK parameters λ_{jk}. Unless the multiplicative structure holds, however, the various SMRs may not be comparable and testing for their equality makes little sense. Our viewpoint is that statistical inference regarding the SMR is best carried out in the context of an explicit model for the data whose validity is checked at each step.

Table 5 shows respiratory cancer SMRs for the Montana smelter workers according to a number of different factors (Breslow, 1984a). Period of hire is of particular interest in view of changes made in the smelting process in 1925 that were designed to reduce the airborne arsenic exposures (Lee-Feldstein, 1983). Since follow-up did not commence until 1938 for workers employed before that time, period of hire is highly correlated with years since first employment. It is less strongly correlated with age, with calendar year and with other time-varying factors. Besides the concern about whether the observed differences in the SMRs reflect parallel differences in the underlying age and calendar year specific rates, therefore, there is an additional question as to whether the differences according to levels of one factor might not be secondary to the effects on the SMR of another related factor. One obvious way to resolve such issues is via regression analysis.

7. REGRESSION ANALYSES OF THE SMR

Regression analyses of the SMR are easily conducted using a multiplicative model for the underlying rates such that

$$SMR_{jk} = \lambda_{jk}/\lambda_j^* = \exp\{\alpha + z_{jk}\beta\} \qquad (16)$$

(Breslow et al., 1983). The fact that the covariables z may depend on j as well as k allows for possible interactions between exposure and stratification variables, or even for systematic variation in the SMRs according to one or more of the stratification variables. Since the lack of comparability between SMRs for different exposure categories is due to confounding with stratification variables, inclusion in the regression equation of covariables representing the main effects of age and/or calendar year may make precisely the adjustment needed to restore comparability.

If we allow a completely arbitrary dependence of the SMRs on the age x year strata, e.g. $\lambda_{jk}/\lambda_j^* = \lambda_j \exp(z_{jk}\beta)$ where the z's now represent only exposure and interaction effects, the result is equivalent to a multiplicative model

$$\lambda_{jk} = \exp\{\alpha_j + z_{jk}\beta\} \quad (17)$$

that makes no reference to an external standard population. Instead both β and the baseline stratum specific rates $\exp(\alpha_j)$ are estimated from the data. Such models are closely related to the notion of internal standardization (Gilbert, 1983).

Table 5. Variations in Respiratory Cancer SMRs Among Montana Smelter Workers

Period Analyzed	Level	No. Deaths	SMR[a] (100)	Test of Significance (Formula 13)
Period of Hire	1885-1924	115	362	$\chi_1^2 = 39.5$
	1925-1955	161	164	$(p < 0.0001)$
Age of Hire (Years)	<24	69	255	$\chi_2^2 = 5.2$
	25-34	116	222	$(p = 0.07)$
	35+	91	184	
Birthplace	US	198	180	$\chi_1^2 = 28.5$
	Foreign	80	381	$(p < 0.0001)$
Years Since First Employed[b]	1-14	101	165	$\chi_2^2 = 24.0$
	15-29	59	185	$(p < 0.0001)$
	30+	116	315	
Years Since Last Employed[b]	None	110	230	$\chi_2^2 = 3.2$
	0+-9	84	227	$(p = 0.20)$
	10+	82	181	
Arsenic Exposure[b]	Light Only	153	160	$\chi_3^2 = 44.4$
	Moderate[c]	91	339	$(p < 0.001)$
	Heavy[c]	32	434	
Age at Follow-up (Years)	40-49	21	166	$\chi_3^2 = 2.5$
	50-59	80	199	$(p = 0.48)$
	60-69	117	228	
	70-79	58	223	
Calendar Year at Follow-up	1938-49	34	403	$\chi_3^2 = 28.4$
	1950-59	65	294	$(p < 0.0001)$
	1960-69	94	211	
	1970-77	83	151	

[a] Calculated with reference to U.S. mortality rates for white males by age and calendar year.

[b] Time-dependent exposure variable lagged two years.

[c] Worked in moderate or heavy arsenic exposure area for at least one year.

Detailed expositions of the maximum likelihood fitting of Poisson regression models to grouped data are given by Holford (1980), Berry (1983), Breslow et al. (1983) and Frome (1983), among others. Multiplicative models of the form (16) and (17) are particularly easy to fit using standard features of the computer program GLIM (Baker and Nelder, 1978). Additive and other non-multiplicative models may also be fitted in GLIM by writing special purpose macros (Frome, 1983). Applying model (16) to the Montana smelter workers data, Breslow (1984a) conducts an analysis of variance of the joint effects of period of hire, birthplace, years since employment and level of arsenic exposure on the respiratory cancer SMR. This shows that the apparent effects of duration of employment are secondary to those of period of hire, whereas the effect of arsenic and birthplace are independent.

Table 6 presents results of further regression analyses based on (16) and (17) that consider period of hire, duration of arsenic exposure and calendar year as explanatory variables. When calendar year is included in the SMR analysis, the regression coefficient for period of hire is much closer to that based on the internally controlled analysis shown in the final column of the table. This confirms that part of the difference between the SMRs for those hired before and after 1925 is due to the confounding effects of calendar year on the ratios of cohort to standard death rates. Appropriate adjustment is made either by including calendar year as covariable in the SMR regression analysis based on (16) or else, and what is nearly equivalent, by conducting a parallel internally controlled regression analysis of death rates based on equation 17.

8. NON-PARAMETRIC ESTIMATION OF THE SMR AS A CONTINUOUS FUNCTION OF TIME

The preceding examples have emphasized the possible dependence of the SMR on such time-dependent factors as age, calendar year, years since initial employment, and duration of heavy exposure. Each of these was treated as a discrete factor with a small number of levels. In some applications we may wish to estimate the SMR as a function of a continuous time variable. This is made possible by considering a continuous time version of the model (16) in the form

$$\lambda(t|z) = \theta(t)\lambda^*(t)\exp\{z(t)\beta\}, \tag{18}$$

where now t is the continuous time variable, $z = z(t)$ is a vector of possibly continuous time-dependent covariables, $\lambda^*(t)$ are the background rates assumed known from the external standard population and $\theta(t)$ represents the unknown SMR considered as a continuous function of t. The identity of this structure with the proportional hazards model of Cox (1972) is made transparent by incorporating the standard rates in the exponential term

$$\lambda(t|z) = \theta(t)\exp\{z_0(t) + z(t)\beta\} , \qquad (19)$$

such that $z_0(t) = \log\lambda^*(t)$ is regarded as an additional covariate with known regression coefficient $\beta_0 = 1$. The unknown SMR $\theta(t)$ plays the role of the unknown hazard function $\lambda_0(t)$ usually associated with this model. This formulation of the proportional hazards model is due to Andersen (1984).

Table 6. Estimated Regression Coefficients (log-SMR ratios) and Standard Errors for Respiratory Cancer SMRs from the Montana Cohort

Variable	Externally Controlled Analysis: Without Calendar Year	Externally Controlled Analysis: With Calendar Year	Internally Controlled Analysis[a]
Hired before 1925	0.708 ± 0.127	0.526 ± 0.140	0.485 ± 0.150
Duration of heavy/ moderate arsenic exposure:			
6 months - 1 year	-------	------	------
1 - 4 years	0.772 ± 0.158	0.796 ± 0.159	0.797 ± 0.158
5 - 14 years	0.564 ± 0.206	0.581 ± 0.206	0.571 ± 0.206
15+ years	0.948 ± 0.180	0.951 ± 0.181	0.951 ± 0.181
Calendar year:			
1938 - 49			
1950 - 59		-0.117 ± 0.215	
1960 - 69		-0.331 ± 0.211	
1970 - 77		-0.612 ± 0.222	

[a]Joint estimation of age and calendar year specific rates in 16 strata.

Using well known results form the theory of survival analysis (Kalbfleisch and Prentice, 1980; Andersen and Gill, 1982) estimates of β are obtained from the partial likelihood (Cox, 1975). Specifically, for each time t_i that an individual dies from the cause of interest, a risk set $R(t_i)$ is constructed of all cohort members who are under observation at t_i and thus at risk of death at that time. The partial likelihood is built up of multiplicative contributions from each risk set of the form

$$\frac{\exp\{s(t_i)\beta\}}{\sum_{\ell \in R(t_i)} \exp\{s_\ell(t_i)\beta\}}$$

where $s(t_i)$ denotes the sum of the covariates for the d_i individuals who die at t_i, ℓ ranges over the subjects of size d_i drawn without replacement from $R(t_i)$ and s_ℓ denotes the sum of the covariates for each subset. Once the maximum partial likelihood $\hat{\beta}$ estimates is obtained, an estimator (Breslow, 1974) of the cumulative SMR function $\Theta(t) = \int_0^t \theta(s)ds$ is given by

$$\hat{\Theta}(t) = \sum_{t_i \le t} \frac{d_i}{\sum_{j \in R(t_i)} \exp\{z_{j0}(t_i) + z_j(t_i)\hat{\beta}\}} \tag{20}$$

Large sample properties of this estimator are derived by Tsiatis (1981).

Figure 1 plots the cumulative SMR for the Montana smelter workers with t defined as years since date of hire. For the upper curve no covariables other than z_0 were used. The lower curve is adjusted for period of hire, birthplace and duration of moderate or heavy arsenic exposure in the sense that covariables $z(t)$ representing these factors are included in the model. There is an apparent change in the slopes at about 35 years, with the unadjusted SMR increasing at that point and the adjusted SMR decreasing slightly. This suggests that the adverse effects of additional years since date of hire are adequately accounted for by the exposure variables considered.

9. ALTERNATIVES TO THE SMR

Two alternatives to the SMR as a measure of comparative mortality have already been mentioned. Synoptic measures such as the CMF are less prone to confounding bias but more sensitive to sampling errors. A better alternative is to conduct an internally controlled analysis in terms of a regression model in which the background rates as well as the exposure effects are estimated directly from the data. We have noted already that use of the SMR as a comparative measure is fully justified only under a multiplicative model such as embodied in equation (16) and we therefore confine our discussion of SMR alternatives to a comparison of internally vs. externally controlled analyses in the multiplicative environment.

Model (16) may be viewed as a special case of (17) in which the background rates are assumed known up to a multiplicative constant $\theta = \exp(\alpha)$. Since fewer assumption are required, the internally controlled approach based on (17) would seem the obvious choice. The only drawback would be if the necessity of estimating a large number of nuisance parameters (the unknown α_j) severely reduced the statistical precision of the β regression coefficients of primary interest. Fortunately, several studies suggest that this is rarely the case for the multiplicative model.

Theoretical calculations (Stewart and Pierce, 1982; Breslow, 1984b) show that the relative efficiency of β parameter estimation for a single covariable z in (16) vs. (17) is determined by the ratio of its within stratum to total variability. The stratum component of variation in z has to be quite large, in fact, before there is a serious loss of precision from use of internally controlled analyses. Gilbert (1983) computes the power of both methods for detecting low level radiation effects using data at the Hanford nuclear reservation, and finds that the power curves are nearly identical. (Her "EXT-adj" method corresponds to the SMR regression analyses proposed here that use an unknown scale factor θ to adjust the background rates.)

It is likely that knowledge of the background rates is of greater value when the effects of exposure are measured by excess disease rates in an additive model (Breslow, 1984b). However, the SMR should probably not be used in such circumstances for reasons already mentioned. See Pocock et al. (1982) for an example of this approach.

In spite of the advantages of comparative analyses involving internal estimation of background rates, we do not believe that the SMR will or should cease entirely to be used as a summary measure. The main reason is that the presentation of observed D_k and expected E_k numbers of deaths for a few occupations or exposure categories is an extremely simple and appealing way to summarize a complicated set of mortality data. Tests for heterogeneity and trend in mortality rates for K different levels of exposure are easily calculated from such summary quantities. The data required for the analogous tests based on internal control, on the other hand, consist of the deaths d_{jk} and person-years n_{jk} in the full J x K cross-classification. These are often sufficiently voluminous as to discourage publication. Furthermore, the comparison of cohort mortality rates with those of an external standard population may itself be of intrinsic interest and may help to identify particular causes of death for further detailed analysis. We would only caution that those who undertake to report their data in terms of the SMR should also undertake the additional work needed to determine whether or not seriously misleading inferences could result.

Figure 1. Estimates of the cumulative SMR for respiratory cancer among Montana smelter workers by years since date of hire, both with (lower curve) and without (upper curve) adjustment for covariables.

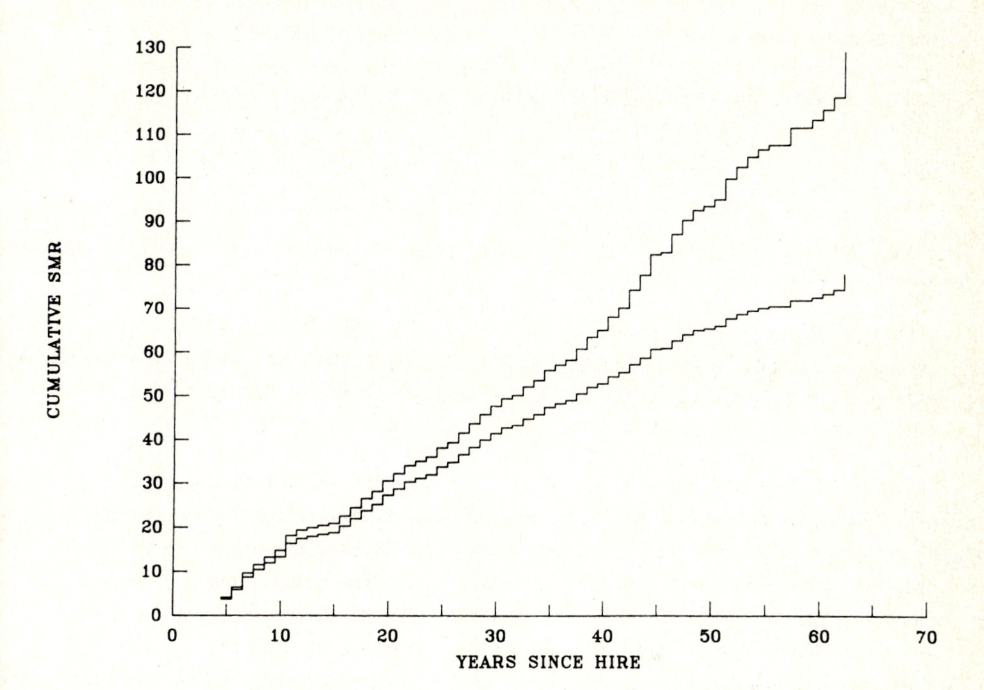

ACKNOWLEDGEMENTS: This research was supported in part by grant 1-K07-CA00723 from the United States Public Health Service. Portions of this paper are drawn from the authors' forthcoming monograph Statistical Methods in Cancer Research II: Design and Analysis of Cohort Studies to be published by IARC, Lyon. We gratefully acknowledge the helpful comments made on an earlier version of this paper by N. Keiding and P.K. Andersen.

REFERENCES

[1] Andersen, E.B., Multiplicative Poisson models with unequal cell rates, Scand. J. Statist. 4:153-158 (1977).

[2] Andersen, P.K., A Cox regression model for the excess mortality in long-term follow-up studies, Proceedings of the XIIth International Biometrics Conference, Tokyo, in press (1984).

[3] Andersen, P.K. and Gill, R.D., Cox's regression model for counting processes. A large sample study, Ann. Statist. 10: 1100-1120 (1982).

[4] Armitage, P., Tests for linear trends in proportions and frequencies, Biometrics 11:375-386 (1955).

[5] Armitage, P., The chi-square test for heterogeneity of proportions, after adjustment for stratification, J. Roy. Statist. Soc. B 26:150-163 (1966).

[6] Armitage, P., Statistical Methods in Medical Research, Oxford, Blackwell Scientific Publications (1971).

[7] Baker, R.O. and Nelder, J.A., The GLIM System. Release 3. Oxford, Numerical Algorithms Group (1978).

[8] Benjamin, B., Health and Vital Statistics, London, Allen and Unwin (1968).

[9] Berry. G., The Analysis of mortality by the subject-years method, Biometrics 39:173-184 (1983).

[10] Birch, M.W., The detection of partial association I: the 2 x 2 case, J. Roy. Statist. Soc. B:313-324 (1964).

[11] Breslow, N., Covariance analysis of censored survival data, Biometrics 30: 89-99 (1974).

[12] Breslow, N.E., Multivariate cohort analysis, J. Nat. Cancer Instit. (in press) (1984a).

[13] Breslow, N.E., Cohort Analysis in Epidemiology, In Fienberg, S.E. and Atkinson, A.C. (eds). A Celebration of Statistics, New York, Springer-Verlag (in press) (1984b).

[14] Breslow, N.E., Lubin, J.H., Marek, P., and Langholz, B., Multiplicative models and the analysis of cohort data, J. Amer. Statist. Assoc. 78:1-12 (1983).

[15] Cox, D.R., Regression models and life tables (with discussion), J. Roy. Statist. Soc. B 34:187-220 (1972).

[16] Cox, D.R., Partial likelihood, Biometrika 62:269-276 (1975).

[17] Frome, E.L., The analysis of rates using Poisson regression models, Biometrics 39:665-674 (1983).

[18] Gail, M., The analysis of heterogeneity for indirectly standardized mortality ratios, J. Roy. Statist. Soc. A 141:224-234 (1978).

[19] Gart, J.J., The analysis of ratios and cross-product ratios of Poisson variates with application to incidence rates, Comm. in Statist., Theory and Methods A7 917-937 (1978).

[20] Gilbert, E.S., An evaluation of several methods for assessing the effects of occupational exposure to radiation, Biometrics 39:161-171 (1983).

[21] Haenszel, W., Loveland, D., and Sirken, M.G., Lung cancer mortality as related to residence and smoking histories, J. Nat. Cancer Instit. 28:947-1001 (1962).

[22] Holford, T.R., The analysis of rates and of survivorship using log-linear models, Biometrics 36:299-305 (1980).

[23] Inskip, H., Beral, V., Fraser, P., and Haskey, J., Methods for age-adjustment of rates, Statist. in Med. 2:455-66 (1983).

[24] Kalbfleisch, J.D. and Prentice, R.L., The Statistical Analysis of Failure Time Data, New York, Wiley (1980).

[25] Keyfitz, N.,Sampling variance of the standardized mortality rates, Hum. Biol. 38:309-317 (1966).

[26] Kilpatrick, S.J., Occupational mortality indices, Pop. Studies 16:175-189 (1962).

[27] Kilpatrick, S.J., Mortality comparisons in occupational groups, Appl. Statist. 17:65-86 (1963).

[28] Kupper, L.L., Some further remarks "on testing hypotheses concerning standardized mortality ratios," Theoretical Pop. Biol. 2:431-436 (1971).

[29] Kupper, L.L. and Kleinbaum, D.G., On testing hypotheses concerning standardized mortality ratios, Theoretical Pop. Biol. 2:290-298 (1971).

[30] Laird, N. and Olivier, D., Covariance analysis of censored survival data using log-linear analysis techniques, J. Amer. Statist. Assoc. 76:231-240 (1981).

[31] Lee, A.M. and Fraumeni, J.F., Arsenic and respiratory cancer in man: an occupational study, J. Nat. Cancer Instit. 42:1045-1052 (1969).

[32] Lee-Feldstein, A., Arsenic and respiratory cancer in humans: follow-up of copper smelter employees in Montana, J. Nat. Cancer Instit. 70:601-610 (1983).

[33] Logan, WPD., Cancer Mortality by Occupation and Social Class 1851-1971, IARC Scientific Publications No. 36, Lyon, IARC (1982).

[34] Mantel, N., Chi-square tests with one degree of freedom. Extensions of the Mantel-Haenszel procedure, J. Amer. Statist. Assoc. 58:690-700 (1963).

[35] Miettinen, O.S., Standardization of risk ratios, Amer. J. Epidemiol. 96: 383-388 (1972).

[36] Monson, R.R., Occupational Epidemiology, Boca Raton, FL, CRC Press (1980).

[37] Mosteller, F. and Tukey, J.W., Data Analysis and Regression: a Second Course in Statistics, Reading, Mass., Addison and Wesley (1977).

[38] Office of Population Surveys and Censuses, Occupational Mortality: The Registrar's Decennial Supplement for England and Wales 1970-72, London, Her Majesty's Stationery Office.

[39] Pearson, E.S. and Hartley, H.O., Biometrika Tables for Statisticians, Vol. I (3rd edition), Cambridge, The University Press (1966).

[40] Pocock, S.J., Gore, S.M., and Kerr, G.R., Long term survival analysis: The curability of breast cancer, Statist. in Med. **1**:93-104 (1982).

[41] Rothman, K.J. and Boice, J., Epidemiologic Analysis with a Programmable Calculator. NIH Publication 79-1649, U.S. Government Printing Office (1979).

[42] Silcock, H., The comparison of occupational mortality rates, Pop. Studies **13**:183-192 (1959).

[43] Simpson, C.H., The interpretation of interaction in contingency tables, J. Roy. Statist. Soc. B **13**:238-241 (1951).

[44] Spiegelman, M. and Marks, H.H., Indexes of mortality and their statistical significance: I Empirical testing of standards for the age adjustment of death rates by the direct method, Hum. Biol. **38**:280-292 (1966).

[45] Stewart, W.H. and Pierce, D.A., Efficiency of Cox's model in estimating regression parameters with grouped survival data, Biometrika **69**:539-545 (1982).

[46] Tsiatis, A.A., A large sample study of Cox's regression model, Ann. Statist. **9**:93-108 (1981).

[47] Whittemore, A.S., Collapsibility of multidimensional contingency tables, J. Roy Statist. Soc. B **40**:328-340 (1978).

[48] Yule, G.U., On some points relating to vital statistics, more especially statistics of occupational mortality, J. Roy. Statist. Soc. 97:1-84 (1934).

BIOSTATISTICS: Statistics in Biomedical, Public Health
and Environmental Sciences, edited by P.K. Sen

CAUSE-OF-DEATH DIFFERENTIALS BY AGE, SEX, AND RACE: A LIFE-TABLE ANALYSIS OF 1978 U.S. DATA

Prithwis Das Gupta

Population Division
U.S. Bureau of the Census
Washington, D.C. 20233

This paper uses the U.S. mortality data by cause of death for 1978 to compute three measures of the effect of the existence of a specified cause of death, viz., (1) gain in expectation of life from the elimination of a cause, (2) fulfillment index for an age group (i.e., percent recovery of the wastage of life potential in an age group) from the elimination of a cause, and (3) probability of eventually dying from a cause. These three measures are shown for 28 causes of death corresponding to 19 age groups and nine race-sex groups (viz., total, and combinations of two sexes with total, white, black, and other races).

KEY WORDS: Life Tables, Mortality by Cause, Differential Mortality

INTRODUCTION

The expectation of life at birth ($\overset{o}{e}_0$) in the United States rose from 70.2 in 1968 to 73.7 in 1980. This increase of 3.5 in $\overset{o}{e}_0$ in 12 years is about six times the corresponding increase of 0.6 (from 69.6 to 70.2) during the preceding 14-year period 1954-68. This rather relatively sharp rise in $\overset{o}{e}_0$ in recent years has been partly due to reduced death rates at older ages (Bayo, 1972; Siegel, 1979; Rosenwaike, et al., 1980; Wilkin, 1981; Crimmins, 1983). The expectation of life at age 65, for example, rose from 14.4 to only 14.6 during 1954-68, whereas the same for 1980 is as high as 16.4.

In order to understand fully the reasons for the recent mortality decline and its implications for the future, a thorough analysis of the mortality statistics by causes of death is essential. The objective of the present work is to construct and analyze the United States life tables by causes of death for the year 1978. Official U.S. life tables by causes of death are available for 1959-61 and 1969-71 (U.S. National Center for Health Statistics, 1968; 1975). The corresponding tables for 1979-81 may not be available for several years. However, because of the changed pattern of mortality decline in recent years, a similar cause-of-death analysis at this time appears useful. Earlier life tables by causes of death for National populations were constructed by Preston, et al. (1972).

THE DATA

The U.S. mortality data for 1978 are obtained from Table 1-26 (U.S. National

Center for Health Statistics, 1982), which gives deaths from 281 selected causes by age, race, and sex (1978 is the latest year for which published cause-of-death data are available at this time). For our analysis of the leading causes of death, these 281 causes are grouped into 24 mutually exclusive and exhaustive categories (the last one being the residual group). In addition, four overlapping sub-total categories are also included. These 28 causes or cause-groups of death are shown in Table A.

The total and all the eight race-sex categories (viz., combinations of two sexes with four race groups - total, white, black, and other races) that are available for deaths are used for life-table analysis. The U.S. National Center for Health Statistics started publishing annual life tables for Blacks separately, beginning in 1979.

The estimates[1] of the U.S. resident population by age, race, and sex for July 1, 1978 are obtained from the U.S. Bureau of the Census (1980, Table 2). Since further breakup of population 85 years and over is not provided by the Census Bureau, we consider the following 19 age groups for the computation of death rates and for subsequent analysis: under 1 year, 1-4, 5-9, 10-14, 15-19, ... , 80-84, 85 years and over.

The cause-specific death rates and the overall death rates for the nine race-sex categories are shown in Table A to facilitate comparison.

THE ABRIDGED LIFE TABLES FOR ALL CAUSES COMBINED

For any of the nine race-sex groups, let D(x,x+n) be the number of deaths and population size in the age group (x,x+n). The 19 age-specific death rates m(x,x+n) are obtained from

$$m(x,x+n) = D(x,x+n)/P(x,x+n). \qquad (1)$$

Our next step is to convert the central death rate m(x,x+n) into the probability q(x,x+n) that a person aged x will die before reaching age x+n. We note that the probability that a person aged x will survive to age x+n is given by

$$p(x,x+n) = \exp\left[- \int_x^{x+n} \mu(t)dt\right], \qquad (2)$$

where $\mu(t)$ is the force of mortality at age t (Chiang, 1978, p. 245).

[1]These estimates were later revised upward based on the 1980 census (U.S. Bureau of the Census, 1982, Table 2), which resulted in an increase of the 1978 expectation of life at birth from 73.3 to 73.5. Since the relative mortality picture by age, sex, and race is not expected to be different in the two cases, we have not revised our computations based on newer estimates of population.

If $\mu(t)$ is the same at all points in the age group $(x,x+n)$ and if this constant value is equal to $m(x,x+n)$, then equation (2) reduces to

$$p(x,x+n) = \exp[\,-nm(x,x+n)]. \tag{3}$$

However, because of uneven distribution of force of mortality within an age group, the relationship in (3) is not, in general, true. A more accurate presentation of equation (3) is given by

$$p(x,x+n) = \exp[\,-A(x,n)m(x,x+n)], \tag{4}$$

where $A(x,n)$'s are constant multipliers that are usually close to n. If mortality is higher in the earlier part in the age group $(x,x+n)$, $A(x,n)$ is less than n, and if it is higher in the latter part, $A(x,n)$ is greater than n.

We choose the values of $A(x,n)$ in such a way that they become consistent with the conversions of m's into p's in the official abridged life tables[2] for the United States, 1978 (U.S. National Center for Health Statistics, 1982, Table 5-1). These values are shown in Table B. The constants $A(x,n)$ are conceptually different from "the fraction of the last age interval of life, a_i" that Chiang (1978, p. 100) used for the construction of abridged life tables. Formula (4) is rather a simplified version of the Reed and Merrell (1939) approach.

The relationship

$$q(x,x+n) = 1 - p(x,x+n) \tag{5}$$

finally gives us the values of $q(x,x+n)$ for the abridged life tables. These values are shown in Table 1, Block 1.

Once the p's and q's are computed, the number $\ell(x)$ of persons who survive to age x out of a cohort of $\ell(0) = 100{,}000$ persons at age 0 is obtained from the simple chain relationship

$$\ell(x+n) = \ell(x)[1 - q(x,x+n)]. \tag{6}$$

The $\ell(x)$ values are shown in Table 1, Block 2.

If deaths are uniformly distributed over the age interval $(x,x+n)$, the number of years lived by $\ell(x)$ persons at age x in the same age interval will be

$$L(x,x+n) = \frac{n}{2}\,[\ell(x) + \ell(x+n)]. \tag{7}$$

[2] These life tables have since been revised based on revised population estimates for 1978 (see footnote 1).

However, because of usually uneven distribution of deaths over an age interval, the relationship (7) may be more accurately written as

$$L(x,x+n) = B(x,n)[\ell(x) + \ell(x+n)], \qquad (8)$$

where the multipliers $B(x,n)$ are close to $n/2$. If more deaths occur early in the age interval, $B(x,n)$ will be less than $n/2$, and if more deaths occur later, $B(x,n)$ will be greater than $n/2$.

Again, the constants $B(x,n)$ are so chosen as to make them consistent with the conversions of ℓ's into L's in the official abridged life tables for males, females, and total population for the United States, 1978. These values are given in Table B.

Formula (8) cannot be used for the age group 85 and over. For this age group, we use the relationship (U.S. National Center for Health Statistics, 1966, p. 4)

$$L(85,\infty) = B(85,\infty)\ \ell(85)/m(85,\infty), \qquad (9)$$

where $B(85,\infty)$ is also chosen appropriately from the official life tables for 1978. The $L(x,x+n)$ values are shown in Table 1, Block 3, as LL's.

The final column in the abridged life table is the expectation of life at age x, $\overset{o}{e}(x)$, which is obtained from the relationships

$$T(x) = L(x,x+n) + \ldots + L(85,\infty),$$

$$\overset{o}{e}(x) = T(x)/\ell(x). \qquad (10)$$

All the $\overset{o}{e}(x)$ values are given in the fourth block of Table 1. Table 1, therefore, shows four selected columns of the abridged life tables for the nine population groups in the United States, 1978.

GAIN IN EXPECTATION OF LIFE FROM THE ELIMINATION OF A SPECIFIED CAUSE OF DEATH

Let $D_i(x,x+n)$ be the observed number of deaths in the age interval $(x,x+n)$ when cause i has been eliminated as a cause of death. Let $q_i(x,x+n)$ be the corresponding probability that a person aged x will die in the interval $(x,x+n)$ when such an elimination has taken place. Combining these with our earlier notations, we can then write (Chiang, 1968, p. 257)

$$q_i(x,x+n) = 1 - [p(x,x+n)]^{\frac{D_i(x,x+n)}{D(x,x+n)}}. \qquad (11)$$

It can be proved that the same value of $q_i(x,x+n)$ is obtained by successive use of the formulas -- similar to (1), (4), and (5) -- as follows:

$$m_i(x,x+n) = D_i(x,x+n)/P(x,x+n),$$

$$p_i(x,x+n) = \exp[-A(x,n)\ m_i(x,x+n)], \tag{12}$$

$$q_i(x,x+n) = 1 - p_i(x,x+n).$$

We found it more convenient to use the set of equations (12), instead of (11), for the computation of $q_i(x,x+n)$ values.

Once $q_i(x,x+n)$ values are obtained from (12), the abridged life table eliminating cause i can be constructed exactly the same way it was done for all causes of death combined by using equations (6) - (10). We have assumed that the values $A(x,n)$ and $B(x,n)$ in Table B do not change when cause i has been eliminated. We also assume that these multipliers depend only on sex and age, and not on race.

The columns of the abridged life tables (when a specified cause has been eliminated) are not shown. Instead, only the gains in expectation of life from the elimination are given in Table 2. For example, for white males, the expectation of life at age 30 will increase from 42.82 to 49.55 from the elimination of heart diseases. Only the gain 6.73 (= 49.55 - 42.82) is shown in Table 2. These gains, however, are not additive, i.e., the sum of the gains from the elimination of causes i and j separately is not, in general, equal to the gain from the elimination of these two causes simultaneously (Keyfitz, 1977).

A comparison of the gains to be achieved at different periods from the elimination of some specified causes is given in Table C.

FULFILLMENT INDEX FOR THE AGE GROUP (x,x+n) FROM THE ELIMINATION OF A SPECIFIED CAUSE

The gain in expectation of life at age x from the elimination of cause i is a function of the change in the mortality condition at all ages above x, and, therefore, it does not say anything specifically about the impact of the elimination on the age group (x,x+n) or any other age group. In order to measure this impact, we define an index which may be called the Fulfillment Index. To develop this index, let us use some notations, as follows:

$L(x,x+n)$ = number of years lived by $\ell(x)$ persons at age x in the age interval (x,x+n), when all causes of death are present,

$L_i(x,x+n)$ = number of years lived by $\ell_i(x)$ persons at age x in the age interval (x,x+n), when cause i has been eliminated,

$\bar{n} = L(x,x+n)/\ell(x)$ = average number of years lived by a person aged x in the age interval (x,x+n), in the presence of all causes of death,

$\bar{n}_i = L_i(x,x+n)/\ell_i(x)$ = average number of years lived by a person aged x in the age interval (x,x+n), when cause i has been eliminated.

Since a person aged x can live up to n years in the interval (x,x+n), for all causes of death combined, a fulfillment index was defined by Case, et al. (1962) as

$$(\bar{n}/n) \times 100. \quad (13)$$

But, in our present context, we define the index in a different way, as follows:

$n - \bar{n}$ = the wastage of life potential per person in the age interval (x,x+n) when all causes of death are present,

$\bar{n}_i - \bar{n}$ = (of the above wastage) the amount of life potential per person recovered from the elimination of cause i in the age interval (x,x+n).

We now define the Fulfillment Index for the age group (x,x+n) from the elimination of cause i as

$$[(\bar{n}_i - \bar{n})/(n - \bar{n})] \times 100. \quad (14)$$

Similar indexes have been used by Arriaga (1984) for measuring temporal changes in life expectancies. A value of 100 for this index means that the fulfillment of life potential is maximum, i.e., a person lives n years in the n-year interval because of the elimination of cause i. On the other hand, a value of 0 implies that the elimination of cause i has not at all reduced the wastage of the life potential in the age interval (x,x+n). Any other value between 0 and 100 gives the percentage of recovery (fulfillment) of the wastage of life potential in the age interval (x,x+n) as a result of the elimination of a specified cause. For example, the fulfillment index for black males in the age group 20-24 from the elimination of homicide is 38.4. This means that on a scale from 0 to 100 (ranging from no impact of the elimination to no deaths in the age group after elimination), the fulfillment index for homicide stands at 38.4 for black males in the age group 20-24. In other words, 38.4 percent of the loss of life potential for black males in the age group 20-24 will be salvaged by the elimination of homicide as a cause of death. This significant effect of the elimination of homicide in recent years was noted by Farley (1980).

As in the case of gains in expectation of life, the fulfillment indexes are not additive, i.e., the recoveries from a set of mutually exclusive and exhaustive disease categories treated separately would not add up to 100. Fulfillment indexes for 28 causes and cause-groups are shown in Table 2 for the nine race-sex categories of population.

PROBABILITY OF EVENTUALLY DYING FROM SPECIFIED CAUSES

Still another way of measuring the effect of the existence of a specified cause of death is to obtain the probability that a person of a specified age will eventually die from that cause.

Denoting by $d(x,x+n)$ the number of deaths in the abridged life table in the age interval $(x,x+n)$, we have

$$d(x,x+n) = \ell(x) - \ell(x+n). \quad (15)$$

Out of this number, the number of deaths from cause i is expected to be

$$d_i(x,x+n) = d(x,x+n)[1 - D_i(x,x+n)/D(x,x+n)]. \quad (16)$$

It should be noted here that $d_i(x,x+n)$ in (16) is not the same as the number of deaths in the age group $(x,x+n)$ in the life table eliminating cause i.

Since all persons $\ell(x)$ at age x in the life table will eventually die, the proportion of deaths to these $\ell(x)$ persons from cause i is obtained by adding all such numbers in (16) in the age group $(x,x+n)$ and above, and finally dividing the sum by $\ell(x)$. Therefore, the probability that a person aged x will eventually die from cause i is given by

$$[d_i(x,x+n) + d_i(x+n,x+n+n') + \ldots]/\ell(x), \quad (17)$$

n, n', ... being the lengths of the age intervals.

These probabilities are given in Table 3. Obviously, at a given age, all these probabilities corresponding to different causes of death should add up to one. Table D gives a comparison of these probabilities at birth for five time periods during 1939-1978.

DISCUSSION OF THE RESULTS

This paper uses the U.S. mortality data by cause of death for 1978 to compute three measures of the effect of the existence of a specified cause of death, viz., (1) gain in expectation of life from the elimination of a cause, (2) fulfillment index for an age group from the elimination of a cause, and (3) probability of eventually dying from a cause. These three measures are shown in Tables 2 and 3 for 28 causes of death corresponding to nine race-sex groups and 19 age groups. They are again shown diagrammatically in Charts 1 - 3 for 15 leading causes of death and four ages (or age groups). Also, for a limited number of causes of death and race-sex categories, the trends of gain in expectation of life at birth due to elimination of a cause and probability at birth of eventually dying from a cause are shown in Tables C and D.

The starting point of the analysis is the computation of expectation of life at different ages for the nine race-sex groups for all causes of death combined, and the results are shown in Table 1. The expectations of life at birth are 70.2, 63.6, and 76.3 for white, black, and other males, respectively. The corresponding numbers for females are 77.8, 72.5, and 86.6. The difference between the life expectancies for whites and others becomes even more pronounced at older ages. For example, at age 65, the life expectancies for white, black, and other males are 14.0, 13.6, and 21.5, and those for females are 18.4, 17.7, and 28.1, respectively. It is known that the Asian-Americans are the longest lived subgroup in the United States (Feinleib, 1980). However, since the "other" race group also includes American Indians, the results for this group are somewhat puzzling.

The expectations of life at ages 75 and above for blacks are higher than the corresponding numbers for whites. This "crossover" phenomenon has been attributed to various factors including errors in the data and natural selection (Siegel, 1980; Manton, 1980). No such phenomenon is observed between blacks and other races or between whites and other races.

About some of the results for different causes of death in Tables 2 and 3, infective and parasitic diseases are no longer a significant threat to human lives. Elimination of these diseases would increase the expectation of life at birth by only 0.17, and less than nine newborn babies per 1,000 would eventually die of these diseases.

The percent recovery of the wastage of life potential from the elimination of malignant neoplasms - which, as a group, is the second leading cause of death - is highest in the ages between 50 and 60. The fulfillment indexes in the age group 50-54 for white, black, and other males are 27.4, 24.9, and 27.4, respectively, the corresponding numbers for females being 47.4, 29.7, and 40.2. Over the whole span of life, other females benefit most in terms of gain in expectation of life from the elimination of malignant neoplasms. These gains at age 0 are 2.9, 3.4, and 3.7 for white, black, and other males, and 3.1, 3.3, and 4.3 for the females of the corresponding race groups. Ironically, the probability that a newborn baby would eventually die of malignant neoplasms is lowest for other females (.204, .206, .170 for white, black, and other males, and .184, .170, .162 for the corresponding females). This anomaly between the gain in expectation of life and the probability of dying is explained by the fact that the latter depends on the total number of deaths in the life table from the specified cause, irrespective of their distribution by age at death. However, this distribution is very important in the computation of the gain in expectation of life, deaths at younger ages from a cause having an inflating effect on the gain, when this cause is eliminated. Tables C and D show how, for this disease group, both the gain in expectation of life from the elimination and the probability of eventually dying from them have increased consistently over the past decades.

Major cardiovascular diseases as a group can certainly claim to be by far the leading cause of death in the United States. All three indexes clearly show that for all race groups, the impact of these diseases increases with age. For example, the fulfillment indexes for them in the age group 65-69 are as high as 50.4, 45.9, and 47.8 for white, black, and other males, and 47.6, 53.7, and 41.0 for the corresponding females. Again, a newborn male baby in the three race categories white, black, and other would gain 10.6, 10.6, and 16.7 years of life from the elimination of major cardiovascular diseases, the corresponding numbers for a female baby being 16.4, 20.3, and 23.0. Similarly, the probabilities of eventually dying from these diseases are .521, .430, and .478 for white, black and other males of age 0, while the corresponding numbers for females are .592, .547, and .514. Of white females of age 85, 73 percent would eventually die of these diseases. Although results showing the trend are not available for major cardiovascular diseases, the probability at birth of dying from the diseases of the heart has decreased over the last two decades and the gain in expectation of life at birth from their elimination has shown a sharp increase in the seventies.

Reduction of infant mortality rate is no longer the primary reason for the increase in the expectation of life at birth. Even if certain diseases of early infancy are completely eliminated, the expectation of life at birth for white, black, and other males would rise only by .44, .96, and .42, and for the corresponding females, by .37, .86, and .33.

Motor vehicle accidents is a leading cause of death, particularly among teenagers. The fulfillment indexes in the age group 15-19 are 55.0, 22.9, and 41.6 for white, black, and other males, and 47.5, 16.1, and 43.5 for the corresponding females. Again, the number of persons out of 100 at age 0 who can expect to eventually die from motor vehicle accidents are 2.4, 2.3, and 3.4 for white, black, and other males, and 1.0, 0.7, and 1.6 for the corresponding females.

The age group which is most susceptible to suicide is 25-29. Elimination of this cause of death from this age group would recover 16.3, 6.8, and 13.8 percents of the wastage of life potential among white, black, and other males (and 13.1, 4.0, and 8.5 among the corresponding females). Since the other two measures, viz., the gain in expectation of life and the probability of dying are composite measures covering the total span of life beyond a specified age, suicide as a cause of death does not produce impressive numbers for them.

As in the case of suicide, the most predominant age group for homicide is 25-29. The fulfillment indexes for this age group are 9.9, 36.2, and 13.0 for white, black, and other males, and 6.7, 18.6, and 11.0 for the females of the corresponding race groups. These indexes for blacks are even higher in the age group 20-24, viz., 38.4 and 24.6 for males and females. Black males aged 0 would survive 1.34 years more, on an average, if homicide is eliminated as a cause of death, and 4 out of every 100 newborn black males would die from homicide (as against 0.6 for white males).

REFERENCES

[1] Arriaga, E.E., Measuring and explaining the change in life expectancies, Demography 21:83-96 (1984).

[2] Bayo, F., Mortality of the aged, Trans. Soc. Actuar. 24:1-24 (1972).

[3] Case, R.A.M., Coghill, C., Harley, J.L., and Pearson, J.T., The Chester Beatty Research Institute Serial Abridged Life Tables, England and Wales 1841-1960, Part I, Institute of Cancer Research, London (1962).

[4] Chiang, C.L., Introduction to Stochastic Processes in Biostatistics, John Wiley, New York (1968).

[5] Chiang, C.L., Life Table and Mortality Analysis, World Health Organization (1978).

[6] Crimmins, E.M., Implications of recent mortality trends for the size and composition of the population over 65, Rev. Public Data Use 11:37-48 (1983).

[7] Farley, R., Homicide trends in the United States, Demography 17:177-188 (1980).

[8] Feinleib, M., Recommendations for future research. In: Haynes, S.G. and Feinleib (eds). Epidemiology of Aging, NIH Publication, No. 80-969, 359-362 (1980).

[9] Keyfitz, N., What difference would it make if cancer were eradicated? An examination of the Taeuber Paradox, Demography 14:411-418 (1977).

[10] Manton, K.G., Sex and race specific mortality differentials in multiple cause of death data, Gerontologist 20:480-493 (1980).

[11] Preston, S.H., Keyfitz, N., and Schoen, R., Causes of Death: Life Tables for National Populations, Seminar Press, New York (1972).

[12] Reed, L.J. and Merrell, M., A short method for constructing an abridged life table, Amer. J. Hygiene 30:33-62 (1939).

[13] Rosenwaike, I., Yaffe, N., and Sagi, P.C., The recent decline in mortality of the extreme aged: an analysis of statistical data, Amer. J. Public Health, 70:1074-1080 (1980).

[14] Siegel, J.S., Prospective Trends in the Size and Structure of the Elderly Population, Impact of Mortality Trends, and Some Implications, Current Population Reports, U.S. Bureau of the Census, Series P-23, 78:7-22 (1979).

[15] Siegel, J.S., On the demography of aging, Demography 17:345-364 (1980).

[16] U.S. Bureau of the Census, Estimates of the Population of the United States by Age, Race, and Sex: 1976 to 1979, Current Population Reports, Series P-25, No. 870 (1980).

[17] U.S. Bureau of the Census, Preliminary Estimates of the Population of the United States, by Age, Sex, and Race: 1970 to 1981, Current Population Reports, Series P-25, No. 917 (1982).

[18] U.S. National Center for Health Statistics, Comparison of Two Methods of Constructing Abridged Life Tables by Reference to a "Standard" Table, Series 2, No. 4 (1966).

[19] U.S. National Center for Health Statistics, United States Life Tables by Causes of Death: 1959-61, Life Tables: 1959-61, Vol. 1, No. 6 (1968).

[20] U.S. National Center for Health Statistics, United States Life Tables by Causes of Death: 1969-71, U.S. Decennial Life Tables for 1969-71, Vol. 1, No. 5 (1975).

[21] U.S. National Center for Health Statistics, Vital Statistics of the United States, 1978, Vol. II - Mortality, Part A (1982).

[22] Wilkin, J.C., Recent trends in the mortality of the aged, Trans. Soc. Actuar. 33:53-86 (1981).

Table A. Death Rates per 100,000 Population for 24 Selected Causes: United States, 1978

Cause of Death[a]	All races			White		Black		Other	
	T	M	F	M	F	M	F	M	F
1. Tuberculosis, all forms (010-019)	1	2	1	1	1	5	2	3	2
2. Other infective and parasitic diseases (000-009, 020-136)	7	7	7	7	6	13	11	7	5
Infective and parasitic diseases (000-136)	8	9	8	8	7	18	13	10	7
3. Malignant neoplasms of digestive organs and peritoneum (150-159)	48	52	45	53	46	52	38	34	20
4. Malignant neoplasms of respiratory system (160-163)	46	70	22	72	24	69	16	24	8
5. Other malignant neoplasms (140-149, 170-209)[b]	88	81	94	82	98	80	78	30	33
Malignant neoplasms (140-209)[b]	182	203	161	207	168	201	132	88	61
6. Diabetes mellitus (250)	15	13	18	13	17	15	24	9	10
7. Diseases of heart (390-398, 402, 404, 410-429)	334	375	295	391	308	295	238	147	69
8. Hypertension (400, 401, 403)	3	2	3	2	2	4	4	1	1
9. Cerebrovascular diseases (430-438)	81	70	91	69	94	78	83	36	27
10. Arteriosclerosis (440)	13	11	15	11	17	7	8	4	2
11. Other diseases of arteries, arterioles, and capillaries (441-448)	12	15	10	16	10	9	8	5	3
Major cardiovascular diseases (390-448)	443	473	414	489	431	393	341	193	102
12. Influenza and pneumonia (470-474, 480-486)	27	29	25	29	26	32	18	19	10
13. Bronchitis, emphysema, and asthma (490-493)	10	15	6	15	7	7	3	5	2
14. Other diseases of the respiratory system (460-466, 500-519)	19	27	11	29	12	19	7	10	5
Diseases of the respiratory system (460-519)	56	71	42	73	45	58	28	34	17
15. Peptic ulcer (531-533)	3	3	2	3	2	3	1	2	1
16. Cirrhosis of liver (571)	14	19	9	18	9	24	13	17	10
17. Nephritis and nephrosis (580-584)	4	4	4	4	3	9	8	4	3
18. Congenital anomalies (740-759)	6	6	5	6	5	9	7	7	6
19. Certain diseases of early infancy (760-769.2, 769.4-772, 774-778)	10	12	8	10	7	31	22	12	8
20. Motor vehicle accidents (E810-E823)	24	36	13	36	13	33	10	42	16
21. All other accidents (E800-E807, E825-E949)	24	34	16	32	15	46	17	34	10
22. Suicide (E950-E959)	13	19	6	20	7	11	3	13	5
23. Homicide (E960-E978)	9	15	4	9	3	59	13	15	4
24. All other causes (residual)	72	77	68	72	65	119	91	54	34
All Causes	883	994	778	1000	797	1029	723	534	294

----

a/ The numbers in parentheses represent the codes in the Eighth Revision International Classification of Diseases, Adapted for Use in the United States, U.S. National Center for Health Statistics, 1967.

b/ Include neoplasms of lymphatic and hematopoietic tissues.

Table B. Multiplying Factors A(x,n) and B(x,n) for Conversion of m(x,x+n) to q(x,x+n) and $\ell(x)$ to L(x,x+n), respectively, in an Abridged Life Table: United States, 1978

Age	Male		Female		Total	
(x,x+n)	A(x,n)	B(x,n)	A(x,n)	B(x,n)	A(x,n)	B(x,n)
0 - 1	.97	.497	.97	.498	.97	.497
1 - 5	3.95	2.000	3.96	2.000	3.96	2.000
5 - 10	5.01	2.500	5.01	2.500	5.01	2.500
10 - 15	5.00	2.501	5.01	2.500	5.01	2.500
15 - 20	5.04	2.501	5.01	2.500	5.03	2.501
20 - 25	5.00	2.500	5.04	2.500	5.02	2.500
25 - 30	5.00	2.500	5.02	2.500	5.01	2.500
30 - 35	5.01	2.500	5.02	2.500	5.01	2.500
35 - 40	5.00	2.501	5.00	2.501	5.00	2.501
40 - 45	4.99	2.502	4.99	2.501	4.99	2.501
45 - 50	5.01	2.502	5.01	2.501	5.01	2.502
50 - 55	5.02	2.504	5.01	2.502	5.01	2.503
55 - 60	5.02	2.505	5.01	2.503	5.02	2.504
60 - 65	5.03	2.506	5.02	2.504	5.02	2.505
65 - 70	5.03	2.506	5.03	2.506	5.03	2.506
70 - 75	5.04	2.504	5.05	2.509	5.04	2.507
75 - 80	5.07	2.495	5.07	2.511	5.07	2.505
80 - 85	5.10	2.469	5.09	2.502	5.09	2.491
85 and over	-	.948	-	.940	-	.944

Table C. Gain in Expectation of Life at Birth Due to Elimination of Specified Causes of Death: United States: 1959-61, 1969-71, 1978.

Cause of Death	Total Population			White Male			White Female		
	1959-61	1969-71	1978	1959-61	1969-71	1978	1959-61	1969-71	1978
1. Tuberculosis, all forms	.10	.04	.02	.10	.03	.02	.05	.02	.01
Infective and parasitic diseases	.22	.17	.17	.20	.13	.14	.14	.12	.14
3. Malignant neoplasms of digestive organs and peritoneum	.66	.60	.71	.63	.55	.63	.68	.62	.72
4. Malignant neoplasms of respiratory system	.32	.50	.73	.49	.69	.92	.11	.22	.43
Malignant neoplasms	2.27	2.47	3.09	2.12	2.31	2.85	2.43	2.57	3.12
6. Diabetes mellitus	.22	.24	.22	.15	.17	.15	.27	.28	.25
7. Diseases of the heart	5.89	5.86	7.01	6.51	6.14	6.49	5.04	5.17	6.94
9. Cerebrovascular diseases	-	1.19	1.14	-	.86	.74	-	1.36	1.42
10. Arteriosclerosis	.18	.13	.16	.15	.09	.10	.21	.17	.21
12. Influenza and pneumonia	.53	.47	.39	.46	.41	.33	.42	.40	.39
13. Bronchitis, emphysema, and asthma	-	.20	.14	-	.26	.17	-	.10	.10
Diseases of the respiratory system	-	.83	.84	-	.86	.85	-	.61	.71
15. Peptic ulcer	.09	.06	.04	.11	.06	.04	.05	.04	.03
16. Cirrhosis of liver	.19	.28	.27	.22	.30	.29	.15	.20	.18
17. Nephritis and nephrosis	-	.07	.06	-	.05	.05	-	.05	.05
18. Congenital anomalies	.36	.29	.25	.37	.30	.25	.36	.30	.25
19. Certain diseases of early infancy	1.12	.82	.49	1.12	.82	.44	.90	.66	.37
20. Motor vehicle accidents	.55	.70	.65	.78	.93	.89	.30	.41	.39
21. All other accidents	.62	.63	.56	.77	.76	.69	.35	.35	.33
22. Suicide	.22	.26	.30	.31	.34	.42	.12	.18	.18
23. Homicide	.13	.23	.26	.09	.16	.22	.04	.06	.09

Sources: U.S. National Center for Health Statistics (1968, 1975); Table 3.

Table D. Probability at Birth of Eventually Dying from Specified Causes: United States, 1939-41, 1949-51, 1959-61, 1969-71, 1978

Cause of Death	Total Population			White Male					White Female				
	1959-61	1969-71	1978	1939-41	1949-51	1959-61	1969-71	1978	1939-41	1949-51	1959-61	1969-71	1978
1. Tuberculosis, all forms	.00568	.00246	.00145	.03126	.02010	.00679	.00271	.00146	.02117	.00961	.00270	.00120	.00089
2. All other infective and parasitic diseases	.00475	.00488	.00716	.03411	-	.00456	.00419	.00618	.02857	-	.00360	.00456	.00681
Infective and parasitic diseases	.01043	.00734	.00861	.06537	-	.01135	.00690	.00764	.04974	-	.00630	.00576	.00770
3. Malignant neoplasms of digestive organs and peritoneum	.05434	.04879	.05379	-	-	.05380	.04816	.05296	-	-	.05660	.05028	.05454
4. Malignant neoplasms of respiratory system	.02145	.03292	.04618	-	-	.03507	.05214	.06872	-	-	.00710	.01333	.02395
5. Other malignant neoplasms	.07575	.08129	.09350	-	-	.06369	.06913	.08268	-	-	.09087	.09563	.10508
Malignant neoplasms	.15154	.16300	.19347	.10780	.13589	.15256	.16943	.20436	.13542	.15452	.15457	.15924	.18357
6. Diabetes mellitus	.01746	.02008	.01776	.01847	.01218	.01267	.01473	.01318	.03693	.02427	.02261	.02444	.02084
7. Diseases of the heart	.42252	.41206	.40606	-	-	.43807	.42233	.40974	-	-	.42368	.42108	.41854
9. Cerebrovascular diseases	-	.12244	.10439	-	-	-	.09505	.07774	-	-	-	.15090	.13098
10. Arteriosclerosis	.02855	.02211	.01936	-	-	.02298	.01624	.01406	-	-	.03668	.03045	.02629
12. Influenza and pneumonia	.03454	.03425	.03440	-	.02859	.03240	.03205	.03249	-	.03041	.03391	.03500	.03653
13. Bronchitis, emphysema, and asthma	-	.01564	.01110	-	-	-	.02461	.01608	-	-	-	.00730	.00715
14. Other diseases of the respiratory system	-	.00921	.02153	-	-	-	.01198	.03009	-	-	-	.00601	.01420
Diseases of the respiratory system	-	.05910	.06703	-	-	-	.06864	.07866	-	-	-	.04831	.05789
15. Peptic ulcer	.00652	.00448	.00295	-	-	.00904	.00558	.00324	-	-	.00403	.00351	.00279
16. Cirrhosis of the liver	.00995	.01339	.01263	-	-	.01284	.01659	.01567	-	-	.00703	.00905	.00834
17. Nephritis and nephrosis	-	.00446	.00465	-	-	-	.00383	.00417	-	-	-	.00379	.00374
18. Congenital anomalies	.00586	.00495	.00412	-	-	.00637	.00519	.00423	-	-	.00560	.00483	.00393
19. Certain diseases of early infancy	.01572	.01148	.00663	.02402	-	.01632	.01191	.00621	.01808	-	.01192	.00869	.00471
20. Motor vehicle accidents	.01657	.01991	.01689	.02836	.02487	.02372	.02750	.02403	.00974	.00868	.00921	.01185	.00999
21. All other accidents	.02804	.02643	.02248	.04485	.03877	.03053	.02987	.02594	.03937	.03485	.02488	.02143	.01764
22. Suicide	.00903	.00950	.00993	.01641	-	.01431	.01413	.01543	.00524	-	.00452	.00600	.00566
23. Homicide	.00351	.00625	.00660	-	-	.00266	.00497	.00628	-	-	.00114	.00169	.00220

Sources: See footnote of Table C; Table 4.

CHART 1A. GAIN IN EXPECTATION OF LIFE AT AGE X (X = 0, 20, 40, 60) FROM THE ELIMINATION OF A SPECIFIED CAUSE, UNITED STATES MALES 1978

(D = mal neo Dig, R = mal neo Res, M = Mal neo, B = diaB, T = hearT, C = Cerebro, A = Arterio, J = maJor cardiov, P = resPiratory, L = cir Liver, I = Infant, V = motor V, O = Other acci, S = Suicide, H = Homicide)

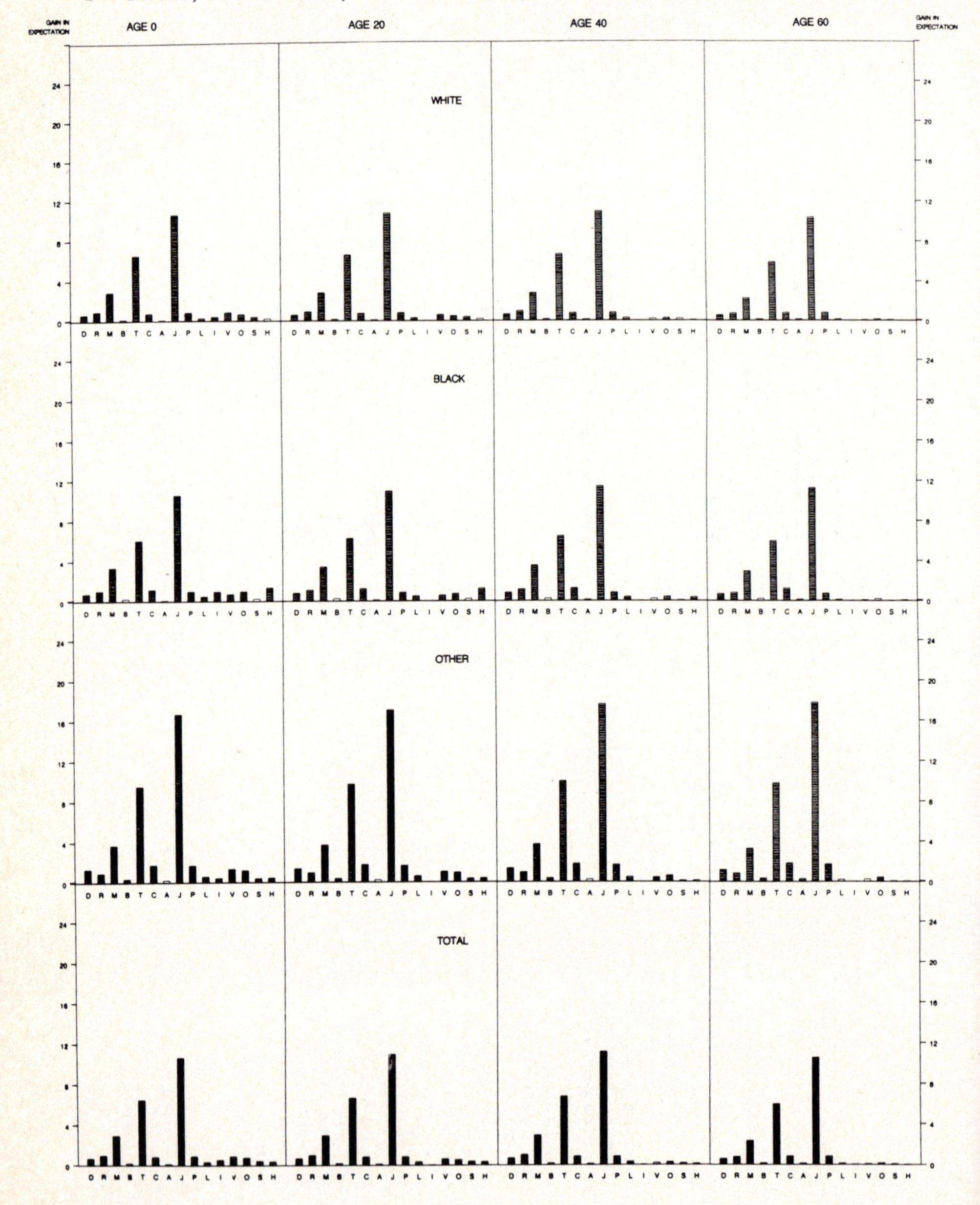

CHART 1B. GAIN IN EXPECTATION OF LIFE AT AGE X (X = 0, 20, 40, 60) FROM THE ELIMINATION OF A SPECIFIED CAUSE, UNITED STATES FEMALES 1978

(D = mal neo Dig, R = mal neo Res, M = Mal neo, B = diaB, T = hearT, C = Cerebro, A = Arterio, J = maJor cardiov, P = resPiratory, L = cir Liver, I = Infant, V = motor V, O = Other acci, S = Suicide, H = Homicide)

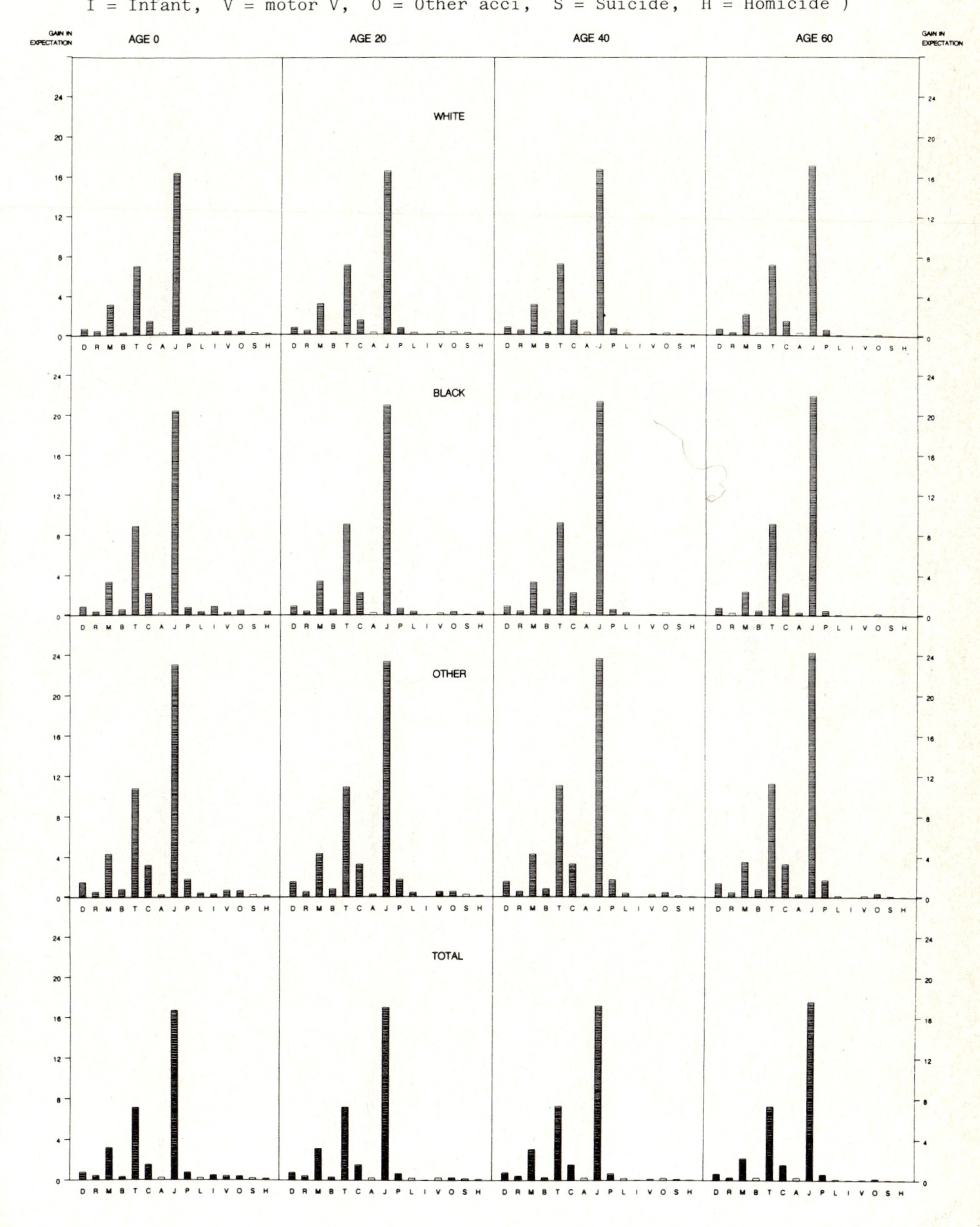

CHART 2A. FULFILLMENT INDEX (PERCENT RECOVERY OF WASTAGE OF LIFE POTENTIAL) IN AN AGE INTERVAL FROM THE ELIMINATION OF A SPECIFIED CAUSE, U.S. MALES 1978

(D = mal neo Dig, R = mal neo Res, M = Mal neo, B = diaB, T = hearT, C = Cerebro, A = Arterio, J = maJor cardiov, P = resPiratory, L = cir Liver, I = Infant, V = motor V, O = Other acci, S = Suicide, H = Homicide)

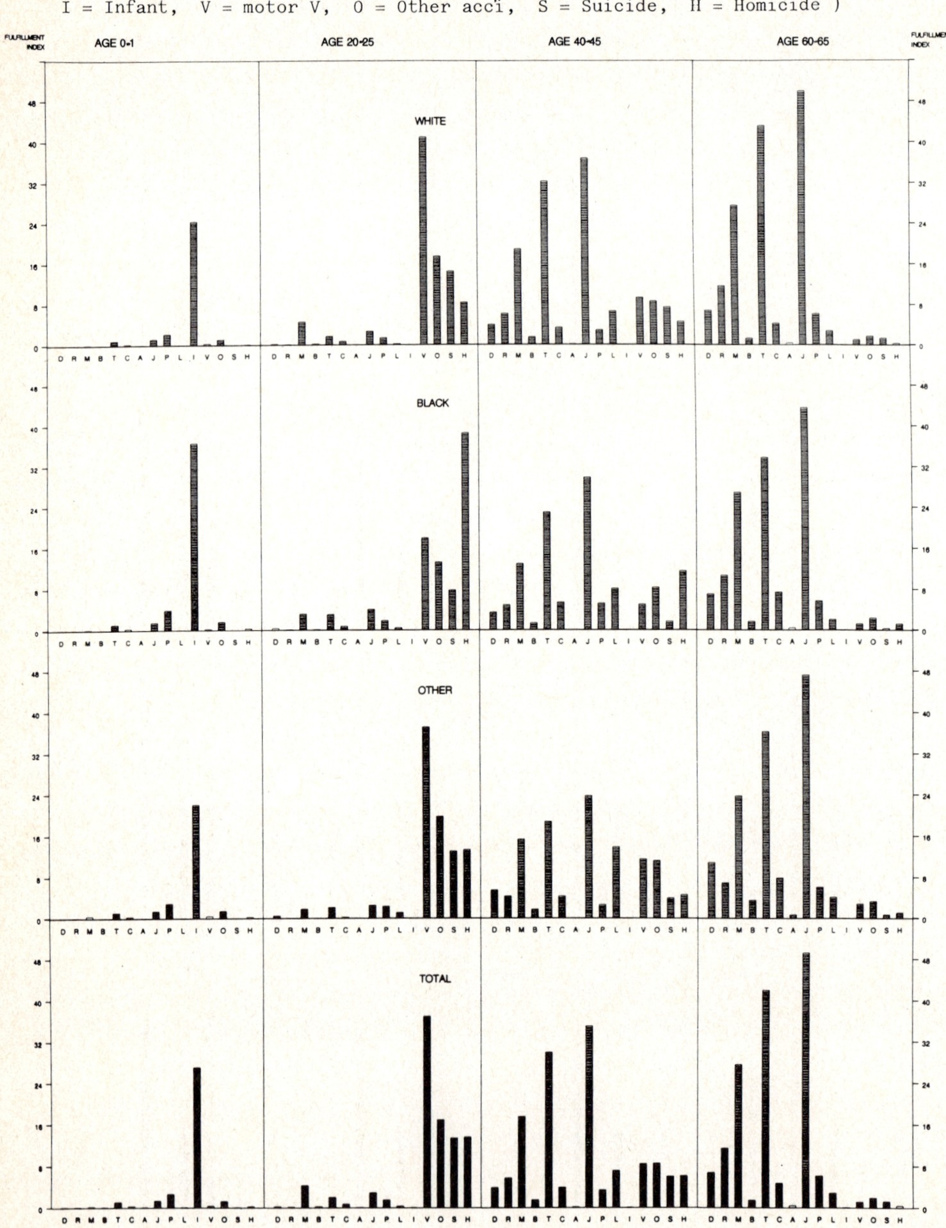

CHART 2B. FULFILLMENT INDEX (PERCENT RECOVERY OF WASTAGE OF LIFE POTENTIAL) IN AN AGE INTERVAL FROM THE ELIMINATION OF A SPECIFIC CAUSE, U.S. FEMALES 1978

(D = mal neo Dig, R = mal neo Res, M = Mal neo, B = diaB, T = hearT, C = Cerebro, A = Arterio, J = maJor cardiov, P = resPiratory, L = cir Liver, I = Infant, V = motor V, O = Other acci, S = Suicide, H = Homicide)

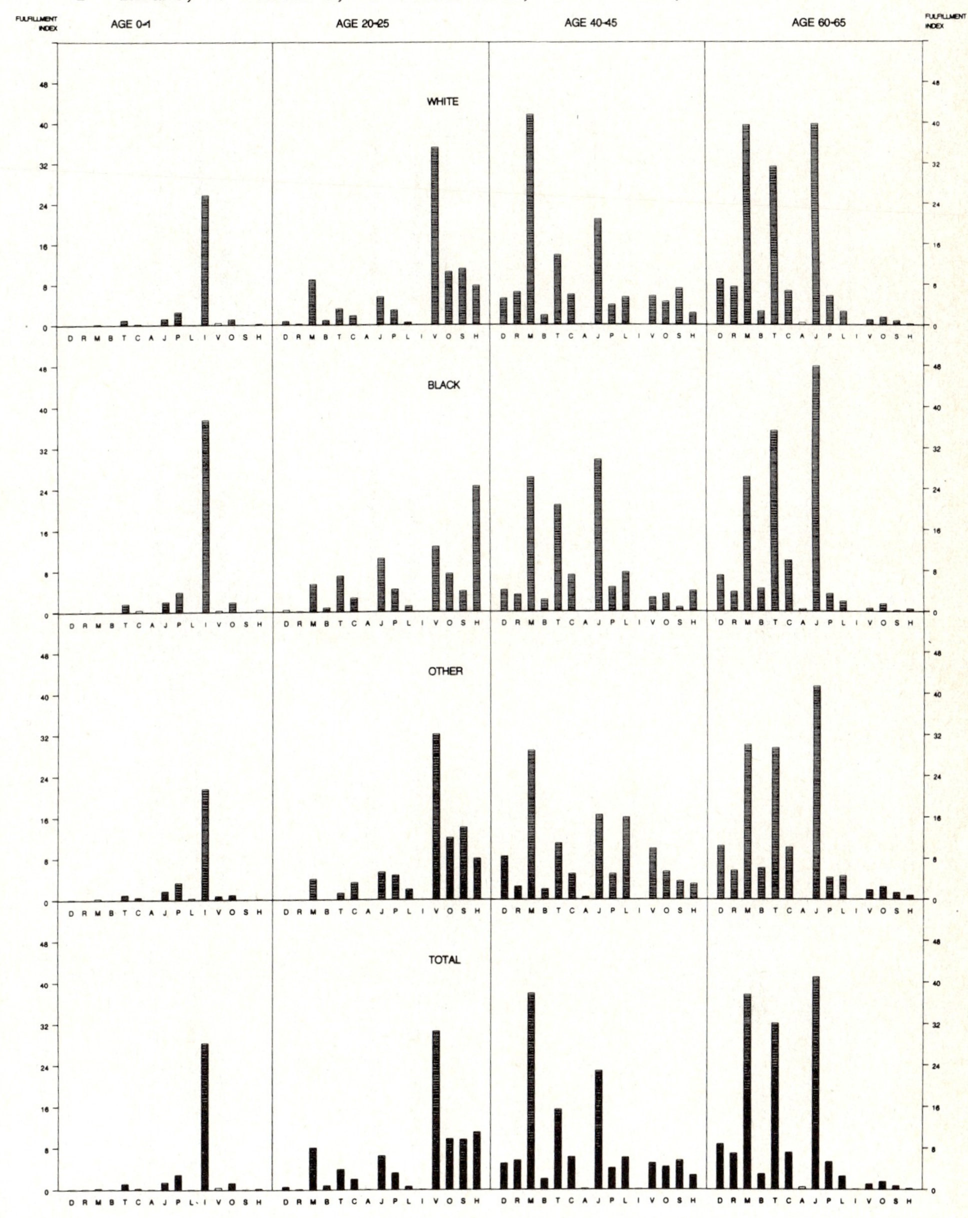

CHART 3A. PROBABILITY THAT A MALE OF AGE X (X = 0, 20, 40, 60) WILL EVENTUALLY DIE FROM A SPECIFIED CAUSE, UNITED STATES 1978

(D = mal neo Dig, R = mal neo Res, M = Mal neo, B = diaB, T = hearT, C = Cerebro, A = Arterio, J = maJor cardiov, P = resPiratory, L = cir Liver, I = Infant, V = motor V, O = Other acci, S = Suicide, H = Homicide)

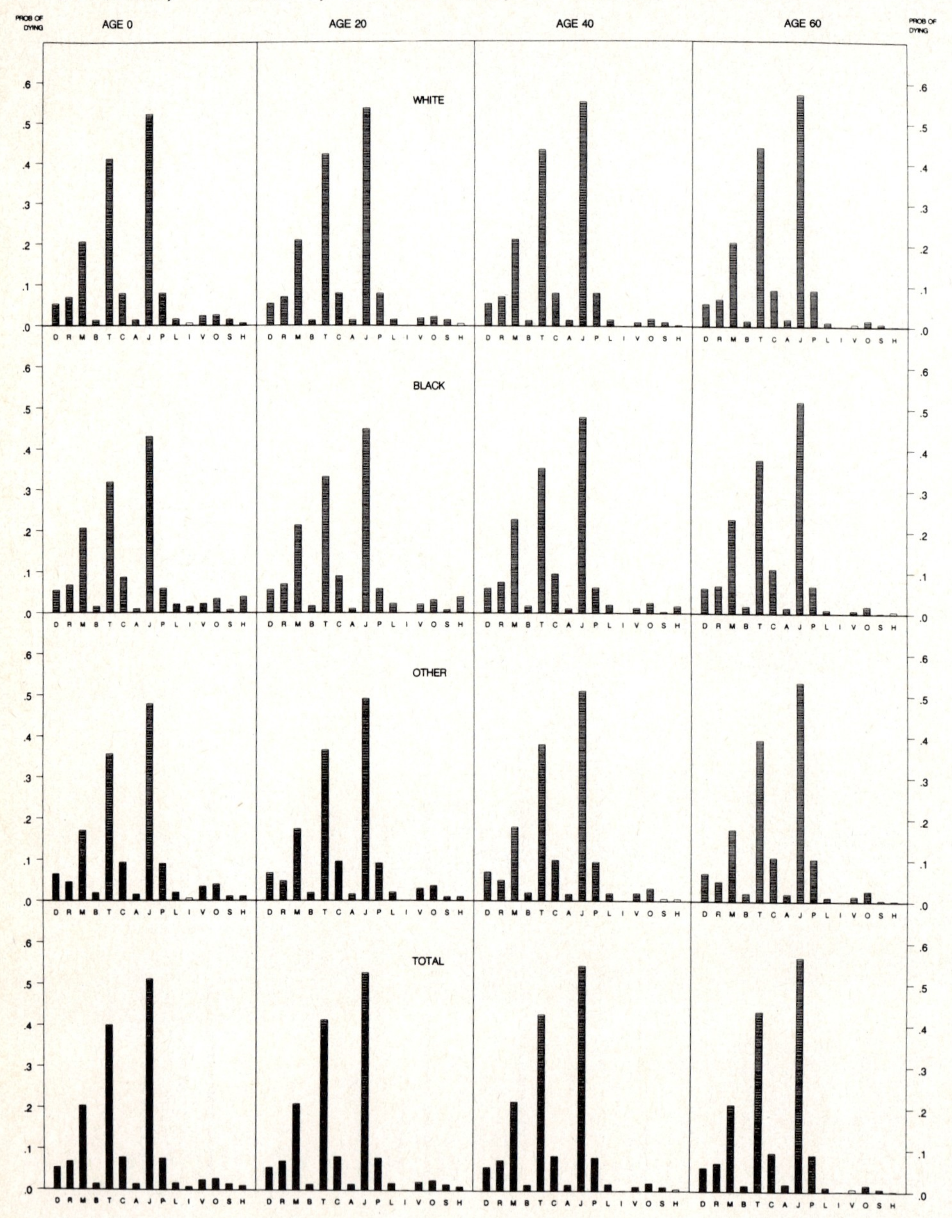

CHART 3B. PROBABILITY THAT A FEMALE OF AGE X (X = 0, 20, 40, 60) WILL EVENTUALLY DIE FROM A SPECIFIED CAUSE, UNITED STATES 1978

(D = mal neo Dig, R = mal neo Res, M = Mal neo, B = diaB, T = hearT, C = Cerebro, A = Arterio, J = maJor cardiov, P = resPiratory, L = cir Liver, I = Infant, V = motor V, O = Other acci, S = Suicide, H = Homicide)

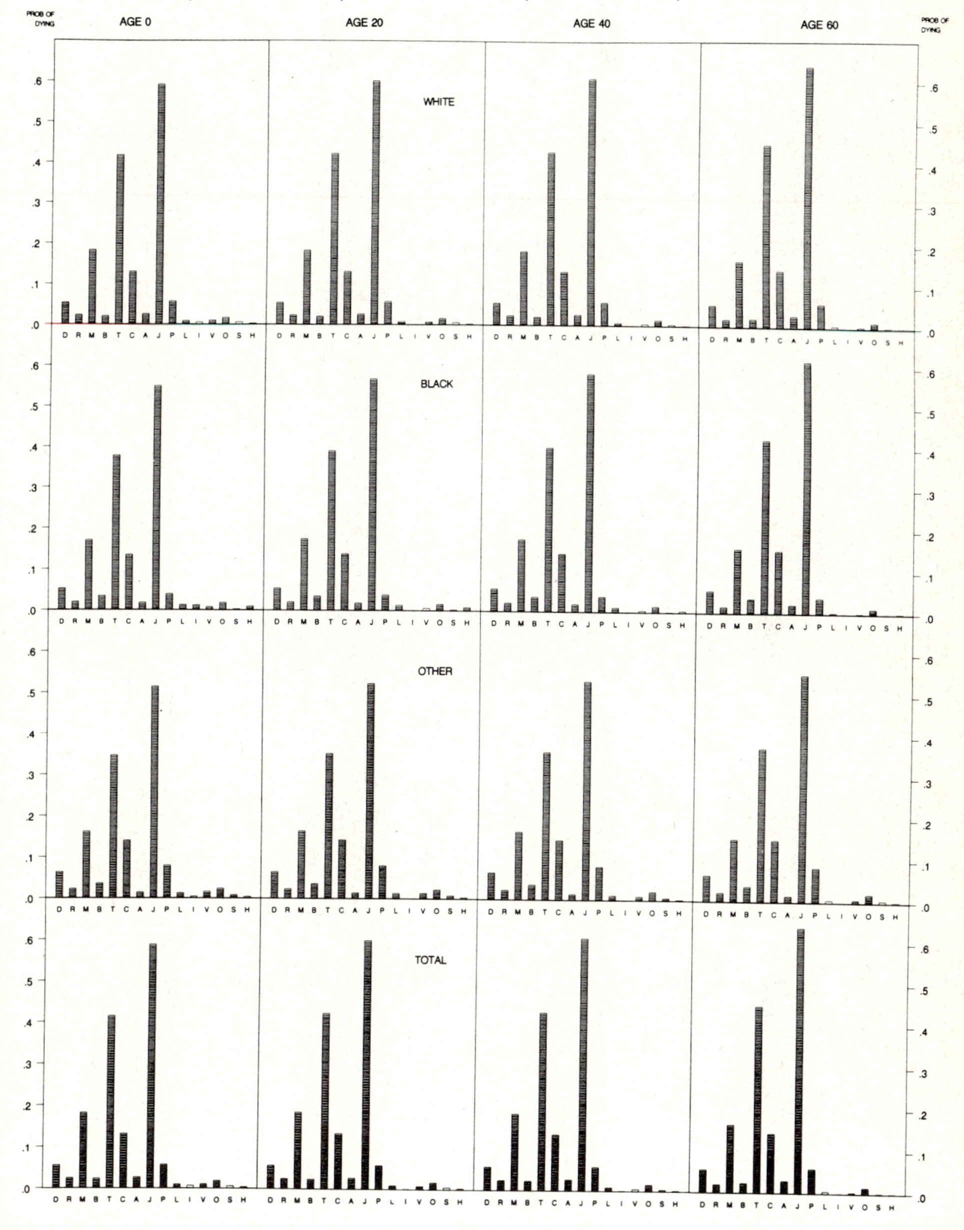

TABLE 1. ABRIDGED LIFE TABLES FOR ALL CAUSES OF DEATH COMBINED: UNITED STATES, 1978

Q(X,X+N) = PROBABILITY THAT A PERSON AGED X WILL DIE IN THE AGE INTERVAL (X,X+N)

AGE (X,X+N)	ALL RACES T	ALL RACES M	ALL RACES F	WHITE M	WHITE F	BLACK M	BLACK F	OTHER M	OTHER F
0 - 1	.01383	.01534	.01224	.01311	.01033	.02831	.02316	.01269	.00989
1 - 5	.00274	.00308	.00237	.00283	.00211	.00446	.00375	.00314	.00239
5 - 10	.00167	.00195	.00139	.00182	.00132	.00270	.00181	.00180	.00115
10 - 15	.00172	.00218	.00123	.00209	.00119	.00272	.00147	.00214	.00117
15 - 20	.00506	.00730	.00277	.00738	.00276	.00672	.00282	.00773	.00263
20 - 25	.00674	.01011	.00335	.00950	.00307	.01437	.00509	.01066	.00361
25 - 30	.00659	.00957	.00363	.00838	.00316	.01913	.00698	.01014	.00372
30 - 35	.00698	.00962	.00440	.00824	.00383	.02222	.00877	.00979	.00446
35 - 40	.00943	.01263	.00637	.01071	.00547	.02915	.01314	.01516	.00674
40 - 45	.01467	.01893	.01062	.01629	.00932	.04151	.02051	.01987	.00926
45 - 50	.02336	.03004	.01697	.02738	.01531	.05474	.03093	.02178	.01327
50 - 55	.03653	.04808	.02575	.04451	.02350	.08426	.04699	.03506	.01901
55 - 60	.05450	.07211	.03806	.06812	.03527	.11530	.06602	.05080	.02928
60 - 65	.08526	.11428	.05910	.11011	.05579	.16364	.09456	.08402	.04277
65 - 70	.11657	.15893	.08129	.15702	.07865	.18108	.10478	.11643	.07147
70 - 75	.17384	.23225	.12859	.22935	.12363	.27748	.19235	.15066	.08793
75 - 80	.26322	.33561	.21261	.33341	.20663	.40043	.31923	.17281	.09154
80 - 85	.36614	.44630	.31773	.45255	.32109	.41611	.29375	.20721	.15438
85 & OVER	1.00000	1.00000	1.00000	1.00000	1.00000	1.00000	1.00000	1.00000	1.00000

L(X) = NUMBER SURVIVING TO AGE X OUT OF A COHORT OF 100,000 PERSONS AT AGE 0

AGE (X,X+N)	ALL RACES T	ALL RACES M	ALL RACES F	WHITE M	WHITE F	BLACK M	BLACK F	OTHER M	OTHER F
0 - 1	100000	100000	100000	100000	100000	100000	100000	100000	100000
1 - 5	98617	98466	98775	98689	98967	97169	97684	98731	99011
5 - 10	98347	98162	98541	98409	98758	96736	97317	98421	98774
10 - 15	98182	97971	98404	98231	98628	96475	97141	98244	98661
15 - 20	98013	97757	98283	98025	98511	96212	96998	98033	98545
20 - 25	97517	97043	98011	97302	98239	95565	96725	97275	98285
25 - 30	96859	96062	97683	96377	97937	94192	96232	96238	97931
30 - 35	96221	95143	97328	95569	97628	92390	95561	95262	97566
35 - 40	95550	94227	96900	94782	97254	90337	94723	94329	97132
40 - 45	94649	93037	96282	93767	96722	87704	93478	92900	96477
45 - 50	93260	91276	95260	92239	95821	84063	91561	91054	95584
50 - 55	91081	88534	93643	89714	94355	79461	88729	89070	94315
55 - 60	87755	84277	91232	85721	92138	72765	84559	85948	92523
60 - 65	82972	78200	87760	79882	88888	64375	78977	81582	89814
65 - 70	75898	69263	82574	71086	83929	53841	71508	74727	85972
70 - 75	67051	58255	75861	59924	77328	44092	64015	66027	79827
75 - 80	55395	44725	66106	46180	67768	31857	51702	56079	72808
80 - 85	40813	29715	52051	30783	53765	19100	35197	46388	66143
85 & OVER	25870	16453	35513	16852	36501	11152	24858	36776	55932

LL(X,X+N) = NUMBER OF YEARS LIVED BY L(X) PERSONS AT AGE X IN THE AGE INTERVAL (X,X+N)

AGE (X,X+N)	ALL RACES T	ALL RACES M	ALL RACES F	WHITE M	WHITE F	BLACK M	BLACK F	OTHER M	OTHER F
0 - 1	98712	98637	98990	98748	99085	97993	98446	98769	99107
1 - 5	393928	393257	394634	394198	395452	387811	390004	394306	395572
5 - 10	491323	490334	492365	491602	493468	483028	486148	491665	493590
10 - 15	490489	489517	491719	490838	492850	481912	485350	490891	493016
15 - 20	489022	487198	490736	488515	491876	479638	484310	488468	492078
20 - 25	485942	482766	489236	484200	490441	474396	482396	483786	490542
25 - 30	482703	478015	487528	479869	488914	466458	479485	478754	488745
30 - 35	479430	473428	485570	475881	487207	456821	475710	473982	486748
35 - 40	475689	468351	483149	471563	485138	445282	470693	468262	484217
40 - 45	469963	461154	479048	465388	481554	429762	462786	460253	480345
45 - 50	461225	449887	472449	455248	475633	409139	450907	450673	474938
50 - 55	447629	432721	462561	439290	466606	381177	433569	438247	467470
55 - 60	427501	407008	448020	414836	453109	343539	409332	419663	456390
60 - 65	397971	369545	426518	378327	432734	296252	376816	391713	440170
65 - 70	358231	319563	397040	328313	404110	245420	339624	352732	415496
70 - 75	306973	257864	356197	265686	364046	190177	290336	305755	382964
75 - 80	241004	185730	296695	192025	305171	127140	218205	255658	348908
80 - 85	166110	113991	219088	117613	225849	74696	150259	205334	305434
85 & OVER	166070	90333	246462	88227	243654	92843	265553	484221	959423

E(X) = EXPECTATION OF LIFE AT AGE X

AGE (X,X+N)	ALL RACES T	ALL RACES M	ALL RACES F	WHITE M	WHITE F	BLACK M	BLACK F	OTHER M	OTHER F
0 - 1	73.30	69.49	77.18	70.20	77.77	63.63	72.50	76.33	86.55
1 - 5	73.33	69.57	77.13	70.14	77.58	64.48	73.21	76.31	86.41
5 - 10	69.52	65.78	73.31	66.33	73.74	60.76	69.48	72.55	82.62
10 - 15	64.63	60.91	68.41	61.45	68.83	55.92	64.60	67.67	77.71
15 - 20	59.74	56.03	63.49	56.57	63.91	51.06	59.69	62.81	72.80
20 - 25	55.03	51.42	58.66	51.97	59.08	46.39	54.85	58.28	67.98
25 - 30	50.39	46.92	53.85	47.44	54.26	42.03	50.12	53.88	63.22
30 - 35	45.70	42.35	49.04	42.82	49.42	37.80	45.46	49.41	58.45
35 - 40	41.01	37.74	44.24	38.16	44.60	33.60	40.84	44.87	53.70
40 - 45	36.37	33.19	39.51	33.54	39.83	29.53	36.34	40.52	49.04
45 - 50	31.88	28.78	34.90	29.05	35.18	25.70	32.05	36.29	44.48
50 - 55	27.57	24.59	30.46	24.79	30.68	22.04	27.99	32.03	40.04
55 - 60	23.52	20.69	26.20	20.82	26.36	18.83	24.24	28.10	35.76
60 - 65	19.72	17.10	22.13	17.15	22.23	15.95	20.78	24.46	31.76
65 - 70	16.32	13.97	18.35	13.95	18.38	13.56	17.68	21.46	28.06
70 - 75	13.13	11.12	14.74	11.07	14.73	11.00	14.44	18.95	25.01
75 - 80	10.35	8.72	11.53	8.62	11.43	9.25	12.26	16.85	22.16
80 - 85	8.14	6.88	8.94	6.69	8.73	8.77	11.81	14.86	19.12
85 & OVER	6.42	5.49	6.94	5.24	6.68	8.32	10.68	13.17	17.15

TABLE 2. GAINS IN EXPECTATION OF LIFE AND FULFILMENT INDEXES FROM THE ELIMINATION OF A SPECIFIED CAUSE OF DEATH: UNITED STATES, 1978

AGE	GAIN IN EXPECTATION OF LIFE AT AGE X									FULFILMENT INDEX FOR AGE (X TO X+N)								
	ALL RACES			WHITE		BLACK		OTHER		ALL RACES			WHITE		BLACK		OTHER	
(X,X+N)	T	M	F	M	F	M	F	M	F	T	M	F	M	F	M	F	M	F
(ELIMINATING TUBERCULOSIS, ALL FORMS)																		
0- 1	.02	.02	.02	.02	.01	.07	.04	.11	.10	.0	.0	.0	.0	.0	.0	.0	.0	.0
1- 5	.02	.02	.02	.02	.01	.08	.04	.11	.11	.1	.1	.2	.1	.2	.1	.2	.0	.0
5-10	.02	.02	.02	.02	.01	.08	.04	.11	.11	.1	.0	.2	.0	.2	.1	.2	.0	.0
10-15	.02	.02	.02	.02	.01	.08	.04	.11	.11	.0	.1	.0	.0	.0	.4	.0	.0	.0
15-20	.02	.02	.02	.02	.01	.08	.04	.11	.11	.1	.1	.1	.1	.0	.0	.4	.4	.0
20-25	.02	.02	.02	.02	.01	.08	.04	.11	.11	.1	.0	.1	.0	.1	.2	.4	.0	.0
25-30	.02	.02	.02	.02	.01	.08	.04	.11	.11	.1	.1	.2	.0	.1	.3	.5	.5	.0
30-35	.02	.02	.01	.02	.01	.08	.04	.11	.11	.2	.2	.2	.1	.1	.6	.4	.0	.6
35-40	.02	.02	.01	.02	.01	.07	.04	.11	.11	.2	.2	.3	.1	.2	.4	.7	.3	.6
40-45	.02	.02	.01	.02	.01	.07	.04	.11	.11	.3	.3	.2	.1	.2	.7	.4	.3	.0
45-50	.02	.02	.01	.02	.01	.06	.03	.11	.11	.2	.3	.2	.2	.1	.6	.5	.8	1.0
50-55	.02	.02	.01	.02	.01	.06	.03	.11	.10	.3	.3	.2	.2	.2	.6	.5	.9	.8
55-60	.02	.02	.01	.01	.01	.05	.03	.10	.10	.2	.2	.2	.2	.1	.6	.2	.6	1.5
60-65	.01	.02	.01	.01	.01	.05	.02	.10	.09	.2	.2	.1	.2	.1	.6	.2	.6	.3
65-70	.01	.01	.01	.01	.01	.04	.02	.10	.09	.2	.2	.1	.2	.1	.4	.2	.6	.8
70-75	.01	.01	.01	.01	.01	.03	.02	.09	.08	.1	.2	.1	.1	.1	.3	.2	.6	.6
75-80	.01	.01	.01	.01	.01	.03	.02	.09	.07	.1	.1	.1	.1	.1	.3	.1	.7	.7
80-85	.01	.01	.01	.01	.01	.03	.02	.08	.07	.1	.1	.1	.1	.1	.3	.2	.4	.5
(ELIMINATING OTHER INFECTIVE AND PARASITIC DISEASES)																		
0- 1	.15	.14	.15	.12	.13	.27	.32	.26	.34	2.6	2.7	2.9	2.4	2.4	3.9	4.5	2.9	3.0
1- 5	.10	.09	.11	.08	.10	.17	.22	.21	.30	4.2	4.0	4.5	4.0	4.5	3.6	4.1	5.5	7.3
5-10	.09	.08	.10	.07	.09	.16	.21	.19	.29	2.3	2.0	2.7	1.9	2.9	2.3	2.0	1.4	2.2
10-15	.09	.08	.10	.07	.09	.16	.21	.19	.28	1.8	2.4	2.3	2.4	2.3	2.3	2.2	4.0	4.8
15-20	.09	.08	.10	.07	.09	.15	.21	.19	.28	1.3	.8	2.0	.7	1.8	1.1	2.6	1.5	4.9
20-25	.09	.08	.10	.07	.08	.15	.20	.18	.27	.7	.5	1.3	.5	1.0	.8	2.4	.0	.7
25-30	.08	.07	.09	.06	.08	.15	.20	.19	.27	.9	.6	1.7	.5	1.6	.9	2.0	1.1	.6
30-35	.08	.07	.09	.06	.08	.14	.19	.18	.27	1.2	.8	2.1	.6	2.0	1.3	2.5	.9	2.3
35-40	.08	.07	.09	.06	.08	.14	.18	.18	.27	1.1	.9	1.6	.8	1.4	1.3	2.1	1.9	1.2
40-45	.08	.07	.08	.06	.07	.13	.17	.17	.26	1.0	.9	1.2	.8	1.0	1.0	1.8	1.2	2.5
45-50	.07	.06	.08	.06	.07	.12	.17	.17	.26	.8	.7	1.0	.6	.8	1.0	1.5	1.3	1.7
50-55	.07	.06	.08	.05	.07	.12	.16	.16	.25	.7	.7	.9	.6	.7	.9	1.4	.9	3.1
55-60	.06	.06	.07	.05	.06	.11	.15	.16	.24	.7	.6	.9	.5	.8	1.0	1.2	1.6	1.3
60-65	.06	.05	.07	.05	.06	.10	.14	.14	.23	.6	.5	.9	.5	.8	.7	1.2	.7	1.3
65-70	.05	.05	.06	.04	.05	.10	.13	.14	.22	.6	.5	.7	.5	.7	.8	1.0	1.1	1.2
70-75	.05	.04	.05	.04	.05	.10	.12	.13	.22	.6	.5	.7	.4	.7	.8	1.0	.7	1.3
75-80	.04	.04	.05	.04	.04	.09	.12	.13	.21	.5	.5	.6	.4	.5	.8	1.0	.8	.7
80-85	.04	.04	.04	.03	.04	.10	.11	.13	.22	.5	.4	.5	.4	.5	.9	.9	.8	1.9

AGE	GAIN IN EXPECTATION OF LIFE AT AGE X									FULFILMENT INDEX FOR AGE (X TO X+N)								
	ALL RACES			WHITE		BLACK		OTHER		ALL RACES			WHITE		BLACK		OTHER	
(X,X+N)	T	M	F	M	F	M	F	M	F	T	M	F	M	F	M	F	M	F
(ELIMINATING INFECTIVE AND PARASITIC DISEASES)																		
0- 1	.17	.17	.17	.14	.14	.34	.36	.37	.45	2.6	2.8	2.9	2.4	2.4	3.9	4.5	2.9	3.0
1- 5	.12	.11	.13	.10	.11	.24	.26	.32	.41	4.3	4.0	4.7	4.1	4.7	3.7	4.3	5.5	7.3
5-10	.11	.11	.12	.09	.10	.24	.25	.31	.39	2.4	2.0	2.9	1.9	3.1	2.5	2.2	1.4	2.2
10-15	.11	.10	.12	.09	.10	.23	.25	.30	.39	1.8	2.5	2.3	2.4	2.3	2.6	2.2	4.0	4.8
15-20	.11	.10	.11	.09	.10	.23	.25	.30	.39	1.3	.8	2.1	.8	1.8	1.1	2.9	1.8	4.9
20-25	.11	.10	.11	.08	.09	.23	.25	.30	.38	.8	.6	1.4	.5	1.1	1.0	2.8	.0	.7
25-30	.11	.10	.11	.08	.09	.23	.24	.30	.38	1.0	.7	1.9	.5	1.7	1.1	2.5	1.6	.6
30-35	.10	.10	.11	.08	.09	.22	.23	.29	.38	1.4	1.0	2.3	.7	2.1	1.8	2.8	.9	2.8
35-40	.10	.09	.10	.08	.09	.21	.22	.29	.38	1.3	1.1	1.9	.8	1.6	1.7	2.8	2.2	1.8
40-45	.10	.09	.10	.08	.08	.20	-.21	.28	.37	1.2	1.2	1.5	1.0	1.2	1.8	2.2	1.4	2.5
45-50	.09	.09	.09	.07	.08	.19	.20	.28	.37	1.1	1.0	1.2	.8	.9	1.7	2.0	2.0	2.7
50-55	.09	.08	.09	.07	.08	.18	.19	.27	.36	1.0	.9	1.1	.8	.9	1.6	1.9	1.9	4.0
55-60	.08	.07	.08	.06	.07	.16	.17	.26	.34	.9	.8	1.0	.7	.9	1.6	1.4	2.2	2.8
60-65	.07	.07	.08	.06	.07	.15	.16	.25	.32	.8	.7	1.0	.6	.9	1.3	1.4	1.4	1.5
65-70	.07	.06	.07	.05	.06	.13	.15	.24	.31	.8	.7	.9	.7	.8	1.2	1.2	1.7	2.0
70-75	.06	.05	.06	.05	.06	.13	.15	.23	.30	.7	.6	.8	.6	.8	1.2	1.2	1.2	1.8
75-80	.05	.05	.05	.04	.05	.12	.14	.22	.29	.7	.6	.7	.6	.6	1.0	1.1	1.5	1.4
80-85	.05	.04	.05	.04	.05	.13	.14	.22	.29	.6	.6	.6	.5	.6	1.2	1.1	1.2	2.4
(ELIMINATING MALIGNANT NEOPLASMS OF DIGESTIVE ORGANS AND PERITONEUM)																		
0- 1	.71	.65	.74	.63	.72	.77	.86	1.30	1.48	.0	.0	.0	.0	.0	.0	.0	.0	.1
1- 5	.72	.66	.74	.63	.73	.79	.88	1.32	1.50	.2	.3	.2	.3	.2	.2	.1	.0	1.0
5-10	.72	.66	.75	.64	.73	.79	.88	1.32	1.50	.2	.1	.3	.1	.2	.1	.7	.0	.0
10-15	.72	.66	.75	.64	.73	.80	.88	1.33	1.50	.2	.4	.2	.5	.2	.0	.5	.0	.0
15-20	.72	.66	.75	.64	.73	.80	.88	1.33	1.50	.3	.1	.5	.1	.5	.3	.6	.0	.0
20-25	.72	.66	.75	.64	.73	.80	.88	1.34	1.50	.3	.2	.6	.2	.7	.4	.4	.5	.0
25-30	.73	.67	.75	.65	.73	.81	.89	1.35	1.51	.7	.6	1.1	.5	1.1	.5	1.0	1.6	3.0
30-35	.73	.67	.75	.65	.73	.82	.89	1.36	1.51	1.7	1.4	2.3	1.6	2.4	1.0	1.8	1.2	2.3
35-40	.73	.68	.75	.65	.73	.83	.89	1.36	1.51	3.0	2.6	3.9	2.9	4.2	1.5	2.8	5.4	4.9
40-45	.73	.67	.74	.65	.73	.84	.89	1.35	1.50	4.2	3.9	5.0	4.0	5.2	3.5	4.3	5.5	8.4
45-50	.72	.66	.73	.64	.72	.84	.88	1.34	1.48	5.6	5.2	6.1	5.0	6.3	5.5	5.0	7.1	8.5
50-55	.70	.64	.71	.62	.70	.81	.86	1.31	1.46	6.8	6.3	7.6	6.2	7.8	6.5	6.5	11.1	10.2
55-60	.66	.60	.68	.58	.66	.76	.81	1.24	1.41	7.3	6.7	8.5	6.5	8.7	7.1	7.3	10.0	11.1
60-65	.60	.55	.62	.54	.61	.68	.75	1.17	1.35	7.5	6.8	8.7	6.8	8.9	7.0	7.1	10.8	10.4
65-70	.53	.48	.55	.47	.54	.60	.68	1.05	1.27	7.4	6.5	8.7	6.5	8.9	6.7	7.3	8.5	11.3
70-75	.45	.40	.47	.39	.46	.53	.62	.94	1.14	6.6	5.8	7.7	5.8	7.9	5.9	6.5	7.9	8.7
75-80	.37	.32	.39	.31	.38	.46	.54	.83	1.05	5.5	4.7	6.2	4.7	6.3	4.7	4.9	7.6	9.4
80-85	.29	.25	.30	.24	.29	.39	.46	.71	.96	4.1	3.6	4.5	3.5	4.5	3.7	4.2	5.3	6.9

TABLE 2 (CONTINUED)

AGE	GAIN IN EXPECTATION OF LIFE AT AGE X									FULFILMENT INDEX FOR AGE (X TO X+N)								
	ALL RACES			WHITE		BLACK		OTHER		ALL RACES			WHITE		BLACK		OTHER	
(X,X+N)	T	M	F	M	F	M	F	M	F	T	M	F	M	F	M	F	M	F
	(ELIMINATING MALIGNANT NEOPLASMS OF RESPIRATORY SYSTEM)																	
0- 1	.73	.94	.42	.92	.43	1.04	.37	.90	.53	.0	.0	.0	.0	.0	.0	.0	.0	.0
1- 5	.74	.95	.42	.93	.43	1.07	.38	.91	.53	.0	.0	.0	.1	.0	.0	.1	.0	.0
5-10	.74	.95	.43	.94	.43	1.07	.38	.92	.53	.0	.0	.1	.0	.1	.0	.0	.0	.0
10-15	.74	.95	.43	.94	.43	1.08	.38	.92	.53	.1	.2	.0	.2	.1	.2	.0	.0	.0
15-20	.74	.96	.43	.94	.43	1.08	.38	.92	.53	.1	.1	.1	.1	.1	.1	.2	.0	.0
20-25	.75	.96	.43	.95	.43	1.09	.39	.93	.54	.1	.1	.1	.1	.2	.1	.1	.0	.0
25-30	.75	.97	.43	.96	.43	1.10	.39	.94	.54	.3	.3	.4	.3	.4	.1	.3	.3	.0
30-35	.75	.98	.43	.96	.44	1.12	.39	.95	.54	1.0	.9	1.3	1.0	1.4	.7	.9	.6	.6
35-40	.76	.99	.43	.97	.43	1.14	.39	.95	.54	2.9	2.8	3.2	3.0	3.8	2.5	1.7	1.6	1.2
40-45	.75	.99	.42	.97	.43	1.15	.39	.96	.54	5.6	5.8	5.6	6.2	6.4	4.9	3.4	4.3	2.5
45-50	.74	.97	.41	.95	.41	1.14	.37	.94	.54	8.5	8.9	7.4	9.1	8.1	8.4	4.9	4.8	3.4
50-55	.70	.93	.37	.91	.38	1.09	.33	.93	.52	10.6	11.7	8.5	11.7	9.2	11.6	5.6	8.0	5.9
55-60	.62	.84	.32	.83	.33	.96	.28	.88	.49	10.7	12.1	8.1	12.2	8.8	11.9	4.4	7.4	4.5
60-65	.53	.73	.26	.72	.26	.79	.23	.83	.46	9.9	11.5	6.9	11.7	7.4	10.6	3.9	6.9	5.6
65-70	.41	.58	.19	.58	.19	.61	.17	.75	.41	8.4	10.1	5.3	10.3	5.7	9.2	3.0	7.9	5.4
70-75	.29	.43	.13	.42	.13	.47	.13	.63	.34	6.2	8.0	3.3	8.1	3.5	7.1	1.7	7.1	2.9
75-80	.19	.28	.09	.28	.08	.33	.11	.50	.31	3.7	5.1	1.9	5.2	1.9	4.0	1.2	4.5	3.6
80-85	.11	.17	.06	.17	.05	.24	.08	.43	.26	2.0	3.0	1.0	3.0	1.0	3.0	.9	3.9	2.8
	(ELIMINATING OTHER MALIGNANT NEOPLASMS INCLUDING NEOPLASMS OF LYM & HEM TISSUES)																	
0- 1	1.45	1.09	1.82	1.08	1.82	1.20	1.90	1.18	2.01	.1	.1	.2	.1	.2	.1	.1	.4	.1
1- 5	1.47	1.11	1.84	1.09	1.84	1.24	1.94	1.19	2.03	6.8	6.7	6.9	7.6	8.4	4.1	3.1	3.9	3.1
5-10	1.46	1.09	1.84	1.08	1.83	1.23	1.94	1.18	2.02	13.6	12.9	14.5	13.8	16.0	9.6	9.4	12.5	8.9
10-15	1.45	1.08	1.82	1.07	1.82	1.22	1.93	1.17	2.02	10.9	15.7	12.8	16.8	13.0	11.3	11.7	14.1	14.3
15-20	1.44	1.07	1.81	1.06	1.81	1.21	1.92	1.16	2.01	6.1	4.9	7.2	4.7	7.4	6.1	6.5	6.2	3.0
20-25	1.43	1.06	1.81	1.05	1.80	1.20	1.91	1.14	2.01	4.8	4.0	7.4	4.3	8.1	2.7	5.0	1.4	4.0
25-30	1.42	1.05	1.80	1.04	1.79	1.20	1.91	1.15	2.01	7.0	5.1	12.2	6.0	13.7	2.1	7.6	3.0	9.1
30-35	1.41	1.04	1.78	1.02	1.77	1.21	1.90	1.14	1.99	10.2	6.4	18.3	7.6	20.3	2.7	12.2	3.3	14.1
35-40	1.39	1.02	1.75	1.00	1.74	1.21	1.87	1.14	1.96	13.0	7.2	24.6	8.4	27.6	3.8	16.8	3.8	17.2
40-45	1.36	1.00	1.70	.98	1.69	1.21	1.81	1.13	1.92	14.6	7.8	27.1	8.9	30.0	4.6	18.6	5.5	17.8
45-50	1.30	.98	1.61	.96	1.61	1.21	1.71	1.11	1.87	15.8	8.3	27.8	8.9	30.4	5.8	18.1	7.1	26.5
50-55	1.22	.94	1.48	.92	1.48	1.19	1.58	1.08	1.74	15.8	8.8	27.9	9.3	30.3	6.5	17.5	8.2	24.1
55-60	1.12	.89	1.32	.86	1.31	1.17	1.42	1.04	1.61	14.9	8.9	25.5	9.2	27.3	7.6	16.9	7.5	19.9
60-65	.99	.82	1.13	.79	1.12	1.14	1.24	.99	1.46	13.5	8.9	21.8	8.9	23.0	8.7	15.1	5.8	14.0
65-70	.85	.74	.91	.71	.90	1.08	1.05	.96	1.34	12.1	8.8	17.6	8.7	18.4	9.8	13.4	6.2	12.5
70-75	.71	.65	.73	.62	.72	1.01	.90	.91	1.18	10.6	8.3	13.7	8.2	14.2	9.7	10.1	6.0	15.4
75-80	.57	.55	.56	.53	.55	.93	.76	.85	.95	8.5	7.5	9.6	7.3	9.8	8.5	7.3	6.3	10.5
80-85	.45	.46	.42	.43	.40	.86	.62	.79	.83	6.3	6.0	6.3	5.9	6.3	7.8	6.2	5.6	6.0

AGE	GAIN IN EXPECTATION OF LIFE AT AGE X									FULFILMENT INDEX FOR AGE (X TO X+N)								
	ALL RACES			WHITE		BLACK		OTHER		ALL RACES			WHITE		BLACK		OTHER	
(X,X+N)	T	M	F	M	F	M	F	M	F	T	M	F	M	F	M	F	M	F
	(ELIMINATING MALIGNANT NEOPLASMS INCLUDING NEOPLASMS OF LYMPH AND HEMAT TISSUES)																	
0- 1	3.09	2.90	3.13	2.85	3.12	3.36	3.32	3.68	4.28	.2	.1	.2	.1	.2	.1	.1	.4	.3
1- 5	3.13	2.94	3.17	2.88	3.15	3.45	3.39	3.72	4.32	7.1	7.0	7.2	7.9	8.6	4.3	3.3	3.9	4.2
5-10	3.12	2.94	3.16	2.88	3.14	3.46	3.40	3.73	4.32	13.8	13.0	14.9	14.0	16.3	9.7	10.1	12.5	8.9
10-15	3.11	2.93	3.15	2.86	3.13	3.45	3.39	3.72	4.32	11.3	16.3	13.0	17.6	13.2	11.5	12.2	14.1	14.3
15-20	3.10	2.92	3.14	2.86	3.12	3.45	3.38	3.71	4.31	6.4	5.1	7.8	4.9	8.0	6.5	7.3	6.2	3.0
20-25	3.10	2.92	3.14	2.86	3.12	3.45	3.38	3.71	4.32	5.2	4.3	8.1	4.6	8.9	3.2	5.5	1.8	4.0
25-30	3.10	2.93	3.13	2.86	3.11	3.48	3.38	3.74	4.32	8.0	5.9	13.7	6.8	15.2	2.7	8.9	4.9	12.2
30-35	3.10	2.93	3.12	2.86	3.09	3.52	3.37	3.75	4.31	13.0	8.8	21.9	10.2	24.2	4.4	14.9	5.1	16.9
35-40	3.08	2.92	3.08	2.85	3.06	3.56	3.34	3.76	4.28	18.9	12.6	31.7	14.2	35.5	7.8	21.3	10.7	23.4
40-45	3.04	2.90	3.02	2.82	3.00	3.59	3.27	3.75	4.23	24.4	17.6	37.8	19.1	41.6	12.9	26.3	15.3	28.8
45-50	2.95	2.85	2.90	2.77	2.88	3.57	3.14	3.70	4.15	30.0	22.4	41.4	23.1	44.9	19.8	28.1	19.0	38.4
50-55	2.80	2.74	2.71	2.66	2.69	3.47	2.95	3.63	3.98	33.2	27.0	44.1	27.4	47.4	24.9	29.7	27.4	40.2
55-60	2.57	2.55	2.44	2.48	2.42	3.25	2.68	3.46	3.75	33.1	27.9	42.3	28.0	45.0	26.9	28.8	25.1	35.6
60-65	2.28	2.30	2.12	2.24	2.10	2.95	2.36	3.27	3.49	31.3	27.6	37.5	27.7	39.5	26.7	26.3	23.7	30.0
65-70	1.92	1.97	1.75	1.92	1.73	2.58	2.02	3.01	3.23	28.3	25.9	31.9	25.8	33.2	26.2	23.9	22.8	29.3
70-75	1.56	1.62	1.41	1.57	1.38	2.26	1.76	2.71	2.84	23.8	22.7	25.0	22.6	25.8	23.3	18.5	21.3	27.2
75-80	1.20	1.27	1.09	1.22	1.06	1.92	1.49	2.39	2.46	18.1	17.8	17.9	17.8	18.3	17.9	13.6	18.6	23.6
80-85	.90	.97	.81	.92	.79	1.64	1.22	2.12	2.20	12.7	13.0	11.9	12.9	11.9	14.9	11.4	15.0	15.9
	(ELIMINATING DIABETES MELLITUS)																	
0- 1	.22	.16	.28	.15	.25	.23	.55	.35	.78	.0	.0	.0	.0	.0	.0	.0	.0	.0
1- 5	.22	.17	.29	.16	.26	.24	.56	.35	.79	.1	.1	.1	.1	.2	.1	.1	.0	.0
5-10	.23	.17	.29	.16	.26	.24	.56	.35	.79	.2	.1	.3	.2	.3	.0	.4	.0	.0
10-15	.23	.17	.29	.16	.26	.24	.56	.35	.79	.3	.3	.6	.3	.6	.2	.5	.0	.0
15-20	.23	.17	.29	.16	.26	.24	.56	.35	.79	.2	.1	.3	.1	.3	.4	.5	.0	1.0
20-25	.23	.17	.29	.16	.26	.24	.56	.36	.79	.4	.2	.8	.2	.9	.1	.8	.0	.0
25-30	.23	.17	.29	.16	.26	.24	.56	.36	.79	.9	.7	1.7	.6	1.8	.8	1.5	.0	.0
30-35	.22	.17	.28	.16	.25	.24	.56	.36	.79	1.5	1.3	1.9	1.4	1.9	1.2	2.0	.9	1.7
35-40	.22	.16	.28	.15	.25	.24	.56	.36	.79	1.8	1.5	2.2	1.5	2.2	1.6	2.1	.9	1.8
40-45	.22	.16	.28	.15	.25	.23	.56	.36	.79	1.6	1.5	2.0	1.5	1.9	1.4	2.3	1.7	2.0
45-50	.21	.15	.27	.14	.24	.22	.55	.36	.79	1.6	1.3	2.0	1.3	1.7	1.2	2.9	2.0	2.7
50-55	.21	.15	.27	.14	.24	.22	.54	.35	.79	1.7	1.4	2.2	1.4	1.8	1.7	3.7	1.7	2.3
55-60	.20	.14	.26	.13	.23	.20	.52	.35	.79	1.9	1.4	2.7	1.4	2.4	1.9	4.0	2.2	4.8
60-65	.19	.13	.24	.12	.22	.19	.49	.34	.77	2.0	1.4	2.9	1.3	2.6	1.7	4.5	3.5	6.1
65-70	.17	.12	.22	.11	.20	.17	.45	.29	.72	2.0	1.4	3.1	1.4	2.8	2.0	4.6	2.5	6.0
70-75	.15	.10	.20	.10	.18	.15	.42	.26	.66	2.0	1.4	3.0	1.3	2.8	1.6	4.5	2.4	7.2
75-80	.13	.09	.17	.08	.16	.14	.36	.23	.56	1.9	1.2	2.6	1.2	2.4	1.5	3.6	1.9	4.7
80-85	.11	.07	.14	.07	.13	.11	.29	.20	.52	1.5	1.0	2.0	.9	1.9	1.1	3.0	1.8	3.9

TABLE 2 (CONTINUED)

AGE	GAIN IN EXPECTATION OF LIFE AT AGE X									FULFILMENT INDEX FOR AGE (X TO X+N)								
	ALL RACES			WHITE		BLACK		OTHER		ALL RACES			WHITE		BLACK		OTHER	
(X,X+N)	T	M	F	M	F	M	F	M	F	T	M	F	M	F	M	F	M	F
				(ELIMINATING DISEASES OF HEART)														
0- 1	7.01	6.47	7.13	6.49	6.94	6.07	8.80	9.48	10.73	1.0	1.0	1.1	.9	1.0	1.1	1.6	1.1	1.0
1- 5	7.09	6.55	7.19	6.56	7.00	6.22	8.97	9.58	10.82	3.3	2.8	4.0	2.5	3.5	3.7	5.0	2.3	8.3
5-10	7.10	6.57	7.20	6.57	7.01	6.23	8.99	9.60	10.82	2.5	2.1	3.2	2.0	3.1	2.3	3.3	2.8	4.4
10-15	7.11	6.58	7.21	6.58	7.01	6.25	9.00	9.61	10.83	3.3	4.3	4.5	4.1	3.9	5.1	7.2	6.0	4.8
15-20	7.12	6.59	7.21	6.59	7.02	6.26	9.01	9.63	10.84	2.4	1.9	2.8	1.6	2.4	3.8	5.2	2.2	4.0
20-25	7.15	6.63	7.23	6.63	7.03	6.29	9.02	9.69	10.86	2.4	2.0	3.9	1.7	3.2	3.1	7.1	2.1	1.3
25-30	7.19	6.69	7.24	6.69	7.05	6.35	9.05	9.78	10.89	4.1	3.7	5.2	3.3	4.6	5.1	6.9	3.0	6.7
30-35	7.22	6.73	7.26	6.73	7.06	6.43	9.08	9.86	10.92	8.2	8.6	7.3	8.5	6.5	9.2	10.4	6.9	5.6
35-40	7.24	6.76	7.27	6.75	7.08	6.49	9.12	9.92	10.95	16.6	18.8	12.1	20.1	11.5	15.5	14.0	12.6	9.2
40-45	7.24	6.75	7.29	6.74	7.09	6.51	9.15	9.98	10.99	23.9	30.0	15.4	32.5	13.7	22.9	20.9	18.7	10.9
45-50	7.21	6.68	7.30	6.66	7.10	6.46	9.16	10.01	11.04	30.6	36.8	18.5	39.2	16.8	27.1	25.3	28.9	13.6
50-55	7.13	6.52	7.30	6.50	7.11	6.37	9.14	9.97	11.09	34.0	40.8	22.0	42.9	20.6	31.1	28.1	32.2	18.1
55-60	7.00	6.28	7.30	6.25	7.12	6.19	9.11	9.90	11.15	37.1	42.7	27.1	44.4	26.1	32.9	32.0	36.5	22.7
60-65	6.82	5.97	7.27	5.93	7.09	5.99	9.05	9.76	11.20	38.5	42.0	31.9	43.2	31.3	33.6	35.3	36.2	29.3
65-70	6.59	5.60	7.19	5.54	7.03	5.77	8.96	9.61	11.20	39.8	41.0	37.2	41.9	37.1	34.1	38.3	35.8	29.7
70-75	6.35	5.24	7.07	5.15	6.91	5.64	8.90	9.48	11.23	40.6	39.7	41.2	40.4	41.5	32.8	38.7	36.9	32.8
75-80	6.12	4.91	6.92	4.80	6.74	5.64	8.92	9.32	11.27	40.4	37.4	43.1	37.9	43.5	32.3	38.5	35.1	38.9
80-85	5.95	4.70	6.75	4.57	6.57	5.70	9.01	9.27	11.24	39.0	34.8	42.1	35.1	42.3	30.5	39.7	30.6	33.9
				(ELIMINATING HYPERTENSION)														
0- 1	.03	.03	.04	.02	.03	.06	.09	.06	.03	.0	.0	.0	.0	.0	.0	.0	.0	.0
1- 5	.03	.03	.04	.02	.03	.07	.09	.06	.03	.0	.0	.0	.0	.0	.0	.0	.0	.0
5-10	.03	.03	.04	.02	.03	.07	.10	.06	.03	.0	.0	.0	.0	.0	.0	.0	.0	.0
10-15	.03	.03	.04	.02	.03	.07	.10	.06	.03	.0	.0	.0	.0	.1	.0	.0	.0	.0
15-20	.03	.03	.04	.02	.03	.07	.10	.06	.03	.0	.0	.0	.0	.0	.0	.0	.0	.0
20-25	.03	.03	.04	.02	.03	.07	.10	.06	.03	.1	.0	.1	.0	.0	.1	.1	.0	.7
25-30	.03	.03	.04	.02	.03	.07	.10	.06	.03	.1	.1	.1	.0	.1	.2	.3	.0	.6
30-35	.03	.03	.04	.02	.03	.07	.10	.06	.03	.1	.1	.1	.0	.1	.3	.2	.0	.0
35-40	.03	.03	.04	.02	.03	.07	.10	.06	.03	.3	.2	.4	.1	.2	.6	.9	.0	.0
40-45	.03	.03	.04	.02	.03	.06	.09	.07	.03	.2	.2	.3	.1	.2	.6	.8	.0	.0
45-50	.03	.03	.04	.02	.03	.06	.09	.07	.03	.2	.2	.2	.1	.1	.5	.6	.3	.0
50-55	.03	.03	.04	.02	.03	.05	.09	.07	.03	.2	.2	.3	.1	.2	.4	.8	.9	.0
55-60	.03	.02	.04	.02	.03	.05	.08	.06	.03	.2	.2	.2	.1	.2	.4	.6	.3	.3
60-65	.03	.02	.04	.02	.03	.05	.08	.06	.03	.2	.2	.3	.1	.2	.5	.7	.4	.3
65-70	.03	.02	.03	.02	.03	.04	.07	.06	.03	.2	.2	.3	.1	.2	.4	.6	.5	.6
70-75	.03	.02	.03	.02	.03	.04	.07	.05	.02	.3	.2	.3	.2	.3	.4	.6	.2	.2
75-80	.03	.02	.03	.02	.03	.04	.06	.05	.02	.3	.2	.3	.2	.3	.3	.6	.2	.2
80-85	.03	.02	.03	.02	.03	.05	.06	.06	.02	.3	.3	.3	.3	.3	.4	.5	.0	.0

AGE	GAIN IN EXPECTATION OF LIFE AT AGE X									FULFILMENT INDEX FOR AGE (X TO X+N)								
	ALL RACES			WHITE		BLACK		OTHER		ALL RACES			WHITE		BLACK		OTHER	
(X,X+N)	T	M	F	M	F	M	F	M	F	T	M	F	M	F	M	F	M	F
				(ELIMINATING CEREBROVASCULAR DISEASES)														
0- 1	1.14	.80	1.49	.74	1.42	1.19	2.18	1.70	3.20	.2	.2	.2	.2	.2	.3	.4	.3	.5
1- 5	1.15	.81	1.51	.75	1.43	1.22	2.22	1.72	3.23	1.2	1.3	1.0	1.4	1.1	1.1	.7	.0	.0
5-10	1.15	.81	1.51	.75	1.43	1.22	2.23	1.72	3.24	1.6	1.4	1.8	1.7	1.6	.7	2.6	.0	2.2
10-15	1.15	.81	1.51	.75	1.43	1.22	2.23	1.73	3.24	1.7	2.6	1.8	2.7	1.7	2.1	1.7	6.0	4.8
15-20	1.15	.81	1.51	.75	1.43	1.23	2.23	1.72	3.24	1.1	.7	1.6	.7	1.6	.6	1.8	1.1	.0
20-25	1.16	.81	1.51	.75	1.44	1.23	2.23	1.73	3.25	1.0	.7	2.0	.7	1.8	.8	2.7	.2	3.3
25-30	1.16	.81	1.51	.76	1.44	1.24	2.24	1.75	3.25	1.6	1.0	3.0	.9	2.5	1.5	4.5	.8	1.8
30-35	1.16	.82	1.51	.76	1.44	1.26	2.24	1.76	3.26	2.4	1.7	3.7	1.5	3.6	2.5	4.2	.6	3.4
35-40	1.16	.82	1.51	.76	1.44	1.26	2.24	1.78	3.26	4.0	2.9	6.2	2.6	6.1	3.8	6.7	3.8	3.7
40-45	1.16	.82	1.51	.76	1.43	1.27	2.24	1.78	3.27	4.6	3.9	6.2	3.4	5.9	5.4	7.2	4.3	5.0
45-50	1.15	.81	1.50	.75	1.43	1.25	2.23	1.78	3.28	4.6	3.7	5.9	3.1	5.5	6.2	7.7	3.8	7.1
50-55	1.15	.80	1.49	.75	1.42	1.24	2.22	1.79	3.28	4.8	3.8	6.6	3.3	6.0	6.1	9.0	4.1	5.9
55-60	1.15	.80	1.48	.75	1.41	1.23	2.21	1.82	3.30	5.0	4.2	6.4	3.7	5.7	6.6	9.1	6.6	10.8
60-65	1.15	.80	1.48	.76	1.41	1.23	2.21	1.82	3.29	5.6	4.7	7.1	4.3	6.6	7.4	10.0	7.9	10.1
65-70	1.16	.81	1.48	.76	1.42	1.22	2.21	1.81	3.30	6.8	5.7	8.6	5.3	8.0	9.0	12.2	8.8	8.4
70-75	1.17	.81	1.47	.77	1.42	1.21	2.21	1.79	3.36	8.5	6.9	10.7	6.6	10.3	9.4	13.3	9.9	11.3
75-80	1.17	.80	1.46	.76	1.40	1.21	2.22	1.75	3.41	10.2	8.0	12.6	7.8	12.4	9.4	13.9	9.4	15.2
80-85	1.16	.79	1.43	.75	1.38	1.22	2.21	1.73	3.40	11.0	8.1	13.6	8.0	13.5	9.0	14.8	8.4	14.3
				(ELIMINATING ARTERIOSCLEROSIS)														
0- 1	.16	.10	.21	.10	.21	.09	.21	.22	.25	.0	.0	.0	.0	.0	.0	.0	.0	.0
1- 5	.16	.10	.22	.10	.21	.10	.21	.22	.25	.0	.0	.0	.0	.0	.0	.0	.0	.0
5-10	.16	.10	.22	.10	.22	.10	.21	.23	.25	.0	.0	.0	.0	.0	.0	.0	.0	.0
10-15	.16	.10	.22	.10	.22	.10	.21	.23	.25	.0	.0	.0	.0	.0	.0	.0	.0	.0
15-20	.16	.10	.22	.10	.22	.10	.22	.23	.25	.0	.0	.0	.0	.0	.0	.0	.0	.0
20-25	.16	.10	.22	.10	.22	.10	.22	.23	.25	.0	.0	.0	.0	.0	.0	.0	.0	.0
25-30	.16	.10	.22	.10	.22	.10	.22	.23	.25	.0	.0	.0	.0	.0	.0	.0	.0	.0
30-35	.16	.11	.22	.11	.22	.10	.22	.23	.26	.0	.0	.0	.0	.0	.0	.0	.0	.0
35-40	.16	.11	.22	.11	.22	.10	.22	.23	.26	.0	.0	.0	.1	.1	.0	.0	.0	.0
40-45	.16	.11	.22	.11	.22	.11	.22	.24	.26	.1	.1	.1	.1	.1	.0	.0	.0	.5
45-50	.17	.11	.22	.11	.22	.11	.23	.24	.26	.1	.1	.1	.1	.1	.1	.1	.0	.0
50-55	.17	.11	.23	.11	.22	.12	.23	.25	.26	.2	.2	.2	.2	.2	.1	.1	.0	.0
55-60	.17	.12	.23	.11	.23	.12	.24	.26	.27	.3	.3	.3	.3	.2	.3	.4	.3	.0
60-65	.18	.12	.24	.12	.23	.13	.26	.27	.28	.4	.4	.4	.4	.4	.4	.5	.7	.0
65-70	.19	.13	.25	.12	.24	.15	.27	.28	.29	.6	.5	.7	.5	.6	.6	.7	.8	.8
70-75	.20	.14	.26	.13	.26	.16	.29	.29	.29	.9	.8	1.1	.8	1.1	.8	1.2	.7	.7
75-80	.22	.15	.28	.15	.27	.19	.32	.32	.31	1.4	1.1	1.6	1.1	1.7	1.0	1.4	.6	.9
80-85	.25	.18	.30	.17	.29	.24	.37	.36	.32	1.9	1.5	2.3	1.5	2.3	1.5	2.0	1.1	2.2

TABLE 2 (CONTINUED)

AGE	GAIN IN EXPECTATION OF LIFE AT AGE X									FULFILMENT INDEX FOR AGE (X TO X+N)								
	ALL RACES			WHITE		BLACK		OTHER		ALL RACES			WHITE		BLACK		OTHER	
(X,X+N)	T	M	F	M	F	M	F	M	F	T	M	F	M	F	M	F	M	F
(ELIMINATING OTHER DISEASES OF ARTERIES, ARTERIOLES, AND CAPILLARIES)																		
0- 1	.16	.17	.15	.17	.14	.13	.19	.19	.24	.0	.1	.0	.0	.0	.1	.0	.1	.1
1- 5	.17	.17	.15	.17	.14	.14	.19	.19	.24	.1	.2	.1	.2	.1	.0	.0	.0	1.0
5-10	.17	.17	.15	.17	.14	.14	.19	.19	.24	.2	.2	.2	.2	.3	.3	.0	.0	.0
10-15	.17	.17	.15	.17	.14	.14	.19	.19	.24	.2	.3	.3	.3	.3	.2	.0	.0	.0
15-20	.17	.17	.15	.17	.14	.14	.19	.20	.24	.2	.2	.3	.2	.3	.2	.5	.0	.0
20-25	.17	.17	.15	.17	.14	.14	.19	.20	.24	.3	.2	.5	.2	.4	.2	.6	.2	.0
25-30	.17	.17	.15	.17	.14	.14	.19	.20	.25	.5	.4	.7	.4	.6	.5	1.1	.0	.0
30-35	.17	.17	.15	.17	.14	.14	.19	.20	.25	.5	.3	.7	.4	.8	.2	.4	.0	.0
35-40	.17	.17	.15	.17	.14	.14	.19	.20	.25	.7	.6	1.0	.6	1.0	.4	1.0	.3	.0
40-45	.17	.17	.15	.17	.14	.14	.19	.20	.25	.8	.8	.9	.8	.9	.8	.8	.6	.0
45-50	.16	.17	.14	.17	.14	.14	.18	.20	.25	.8	.7	.8	.7	.8	.9	.7	.3	.3
50-55	.16	.17	.14	.17	.14	.13	.18	.20	.25	1.0	1.0	1.0	1.0	1.0	.7	.9	1.1	.3
55-60	.16	.17	.14	.17	.13	.13	.18	.20	.26	1.2	1.2	1.1	1.3	1.1	1.0	1.1	1.4	1.0
60-65	.15	.16	.13	.17	.13	.12	.18	.19	.25	1.3	1.5	1.0	1.6	1.1	1.0	.9	1.5	1.5
65-70	.15	.15	.13	.15	.13	.12	.18	.18	.25	1.6	1.8	1.3	1.9	1.3	.9	1.1	1.4	1.2
70-75	.13	.14	.12	.14	.12	.11	.18	.17	.24	1.7	1.8	1.4	1.9	1.4	1.0	1.3	1.1	1.3
75-80	.12	.12	.11	.12	.11	.11	.17	.16	.24	1.5	1.6	1.3	1.7	1.2	1.0	1.2	1.5	1.4
80-85	.10	.10	.11	.09	.10	.11	.17	.14	.23	1.3	1.3	1.2	1.3	1.2	1.0	1.3	1.2	1.7
(ELIMINATING MAJOR CARDIOVASCULAR DISEASES)																		
0- 1	13.87	10.63	16.74	10.57	16.38	10.58	20.29	16.65	22.96	1.2	1.3	1.4	1.2	1.2	1.5	2.0	1.4	1.7
1- 5	14.04	10.77	16.92	10.68	16.52	10.84	20.72	16.83	23.15	4.6	4.3	5.2	4.2	4.8	4.8	5.7	2.3	9.4
5-10	14.07	10.79	16.95	10.70	16.55	10.87	20.77	16.88	23.18	4.4	3.8	5.2	3.9	5.0	3.3	5.9	2.8	6.7
10-15	14.09	10.81	16.97	10.72	16.57	10.90	20.80	16.91	23.20	5.3	7.2	6.6	7.1	6.0	7.3	9.0	12.1	9.5
15-20	14.10	10.83	16.98	10.73	16.58	10.92	20.82	16.93	23.22	3.7	2.8	4.8	2.6	4.3	4.6	7.6	3.3	4.0
20-25	14.16	10.89	17.02	10.80	16.62	10.98	20.86	17.04	23.27	3.8	2.9	6.5	2.7	5.5	4.1	10.5	2.5	5.3
25-30	14.24	10.99	17.06	10.89	16.65	11.10	20.93	17.21	23.34	6.3	5.2	9.0	4.7	7.8	7.3	12.8	3.8	9.1
30-35	14.31	11.07	17.10	10.96	16.69	11.25	21.02	17.36	23.40	11.2	10.8	11.9	10.5	11.1	12.2	15.3	7.5	9.0
35-40	14.37	11.12	17.14	11.01	16.73	11.37	21.12	17.48	23.47	21.6	22.6	19.7	23.4	18.8	20.3	22.7	16.7	12.9
40-45	14.40	11.14	17.18	11.02	16.77	11.45	21.23	17.60	23.57	29.7	35.1	22.8	37.0	20.8	29.7	29.8	23.7	16.4
45-50	14.41	11.09	17.25	10.96	16.83	11.46	21.34	17.71	23.69	36.4	41.6	25.6	43.3	23.4	35.1	34.4	33.2	21.0
50-55	14.40	10.97	17.33	10.83	16.92	11.41	21.46	17.75	23.83	40.3	46.0	30.1	47.6	28.1	38.6	38.9	38.4	24.3
55-60	14.37	10.78	17.45	10.64	17.04	11.35	21.62	17.78	24.03	43.9	48.7	35.1	49.9	33.4	41.7	43.4	45.3	34.8
60-65	14.33	10.54	17.59	10.39	17.18	11.27	21.82	17.73	24.19	46.3	49.2	40.9	49.9	39.7	43.4	47.8	47.1	41.3
65-70	14.30	10.28	17.75	10.10	17.35	11.22	22.06	17.66	24.33	49.5	49.9	48.4	50.4	47.6	45.9	53.7	47.8	41.0
70-75	14.31	10.05	17.91	9.85	17.51	11.26	22.29	17.63	24.67	52.9	50.6	55.6	51.1	55.4	45.9	56.6	49.4	46.8
75-80	14.38	9.91	18.06	9.67	17.65	11.52	22.73	17.58	24.99	55.7	50.6	60.8	50.9	60.9	46.8	58.9	47.7	57.3
80-85	14.58	9.92	18.26	9.65	17.83	11.90	23.29	17.68	25.07	57.1	49.6	63.1	49.9	63.2	45.5	61.5	42.4	53.3

AGE	GAIN IN EXPECTATION OF LIFE AT AGE X									FULFILMENT INDEX FOR AGE (X TO X+N)								
	ALL RACES			WHITE		BLACK		OTHER		ALL RACES			WHITE		BLACK		OTHER	
(X,X+N)	T	M	F	M	F	M	F	M	F	T	M	F	M	F	M	F	M	F
(ELIMINATING INFLUENZA AND PNEUMONIA)																		
0- 1	.39	.36	.41	.33	.39	.54	.49	.94	1.24	1.8	1.8	2.1	1.5	1.7	2.9	3.1	2.0	2.2
1- 5	.36	.33	.38	.31	.37	.48	.42	.91	1.22	4.2	3.7	4.9	3.3	4.7	5.0	5.4	3.9	7.3
5-10	.36	.32	.37	.30	.36	.46	.41	.90	1.21	2.8	2.4	3.2	2.2	3.2	3.2	2.8	2.8	6.7
10-15	.35	.32	.37	.30	.36	.46	.41	.90	1.20	2.3	3.0	3.0	3.0	2.9	3.2	3.7	2.0	2.4
15-20	.35	.32	.37	.30	.35	.46	.41	.90	1.20	1.4	1.1	1.6	1.0	1.3	1.7	3.4	.7	1.0
20-25	.35	.32	.36	.29	.35	.46	.40	.91	1.21	1.1	.9	2.0	.8	1.8	1.2	2.5	1.4	3.3
25-30	.35	.32	.36	.29	.35	.45	.40	.91	1.20	1.5	1.3	2.0	1.1	1.8	1.8	3.1	1.6	1.2
30-35	.35	.31	.36	.29	.35	.45	.39	.91	1.20	1.7	1.5	2.2	1.3	1.8	2.4	3.2	1.8	2.8
35-40	.34	.31	.36	.29	.35	.44	.38	.91	1.20	2.0	1.7	2.5	1.4	2.4	2.6	3.0	1.3	1.2
40-45	.34	.31	.35	.29	.34	.43	.37	.91	1.21	1.9	2.0	1.9	1.7	1.8	3.1	2.4	2.3	1.5
45-50	.34	.30	.35	.29	.34	.41	.36	.92	1.21	1.8	1.7	1.9	1.4	1.8	2.9	2.3	.8	2.0
50-55	.33	.30	.34	.28	.34	.39	.35	.93	1.22	1.7	1.7	1.6	1.5	1.5	2.8	1.8	2.6	3.4
55-60	.33	.29	.34	.28	.34	.37	.35	.94	1.22	1.7	1.7	1.7	1.6	1.7	2.4	1.8	3.0	2.0
60-65	.33	.29	.34	.28	.34	.37	.34	.95	1.23	1.7	1.8	1.7	1.6	1.7	2.5	1.7	2.7	1.5
65-70	.33	.29	.34	.28	.34	.36	.34	.98	1.27	1.9	1.9	1.8	1.8	1.8	2.6	1.6	2.8	2.0
70-75	.33	.30	.34	.29	.34	.36	.35	1.02	1.32	2.3	2.3	2.2	2.2	2.2	2.6	1.9	2.9	3.1
75-80	.34	.31	.35	.30	.35	.38	.37	1.10	1.38	2.7	2.7	2.7	2.7	2.8	2.7	2.0	3.8	4.0
80-85	.36	.34	.36	.32	.35	.42	.40	1.18	1.43	3.3	3.3	3.1	3.3	3.1	3.2	2.5	6.0	4.6
(ELIMINATING BRONCHITIS, EMPHYSEMA, AND ASTHMA)																		
0- 1	.14	.16	.10	.17	.10	.12	.08	.18	.13	.1	.1	.1	.1	.1	.1	.1	.0	.0
1- 5	.14	.16	.10	.17	.10	.12	.08	.18	.13	.6	.6	.5	.4	.5	1.3	.7	.0	.0
5-10	.14	.16	.10	.17	.10	.11	.07	.18	.13	.3	.4	.2	.3	.2	.6	.2	1.4	.0
10-15	.14	.16	.10	.17	.10	.11	.07	.18	.13	.7	1.2	.6	1.3	.6	.8	1.0	2.0	.0
15-20	.14	.16	.10	.17	.10	.11	.07	.18	.13	.3	.2	.5	.2	.4	.6	.9	.0	1.0
20-25	.14	.16	.10	.17	.10	.11	.07	.18	.13	.3	.2	.5	.1	.5	.3	.7	.2	1.3
25-30	.14	.16	.10	.17	.10	.11	.07	.18	.12	.3	.2	.5	.2	.4	.3	1.0	.3	.6
30-35	.14	.16	.10	.17	.10	.11	.07	.18	.12	.3	.3	.5	.2	.4	.4	.9	.3	.0
35-40	.14	.16	.10	.17	.10	.11	.06	.18	.12	.4	.3	.6	.3	.6	.6	.6	.0	.0
40-45	.14	.17	.10	.17	.10	.11	.06	.19	.12	.5	.4	.8	.4	.8	.5	.7	.0	1.5
45-50	.14	.17	.09	.17	.10	.11	.06	.19	.12	.8	.7	1.0	.7	1.0	.6	.7	.3	.7
50-55	.14	.17	.09	.17	.10	.10	.05	.19	.12	.9	.8	1.0	.9	1.1	.8	.6	.9	.3
55-60	.13	.17	.09	.17	.09	.10	.05	.19	.12	1.2	1.2	1.3	1.3	1.4	.8	.6	.8	.5
60-65	.13	.16	.08	.17	.08	.09	.04	.19	.12	1.5	1.6	1.4	1.7	1.6	1.0	.7	.9	.5
65-70	.11	.15	.06	.15	.07	.08	.03	.19	.12	1.7	1.9	1.2	2.0	1.4	.9	.3	1.0	.8
70-75	.10	.13	.05	.13	.05	.07	.03	.19	.11	1.6	1.9	1.0	2.0	1.2	.9	.3	2.2	.4
75-80	.07	.10	.04	.11	.04	.06	.03	.14	.11	1.3	1.7	.7	1.8	.8	.7	.2	1.2	1.6
80-85	.05	.08	.03	.08	.03	.05	.03	.13	.09	.8	1.2	.4	1.2	.4	.5	.3	1.3	.6

TABLE 2 (CONTINUED)

AGE	GAIN IN EXPECTATION OF LIFE AT AGE X									FULFILMENT INDEX FOR AGE (X TO X+N)								
	ALL RACES			WHITE		BLACK		OTHER		ALL RACES			WHITE		BLACK		OTHER	
(X,X+N)	T	M	F	M	F	M	F	M	F	T	M	F	M	F	M	F	M	F
(ELIMINATING OTHER DISEASES OF THE RESPIRATORY SYSTEM)																		
0- 1	.28	.32	.20	.32	.20	.29	.17	.46	.33	.7	.7	.8	.6	.7	.9	.8	.9	1.0
1- 5	.27	.31	.19	.32	.19	.28	.16	.45	.32	2.5	2.6	2.3	2.6	2.7	2.6	1.1	2.3	2.1
5-10	.27	.31	.19	.31	.19	.27	.16	.44	.32	1.3	1.2	1.4	1.0	1.4	1.9	1.5	.0	.0
10-15	.27	.31	.19	.31	.19	.27	.15	.45	.32	.8	1.0	1.2	1.1	1.2	.9	1.5	.0	.0
15-20	.27	.31	.19	.31	.19	.27	.15	.45	.32	.7	.5	1.0	.4	.8	.9	2.2	.0	.0
20-25	.26	.31	.18	.31	.19	.27	.15	.45	.32	.5	.4	.7	.4	.6	.5	1.3	.7	.0
25-30	.26	.31	.18	.31	.19	.27	.15	.45	.32	.7	.5	1.0	.5	1.0	.6	1.0	.3	.6
30-35	.26	.31	.18	.31	.18	.27	.14	.45	.32	.7	.6	1.0	.6	.9	.9	1.4	.0	.6
35-40	.26	.31	.18	.31	.18	.27	.14	.46	.32	.9	.9	1.0	.7	.9	1.3	1.4	.6	1.2
40-45	.26	.31	.18	.32	.18	.26	.14	.46	.32	1.1	1.0	1.3	.8	1.2	1.7	1.7	.3	2.0
45-50	.26	.31	.18	.32	.18	.25	.13	.47	.31	1.3	1.2	1.3	1.2	1.4	1.2	1.0	1.5	.3
50-55	.26	.31	.17	.32	.18	.25	.12	.47	.31	1.7	1.7	1.7	1.7	1.9	1.8	1.3	.9	.3
55-60	.25	.31	.16	.32	.17	.24	.11	.48	.32	2.2	2.2	2.1	2.2	2.3	1.8	1.2	1.6	2.5
60-65	.24	.30	.15	.31	.16	.23	.10	.48	.30	2.6	2.8	2.2	2.9	2.4	2.2	1.0	2.5	2.3
65-70	.22	.28	.13	.29	.14	.21	.09	.47	.29	3.0	3.4	2.3	3.6	2.5	2.4	.9	2.4	2.2
70-75	.19	.25	.11	.25	.11	.19	.08	.47	.27	2.9	3.4	2.0	3.6	2.2	1.9	.8	2.8	2.9
75-80	.15	.21	.09	.21	.09	.17	.08	.45	.23	2.4	3.1	1.5	3.3	1.6	1.6	.6	2.6	1.4
80-85	.11	.16	.07	.16	.07	.16	.07	.45	.22	1.6	2.3	.9	2.4	.9	1.6	.6	1.4	2.0
(ELIMINATING DISEASES OF THE RESPIRATORY SYSTEM)																		
0- 1	.84	.88	.72	.85	.71	.97	.74	1.67	1.76	2.5	2.6	2.9	2.2	2.5	3.9	3.9	2.9	3.3
1- 5	.80	.84	.68	.82	.68	.89	.67	1.64	1.72	7.2	6.9	7.7	6.3	7.9	8.9	7.2	6.2	9.4
5-10	.79	.83	.67	.81	.66	.87	.65	1.63	1.71	4.3	4.0	4.8	3.5	4.8	5.6	4.6	4.2	6.7
10-15	.78	.83	.67	.81	.66	.86	.65	1.62	1.70	3.9	5.3	4.9	5.4	4.6	4.9	6.2	4.0	2.4
15-20	.78	.82	.66	.81	.66	.86	.64	1.62	1.70	2.4	1.8	3.1	1.6	2.5	3.1	6.5	.7	2.0
20-25	.78	.82	.66	.81	.66	.86	.63	1.63	1.70	1.9	1.5	3.2	1.4	2.9	1.9	4.4	2.3	4.7
25-30	.78	.82	.66	.81	.65	.86	.62	1.64	1.70	2.4	2.0	3.6	1.8	3.2	2.8	5.1	2.2	2.4
30-35	.78	.82	.65	.81	.65	.85	.61	1.64	1.70	2.8	2.4	3.7	2.1	3.1	3.6	5.6	2.1	3.4
35-40	.77	.82	.65	.81	.65	.84	.59	1.65	1.70	3.3	2.9	4.1	2.4	3.9	4.5	4.9	1.9	2.5
40-45	.77	.82	.64	.81	.64	.82	.58	1.66	1.70	3.6	3.4	4.1	2.9	3.9	5.2	4.8	2.6	5.0
45-50	.76	.82	.63	.81	.64	.80	.56	1.67	1.70	3.8	3.5	4.2	3.3	4.2	4.8	4.0	2.5	3.0
50-55	.75	.81	.62	.81	.62	.77	.54	1.69	1.70	4.3	4.2	4.4	4.0	4.5	5.4	3.8	4.5	4.0
55-60	.74	.80	.61	.80	.61	.74	.51	1.71	1.71	5.1	5.1	5.1	5.1	5.4	5.0	3.6	5.3	5.0
60-65	.72	.79	.58	.79	.59	.72	.49	1.72	1.71	5.9	6.1	5.3	6.2	5.6	5.7	3.5	6.1	4.3
65-70	.69	.76	.55	.76	.55	.68	.47	1.74	1.73	6.6	7.3	5.3	7.5	5.7	5.9	2.8	6.2	5.0
70-75	.64	.72	.52	.71	.52	.65	.47	1.79	1.76	6.8	7.6	5.2	7.9	5.5	5.4	3.0	8.0	6.5
75-80	.59	.66	.49	.65	.49	.64	.48	1.81	1.77	6.5	7.6	4.9	7.8	5.1	5.1	2.8	7.6	7.0
80-85	.55	.61	.47	.59	.46	.66	.51	1.88	1.80	5.8	6.9	4.5	7.0	4.5	5.2	3.4	8.8	7.2

AGE	GAIN IN EXPECTATION OF LIFE AT AGE X									FULFILMENT INDEX FOR AGE (X TO X+N)								
	ALL RACES			WHITE		BLACK		OTHER		ALL RACES			WHITE		BLACK		OTHER	
(X,X+N)	T	M	F	M	F	M	F	M	F	T	M	F	M	F	M	F	M	F
(ELIMINATING PEPTIC ULCER)																		
0- 1	.04	.04	.03	.04	.03	.04	.03	.07	.06	.0	.0	.0	.0	.0	.0	.1	.0	.0
1- 5	.04	.04	.03	.04	.03	.04	.03	.07	.06	.1	.1	.1	.1	.0	.1	.1	.0	.0
5-10	.04	.04	.03	.04	.03	.04	.03	.07	.06	.0	.0	.1	.0	.1	.0	.2	.0	.0
10-15	.04	.04	.03	.04	.03	.04	.03	.07	.06	.1	.2	.1	.1	.1	.2	.2	2.0	.0
15-20	.04	.04	.03	.04	.03	.04	.03	.07	.06	.1	.1	.1	.1	.1	.1	.2	.0	.0
20-25	.04	.04	.03	.04	.03	.04	.03	.07	.06	.1	.0	.1	.0	.1	.1	.1	.0	.7
25-30	.04	.04	.03	.04	.03	.04	.03	.07	.06	.1	.1	.1	.1	.1	.1	.0	.0	.0
30-35	.04	.04	.03	.04	.03	.04	.03	.07	.06	.2	.2	.2	.2	.2	.2	.1	.0	.0
35-40	.04	.04	.03	.04	.03	.04	.03	.08	.06	.2	.3	.2	.3	.2	.1	.2	.0	.0
40-45	.04	.04	.03	.04	.03	.04	.03	.08	.06	.3	.3	.3	.3	.3	.3	.1	.6	.5
45-50	.03	.04	.03	.03	.03	.04	.03	.07	.06	.3	.3	.3	.3	.3	.3	.1	.5	.3
50-55	.03	.03	.03	.03	.03	.03	.03	.07	.05	.3	.3	.2	.3	.3	.3	.2	.7	.0
55-60	.03	.03	.03	.03	.03	.03	.03	.07	.06	.3	.3	.3	.3	.3	.3	.2	.8	.5
60-65	.03	.03	.03	.03	.03	.03	.03	.06	.05	.3	.3	.3	.3	.3	.3	.2	.5	.8
65-70	.03	.03	.03	.03	.03	.03	.02	.06	.04	.3	.3	.3	.3	.3	.3	.1	.8	.4
70-75	.03	.02	.02	.02	.02	.02	.02	.04	.04	.3	.3	.3	.3	.3	.3	.2	.2	.6
75-80	.02	.02	.02	.02	.02	.02	.02	.04	.03	.3	.3	.3	.3	.3	.2	.2	.3	.2
80-85	.02	.02	.02	.02	.02	.02	.02	.04	.03	.2	.2	.2	.2	.3	.2	.1	.4	.5
(ELIMINATING CIRRHOSIS OF LIVER)																		
0- 1	.27	.31	.20	.29	.18	.48	.33	.57	.43	.0	.0	.0	.0	.0	.0	.0	.0	.3
I- 5	.27	.32	.21	.29	.18	.49	.34	.58	.43	.2	.1	.3	.1	.3	.2	.1	.0	1.0
5-10	.27	.32	.21	.29	.18	.49	.34	.58	.43	.1	.1	.0	.2	.1	.1	.0	.0	.0
10-15	.27	.32	.21	.29	.18	.50	.34	.58	.43	.1	.2	.0	.2	.1	.0	.0	4.0	.0
15-20	.27	.32	.21	.29	.18	.50	.34	.58	.43	.1	.1	.2	.0	.1	.2	.6	.0	.0
20-25	.28	.32	.21	.29	.18	.50	.34	.59	.43	.4	.3	.6	.2	.5	.5	1.1	1.1	2.0
25-30	.28	.32	.21	.30	.18	.50	.34	.59	.43	1.6	1.4	1.9	1.2	1.4	2.4	3.4	1.9	2.4
30-35	.27	.32	.20	.29	.18	.50	.33	.58	.43	3.6	3.6	3.7	2.7	2.3	6.3	7.7	7.8	10.7
35-40	.26	.31	.20	.29	.18	.45	.31	.55	.40	6.2	6.2	6.1	5.4	5.1	8.2	8.6	12.3	11.7
40-45	.24	.29	.18	.27	.17	.39	.27	.48	.37	6.6	7.2	6.1	6.7	5.4	8.1	7.7	13.8	15.9
45-50	.22	.25	.16	.24	.15	.31	.22	.39	.31	6.2	6.4	5.7	6.2	5.4	6.4	6.8	12.7	8.1
50-55	.18	.21	.13	.20	.13	.24	.16	.31	.27	5.2	5.2	5.0	5.3	5.0	4.9	4.5	6.5	9.9
55-60	.14	.16	.10	.16	.10	.17	.11	.25	.21	3.7	3.9	3.4	3.9	3.5	3.6	2.8	5.3	6.0
60-65	.10	.12	.07	.12	.07	.11	.07	.19	.15	2.7	2.8	2.5	2.8	2.6	2.1	2.0	4.1	4.6
65-70	.07	.08	.05	.08	.05	.06	.04	.13	.10	1.8	1.9	1.5	1.9	1.6	1.4	1.1	2.3	2.8
70-75	.04	.05	.03	.05	.03	.04	.03	.08	.06	.9	1.0	.9	1.0	.9	.7	.5	1.2	.7
75-80	.02	.03	.02	.03	.02	.02	.01	.06	.04	.5	.5	.4	.5	.5	.3	.2	.9	.9
80-85	.01	.01	.01	.01	.01	.02	.00	.03	.03	.2	.2	.2	.2	.2	.1	.1	.6	.5

TABLE 2 (CONTINUED)

AGE	GAIN IN EXPECTATION OF LIFE AT AGE X									FULFILMENT INDEX FOR AGE (X TO X+N)								
	ALL RACES			WHITE		BLACK		OTHER		ALL RACES			WHITE		BLACK		OTHER	
(X,X+N)	T	M	F	M	F	M	F	M	F	T	M	F	M	F	M	F	M	F
(ELIMINATING NEPHRITIS AND NEPHROSIS)																		
0- 1	.06	.06	.06	.05	.05	.14	.19	.15	.23	.1	.1	.1	.1	.1	.1	.1	.1	.1
1- 5	.06	.05	.06	.04	.05	.14	.19	.15	.23	.3	.3	.3	.2	.3	.5	.1	.0	.0
5-10	.06	.05	.06	.04	.05	.14	.19	.15	.23	.3	.2	.4	.2	.5	.1	.0	.0	.0
10-15	.06	.05	.06	.04	.05	.14	.19	.15	.23	.3	.3	.4	.3	.4	.2	.3	.0	.0
15-20	.06	.05	.06	.04	.05	.14	.19	.15	.23	.2	.1	.4	.1	.3	.3	.6	.0	1.0
20-25	.06	.05	.06	.04	.05	.14	.19	.16	.23	.2	.2	.4	.2	.3	.3	1.0	.0	.0
25-30	.06	.05	.06	.04	.05	.14	.19	.16	.23	.4	.3	.6	.3	.5	.5	1.2	.3	.0
30-35	.06	.05	.06	.04	.04	.14	.19	.16	.23	.5	.4	.7	.2	.5	.8	1.3	1.2	.0
35-40	.06	.05	.06	.04	.04	.13	.18	.15	.23	.5	.4	.7	.3	.6	.8	1.0	.3	1.8
40-45	.06	.05	.06	.04	.04	.13	.18	.15	.22	.4	.5	.5	.3	.3	.9	.9	.3	.5
45-50	.05	.05	.06	.04	.04	.12	.18	.15	.22	.5	.4	.6	.3	.4	.9	1.3	.3	1.0
50-55	.05	.05	.05	.04	.04	.12	.17	.16	.22	.5	.4	.6	.3	.4	.8	1.4	.6	2.0
55-60	.05	.05	.05	.04	.04	.11	.16	.16	.21	.5	.4	.7	.3	.4	1.0	1.8	.2	1.5
60-65	.05	.04	.05	.04	.04	.11	.14	.16	.20	.5	.4	.6	.3	.5	.8	1.3	1.1	1.0
65-70	.04	.04	.04	.04	.03	.10	.13	.15	.20	.5	.4	.6	.3	.5	.9	1.3	.8	1.6
70-75	.04	.04	.04	.03	.03	.10	.12	.15	.18	.5	.4	.5	.4	.4	1.0	1.1	.6	.9
75-80	.03	.04	.03	.03	.03	.09	.11	.16	.18	.4	.4	.4	.4	.4	.9	1.0	1.1	.7
80-85	.03	.03	.03	.03	.02	.09	.10	.15	.19	.4	.4	.4	.4	.3	.8	1.0	.8	1.4
(ELIMINATING CONGENITAL ANOMALIES)																		
0- 1	.25	.25	.25	.25	.25	.25	.26	.22	.27	9.7	9.6	11.8	10.3	12.9	7.2	8.3	8.1	14.1
1- 5	.07	.06	.07	.07	.07	.06	.07	.07	.05	12.2	11.5	13.1	13.0	14.0	6.5	11.2	9.4	7.3
5-10	.04	.04	.04	.04	.04	.04	.04	.04	.04	6.2	5.5	7.4	5.6	7.3	4.8	8.1	6.9	2.2
10-15	.04	.03	.04	.03	.04	.04	.03	.04	.04	4.7	6.1	6.2	6.3	6.6	5.5	5.2	6.0	2.4
15-20	.03	.03	.03	.03	.03	.03	.03	.03	.03	2.1	1.5	2.8	1.6	2.8	1.6	3.2	.7	.0
20-25	.03	.02	.03	.02	.03	.03	.02	.03	.03	1.0	.7	1.8	.8	1.9	.7	1.4	.5	1.3
25-30	.02	.02	.02	.02	.02	.02	.02	.02	.03	.9	.8	1.3	.9	1.4	.5	1.1	1.1	1.2
30-35	.02	.02	.02	.02	.02	.02	.02	.02	.03	.9	.6	1.4	.7	1.6	.5	.7	.0	1.1
35-40	.02	.01	.02	.01	.02	.02	.01	.02	.03	.6	.5	.9	.6	1.1	.2	.5	.3	.6
40-45	.01	.01	.02	.01	.02	.01	.01	.02	.02	.5	.5	.6	.5	.6	.2	.3	.3	1.5
45-50	.01	.01	.01	.01	.01	.01	.01	.02	.02	.4	.3	.5	.3	.5	.2	.2	.3	.3
50-55	.01	.01	.01	.01	.01	.01	.01	.01	.02	.2	.2	.3	.2	.4	.1	.1	.2	.6
55-60	.01	.01	.01	.01	.01	.01	.01	.01	.01	.2	.2	.2	.2	.3	.1	.1	.3	.0
60-65	.01	.01	.01	.00	.01	.01	.00	.01	.01	.1	.1	.2	.1	.2	.1	.1	.1	.5
65-70	.00	.00	.01	.00	.01	.01	.00	.01	.01	.1	.1	.2	.1	.2	.1	.1	.1	.2
70-75	.00	.00	.00	.00	.00	.00	.00	.01	.00	.1	.0	.1	.0	.1	.0	.0	.2	.2
75-80	.00	.00	.00	.00	.00	.00	.00	.00	.00	.0	.0	.1	.0	.1	.0	.0	.1	.0
80-85	.00	.00	.00	.00	.00	.00	.00	.00	.00	.0	.0	.0	.0	.0	.0	.0	.0	.0

AGE	GAIN IN EXPECTATION OF LIFE AT AGE X									FULFILMENT INDEX FOR AGE (X TO X+N)								
	ALL RACES			WHITE		BLACK		OTHER		ALL RACES			WHITE		BLACK		OTHER	
(X,X+N)	T	M	F	M	F	M	F	M	F	T	M	F	M	F	M	F	M	F
(ELIMINATING CERTAIN CAUSES OF MORTALITY IN EARLY INFANCY)																		
0- 1	.49	.52	.45	.44	.37	.96	.86	.42	.33	25.5	27.1	28.3	24.5	25.6	36.4	37.5	22.0	21.5
1- 5	.00	.00	.00	.00	.00	.00	.00	.00	.00	.3	.4	.2	.4	.1	.4	.5	.0	.0
5-10	.00	.00	.00	.00	.00	.00	.00	.00	.00	.1	.1	.0	.1	.0	.0	.0	1.4	.0
10-15	.00	.00	.00	.00	.00	.00	.00	.00	.00	.0	.0	.0	.0	.0	.0	.0	.0	.0
15-20	.00	.00	.00	.00	.00	.00	.00	.00	.00	.0	.0	.0	.0	.0	.1	.0	.0	.0
20-25	.00	.00	.00	.00	.00	.00	.00	.00	.00	.0	.0	.0	.0	.0	.0	.0	.0	.0
25-30	.00	.00	.00	.00	.00	.00	.00	.00	.00	.0	.0	.0	.0	.0	.0	.0	.0	.0
30-35	.00	.00	.00	.00	.00	.00	.00	.00	.00	.0	.0	.0	.0	.0	.0	.0	.0	.0
35-40	.00	.00	.00	.00	.00	.00	.00	.00	.00	.0	.0	.0	.0	.0	.0	.0	.0	.0
40-45	.00	.00	.00	.00	.00	.00	.00	.00	.00	.0	.0	.0	.0	.0	.0	.0	.0	.0
45-50	.00	.00	.00	.00	.00	.00	.00	.00	.00	.0	.0	.0	.0	.0	.0	.0	.0	.0
50-55	.00	.00	.00	.00	.00	.00	.00	.00	.00	.0	.0	.0	.0	.0	.0	.0	.0	.0
55-60	.00	.00	.00	.00	.00	.00	.00	.00	.00	.0	.0	.0	.0	.0	.0	.0	.0	.0
60-65	.00	.00	.00	.00	.00	.00	.00	.00	.00	.0	.0	.0	.0	.0	.0	.0	.0	.0
65-70	.00	.00	.00	.00	.00	.00	.00	.00	.00	.0	.0	.0	.0	.0	.0	.0	.0	.0
70-75	.00	.00	.00	.00	.00	.00	.00	.00	.00	.0	.0	.0	.0	.0	.0	.0	.0	.0
75-80	.00	.00	.00	.00	.00	.00	.00	.00	.00	.0	.0	.0	.0	.0	.0	.0	.0	.0
80-85	.00	.00	.00	.00	.00	.00	.00	.00	.00	.0	.0	.0	.0	.0	.0	.0	.0	.0
(ELIMINATING MOTOR VEHICLE ACCIDENTS)																		
0- 1	.65	.87	.38	.89	.39	.71	.28	1.32	.67	.3	.3	.4	.3	.4	.2	.3	.4	.7
1- 5	.65	.88	.38	.90	.39	.72	.27	1.32	.67	15.2	14.9	15.7	15.3	16.2	12.3	13.8	23.4	20.8
5-10	.62	.85	.35	.87	.36	.69	.24	1.27	.63	25.7	26.8	24.1	27.4	24.8	24.2	19.9	30.5	37.8
10-15	.60	.82	.33	.84	.34	.65	.21	1.24	.59	26.3	43.3	24.2	48.6	26.4	23.5	14.7	38.3	19.0
15-20	.57	.78	.31	.80	.32	.63	.20	1.20	.58	53.0	50.8	43.0	55.0	47.5	22.9	16.1	41.6	43.5
20-25	.44	.61	.23	.61	.24	.57	.18	1.03	.50	35.4	37.0	30.5	41.1	35.1	17.9	12.8	37.1	32.0
25-30	.31	.43	.18	.42	.18	.46	.14	.82	.42	24.5	26.9	18.3	30.4	21.5	13.5	6.8	32.9	28.6
30-35	.24	.31	.14	.30	.14	.36	.12	.65	.36	17.3	19.6	12.3	22.3	14.0	10.2	5.3	23.8	20.9
35-40	.19	.24	.12	.23	.12	.29	.10	.54	.31	11.9	13.6	8.5	15.5	10.2	7.4	3.1	16.1	16.0
40-45	.15	.18	.10	.18	.10	.22	.09	.45	.26	7.0	8.5	5.0	9.4	5.6	5.0	2.8	11.5	9.9
45-50	.12	.14	.08	.13	.08	.17	.07	.37	.22	4.1	4.6	2.9	4.9	3.2	3.3	1.7	8.6	4.7
50-55	.09	.11	.07	.10	.07	.14	.06	.32	.20	2.5	2.8	2.0	2.9	2.2	2.4	1.1	5.8	2.8
55-60	.07	.08	.05	.08	.06	.11	.04	.27	.18	1.7	1.8	1.5	1.8	1.6	1.6	1.0	4.1	3.5
60-65	.06	.06	.04	.06	.04	.08	.03	.23	.16	1.1	1.1	1.0	1.1	1.0	1.2	.6	2.8	1.8
65-70	.04	.05	.03	.05	.03	.06	.02	.19	.14	.8	.8	.8	.7	.9	.9	.4	3.2	2.6
70-75	.03	.04	.03	.04	.03	.05	.02	.13	.10	.6	.6	.6	.6	.6	.6	.2	1.2	.7
75-80	.02	.03	.02	.03	.02	.04	.01	.12	.09	.4	.5	.4	.5	.4	.6	.2	1.5	.9
80-85	.02	.02	.01	.02	.01	.03	.01	.08	.09	.3	.3	.2	.3	.2	.3	.1	.7	.5

TABLE 2 (CONTINUED)

AGE	GAIN IN EXPECTATION OF LIFE AT AGE X									FULFILMENT INDEX FOR AGE (X TO X+N)								
	ALL RACES			WHITE		BLACK		OTHER		ALL RACES			WHITE		BLACK		OTHER	
(X,X+N)	T	M	F	M	F	M	F	M	F	T	M	F	M	F	M	F	M	F
	(ELIMINATING ALL OTHER ACCIDENTS)																	
0- 1	.56	.73	.36	.69	.33	.97	.48	1.18	.64	1.2	1.2	1.3	1.1	1.1	1.6	1.9	1.4	.9
1- 5	.55	.71	.34	.67	.32	.95	.45	1.17	.63	26.3	28.0	23.9	28.1	23.1	27.6	26.2	31.2	25.0
5-10	.50	.66	.30	.62	.29	.88	.38	1.10	.58	24.6	28.3	19.1	27.6	18.0	30.8	23.4	31.9	20.0
10-15	.47	.62	.28	.59	.27	.83	.35	1.06	.56	25.1	46.0	17.4	45.0	16.9	50.8	19.7	38.3	19.0
15-20	.45	.59	.27	.56	.26	.78	.33	1.03	.55	18.4	19.9	9.6	19.5	9.3	22.5	10.6	21.9	17.8
20-25	.40	.52	.25	.49	.24	.72	.32	.94	.52	15.2	17.0	9.8	17.7	10.4	13.2	7.5	19.7	12.0
25-30	.35	.44	.23	.41	.22	.64	.30	.83	.49	13.7	15.7	8.5	17.0	8.9	11.2	7.7	13.5	4.9
30-35	.31	.38	.22	.35	.21	.56	.28	.77	.48	11.9	14.3	6.7	15.5	7.2	10.0	4.9	16.3	6.2
35-40	.27	.32	.20	.30	.20	.49	.26	.70	.47	10.6	12.7	6.3	13.6	6.6	9.8	5.2	12.0	6.1
40-45	.24	.27	.19	.25	.18	.42	.24	.63	.45	6.8	8.6	4.3	8.7	4.5	8.3	3.5	11.2	5.4
45-50	.21	.23	.18	.21	.17	.33	.22	.56	.43	4.7	5.5	3.1	5.4	3.1	5.5	2.6	7.3	4.7
50-55	.18	.19	.16	.18	.16	.28	.20	.52	.41	3.2	3.8	2.1	3.7	2.1	4.0	1.9	7.2	3.7
55-60	.16	.16	.15	.15	.15	.23	.18	.47	.40	2.3	2.6	1.7	2.5	1.7	2.8	1.6	5.2	2.0
60-65	.14	.14	.14	.13	.14	.19	.17	.42	.39	1.8	1.9	1.5	1.8	1.5	2.4	1.5	3.3	2.5
65-70	.13	.11	.13	.11	.13	.16	.16	.39	.37	1.4	1.4	1.3	1.3	1.2	1.6	1.4	2.9	1.4
70-75	.12	.10	.13	.10	.12	.14	.15	.36	.37	1.1	1.1	1.2	1.0	1.2	1.5	1.2	2.2	1.1
75-80	.11	.09	.12	.09	.12	.13	.14	.35	.39	1.1	1.0	1.2	1.0	1.2	1.2	1.0	2.0	1.4
80-85	.11	.09	.12	.09	.12	.13	.14	.35	.40	1.1	1.0	1.2	1.0	1.2	1.0	1.1	2.3	1.1
	(ELIMINATING SUICIDE)																	
0- 1	.30	.40	.17	.42	.18	.24	.07	.40	.24	.0	.0	.0	.0	.0	.0	.0	.0	.0
1- 5	.30	.40	.17	.42	.19	.24	.08	.41	.25	.0	.0	.0	.0	.0	.0	.0	.0	.0
5-10	.30	.40	.17	.43	.19	.24	.08	.41	.25	.0	.1	.0	.0	.0	.1	.0	.0	.0
10-15	.30	.41	.17	.43	.19	.24	.08	.41	.25	2.4	4.5	1.5	5.2	1.5	1.3	1.2	6.0	4.8
15-20	.30	.40	.17	.42	.19	.24	.08	.41	.24	9.4	9.9	5.6	10.5	6.1	4.8	2.4	14.9	5.9
20-25	.28	.37	.16	.39	.18	.23	.07	.35	.23	12.5	13.5	9.7	14.7	11.0	7.8	4.1	13.0	14.0
25-30	.24	.31	.15	.32	.16	.18	.06	.27	.20	13.3	14.2	10.9	16.3	13.1	6.8	4.0	13.8	8.5
30-35	.19	.25	.13	.26	.14	.13	.05	.20	.18	11.2	12.1	9.4	14.4	11.4	4.8	3.1	6.3	6.2
35-40	.16	.20	.11	.21	.12	.10	.04	.17	.17	9.0	9.3	8.3	11.3	10.5	3.3	2.3	6.9	4.3
40-45	.13	.16	.09	.18	.10	.07	.03	.13	.15	5.7	6.1	5.5	7.5	7.1	1.7	.8	4.0	3.5
45-50	.11	.13	.07	.14	.07	.05	.02	.10	.14	3.8	3.9	3.4	4.5	4.2	1.3	.5	2.8	2.4
50-55	.08	.11	.05	.12	.05	.04	.02	.09	.13	2.5	2.6	2.3	3.0	2.7	.7	.5	1.1	2.3
55-60	.06	.08	.04	.09	.04	.03	.01	.08	.12	1.6	1.7	1.5	2.0	1.7	.4	.2	1.3	1.5
60-65	.05	.07	.02	.07	.02	.02	.01	.07	.11	1.0	1.2	.7	1.4	.8	.3	.2	.7	1.3
65-70	.03	.05	.02	.05	.02	.02	.01	.06	.10	.7	.8	.5	.9	.6	.3	.1	.4	1.0
70-75	.03	.04	.01	.04	.01	.01	.00	.05	.08	.5	.6	.3	.7	.3	.2	.1	.7	.0
75-80	.02	.03	.01	.03	.01	.01	.00	.04	.09	.3	.5	.2	.5	.2	.1	.0	.3	.4
80-85	.01	.02	.00	.02	.00	.01	.00	.04	.09	.2	.3	.1	.3	.1	.1	.0	.1	.2

AGE	GAIN IN EXPECTATION OF LIFE AT AGE X									FULFILMENT INDEX FOR AGE (X TO X+N)								
	ALL RACES			WHITE		BLACK		OTHER		ALL RACES			WHITE		BLACK		OTHER	
(X,X+N)	T	M	F	M	F	M	F	M	F	T	M	F	M	F	M	F	M	F
	(ELIMINATING HOMICIDE)																	
0- 1	.26	.36	.13	.22	.09	1.34	.37	.44	.16	.2	.2	.2	.1	.2	.3	.4	.2	.1
1- 5	.26	.37	.12	.22	.09	1.37	.37	.45	.16	3.7	3.5	3.9	2.4	2.7	7.6	8.0	3.1	1.0
5-10	.25	.36	.12	.22	.08	1.36	.35	.44	.16	3.0	2.7	3.5	2.0	2.5	5.2	6.8	4.2	8.9
10-15	.25	.36	.11	.22	.08	1.35	.34	.44	.15	4.5	6.2	5.5	5.1	4.4	10.3	10.2	6.0	9.5
15-20	.24	.35	.11	.21	.08	1.35	.33	.43	.14	10.5	10.1	8.2	6.7	6.3	33.8	19.7	12.0	6.9
20-25	.22	.32	.09	.19	.06	1.25	.30	.38	.13	13.1	13.7	11.1	8.5	7.7	38.4	24.6	13.3	8.0
25-30	.17	.25	.07	.15	.05	1.02	.23	.31	.11	14.0	15.6	9.6	9.9	6.7	36.2	18.6	13.0	11.0
30-35	.13	.19	.06	.12	.04	.75	.17	.24	.09	10.9	13.0	6.5	8.4	4.8	28.2	12.2	13.3	4.5
35-40	.10	.14	.04	.09	.03	.53	.13	.18	.07	8.3	10.1	4.8	7.1	3.6	19.3	8.4	7.2	4.9
40-45	.07	.10	.03	.06	.03	.36	.09	.14	.06	4.8	6.2	2.7	4.6	2.3	11.5	4.1	4.6	3.0
45-50	.05	.06	.02	.04	.02	.24	.06	.11	.05	2.4	3.0	1.2	2.1	1.1	6.8	1.6	4.0	1.0
50-55	.03	.04	.02	.03	.01	.16	.05	.08	.04	1.3	1.7	.7	1.1	.6	4.0	1.2	2.6	.6
55-60	.02	.03	.01	.02	.01	.10	.04	.06	.04	.7	.8	.4	.6	.3	2.1	.7	1.1	.0
60-65	.01	.02	.01	.01	.01	.06	.03	.05	.04	.4	.4	.2	.3	.2	1.2	.5	1.1	.8
65-70	.01	.01	.01	.01	.01	.04	.02	.03	.03	.2	.2	.1	.2	.1	.8	.4	.1	.6
70-75	.01	.01	.01	.00	.00	.02	.02	.03	.02	.1	.1	.1	.1	.1	.4	.2	.5	.4
75-80	.00	.00	.00	.00	.00	.02	.01	.02	.02	.1	.1	.1	.0	.1	.2	.1	.2	.2
80-85	.00	.00	.00	.00	.00	.01	.01	.01	.02	.0	.0	.0	.0	.0	.1	.1	.1	.0
	(ELIMINATING ALL OTHER CAUSES)																	
0- 1	1.35	1.28	1.37	1.13	1.23	2.32	2.53	2.15	2.74	9.7	10.5	10.5	9.4	9.3	14.2	14.4	11.2	9.0
1- 5	1.18	1.09	1.22	.98	1.10	1.99	2.24	1.96	2.62	18.3	18.8	17.5	17.7	17.1	23.0	19.2	14.8	14.6
5-10	1.14	1.06	1.19	.94	1.08	1.93	2.20	1.93	2.60	14.8	13.2	17.2	13.5	17.1	13.3	18.4	4.2	6.7
10-15	1.13	1.04	1.18	.93	1.06	1.92	2.18	1.93	2.60	14.0	19.3	17.2	18.2	17.1	23.1	18.0	24.2	14.3
15-20	1.12	1.03	1.16	.92	1.05	1.90	2.16	1.91	2.59	10.7	8.8	12.0	8.5	10.4	11.3	21.7	8.0	9.9
20-25	1.10	1.00	1.15	.89	1.04	1.87	2.13	1.89	2.58	9.7	7.6	15.8	7.1	13.9	10.3	23.2	8.2	15.3
25-30	1.07	.98	1.12	.87	1.01	1.83	2.08	1.86	2.55	12.6	10.2	18.8	9.2	16.5	13.5	26.2	10.8	18.9
30-35	1.03	.94	1.09	.84	.99	1.76	2.00	1.82	2.51	13.3	11.5	17.4	10.4	15.5	14.8	23.9	14.2	16.4
35-40	1.00	.90	1.06	.81	.97	1.67	1.92	1.77	2.48	14.5	12.5	18.6	11.0	16.9	16.4	22.8	17.3	25.2
40-45	.96	.86	1.02	.78	.94	1.56	1.83	1.68	2.42	12.5	11.8	14.5	10.7	13.5	15.1	17.6	16.7	14.4
45-50	.91	.81	.97	.74	.90	1.44	1.74	1.59	2.38	11.2	10.1	12.4	9.3	11.6	13.4	15.3	11.6	15.3
50-55	.86	.75	.92	.69	.86	1.33	1.65	1.54	2.32	9.3	8.6	10.5	8.0	10.0	11.3	12.6	10.0	11.9
55-60	.80	.69	.87	.64	.81	1.23	1.56	1.49	2.28	8.2	7.5	9.5	7.0	9.0	10.2	11.5	8.2	9.6
60-65	.74	.64	.81	.59	.76	1.13	1.47	1.46	2.25	7.6	6.9	8.8	6.5	8.5	9.3	10.1	7.1	9.9
65-70	.68	.57	.75	.53	.70	1.05	1.41	1.43	2.22	7.1	6.2	8.4	6.0	8.1	8.4	10.0	8.1	10.2
70-75	.62	.52	.69	.48	.65	1.02	1.36	1.39	2.17	6.7	5.8	7.9	5.5	7.6	8.4	9.6	7.4	11.3
75-80	.57	.47	.63	.44	.59	1.01	1.33	1.37	2.12	6.4	5.5	7.4	5.2	7.1	7.4	9.0	7.8	8.5
80-85	.53	.44	.58	.40	.54	1.07	1.31	1.34	2.15	5.8	5.0	6.3	4.8	6.2	8.1	8.4	8.0	8.8

TABLE 3. PROBABILITY OF EVENTUALLY DYING FROM SPECIFIED CAUSES BY AGE, SEX, AND RACE: UNITED STATES, 1978

WHITE MALE

AGE	TUBERC ULOSIS (1)	OTHER INFECT (2)	INFECT & PARA	MA NEO OF DIG (3)	MA NEO OF RES (4)	OTHER MA NEO (5)	MALIGN NEOPLA	DIABET MELLIT (6)	HEART DISEASE (7)	HYPERT ENSION (8)	CEREBRO VASCUL (9)	ARTERIO SCLERO (10)	OTHER ARTERI (11)	MAJOR CARDIOV
0	.00146	.00618	.00764	.05296	.06872	.08268	.20436	.01318	.40974	.00252	.07774	.01406	.01659	.52065
1	.00148	.00564	.00712	.05366	.06963	.08375	.20704	.01335	.41494	.00255	.07873	.01424	.01680	.52727
5	.00148	.00554	.00702	.05380	.06983	.08377	.20740	.01339	.41604	.00256	.07892	.01428	.01684	.52865
10	.00148	.00552	.00700	.05390	.06996	.08367	.20753	.01341	.41676	.00257	.07903	.01431	.01687	.52954
15	.00148	.00550	.00698	.05401	.07010	.08363	.20774	.01343	.41758	.00257	.07916	.01434	.01690	.53056
20	.00149	.00549	.00698	.05440	.07062	.08394	.20895	.01353	.42058	.00259	.07970	.01444	.01701	.53433
25	.00150	.00549	.00700	.05490	.07129	.08433	.21052	.01363	.42445	.00261	.08040	.01458	.01716	.53920
30	.00151	.00550	.00701	.05532	.07186	.08454	.21172	.01369	.42776	.00263	.08100	.01470	.01727	.54336
35	.00152	.00549	.00701	.05565	.07238	.08460	.21263	.01369	.43060	.00265	.08154	.01482	.01738	.54700
40	.00153	.00547	.00701	.05597	.07286	.08467	.21350	.01369	.43325	.00266	.08216	.01498	.01751	.55057
45	.00154	.00544	.00698	.05630	.07314	.08473	.21416	.01369	.43554	.00269	.08301	.01520	.01768	.55412
50	.00153	.00543	.00696	.05652	.07276	.08473	.21401	.01371	.43731	.00273	.08451	.01560	.01800	.55815
55	.00152	.00542	.00693	.05641	.07097	.08456	.21194	.01375	.43889	.00280	.08700	.01623	.01838	.56331
60	.00151	.00544	.00694	.05589	.06752	.08423	.20764	.01378	.43981	.00291	.09073	.01723	.01882	.56950
65	.00148	.00552	.00701	.05431	.06133	.08348	.19912	.01380	.44133	.00309	.09659	.01889	.01917	.57906
70	.00144	.00558	.00702	.05169	.05269	.08208	.18646	.01368	.44401	.00338	.10419	.02133	.01908	.59199
75	.00143	.00575	.00718	.04781	.04155	.07930	.16865	.01328	.44815	.00366	.11321	.02506	.01851	.60859
80	.00134	.00588	.00722	.04273	.03005	.07397	.14675	.01241	.45607	.00409	.12211	.03078	.01722	.63026
85	.00131	.00588	.00720	.03625	.01984	.06615	.12224	.01152	.46649	.00447	.12992	.03832	.01580	.65500

AGE	INFLU & PNEUMO (12)	BRONCH EM & AS (13)	OTHER RESP D (14)	RESPIR DISEASE	PEPTIC ULCER (15)	CIRRHO OF LIV (16)	NEPHRI NEPHRO (17)	CONGEN ANOMAL (18)	INFANT DISEASE (19)	MOTOR V ACCID (20)	OTHER ACCID (21)	SUI CIDE (22)	HOMI CIDE (23)	OTHER CAUSES (24)
0	.03249	.01608	.03009	.07866	.00324	.01567	.00417	.00423	.00621	.02403	.02594	.01543	.00628	.07032
1	.03254	.01627	.03033	.07914	.00328	.01587	.00420	.00163	.00001	.02427	.02600	.01563	.00633	.06885
5	.03254	.01631	.03034	.07919	.00328	.01591	.00421	.00127	.00000	.02391	.02528	.01568	.00628	.06854
10	.03256	.01633	.03038	.07927	.00329	.01593	.00421	.00117	.00000	.02345	.02482	.01571	.00626	.06842
15	.03259	.01635	.03043	.07936	.00330	.01596	.00421	.00109	.00000	.02287	.02429	.01567	.00620	.06833
20	.03276	.01646	.03063	.07985	.00332	.01608	.00424	.00099	.00000	.01939	.02318	.01509	.00581	.06827
25	.03300	.01660	.03088	.08048	.00334	.01621	.00426	.00093	.00000	.01562	.02169	.01382	.00504	.06825
30	.03319	.01673	.03110	.08101	.00336	.01625	.00428	.00086	.00000	.01317	.02044	.01255	.00424	.06804
35	.03336	.01685	.03131	.08151	.00338	.01616	.00429	.00081	.00000	.01142	.01931	.01146	.00358	.06774
40	.03358	.01700	.03158	.08216	.00338	.01579	.00431	.00076	.00000	.00999	.01815	.01044	.00290	.06736
45	.03388	.01723	.03198	.08308	.00339	.01505	.00433	.00069	.00000	.00873	.01715	.00949	.00226	.06687
50	.03445	.01753	.03256	.08455	.00340	.01379	.00437	.00063	.00000	.00767	.01617	.00856	.00176	.06627
55	.03541	.01797	.03335	.08673	.00341	.01211	.00443	.00057	.00000	.00675	.01530	.00763	.00133	.06583
60	.03688	.01838	.03419	.08945	.00342	.01022	.00455	.00049	.00000	.00596	.01461	.00677	.00103	.06563
65	.03939	.01854	.03484	.09278	.00342	.00793	.00472	.00044	.00000	.00529	.01416	.00591	.00075	.06563
70	.04317	.01798	.03433	.09548	.00343	.00563	.00492	.00038	.00000	.00486	.01416	.00522	.00057	.06619
75	.04857	.01669	.03251	.09778	.00346	.00391	.00511	.00036	.00000	.00432	.01490	.00450	.00044	.06752
80	.05635	.01405	.02861	.09901	.00336	.00249	.00520	.00033	.00000	.00363	.01615	.00369	.00038	.06912
85	.06426	.01085	.02438	.09949	.00331	.00169	.00512	.00029	.00000	.00291	.01775	.00293	.00035	.07019

WHITE FEMALE

AGE	TUBERC ULOSIS (1)	OTHER INFECT (2)	INFECT & PARA	MA NEO OF DIG (3)	MA NEO OF RES (4)	OTHER MA NEO (5)	MALIGN NEOPLA	DIABET MELLIT (6)	HEART DISEASE (7)	HYPERT ENSION (8)	CEREBRO VASCUL (9)	ARTERIO SCLERO (10)	OTHER ARTERI (11)	MAJOR CARDIOV
0	.00089	.00681	.00770	.05454	.02395	.10508	.18357	.02084	.41854	.00337	.13098	.02629	.01321	.59239
1	.00090	.00643	.00732	.05511	.02420	.10614	.18544	.02105	.42272	.00340	.13231	.02656	.01334	.59834
5	.00089	.00635	.00724	.05522	.02425	.10618	.18565	.02109	.42354	.00341	.13257	.02662	.01337	.59950
10	.00089	.00631	.00721	.05529	.02428	.10611	.18568	.02112	.42405	.00341	.13272	.02665	.01338	.60023
15	.00089	.00630	.00719	.05535	.02431	.10609	.18575	.02113	.42451	.00342	.13286	.02669	.01340	.60087
20	.00090	.00626	.00716	.05549	.02437	.10617	.18604	.02118	.42562	.00342	.13318	.02676	.01343	.60242
25	.00090	.00625	.00715	.05564	.02444	.10625	.18634	.02122	.42684	.00343	.13354	.02684	.01345	.60410
30	.00090	.00622	.00711	.05579	.02451	.10615	.18645	.02123	.42804	.00344	.13388	.02693	.01348	.60577
35	.00090	.00617	.00706	.05591	.02455	.10578	.18623	.02124	.42944	.00345	.13426	.02703	.01350	.60767
40	.00089	.00613	.00703	.05602	.02450	.10507	.18559	.02125	.43126	.00346	.13471	.02718	.01352	.61013
45	.00089	.00610	.00699	.05610	.02418	.10347	.18374	.02128	.43413	.00348	.13546	.02743	.01357	.61407
50	.00089	.00608	.00696	.05603	.02335	.10058	.17996	.02136	.43839	.00351	.13676	.02783	.01366	.62015
55	.00087	.00605	.00692	.05561	.02183	.09614	.17358	.02146	.44426	.00356	.13868	.02846	.01376	.62872
60	.00086	.00600	.00686	.05464	.01957	.09022	.16443	.02143	.45150	.00363	.14176	.02942	.01389	.64020
65	.00084	.00590	.00674	.05275	.01652	.08247	.15174	.02119	.46037	.00373	.14638	.03092	.01411	.65550
70	.00080	.00581	.00661	.04987	.01319	.07429	.13736	.02066	.46916	.00384	.15219	.03303	.01420	.67241
75	.00077	.00570	.00647	.04577	.01015	.06480	.12073	.01965	.47794	.00398	.15906	.03616	.01416	.69130
80	.00077	.00567	.00644	.04006	.00745	.05453	.10204	.01797	.48653	.00411	.16621	.04092	.01436	.71214
85	.00070	.00555	.00626	.03365	.00551	.04453	.08369	.01565	.49514	.00423	.16970	.04733	.01453	.73094

AGE	INFLU & PNEUMO (12)	BRONCH EM & AS (13)	OTHER RESP D (14)	RESPIR DISEASE	PEPTIC ULCER (15)	CIRRHO OF LIV (16)	NEPHRI NEPHRO (17)	CONGEN ANOMAL (18)	INFANT DISEASE (19)	MOTOR V ACCID (20)	OTHER ACCID (21)	SUI CIDE (22)	HOMI CIDE (23)	OTHER CAUSES (24)
0	.03653	.00715	.01420	.05789	.00279	.00834	.00374	.00393	.00471	.00999	.01764	.00566	.00220	.07862
1	.03659	.00721	.01421	.05802	.00282	.00842	.00377	.00158	.00001	.01001	.01761	.00572	.00219	.07771
5	.03657	.00722	.01419	.05797	.00282	.00843	.00377	.00129	.00000	.00969	.01716	.00573	.00214	.07751
10	.03658	.00722	.01419	.05799	.00283	.00844	.00377	.00119	.00000	.00937	.01694	.00574	.00211	.07739
15	.03658	.00723	.01419	.05800	.00283	.00845	.00377	.00111	.00000	.00907	.01676	.00573	.00206	.07728
20	.03665	.00724	.01421	.05809	.00284	.00847	.00377	.00104	.00000	.00778	.01655	.00557	.00189	.07720
25	.03671	.00724	.01423	.05818	.00284	.00848	.00377	.00098	.00000	.00672	.01628	.00525	.00166	.07701
30	.03677	.00725	.01424	.05826	.00285	.00846	.00377	.00094	.00000	.00606	.01605	.00485	.00145	.07674
35	.03684	.00727	.01426	.05837	.00285	.00841	.00376	.00089	.00000	.00554	.01584	.00444	.00127	.07643
40	.03693	.00728	.01430	.05851	.00286	.00822	.00375	.00084	.00000	.00509	.01561	.00397	.00111	.07606
45	.03712	.00727	.01433	.05872	.00286	.00783	.00376	.00079	.00000	.00466	.01537	.00339	.00092	.07560
50	.03743	.00723	.01434	.05901	.00286	.00715	.00376	.00073	.00000	.00426	.01514	.00283	.00077	.07506
55	.03799	.00715	.01427	.05941	.00286	.00618	.00376	.00066	.00000	.00386	.01502	.00229	.00065	.07460
60	.03881	.00691	.01399	.05971	.00288	.00519	.00375	.00059	.00000	.00346	.01497	.00178	.00057	.07420
65	.04015	.00642	.01346	.06003	.00289	.00401	.00370	.00051	.00000	.00306	.01499	.00142	.00050	.07371
70	.04207	.00581	.01253	.06040	.00287	.00304	.00363	.00040	.00000	.00260	.01523	.00107	.00046	.07326
75	.04492	.00499	.01118	.06109	.00284	.00218	.00351	.00035	.00000	.00209	.01570	.00081	.00038	.07290
80	.04886	.00415	.00975	.06275	.00279	.00145	.00338	.00027	.00000	.00152	.01641	.00052	.00030	.07203
85	.05413	.00356	.00892	.06661	.00265	.00097	.00323	.00021	.00000	.00088	.01708	.00037	.00021	.07124

TABLE 3 (CONTINUED)

BLACK MALE

AGE	TUBERC ULOSIS (1)	OTHER INFECT (2)	INFECT & PARA	MA NEO OF DIG (3)	MA NEO OF RES (4)	OTHER MA NEO (5)	MALIGN NEOPLA	DIABET MELLIT (6)	HEART DISEASE (7)	HYPERT ENSION (8)	CEREBRO VASCUL (9)	ARTERIO SCLERO (10)	OTHER ARTERI (11)	MAJOR CARDIOV
0	.00456	.01150	.01606	.05455	.06783	.08389	.20627	.01600	.31859	.00441	.08736	.00961	.01012	.43008
1	.00468	.01020	.01487	.05614	.06980	.08629	.21224	.01645	.32740	.00453	.08976	.00989	.01038	.44196
5	.00469	.01008	.01477	.05638	.07012	.08650	.21299	.01652	.32870	.00456	.09012	.00993	.01042	.44373
10	.00470	.01005	.01475	.05653	.07031	.08647	.21331	.01657	.32953	.00457	.09034	.00996	.01044	.44484
15	.00471	.01003	.01474	.05669	.07049	.08649	.21367	.01661	.33033	.00458	.09055	.00999	.01047	.44591
20	.00474	.01003	.01477	.05705	.07096	.08671	.21472	.01670	.33233	.00461	.09112	.01005	.01053	.44865
25	.00478	.01006	.01485	.05782	.07198	.08759	.21739	.01692	.33673	.00467	.09234	.01020	.01066	.45459
30	.00482	.01009	.01491	.05884	.07336	.08888	.22109	.01709	.34229	.00471	.09385	.01039	.01078	.46202
35	.00480	.01003	.01483	.05994	.07487	.09029	.22510	.01721	.34797	.00475	.09541	.01063	.01097	.46973
40	.00481	.00996	.01477	.06130	.07639	.09187	.22956	.01724	.35384	.00472	.09715	.01094	.01117	.47783
45	.00471	.00995	.01466	.06249	.07762	.09391	.23402	.01738	.35949	.00469	.09909	.01140	.01132	.48598
50	.00462	.00993	.01455	.06294	.07725	.09601	.23620	.01768	.36474	.00465	.10123	.01197	.01144	.49404
55	.00445	.00997	.01441	.06269	.07365	.09883	.23516	.01776	.36994	.00471	.10494	.01299	.01182	.50439
60	.00418	.00994	.01412	.06140	.06742	.10163	.23046	.01759	.37498	.00477	.10975	.01426	.01199	.51576
65	.00381	.01038	.01419	.05902	.05879	.10353	.22134	.01752	.38069	.00473	.11603	.01615	.01228	.52988
70	.00367	.01074	.01441	.05618	.05008	.10330	.20956	.01674	.38637	.00484	.12033	.01839	.01278	.54271
75	.00358	.01106	.01464	.05163	.03801	.10034	.18998	.01580	.39608	.00512	.12514	.02187	.01301	.56121
80	.00365	.01177	.01542	.04530	.02826	.09448	.16804	.01328	.40302	.00569	.12842	.02760	.01327	.57800
85	.00351	.01121	.01473	.04107	.01891	.08633	.14631	.01189	.41529	.00527	.13267	.03202	.01324	.59849

AGE	INFLU & PNEUMO (12)	BRONCH EM & AS (13)	OTHER RESP D (14)	RESPIR DISEASE	PEPTIC ULCER (15)	CIRRHO OF LIV (16)	NEPHRI NEPHRO (17)	CONGEN ANOMAL (18)	INFANT DISEASE (19)	MOTOR V ACCID (20)	OTHER ACCID (21)	SUI CIDE (22)	HOMI CIDE (23)	OTHER CAUSES (24)
0	.03272	.00777	.01918	.05966	.00262	.02180	.00945	.00466	.01480	.02300	.03527	.00768	.04055	.11210
1	.03246	.00795	.01936	.05976	.00268	.02242	.00970	.00177	.00002	.02360	.03562	.00791	.04160	.10940
5	.03238	.00792	.01933	.05963	.00269	.02251	.00972	.00148	.00000	.02316	.03454	.00794	.04145	.10885
10	.03238	.00793	.01933	.05964	.00269	.02257	.00975	.00136	.00000	.02257	.03380	.00796	.04142	.10879
15	.03241	.00794	.01936	.05971	.00270	.02263	.00977	.00126	.00000	.02217	.03291	.00796	.04133	.10864
20	.03253	.00796	.01944	.05993	.00271	.02277	.00982	.00117	.00000	.02095	.03179	.00773	.03959	.10870
25	.03283	.00804	.01965	.06052	.00274	.02303	.00991	.00109	.00000	.01864	.03031	.00670	.03454	.10878
30	.03311	.00813	.01991	.06116	.00277	.02302	.01001	.00102	.00000	.01635	.02870	.00550	.02812	.10825
35	.03332	.00824	.02016	.06172	.00278	.02211	.01005	.00093	.00000	.01438	.02707	.00452	.02229	.10731
40	.03356	.00832	.02037	.06224	.00283	.02034	.01010	.00090	.00000	.01261	.02498	.00367	.01726	.10567
45	.03368	.00848	.02054	.06271	.00281	.01780	.01016	.00084	.00000	.01101	.02254	.00311	.01313	.10385
50	.03394	.00861	.02103	.06359	.00280	.01513	.01020	.00077	.00000	.00976	.02068	.00251	.00995	.10214
55	.03448	.00867	.02133	.06447	.00282	.01201	.01045	.00074	.00000	.00849	.01892	.00209	.00715	.10112
60	.03572	.00878	.02175	.06624	.00276	.00882	.01052	.00069	.00000	.00749	.01769	.00188	.00524	.10073
65	.03762	.00838	.02151	.06751	.00272	.00610	.01083	.00063	.00000	.00645	.01620	.00160	.00373	.10129
70	.03968	.00818	.02052	.06838	.00268	.00413	.01118	.00054	.00000	.00579	.01592	.00134	.00271	.10390
75	.04337	.00727	.02012	.07076	.00258	.00277	.01114	.00055	.00000	.00529	.01539	.00118	.00193	.10676
80	.04865	.00603	.01950	.07418	.00233	.00196	.01069	.00050	.00000	.00360	.01517	.00098	.00125	.11460
85	.05201	.00567	.01797	.07566	.00189	.00216	.01027	.00041	.00000	.00297	.01648	.00068	.00068	.11740

BLACK FEMALE

AGE	TUBERC ULOSIS (1)	OTHER INFECT (2)	INFECT & PARA	MA NEO OF DIG (3)	MA NEO OF RES (4)	OTHER MA NEO (5)	MALIGN NEOPLA	DIABET MELLIT (6)	HEART DISEASE (7)	HYPERT ENSION (8)	CEREBRO VASCUL (9)	ARTERIO SCLERO (10)	OTHER ARTERI (11)	MAJOR CARDIOV
0	.00226	.01291	.01517	.05274	.01947	.09734	.16956	.03449	.37697	.00581	.13434	.01729	.01286	.54726
1	.00231	.01176	.01407	.05399	.01993	.09962	.17355	.03530	.38540	.00594	.13741	.01770	.01315	.55960
5	.00231	.01165	.01396	.05419	.02000	.09988	.17408	.03543	.38666	.00597	.13790	.01776	.01320	.56149
10	.00231	.01164	.01395	.05428	.02004	.09989	.17421	.03548	.38730	.00598	.13811	.01780	.01322	.56240
15	.00231	.01162	.01394	.05435	.02007	.09986	.17429	.03553	.38777	.00599	.13828	.01782	.01324	.56310
20	.00231	.01158	.01389	.05449	.02012	.09996	.17457	.03561	.38871	.00600	.13862	.01787	.01326	.56448
25	.00230	.01152	.01382	.05474	.02022	.10022	.17518	.03575	.39034	.00603	.13920	.01796	.01330	.56683
30	.00229	.01146	.01374	.05505	.02034	.10039	.17578	.03590	.39259	.00605	.13986	.01809	.01332	.56991
35	.00227	.01134	.01361	.05538	.02044	.10020	.17601	.03604	.39514	.00609	.14072	.01825	.01340	.57360
40	.00222	.01122	.01344	.05576	.02050	.09942	.17568	.03626	.39863	.00605	.14176	.01849	.01344	.57838
45	.00218	.01109	.01327	.05605	.02024	.09773	.17402	.03655	.40274	.00601	.14327	.01887	.01357	.58446
50	.00210	.01097	.01307	.05626	.01933	.09516	.17075	.03681	.40763	.00601	.14543	.01946	.01378	.59231
55	.00196	.01083	.01279	.05588	.01755	.09134	.16477	.03681	.41410	.00594	.14824	.02038	.01404	.60269
60	.00193	.01076	.01269	.05466	.01567	.08597	.15630	.03659	.42103	.00595	.15230	.02156	.01428	.61512
65	.00187	.01065	.01253	.05290	.01314	.07903	.14507	.03565	.42821	.00585	.15762	.02329	.01485	.62982
70	.00189	.01072	.01260	.05046	.01119	.07256	.13421	.03445	.43401	.00578	.16176	.02515	.01527	.64196
75	.00181	.01065	.01247	.04609	.00943	.06427	.11979	.03134	.44264	.00570	.16679	.02804	.01554	.65871
80	.00202	.01035	.01237	.04083	.00711	.05460	.10253	.02627	.45312	.00518	.17025	.03319	.01592	.67765
85	.00201	.01032	.01233	.03715	.00577	.04712	.09004	.02238	.45992	.00498	.17021	.03707	.01635	.68852

AGE	INFLU & PNEUMO (12)	BRONCH EM & AS (13)	OTHER RESP D (14)	RESPIR DISEASE	PEPTIC ULCER (15)	CIRRHO OF LIV (16)	NEPHRI NEPHRO (17)	CONGEN ANOMAL (18)	INFANT DISEASE (19)	MOTOR V ACCID (20)	OTHER ACCID (21)	SUI CIDE (22)	HOMI CIDE (23)	OTHER CAUSES (24)
0	.02679	.00377	.00856	.03912	.00194	.01241	.01117	.00408	.01180	.00721	.01798	.00217	.00925	.11640
1	.02643	.00383	.00851	.03877	.00197	.01270	.01140	.00148	.00002	.00727	.01781	.00222	.00933	.11452
5	.02633	.00382	.00850	.03865	.00197	.01274	.01143	.00107	.00000	.00678	.01689	.00223	.00907	.11423
10	.02633	.00382	.00849	.03863	.00197	.01276	.01146	.00092	.00000	.00643	.01650	.00223	.00896	.11410
15	.02631	.00381	.00848	.03860	.00197	.01278	.01147	.00085	.00000	.00622	.01623	.00221	.00882	.11400
20	.02629	.00380	.00844	.03852	.00197	.01280	.01148	.00076	.00000	.00578	.01598	.00215	.00829	.11371
25	.02630	.00378	.00842	.03849	.00197	.01281	.01149	.00069	.00000	.00516	.01567	.00195	.00707	.11311
30	.02626	.00374	.00840	.03841	.00199	.01266	.01149	.00062	.00000	.00471	.01524	.00168	.00581	.11206
35	.02621	.00369	.00835	.03825	.00200	.01209	.01148	.00056	.00000	.00428	.01495	.00142	.00478	.11093
40	.02618	.00367	.00829	.03814	.00201	.01117	.01150	.00051	.00000	.00395	.01449	.00115	.00379	.10954
45	.02624	.00360	.00813	.03797	.00202	.00984	.01156	.00047	.00000	.00346	.01408	.00101	.00304	.10825
50	.02636	.00350	.00808	.03793	.00205	.00801	.01152	.00041	.00000	.00302	.01371	.00087	.00263	.10690
55	.02678	.00336	.00783	.03797	.00206	.00620	.01139	.00037	.00000	.00262	.01344	.00065	.00218	.10605
60	.02739	.00319	.00755	.03813	.00206	.00465	.01096	.00033	.00000	.00213	.01324	.00053	.00184	.10544
65	.02841	.00276	.00724	.03842	.00203	.00306	.01072	.00028	.00000	.00168	.01303	.00042	.00155	.10574
70	.02985	.00275	.00703	.03962	.00210	.00208	.01042	.00021	.00000	.00145	.01287	.00038	.00131	.10632
75	.03201	.00276	.00671	.04147	.00202	.00125	.01010	.00022	.00000	.00124	.01280	.00028	.00107	.10725
80	.03583	.00312	.00642	.04537	.00192	.00046	.00954	.00017	.00000	.00093	.01327	.00017	.00096	.10840
85	.03820	.00297	.00629	.04747	.00227	.00035	.00857	.00009	.00000	.00061	.01355	.00009	.00096	.11277

TABLE 3 (CONTINUED)

OTHER RACES MALE

AGE	TUBERC ULOSIS (1)	OTHER INFECT (2)	INFECT & PARA	MA NEO OF DIG (3)	MA NEO OF RES (4)	OTHER MA NEO (5)	MALIGN NEOPLA	DIABET MELLIT (6)	HEART DISEASE (7)	HYPERT ENSION (8)	CEREBRO VASCUL (9)	ARTERIO SCLERO (10)	OTHER ARTERI (11)	MAJOR CARDIOV
0	.00599	.01038	.01637	.06557	.04587	.05859	.17003	.01844	.35535	.00352	.09309	.01506	.01079	.47780
1	.00607	.00978	.01585	.06642	.04646	.05925	.17212	.01868	.35964	.00356	.09421	.01525	.01091	.48358
5	.00609	.00964	.01573	.06662	.04660	.05932	.17254	.01873	.36070	.00357	.09451	.01530	.01094	.48503
10	.00610	.00963	.01573	.06675	.04669	.05920	.17263	.01877	.36130	.00358	.09468	.01533	.01096	.48585
15	.00611	.00960	.01571	.06689	.04679	.05913	.17281	.01881	.36200	.00359	.09480	.01536	.01099	.48674
20	.00614	.00957	.01571	.06741	.04715	.05916	.17372	.01896	.36466	.00362	.09547	.01548	.01107	.49030
25	.00620	.00968	.01588	.06809	.04766	.05965	.17539	.01916	.36837	.00365	.09647	.01565	.01117	.49531
30	.00621	.00966	.01587	.06862	.04812	.05995	.17669	.01936	.37184	.00369	.09737	.01581	.01128	.49999
35	.00627	.00967	.01594	.06918	.04854	.06022	.17793	.01946	.37483	.00373	.09828	.01596	.01139	.50419
40	.00632	.00954	.01586	.06946	.04905	.06059	.17910	.01962	.37875	.00379	.09924	.01621	.01152	.50950
45	.00640	.00952	.01591	.06983	.04923	.06079	.17985	.01969	.38290	.00386	.10043	.01654	.01165	.51538
50	.00638	.00947	.01585	.06991	.04933	.06066	.17990	.01971	.38541	.00390	.10188	.01691	.01185	.51995
55	.00630	.00950	.01580	.06871	.04844	.06013	.17728	.01986	.38863	.00373	.10421	.01752	.01191	.52600
60	.00632	.00921	.01553	.06731	.04731	.05954	.17416	.01981	.39112	.00377	.10646	.01830	.01183	.53148
65	.00634	.00939	.01573	.06376	.04538	.05974	.16888	.01850	.39478	.00378	.10906	.01931	.01158	.53850
70	.00639	.00919	.01558	.06080	.04078	.05926	.16084	.01754	.39965	.00362	.11168	.02081	.01128	.54704
75	.00643	.00957	.01600	.05672	.03472	.05850	.14995	.01612	.40313	.00395	.11303	.02325	.01124	.55460
80	.00615	.00976	.01592	.05092	.03135	.05596	.13823	.01498	.40774	.00423	.11485	.02666	.00999	.56348
85	.00641	.00962	.01603	.04701	.02671	.05235	.12607	.01282	.41774	.00534	.11752	.02991	.00855	.57906

AGE	INFLU & PNEUMO (12)	BRONCH EM & AS (13)	OTHER RESP D (14)	RESPIR DISEASE	PEPTIC ULCER (15)	CIRRHO OF LIV (16)	NEPHRI NEPHRO (17)	CONGEN ANOMAL (18)	INFANT DISEASE (19)	MOTOR V ACCID (20)	OTHER ACCID (21)	SUI CIDE (22)	HOMI CIDE (23)	OTHER CAUSES (24)
0	.05325	.01029	.02647	.09001	.00366	.01976	.00907	.00346	.00549	.03365	.03909	.01058	.01079	.09180
1	.05344	.01042	.02659	.09044	.00371	.02001	.00916	.00147	.00002	.03396	.03922	.01071	.01089	.09016
5	.05348	.01045	.02660	.09053	.00372	.02007	.00919	.00118	.00003	.03333	.03837	.01075	.01082	.08998
10	.05353	.01044	.02665	.09062	.00373	.02011	.00921	.00105	.00000	.03284	.03786	.01077	.01077	.09007
15	.05362	.01044	.02670	.09076	.00371	.02010	.00923	.00097	.00000	.03240	.03742	.01071	.01071	.08994
20	.05398	.01052	.02691	.09141	.00374	.02026	.00930	.00093	.00000	.02974	.03618	.00974	.00995	.09007
25	.05441	.01061	.02713	.09215	.00378	.02035	.00940	.00089	.00000	.02605	.03444	.00844	.00862	.09015
30	.05481	.01069	.02738	.09287	.00382	.02036	.00947	.00079	.00000	.02292	.03341	.00711	.00737	.08997
35	.05517	.01077	.02765	.09358	.00385	.01979	.00944	.00080	.00000	.02078	.03212	.00655	.00613	.08945
40	.05583	.01093	.02798	.09474	.00391	.01829	.00954	.00076	.00000	.01875	.03086	.00563	.00516	.08828
45	.05653	.01115	.02849	.09618	.00388	.01606	.00968	.00072	.00000	.01696	.02937	.00499	.00440	.08693
50	.05763	.01135	.02881	.09779	.00386	.01378	.00984	.00069	.00000	.01554	.02849	.00452	.00365	.08644
55	.05885	.01145	.02955	.09985	.00376	.01210	.01001	.00065	.00000	.01417	.02710	.00431	.00291	.08621
60	.06050	.01167	.03034	.10250	.00356	.01005	.01047	.00053	.00000	.01287	.02593	.00390	.00251	.08670
65	.06358	.01195	.03088	.10642	.00344	.00728	.01042	.00046	.00000	.01148	.02529	.00359	.00174	.08828
70	.06817	.01222	.03168	.11208	.00285	.00510	.01075	.00039	.00000	.00868	.02470	.00354	.00183	.08907
75	.07479	.01017	.03199	.11695	.00288	.00382	.01156	.00015	.00000	.00803	.02486	.00292	.00122	.09095
80	.08159	.00941	.03255	.12355	.00276	.00245	.01146	.00000	.00000	.00611	.02537	.00281	.00111	.09176
85	.08333	.00748	.03632	.12714	.00214	.00107	.01175	.00000	.00000	.00534	.02457	.00321	.00107	.08974

OTHER RACES FEMALE

AGE	TUBERC ULOSIS (1)	OTHER INFECT (2)	INFECT & PARA	MA NEO OF DIG (3)	MA NEO OF RES (4)	OTHER MA NEO (5)	MALIGN NEOPLA	DIABET MELLIT (6)	HEART DISEASE (7)	HYPERT ENSION (8)	CEREBRO VASCUL (9)	ARTERIO SCLERO (10)	OTHER ARTERI (11)	MAJOR CARDIOV
0	.00466	.01306	.01772	.06370	.02223	.07584	.16177	.03512	.34636	.00148	.14020	.01352	.01213	.51369
1	.00470	.01265	.01735	.06431	.02245	.07658	.16334	.03547	.34963	.00149	.14151	.01365	.01223	.51851
5	.00471	.01250	.01722	.06444	.02251	.07669	.16363	.03556	.35027	.00149	.14185	.01369	.01223	.51953
10	.00472	.01249	.01721	.06451	.02253	.07667	.16372	.03560	.35062	.00150	.14198	.01370	.01225	.52005
15	.00472	.01245	.01718	.06459	.02256	.07659	.16374	.03564	.35097	.00150	.14210	.01372	.01226	.52055
20	.00474	.01235	.01709	.06476	.02262	.07672	.16410	.03571	.35180	.00150	.14247	.01375	.01229	.52182
25	.00475	.01237	.01713	.06500	.02270	.07685	.16455	.03584	.35302	.00148	.14287	.01380	.01234	.52352
30	.00477	.01240	.01717	.06512	.02278	.07680	.16470	.03597	.35409	.00147	.14333	.01386	.01239	.52513
35	.00477	.01235	.01712	.06532	.02286	.07651	.16468	.03606	.35542	.00147	.14382	.01392	.01244	.52707
40	.00476	.01236	.01713	.06546	.02294	.07599	.16440	.03619	.35728	.00148	.14458	.01401	.01252	.52988
45	.00481	.01227	.01707	.06535	.02294	.07517	.16347	.03636	.35969	.00150	.14550	.01410	.01264	.53343
50	.00474	.01221	.01696	.06515	.02282	.07282	.16080	.03651	.36280	.00152	.14655	.01429	.01277	.53793
55	.00468	.01189	.01658	.06460	.02220	.06993	.15673	.03681	.36659	.00155	.14833	.01457	.01297	.54400
60	.00440	.01190	.01630	.06344	.02160	.06646	.15150	.03658	.37129	.00152	.14977	.01501	.01307	.55066
65	.00449	.01190	.01639	.06189	.02021	.06355	.14566	.03564	.37559	.00148	.15218	.01568	.01302	.55795
70	.00424	.01192	.01616	.05830	.01774	.05920	.13524	.03391	.38257	.00115	.15763	.01629	.01312	.57076
75	.00414	.01187	.01601	.05590	.01672	.05074	.12336	.03053	.38942	.00109	.16242	.01717	.01319	.58330
80	.00387	.01238	.01625	.05260	.01497	.04590	.11348	.02914	.39210	.00103	.16437	.01805	.01315	.58869
85	.00365	.01095	.01460	.04866	.01217	.04258	.10341	.02676	.39903	.00122	.16667	.01703	.01217	.59611

AGE	INFLU & PNEUMO (12)	BRONCH EM & AS (13)	OTHER RESP D (14)	RESPIR DISEASE	PEPTIC ULCER (15)	CIRRHO OF LIV (16)	NEPHRI NEPHRO (17)	CONGEN ANOMAL (18)	INFANT DISEASE (19)	MOTOR V ACCID (20)	OTHER ACCID (21)	SUI CIDE (22)	HOMI CIDE (23)	OTHER CAUSES (24)
0	.05943	.00581	.01464	.07988	.00243	.01239	.01042	.00363	.00386	.01564	.02325	.00772	.00374	.10875
1	.05962	.00587	.01459	.08008	.00245	.01247	.01050	.00110	.00000	.01567	.02331	.00779	.00375	.10820
5	.05959	.00589	.01458	.08005	.00246	.01247	.01052	.00092	.00000	.01521	.02277	.00781	.00374	.10811
10	.05958	.00589	.01459	.08006	.00246	.01249	.01053	.00090	.00000	.01479	.02257	.00782	.00364	.10815
15	.05962	.00590	.01461	.08013	.00246	.01250	.01055	.00087	.00000	.01459	.02237	.00777	.00353	.10811
20	.05975	.00589	.01465	.08029	.00247	.01254	.01055	.00087	.00000	.01348	.02196	.00764	.00336	.10814
25	.05985	.00586	.01470	.08041	.00246	.01251	.01059	.00083	.00000	.01237	.02160	.00716	.00308	.10797
30	.06002	.00586	.01473	.08062	.00247	.01247	.01063	.00079	.00000	.01134	.02150	.00687	.00268	.10767
35	.06017	.00589	.01477	.08083	.00248	.01204	.01067	.00074	.00000	.01046	.02132	.00662	.00249	.10742
40	.06050	.00593	.01480	.08123	.00249	.01142	.01063	.00071	.00000	.00957	.02110	.00641	.00221	.10664
45	.06094	.00585	.01477	.08156	.00247	.01017	.01069	.00059	.00000	.00881	.02083	.00617	.00198	.10640
50	.06150	.00585	.01493	.08227	.00246	.00927	.01071	.00055	.00000	.00832	.02050	.00595	.00188	.10589
55	.06209	.00591	.01516	.08316	.00251	.00768	.01056	.00046	.00000	.00798	.02024	.00566	.00181	.10582
60	.06339	.00595	.01491	.08425	.00245	.00622	.01045	.00048	.00000	.00723	.02029	.00541	.00187	.10633
65	.06558	.00600	.01462	.08620	.00223	.00457	.01049	.00028	.00000	.00681	.02012	.00511	.00163	.10691
70	.06914	.00586	.01410	.08911	.00211	.00283	.01011	.00016	.00000	.00539	.02063	.00476	.00131	.10753
75	.07291	.00609	.01273	.09173	.00180	.00242	.01023	.00000	.00000	.00523	.02159	.00522	.00109	.10749
80	.07647	.00516	.01264	.09427	.00181	.00181	.01057	.00000	.00000	.00490	.02240	.00540	.00103	.11025
85	.08151	.00487	.01095	.09732	.00122	.00122	.00973	.00000	.00000	.00487	.02433	.00608	.00122	.11314

TABLE 3 (CONTINUED)

ALL RACES MALE

AGE	TUBERC ULOSIS (1)	OTHER INFECT (2)	INFECT & PARA	MA NEO OF DIG (3)	MA NEO OF RES (4)	OTHER MA NEO (5)	MALIGN NEOPLA	DIABET MELLIT (6)	HEART DISEASE (7)	HYPERT ENSION (8)	CEREBRO VASCUL (9)	ARTERIO SCLERO (10)	OTHER ARTERI (11)	MAJOR CARDIOV
0	.00182	.00682	.00865	.05320	.06827	.08250	.20396	.01352	.39923	.00273	.07895	.01362	.01584	.51037
1	.00185	.00616	.00801	.05402	.06933	.08375	.20710	.01373	.40518	.00277	.08013	.01383	.01607	.51797
5	.00185	.00606	.00791	.05418	.06954	.08380	.20753	.01377	.40634	.00278	.08033	.01387	.01612	.51944
10	.00186	.00603	.00788	.05429	.06968	.08371	.20768	.01380	.40710	.00279	.08046	.01390	.01614	.52038
15	.00186	.00601	.00787	.05440	.06983	.08368	.20790	.01382	.40793	.00279	.08060	.01393	.01617	.52142
20	.00187	.00600	.00787	.05479	.07034	.08397	.20910	.01392	.41080	.00281	.08115	.01403	.01628	.52507
25	.00188	.00601	.00789	.05533	.07105	.08442	.21079	.01404	.41479	.00284	.08190	.01417	.01643	.53013
30	.00189	.00601	.00790	.05581	.07171	.08475	.21226	.01411	.41844	.00286	.08259	.01431	.01655	.53475
35	.00189	.00599	.00788	.05621	.07231	.08494	.21346	.01411	.42167	.00287	.08323	.01445	.01668	.53889
40	.00190	.00596	.00785	.05662	.07290	.08516	.21467	.01411	.42480	.00288	.08394	.01462	.01682	.54307
45	.00188	.00591	.00780	.05702	.07326	.08540	.21568	.01412	.42766	.00290	.08486	.01489	.01700	.54731
50	.00186	.00589	.00775	.05725	.07288	.08558	.21571	.01416	.42999	.00292	.08639	.01531	.01732	.55193
55	.00182	.00586	.00769	.05708	.07092	.08564	.21364	.01419	.43218	.00299	.08893	.01599	.01772	.55781
60	.00178	.00586	.00764	.05645	.06725	.08552	.20923	.01419	.43373	.00308	.09268	.01702	.01815	.56467
65	.00172	.00596	.00767	.05477	.06089	.08491	.20058	.01417	.43587	.00324	.09848	.01871	.01852	.57480
70	.00165	.00601	.00766	.05210	.05229	.08352	.18791	.01396	.43916	.00350	.10564	.02115	.01851	.58796
75	.00162	.00617	.00779	.04814	.04113	.08062	.16989	.01350	.44403	.00377	.11419	.02487	.01804	.60490
80	.00152	.00630	.00782	.04292	.02990	.07515	.14797	.01250	.45223	.00420	.12253	.03060	.01690	.62645
85	.00149	.00625	.00774	.03663	.01984	.06730	.12377	.01155	.46292	.00453	.12999	.03786	.01558	.65088

AGE	INFLU & PNEUMO (12)	BRONCH EM & AS (13)	OTHER RESP D (14)	RESPIR DISEASE	PEPTIC ULCER (15)	CIRRHO OF LIV (16)	NEPHRI NEFHRO (17)	CONGEN ANOMAL (18)	INFANT DISEASE (19)	MOTOR V ACCID (20)	OTHER ACCID (21)	SUI CIDE (22)	HOMI CIDE (23)	OTHER CAUSES (24)
0	.03274	.01512	.02884	.07670	.00318	.01637	.00476	.00427	.00746	.02393	.02712	.01451	.01012	.07506
1	.03274	.01533	.02910	.07717	.00322	.01661	.00481	.00164	.00002	.02423	.02721	.01474	.01023	.07330
5	.03273	.01536	.02911	.07720	.00323	.01666	.00482	.00129	.00000	.02385	.02642	.01478	.01016	.07295
10	.03275	.01538	.02914	.07727	.00324	.01669	.00482	.00119	.00000	.02337	.02592	.01481	.01012	.07283
15	.03278	.01540	.02919	.07737	.00324	.01672	.00483	.00110	.00000	.02282	.02534	.01478	.01006	.07272
20	.03295	.01550	.02937	.07782	.00326	.01684	.00486	.00101	.00000	.01966	.02422	.01424	.00947	.07268
25	.03319	.01564	.02963	.07846	.00329	.01698	.00489	.00094	.00000	.01606	.02272	.01300	.00815	.07263
30	.03339	.01577	.02986	.07903	.00331	.01701	.00491	.00088	.00000	.01362	.02142	.01175	.00672	.07235
35	.03357	.01590	.03009	.07956	.00332	.01682	.00492	.00082	.00000	.01183	.02023	.01069	.00552	.07193
40	.03379	.01606	.03037	.08023	.00334	.01629	.00493	.00078	.00000	.01035	.01897	.00970	.00438	.07135
45	.03408	.01630	.03078	.08116	.00334	.01532	.00494	.00071	.00000	.00904	.01779	.00881	.00335	.07062
50	.03463	.01661	.03138	.08262	.00335	.01391	.00497	.00065	.00000	.00794	.01672	.00793	.00255	.06980
55	.03555	.01705	.03216	.08476	.00336	.01209	.00502	.00059	.00000	.00698	.01574	.00708	.00188	.06918
60	.03702	.01746	.03300	.08748	.00337	.01009	.00512	.00051	.00000	.00616	.01498	.00631	.00142	.06884
65	.03950	.01761	.03363	.09074	.00337	.00776	.00527	.00045	.00000	.00545	.01444	.00551	.00103	.06875
70	.04315	.01714	.03314	.09342	.00337	.00551	.00545	.00039	.00000	.00497	.01440	.00489	.00075	.06932
75	.04847	.01591	.03151	.09589	.00340	.00382	.00560	.00038	.00000	.00443	.01504	.00423	.00056	.07057
80	.05612	.01346	.02797	.09754	.00330	.00245	.00561	.00034	.00000	.00365	.01617	.00349	.00044	.07227
85	.06365	.01051	.02407	.09823	.00322	.00172	.00549	.00029	.00000	.00293	.01773	.00279	.00038	.07328

ALL RACES FEMALE

AGE	TUBERC ULOSIS (1)	OTHER INFECT (2)	INFECT & PARA	MA NEO OF DIG (3)	MA NEO OF RES (4)	OTHER MA NEO (5)	MALIGN NEOPLA	DIABET MELLIT (6)	HEART DISEASE (7)	HYPERT ENSION (8)	CEREBRO VASCUL (9)	ARTERIO SCLERO (10)	OTHER ARTERI (11)	MAJOR CARDIOV
0	.00105	.00748	.00853	.05431	.02340	.10387	.18158	.02219	.41396	.00360	.13124	.02542	.01314	.58736
1	.00106	.00697	.00803	.05498	.02369	.10511	.18378	.02246	.41885	.00364	.13282	.02573	.01330	.59435
5	.00106	.00688	.00794	.05511	.02374	.10520	.18405	.02251	.41975	.00365	.13311	.02579	.01333	.59563
10	.00106	.00685	.00791	.05518	.02378	.10514	.18410	.02253	.42029	.00365	.13327	.02583	.01334	.59639
15	.00106	.00683	.00789	.05524	.02380	.10511	.18416	.02256	.42076	.00366	.13342	.02586	.01336	.59704
20	.00106	.00680	.00786	.05538	.02387	.10521	.18446	.02261	.42185	.00367	.13374	.02593	.01338	.59857
25	.00106	.00678	.00783	.05555	.02394	.10531	.18480	.02266	.42313	.00368	.13412	.02602	.01341	.60036
30	.00105	.00674	.00779	.05571	.02402	.10525	.18498	.02268	.42448	.00368	.13450	.02611	.01344	.60222
35	.00105	.00668	.00773	.05585	.02407	.10491	.18483	.02269	.42603	.00369	.13493	.02623	.01346	.60435
40	.00104	.00663	.00767	.05600	.02404	.10420	.18423	.02271	.42809	.00370	.13545	.02639	.01349	.60712
45	.00103	.00658	.00761	.05610	.02374	.10261	.18244	.02276	.43115	.00371	.13629	.02667	.01355	.61137
50	.00102	.00653	.00755	.05605	.02292	.09978	.17875	.02282	.43553	.00373	.13765	.02711	.01365	.61768
55	.00099	.00648	.00746	.05563	.02140	.09544	.17247	.02288	.44152	.00376	.13964	.02779	.01377	.62648
60	.00096	.00641	.00737	.05463	.01920	.08961	.16344	.02277	.44881	.00382	.14276	.02879	.01391	.63809
65	.00094	.00629	.00723	.05275	.01622	.08199	.15096	.02241	.45764	.00389	.14738	.03034	.01415	.65340
70	.00089	.00620	.00709	.04990	.01305	.07402	.13697	.02174	.46629	.00397	.15297	.03246	.01427	.66997
75	.00086	.00605	.00691	.04579	.01013	.06463	.12055	.02047	.47529	.00408	.15965	.03562	.01426	.68889
80	.00085	.00597	.00683	.04012	.00746	.05444	.10201	.01852	.48423	.00417	.16648	.04044	.01445	.70977
85	.00079	.00584	.00663	.03391	.00555	.04466	.08412	.01606	.49279	.00426	.16972	.04664	.01462	.72804

AGE	INFLU & PNEUMO (12)	BRONCH EM & AS (13)	OTHER RESP D (14)	RESPIR DISEASE	PEPTIC ULCER (15)	CIRRHO OF LIV (16)	NEPHRI NEFHRO (17)	CONGEN ANOMAL (18)	INFANT DISEASE (19)	MOTOR V ACCID (20)	OTHER ACCID (21)	SUI CIDE (22)	HOMI CIDE (23)	OTHER CAUSES (24)
0	.03580	.00680	.01365	.05625	.00271	.00887	.00448	.00397	.00576	.00972	.01780	.00526	.00306	.08248
1	.03581	.00687	.01367	.05636	.00274	.00897	.00452	.00157	.00001	.00975	.01775	.00532	.00305	.08135
5	.03578	.00688	.01364	.05631	.00274	.00899	.00453	.00126	.00000	.00940	.01722	.00533	.00296	.08112
10	.03579	.00688	.01364	.05632	.00274	.00900	.00453	.00116	.00000	.00908	.01698	.00534	.00292	.08100
15	.03579	.00688	.01365	.05633	.00275	.00901	.00453	.00109	.00000	.00879	.01678	.00533	.00285	.08089
20	.03585	.00689	.01366	.05640	.00275	.00903	.00453	.00101	.00000	.00762	.01656	.00519	.00263	.08078
25	.03590	.00690	.01368	.05648	.00276	.00904	.00453	.00096	.00000	.00662	.01629	.00488	.00227	.08052
30	.03596	.00690	.01369	.05655	.00277	.00900	.00453	.00091	.00000	.00598	.01604	.00450	.00193	.08012
35	.03602	.00691	.01371	.05664	.00277	.00888	.00451	.00085	.00000	.00546	.01581	.00411	.00165	.07971
40	.03611	.00692	.01374	.05677	.00278	.00859	.00450	.00081	.00000	.00501	.01556	.00367	.00139	.07917
45	.03631	.00691	.01375	.05697	.00278	.00807	.00451	.00076	.00000	.00457	.01530	.00316	.00113	.07857
50	.03662	.00687	.01377	.05727	.00278	.00726	.00449	.00070	.00000	.00417	.01506	.00265	.00095	.07788
55	.03719	.00679	.01370	.05768	.00279	.00620	.00445	.00064	.00000	.00377	.01492	.00214	.00079	.07730
60	.03802	.00658	.01344	.05804	.00281	.00515	.00438	.00057	.00000	.00336	.01486	.00167	.00068	.07679
65	.03939	.00611	.01296	.05846	.00282	.00394	.00428	.00049	.00000	.00296	.01487	.00135	.00060	.07624
70	.04133	.00558	.01213	.05903	.00281	.00296	.00416	.00038	.00000	.00252	.01508	.00103	.00052	.07573
75	.04421	.00485	.01088	.05994	.00278	.00212	.00398	.00034	.00000	.00204	.01553	.00078	.00043	.07524
80	.04825	.00409	.00956	.06190	.00273	.00139	.00378	.00026	.00000	.00149	.01625	.00051	.00034	.07423
85	.05335	.00353	.00878	.06567	.00262	.00094	.00356	.00020	.00000	.00089	.01691	.00038	.00026	.07372

TABLE 3 (CONTINUED)

ALL RACES TOTAL

AGE	TUBERC ULOSIS (1)	OTHER INFECT (2)	INFECT & PARA	MA NEO OF DIG (3)	MA NEO OF RES (4)	OTHER MA NEO (5)	MALIGN NEOPLA	DIABET MELLIT (6)	HEART DISEASE (7)	HYPERT ENSION (8)	CEREBRO VASCUL (9)	ARTERIO SCLERO (10)	OTHER ARTERI (11)	MAJOR CARDIOV
0	.00145	.00716	.00861	.05379	.04618	.09350	.19347	.01776	.40606	.00316	.10439	.01936	.01452	.54750
1	.00146	.00657	.00804	.05454	.04683	.09477	.19614	.01801	.41150	.00321	.10580	.01963	.01471	.55486
5	.00147	.00647	.00794	.05469	.04695	.09484	.19648	.01805	.41254	.00322	.10606	.01968	.01475	.55625
10	.00147	.00645	.00791	.05477	.04703	.09477	.19658	.01808	.41319	.00322	.10621	.01972	.01477	.55711
15	.00147	.00643	.00789	.05486	.04711	.09475	.19672	.01811	.41384	.00323	.10637	.01975	.01479	.55798
20	.00147	.00640	.00788	.05513	.04735	.09497	.19745	.01819	.41585	.00324	.10686	.01985	.01486	.56066
25	.00148	.00640	.00788	.05548	.04766	.09529	.19843	.01829	.41850	.00326	.10752	.01998	.01494	.56421
30	.00148	.00638	.00786	.05580	.04796	.09545	.19921	.01835	.42101	.00328	.10813	.02012	.01501	.56753
35	.00148	.00634	.00782	.05608	.04822	.09540	.19970	.01837	.42339	.00329	.10872	.02026	.01508	.57073
40	.00147	.00630	.00778	.05635	.04842	.09518	.19995	.01839	.42596	.00330	.10940	.02044	.01516	.57427
45	.00146	.00626	.00772	.05659	.04835	.09453	.19947	.01843	.42893	.00331	.11038	.02074	.01528	.57863
50	.00144	.00622	.00766	.05668	.04758	.09322	.19749	.01851	.43231	.00334	.11197	.02120	.01547	.58430
55	.00140	.00619	.00759	.05637	.04559	.09108	.19304	.01860	.43649	.00339	.11449	.02194	.01570	.59202
60	.00136	.00616	.00752	.05554	.04226	.08805	.18586	.01862	.44113	.00347	.11832	.02305	.01595	.60193
65	.00131	.00614	.00745	.05373	.03703	.08376	.17453	.01854	.44705	.00359	.12417	.02483	.01618	.61583
70	.00124	.00612	.00736	.05092	.03061	.07863	.16016	.01823	.45377	.00376	.13138	.02731	.01616	.63239
75	.00119	.00611	.00730	.04682	.02321	.07169	.14172	.01751	.46178	.00396	.14009	.03100	.01584	.65267
80	.00112	.00611	.00723	.04124	.01624	.06277	.12025	.01615	.47148	.00418	.14900	.03653	.01539	.67658
85	.00104	.00599	.00703	.03491	.01078	.05295	.09864	.01441	.48185	.00436	.15517	.04343	.01498	.69979

AGE	INFLU & PNEUMO (12)	BRONCH EM & AS (13)	OTHER RESP D (14)	RESPIR DISEASE	PEPTIC ULCER (15)	CIRRHO OF LIV (16)	NEPHRI NEPHRO (17)	CONGEN ANOMAL (18)	INFANT DISEASE (19)	MOTOR V ACCID (20)	OTHER ACCID (21)	SUI CIDE (22)	HOMI CIDE (23)	OTHER CAUSES (24)
0	.03440	.01110	.02153	.06703	.00295	.01263	.00465	.00412	.00663	.01689	.02248	.00993	.00660	.07874
1	.03441	.01124	.02166	.06731	.00299	.01280	.00470	.00161	.00001	.01705	.02249	.01007	.00664	.07729
5	.03439	.01125	.02165	.06729	.00300	.01283	.00470	.00128	.00000	.01667	.02183	.01010	.00656	.07701
10	.03440	.01127	.02166	.06733	.00300	.01285	.00471	.00118	.00000	.01627	.02146	.01012	.00652	.07689
15	.03442	.01127	.02169	.06738	.00300	.01287	.00471	.00110	.00000	.01585	.02106	.01009	.00645	.07678
20	.03454	.01132	.02177	.06762	.00302	.01293	.00472	.00101	.00000	.01365	.02038	.00974	.00603	.07671
25	.03470	.01137	.02188	.06795	.00303	.01299	.00474	.00095	.00000	.01133	.01948	.00895	.00518	.07657
30	.03483	.01143	.02198	.06824	.00305	.01298	.00474	.00089	.00000	.00978	.01870	.00813	.00429	.07624
35	.03495	.01149	.02208	.06852	.00305	.01281	.00474	.00084	.00000	.00863	.01799	.00740	.00355	.07584
40	.03511	.01156	.02221	.06889	.00306	.01239	.00474	.00079	.00000	.00767	.01724	.00668	.00285	.07529
45	.03536	.01166	.02239	.06940	.00307	.01164	.00475	.00074	.00000	.00680	.01653	.00597	.00222	.07464
50	.03580	.01176	.02264	.07020	.00307	.01051	.00475	.00068	.00000	.00604	.01587	.00527	.00173	.07391
55	.03655	.01187	.02288	.07131	.00308	.00905	.00476	.00061	.00000	.00536	.01532	.00457	.00132	.07336
60	.03771	.01187	.02300	.07258	.00308	.00750	.00476	.00054	.00000	.00472	.01493	.00392	.00103	.07300
65	.03961	.01154	.02276	.07391	.00308	.00570	.00477	.00047	.00000	.00414	.01468	.00331	.00079	.07278
70	.04231	.01081	.02168	.07480	.00307	.00410	.00476	.00039	.00000	.00364	.01480	.00278	.00063	.07290
75	.04617	.00955	.01970	.07543	.00304	.00284	.00469	.00036	.00000	.00307	.01534	.00225	.00049	.07331
80	.05145	.00778	.01685	.07608	.00296	.00181	.00451	.00029	.00000	.00235	.01624	.00169	.00038	.07348
85	.05712	.00608	.01438	.07759	.00284	.00122	.00426	.00024	.00000	.00163	.01721	.00126	.00030	.07356

BIOSTATISTICS: Statistics in Biomedical, Public Health
and Environmental Sciences, edited by P.K. Sen

PATTERNS AND TRENDS IN MORTALITY FROM DIFFERENT CAUSES IN U.S. WHITE MALES AND FEMALES OVER THE DECADE 1968-77

Regina C. Elandt-Johnson

Department of Biostatistics
University of North Carolina
Chapel Hill, NC 27514

Six types of disease groups as "causes" of death are analyzed by multiple decrement life table (MDLT) and truncated MDLT techniques searching for patterns and trends in mortality from these causes. The target populations are adult (aged ≥ 30) U.S. white males and females over the decade 1968-77. It is shown that over this period mortality from Diseases of Circulatory System, especially from Ischemic Heart Disease, steadily decreases, while from Malignant Neoplasms increases over this period. Mortality from a variety of remaining causes decrease, too, but exhibits some variation in specific age groups, presumably due to seasonal and secular variation of infectious diseases and external causes.

KEY WORDS: Multiple decrement life tables (MDLT); Truncated MDLT; Expected deaths; Proportionate distributions.

1. INTRODUCTION

In analysis of patterns and trends in mortality from specific causes, standardized rates are commonly employed as comparative indices. These are often directly age adjusted rates as used, for example, in studying the decline of Coronary Heart Disease (CHD) mortality (Proceedings (1979)) or morbidity of and mortality from cancer by Doll and Peto (1981). Analysis of rate functions from grouped data by X^2-ANOVA-like method, and by fitting Poisson multiplicative models are discussed by Elandt-Johnson (1982). Of course, to appreciate fully the meaning of a decline (or increase) from a given cause over a specified calendar period, it should be considered against the background of mortality patterns from all other causes acting simultaneously in the population. Therefore, it would be desirable to carry out age adjusted rate analyses or fitting multiplicative models *separately* for each of several other causes.

In this article, methods employing techniques of multiple decrement life tables (MDLT) in which several causes of death are analyzed *simultaneously* are presented. Proportionate distributions of life table deaths from different causes over different age ranges will be analyzed to:

(a) detect age intervals in which mortality from specified causes are most (least) frequent;

(b) investigate the critical ages at which the patterns of proportionate distributions of expected deaths from different causes change (shift analysis);

(c) analyze these patterns over a certain calendar period to detect changes (if they exist) in the patterns, as well as secular trends (if the exist) in mortality from specific causes.

The data analyzed in this paper are population mortality data for U.S. white males and females aged 30 and over, for the period 1968-77, obtained from *Vital Statistics* (1972-82). Six "causes" are studied. They are defined using ICDA, Eighth Revision (1968), as follows:

(1) Lung Cancer (162);

(2) Other Cancers (i.e., Malignant Neoplasms of organs other than lungs (140-209, except 162);

(3) Ischemic Heart Disease (IHD) (410-413);

(4) Other DCS (Diseases of Circulatory System other than IHD) (390-458, except 410-413);

(5) External Causes (E800-#999);

(6) Other Causes (all causes different than those listed in (1) through (5).

Special consideration is given to Lung Cancer (cause (1)), and IHD (cause (2)) mortality.

2. TRENDS IN MORTALITY FROM ALL CAUSES: LIFE TABLES

It is well known that general mortality (from all causes) in several developed countries and, in particular, in the United States, was decreasing over the period 1968-77.

Table 1.A. U.S. White Males. Survival Distributions

Age X_i	1968	1969	1970	1971	1972	1973	1974	1975	1976	1977
30	1.0000	1.00	1.00	1.00	1.00	1.00	1.00	1.00	1.00	1.00
35	.9909	.9908	.9908	.9909	.9911	.9909	.9913	.9916	.9918	.9918
45	.9575	.9572	.9576	.9586	.9596	.9594	.9615	.9626	.9639	.9645
55	.8735	.8752	.8759	.8790	.8802	.8820	.8858	.8896	.8930	.8952
65	.6928	.6978	.7001	.7068	.7066	.7114	.7211	.7291	.7346	.7411
75	.4122	.4207	.4273	.4357	.4324	.4394	.4519	.4637	.4712	.4804
85+	.1440	.1491	.1473	.1517	.1480	.1513	.1607	.1698	.1731	.1794

Table 1.B. U.S. White Females. Survival Distributions

Age X_i	1968	1969	1970	1971	1972	1973	1974	1975	1976	1977
30	1.0000	1.00	1.00	1.00	1.00	1.00	1.00	1.00	1.00	1.00
35	.9952	.9951	.9952	.9953	.9953	.9954	.9957	.9958	.9960	.9961
45	.9759	.9757	.9763	.9768	.9769	.9776	.9787	.9795	.9805	.9809
55	.9309	.9320	.9318	.9335	.9344	.9355	.9379	.9398	.9417	.9429
65	.8378	.9412	.8409	.8435	.8446	.8457	.8501	.8542	.8560	.8588
75	.6403	.6472	.6529	.6603	.6585	.6654	.6736	.6834	.6879	.6938
85	.3067	.3166	.3207	.3301	.3284	.3345	.3468	.3637	.3684	.3789

Tables 1A and 1B represent the survival distributions, $S_X(x)$, where x is age, for white males and females aged ≥ 30, respectively. ($S_X(x)$ corresponds to the ℓ_x column of the life table, with $\ell_0 = 1$.) These tables were calculated from cross-sectional population data grouped in 5 year intervals, by piecewise exponential fitting techniques. If μ_i is the age specific rate in the age group x_i to x_{i+1} with $x_{i+1} - x_i = h_i$ (= 5 years), then

$$S_X(x) = Pr\{X>x\} = \exp[-\sum_{j=0}^{i-1} \mu_j h_j - \mu_i(x-x_i)] \quad x_i \le x < x_{i+1} \tag{1}$$

(To save space, values $S_X(x)$ are given only at x_i = 30,40,...,80, ∞.)

Table 2.A. U.S. White Males. Life Expectancies

Age X_i	1968	1969	1970	1971	1972	1973	1974	1975	1976	1977
30	40.72	40.94	41.09	41.25	41.21	41.36	41.79	42.21	42.40	42.70
35	36.07	36.30	36.45	36.61	36.56	36.72	37.14	37.55	37.73	38.03
45	27.14	27.38	27.52	27.66	27.58	27.75	28.12	28.52	28.66	28.95
55	19.22	19.43	19.57	19.66	19.57	19.70	20.05	20.40	20.50	20.76
65	12.83	13.00	13.14	13.15	13.06	13.14	13.40	13.70	13.75	13.95
75	8.09	8.21	8.31	8.18	8.11	8.12	8.34	8.60	8.56	8.71
85	4.64	4.79	5.39	4.90	4.93	4.89	5.12	5.48	5.33	5.54

Table 2.B. U.S. White Females. Life Expectancies

Age X_i	1968	1969	1970	1971	1972	1973	1974	1975	1976	1977
30	47.04	47.28	47.75	47.90	47.87	48.03	48.45	49.04	49.13	49.50
35	42.26	42.50	42.97	43.11	43.09	43.24	43.65	44.23	44.32	44.68
45	32.98	33.24	33.69	33.83	33.79	33.93	34.31	34.88	34.93	35.29
55	24.31	24.54	25.04	25.14	25.08	25.21	25.57	26.12	26.14	26.49
65	16.41	16.60	17.16	17.25	17.18	17.31	17.65	18.19	18.21	18.55
75	9.81	9.95	10.55	10.53	10.50	10.54	10.84	11.37	11.33	11.67
85	5.00	5.05	6.26	6.01	6.00	5.99	6.31	6.90	6.75	7.12

Tables 2A and 2B give the corresponding life expectancies. It is clear from these tables that mortality steadily decreases, especially for age ranges 55-75. Life expectancy at age 30, over this decade, increased by about 2 years for males and about 2.5 years for females. Having this background, we now analyze mortality functions from the six causes mentioned above.

3. MULTIPLE DECREMENT LIFE TABLE TECHNIQUES

Let D_{ki} denote the number of observed deaths in age interval $[x_i, x_{i+1})$, i = 0,1, ...,I, from the kth cause, k = 1,2,...,K; $D_{.i} = \sum_{k=1}^{K} D_{ki}$ - the number of all deaths and N_i - the midyear population size in age interval $[x_i, x_{i+1})$. Having these data, multiple decrement life tables (MDLT) can be computed using actuarial or piecewise exponential fitting methods. Details of such methods can be found in various actuarial textbooks or in books or articles on survival analyses. In this paper, piecewise exponential fitting is used as described in Elandt-Johnson (1984).

Let d_{ki} and $d_{\cdot i}$ be the life table deaths in age interval $[x_i, x_{i+1})$ from the kth cause, and for all causes, respectively. In this setting, these are, in fact, the corresponding *expected* numbers of deaths in a cohort of ℓ_{30} individuals who started their life experience at age 30.

Let $d_{k\cdot} = \sum_{i=0}^{I} d_{ki} = \sum_{i=0}^{\infty} d_{ki}$ be the total number of deaths from the dth cause which are expected to occur over the whole life span of this cohort, *in the presence* of all other causes. Similarly, $d_{\cdot\cdot} = \sum_{k=1}^{K} d_{k\cdot} = \sum_{i=0}^{\infty} d_{\cdot i}$, is the total number of expected deaths. Then

$$\pi_{k0} = d_{k\cdot}/d_{\cdot\cdot} = \Big(\sum_{i=0}^{\infty} d_{ji}\Big)/\Big(\sum_{i=0}^{\infty} d_{\cdot i}\Big), \quad (k=1,2,\ldots,K) \tag{2}$$

represents the (ultimate) distribution of deaths from different causes over the whole life span for this cohort.

The corresponding *observed* proportionate distribution in the population is

$$D_{k\cdot}/D_{\cdot\cdot} = \Big(\sum_{i=0}^{\infty} D_{ki}\Big)/\Big(\sum_{i=0}^{\infty} D_{\cdot i}\Big), \quad k=1,2,\ldots,K. \tag{3}$$

Of course, the observed and expected distributions do not need to be the same, but for large population data, they often are similar, as in the case for white males (Table 3A). But for white females they are different (Table 3B).

Analyzing Table 3A over the decade 1968-77, we notice that in white males, mortality from cancers (cause (1) and (2)) increases, while mortality from DCS (causes (3) and (4)), especially from CHD (cause (3)), decreases. In white females (Table 3A), cancer mortality increases, too, but lung cancer is much less frequent in females (1-2%) than in males (5-6.5%). On the other hand, mortality from DCS, though it decreases, too, and it is generally lower (53-55%) than in males (56-59%), is much higher in the DCS-CHD group, in females (25.5-29%) than in males (17-18%).

Similar proportionate distributions can be studied at different ages, x_i.

4. TRUNCATED MDLT: SHIFT ANALYSIS

The expected ratios, $d_{k\cdot}/d_{\cdot\cdot}$, are indices which are rather sensitive to mortality data in the last age intervals. Since these data are often sparse, these ratios might sometimes be misleading. Also, it would be of interest to analyze propor-

Table 3.A. U.S. White Males. Observed and Expected Distributions of Deaths from Different Causes

Year		(1) Lung Cancer	(2) Other Cancers	(3) IHD	(4) Other DCS	(5) External Causes	(6) Other Causes	All Deaths	Midyear Population
	$D_{k\cdot}$	43968	107017	358080	149194	38247	174225	870731	40,250,000
1968	$D_{k\cdot}/D_{\cdot\cdot}$	.0505	.1229	.4112	.1713	.0439	.2001		
	$d_{k\cdot}/d_{\cdot\cdot}$	.0478	.1210	.4139	.1797	.0408	.1967		
	$D_{k\cdot}$	45261	107585	354489	146655	38483	171082	863666	40,514,000
1969	$D_{k\cdot}/D_{\cdot\cdot}$	.0524	.1246	.4105	.1698	.0446	.1981		
	$d_{k\cdot}/d_{\cdot\cdot}$	.0493	.1226	.4139	.1789	.0412	.1942		
	$D_{k\cdot}$	47323	108978	350699	145645	37985	170462	861092	40,597,954
1970	$D_{k\cdot}/D_{\cdot\cdot}$	.0550	.1266	.4073	.1691	.0441	.1980		
	$d_{k\cdot}/d_{\cdot\cdot}$	.0515	.1246	.4111	.1784	.0407	.1938		
	$D_{k\cdot}$	49269	110432	352446	146208	36729	167973	863057	41,053,000
1971	$D_{k\cdot}/D_{\cdot\cdot}$	.0571	.1280	.4084	.1694	.0426	.1946		
	$d_{k\cdot}/d_{\cdot\cdot}$	.0536	.1259	.4121	.1786	.0392	.1905		
	$D_{k\cdot}$	51490	112136	355206	149476	36800	174800	879914	41,453,000
1972	$D_{k\cdot}/D$	.0585	.1274	.4037	.1699	.0418	.1987		
	$d_{k\cdot}/d_{\cdot\cdot}$	.0552	.1257	.4071	.1783	.0387	.1950		
	$D_{k\cdot}$	52671	112863	353851	148185	36641	176347	880558	42,002,000
1973	$D_{k\cdot}/D$	.0598	.1281	.4018	.1683	.0416	.2003		
	$d_{k\cdot}/d_{\cdot\cdot}$	.0564	.1265	.4055	.1768	.0382	.1966		
	$D_{k\cdot}$	54716	115679	341658	145171	33171	171737	862132	42,454,000
1974	$D_{k\cdot}/D_{\cdot\cdot}$	.0635	.1342	.3963	.1684	.0385	.1992		
	$d_{k\cdot}/d_{\cdot\cdot}$	.0592	.1319	.4006	.1783	.0349	.1950		
	$D_{k\cdot}$	56292	116177	330800	137327	32484	171976	845056	42,876,000
1975	$D_{k\cdot}/D_{\cdot\cdot}$	.0666	.1375	.3915	.1625	.0384	.2035		
	$d_{k\cdot}/d_{\cdot\cdot}$	.0615	.1350	.3962	.1735	.0346	.1993		
	$D_{k\cdot}$	59179	118259	329226	136379	31333	173998	848374	43,314,000
1976	$D_{k\cdot}/D_{\cdot\cdot}$	.0698	.1394	.3881	.1608	.0369	.2051		
	$d_{k\cdot}/d_{\cdot\cdot}$	.0650	.1364	.3927	.1707	.0332	.2021		
	$D_{k\cdot}$	60428	120986	323503	134123	31946	170089	841075	44,245,000
1977	$D_{k\cdot}/D$	.0718	.1438	.3846	.1595	.0380	.2022		
	$d_{k\cdot}/d_{\cdot\cdot}$	.0662	.1409	.3904	.1705	.0335	.1985		

tionate distributions over specified age ranges, to detect age intervals in which the proportionate distributions may change (shift) their patterns, for example, some causes may become more (or less) frequent than in earlier age groups. This leads to construction of *truncated* MDLT's over restricted age ranges, in which the

Table 3.B. U.S. White Females. Observed and Expected Distributions of Deaths from Different Causes

Year		(1) Lung Cancer	(2) Other Cancers	(3) IHD	(4) Other DCS	(5) External Causes	(6) Other Causes	All Deaths	Midyear Population
1968	$D_{k\cdot}$	9359	117900	256635	164841	21433	122301	692649	44,765,000
	$D_{k\cdot}/D_{\cdot\cdot}$	.0138	.1702	.3705	.2380	.0309	.1766		
	$d_{k\cdot}/d_{\cdot\cdot}$	.0112	.1470	.3906	.2556	.0291	.1665		
1969	$D_{k\cdot}$	10261	119256	256511	162852	21455	121870	692205	45,214,000
	$D_{k\cdot}/D_{\cdot\cdot}$	.0148	.1723	.3706	.2353	.0310	.1761		
	$d_{k\cdot}/d_{\cdot\cdot}$	.0119	.1484	.3923	.2533	.0287	.1654		
1970	$D_{k\cdot}$	11167	121759	256972	162574	20772	120493	693737	45,585,237
	$D_{k\cdot}/D_{\cdot\cdot}$	.0161	.1755	.3704	.2343	.0299	.1737		
	$d_{k\cdot}/d_{\cdot\cdot}$	.0128	.1507	.3932	.2536	.0276	.1621		
1971	$D_{k\cdot}$	12335	122784	262962	164316	20796	120128	703321	46,291,000
	$D_{k\cdot}/D_{\cdot\cdot}$	.0175	.1746	.3739	.2336	.0296	.1708		
	$d_{k\cdot}/d_{\cdot\cdot}$	.0140	.1503	.3965	.2522	.0273	.1598		
1972	$D_{k\cdot}$	13596	125050	269048	168666	21192	125058	722610	46,798,000
	$D_{k\cdot}/D_{\cdot\cdot}$	.0188	.1731	.3723	.2334	.0293	.1731		
	$d_{k\cdot}/d_{\cdot\cdot}$	.0150	.1503	.3942	.2508	.0270	.1627		
1973	$D_{k\cdot}$	14206	126231	268822	172418	20920	128027	730624	47,467,000
	$D_{k\cdot}/D_{\cdot\cdot}$	.0194	.1728	.3679	.2360	.0286	.1752		
	$d_{k\cdot}/d_{\cdot\cdot}$	.0156	.1500	.3891	.2534	.0265	.1654		
1974	$D_{k\cdot}$	15615	128687	264018	169933	18670	125935	722858	48,048,000
	$D_{k\cdot}/D_{\cdot\cdot}$	.0216	.1780	.3652	.2351	.0258	.1742		
	$d_{k\cdot}/d_{\cdot\cdot}$	.0172	.1537	.3879	.2534	.0241	.1638		
1975	$D_{k\cdot}$	16892	129864	254933	161260	18050	125057	706056	48,593,000
	$D_{k\cdot}/d_{\cdot\cdot}$	.0239	.1839	.3611	.2284	.0256	.1771		
	$d_{k\cdot}/d_{\cdot\cdot}$	.0186	.1577	.3854	.2477	.0236	.1670		
1976	$D_{k\cdot}$	18598	133188	259117	161987	17760	130270	720920	49,123,000
	$D_{k\cdot}/D_{\cdot\cdot}$	.0258	.1847	.3594	.2247	.0246	.1807		
	$d_{k\cdot}/d_{\cdot\cdot}$	.0203	.1592	.3827	.2423	.0225	.1728		
1977	$D_{k\cdot}$	19900	135549	256466	159417	18055	126997	716384	50,143,000
	$D_{k\cdot}/D_{\cdot\cdot}$	.0278	.1892	.3580	.2225	.0252	.1773		
	$d_{k\cdot}/d_{\cdot\cdot}$	.0217	.1624	.3830	.2415	.0228	.1687		

mortality from different causes will be analyzed (Elandt-Johnson (1984)).

Let x_i be the exact age of a cohort of survivors whose mortality experience is under consideration until the cohort reaches exace age x_a, that is, over the interval $[x_i,x_a)$. The proportionate distribution of expected deaths from different causes over this age range is

$$\pi'_{ki;a} = (\sum_{j=i}^{a-1} d_{kj})/(\sum_{j=1}^{a-1} d_{\cdot j}) \ . \quad (4)$$

We may fix age x_a, and analyze the behavior of distributions (4) for different age (x_i) cohorts. We also may fix x_i and consider different truncation ages x_a. Finally, we may make both x_i and x_a flexible; in particular, if $x_a = x_{i+1}$, then we consider the proportionate distributions of expected deaths in the grouping intervals $[x_i, x_{i+1}]$, and we have

$$\pi_{ki;i+1} = d_{ki}/d_{\cdot i} = D_{ki}/D_{\cdot i} \ . \quad (5)$$

In this paper, we present the analysis for the cohort aged 30 ($x_i = x_0 = 30$ fixed) using different truncation ages, x_a = 45, 55, 65, 75 and ∞ (Tables 4A and 4B). (Note that for $x_a = \infty$, $\pi'_{ki;\infty} = \pi_{ki}$ and this is already given in Tables 3A and 3B; here it is repeated for comparison.)

For a *given calendar year,* the mortality from cardiovascular diseases (IHD and other DCS) increases, while from External and Other Causes it decreases with age. Of special interest is the pattern in Malignant Neoplasms (cancers). In males, deaths from Lung Cancer increase with age up to age about 75, and then decrease. For females, the same phenomenon is observed but with corresponding threshold ages 60 and 55 (or 60), probably because of the larger frequency of breast cancer in females. More detailed analysis of this phenomenon, for ages 50 to 70, is presented in Tables 5A and 5B.

Analyzing the secular trends over calendar time, we observe that mortality from Disease of Circulatory System, and in particular from IHD, steadily decreases and from cancers it increases, at *each given age.*

Mortality from External and Other Causes also decrease, but the patterns are not so clear, and vary from age to age. This is possibly caused by different intensities of epidemics and environmental conditions occurring in different calendar years together, of course, with random variation.

4. CONCLUSION

4.1. Over the decade 1968-77, general mortality in the U.S. white population decreased. There were increases in life expectancy of about 2 years for males and about 2.5 years for females.

4.2. For a *given calendar year,* mortality from Cancers increases until age 70 (or 75) in males, and 60 (or even 55) in females; from Diseases of Circulatory System, it decreases steadily with age; and from all other causes (including accidents)

Table 4.A. U.S. White Males. Truncated MDLT (x_0=30, fixed)

Year	Age x_a	(1) Lung Cancer	(2) Other Cancers	(3) IHD	(4) Other DCS	(5) External Causes	(6) Other Causes
	45	.0336	.1055	.2127	.0836	.2364	.3280
	55	.0560	.1164	.3265	.0956	.1342	.2714
1968	65	.0694	.1277	.3817	.1089	.0790	.2333
	75	.0661	.1318	.4030	.1339	.0530	.2122
	∞	.0478	.1210	.4139	.1797	.0408	.1967
	45	.0351	.1026	.2064	.0822	.2447	.3290
	55	.0574	.1153	.3227	.0931	.1390	.2725
1969	65	.0717	.1286	.3782	.1081	.0814	.2319
	75	.0684	.1330	.4010	.1337	.0541	.2098
	∞	.0493	.1226	.4139	.1789	.0412	.1942
	45	.0355	.1006	.2012	.0826	.2412	.3389
	55	.0594	.1155	.3170	.0932	.1373	.2775
1970	65	.0743	.1307	.3721	.1080	.0807	.2343
	75	.0710	.1350	.3958	.1330	.0540	.2113
	∞	.0515	.1246	.4111	.1784	.0407	.1938
	45	.0349	.1042	.2045	.0815	.2381	.3367
	55	.0605	.1170	.3193	.0936	.1343	.2753
1971	65	.0753	.1317	.3751	.1081	.0787	.2310
	75	.0736	.1369	.3971	.1328	.0524	.2073
	∞	.0536	.1259	.4121	.1786	.0392	.1905
	45	.0353	.1009	.1989	.0803	.2333	.3513
	55	.0631	.1133	.3154	.0925	.1325	.2833
1972	65	.0781	.1307	.3706	.1098	.0772	.2336
	75	.0756	.1364	.3926	.1344	.0509	.2101
	∞	.0552	.1257	.4071	.1783	.0387	.1950
	45	.0350	.1032	.1937	.0785	.2410	.3487
	55	.0633	.1174	.3108	.0913	.1359	.2813
1973	65	.0797	.1319	.3691	.1065	.0784	.2344
	75	.0776	.1371	.3901	.1320	.0512	.2119
	∞	.0564	.1265	.4055	.1768	.0382	.1967
	45	.0363	.1034	.1904	.0779	.2279	.3641
	55	.0679	.1231	.3077	.0895	.1264	.2852
1974	65	.0852	.1391	.3633	.1059	.0725	.2341
	75	.0817	.1440	.3847	.1320	.0471	.2105
	∞	.0592	.1319	.4006	.1783	.0349	.1950
	45	.0356	.1046	.1897	.0736	.2270	.3695
	55	.0700	.1237	.3047	.0867	.1266	.2884
1975	65	.0876	.1396	.3603	.1030	.0721	.2371
	75	.0856	.1465	.3800	.1276	.0468	.2135
	∞	.0615	.1350	.3962	.1735	.0346	.1993
	45	.0373	.1050	.1912	.0769	.2255	.3641
	55	.0724	.1271	.3029	.0899	.1239	.2837
1976	65	.0910	.1443	.3566	.1048	.0704	.2331
	75	.0888	.1503	.3765	.1277	.0452	.2116
	∞	.0650	.1364	.3927	.1707	.0332	.2021
	45	.0356	.1078	.1824	.0772	.2323	.3647
	55	.0740	.1290	.2974	.0888	.1284	.2825
1977	65	.0936	.1488	.3506	.1032	.0728	.2309
	75	.0922	.1543	.3718	.1257	.0464	.2096
	∞	.0662	.1409	.3904	.1705	.0335	.1985

Table 4.B. U.S. White Females. Truncated MDLT (x_0=30, fixed)

Year	Age x_a	(1) Lung Cancer	(2) Other Cancers	(3) IHD	(4) Other DCS	(5) External Causes	(6) Other Causes
	45	.0239	.2851	.0725	.1320	.1189	.3676
	55	.0327	.3213	.1266	.1392	.0766	.3036
1968	65	.0304	.3035	.2131	.1494	.0513	.2522
	75	.0221	.2458	.3103	.1791	.0358	.2069
	∞	.0112	.1470	.3906	.2556	.0291	.1665
	45	.0244	.2781	.0700	.1268	.1261	.3747
	55	.0349	.3166	.1238	.1333	.0816	.3098
1969	65	.0340	.3049	.2065	.1452	.0545	.2549
	75	.0245	.2481	.3051	.1770	.0371	.2081
	∞	.0119	.1484	.3923	.2533	.0287	.1654
	45	.0256	.2765	.0686	.1271	.1229	.3793
	55	.0383	.3166	.1229	.1331	.0801	.3090
1970	65	.0373	.3050	.2063	.1432	.0529	.2555
	75	.0264	.2515	.3030	.1746	.0363	.2082
	∞	.0128	.1507	.3932	.2536	.0276	.1621
	45	.0269	.2744	.0710	.1274	.1253	.3750
	55	.0416	.3143	.1254	.1322	.0796	.3068
1971	65	.0403	.3051	.2082	.1429	.0528	.2506
	75	.0294	.2536	.3020	.1725	.0365	.2060
	∞	.0140	.1503	.3965	.2522	.0273	.1598
	45	.0302	.2727	.0702	.1251	.1264	.3753
	55	.0461	.3133	.1202	.1320	.0823	.3061
1972	65	.0449	.3038	.2016	.1421	.0541	.2534
	75	.0324	.2530	.2977	.1715	.0368	.2085
	∞	.0150	.1503	.3942	.2508	.0270	.1627
	45	.0290	.2707	.0704	.1250	.1242	.3806
	55	.0449	.3143	.1219	.1307	.0788	.3094
1973	65	.0463	.3064	.2004	.1407	.0522	.2540
	75	.0339	.2566	.2915	.1712	.0361	.2107
	∞	.0156	.1500	.3891	.2534	.0265	.1654
	45	.0314	.2842	.0676	.1245	.1120	.3803
	55	.0487	.3252	.1183	.1283	.0703	.3093
1974	65	.0516	.3162	.1958	.1378	.0455	.2531
	75	.0381	.2641	.2863	.1692	.0315	.2108
	∞	.0172	.1537	.3879	.2534	.0411	.0241
	45	.0350	.2832	.0678	.1193	.1181	.3766
	55	.0541	.3274	.1191	.1242	.0723	.3029
1975	65	.0572	.3186	.1934	.1344	.0458	.2505
	75	.0423	.2696	.2810	.1646	.0318	.2106
	∞	.0186	.1577	.3854	.2477	.0236	.1670
	45	.0358	.2935	.0657	.1185	.1173	.3693
	55	.0581	.3328	.1174	.1227	.0719	.2970
1976	65	.0614	.3238	.1914	.1322	.0457	.2455
	75	.0465	.2759	.2756	.1609	.0314	.2096
	∞	.0203	.1592	.3827	.2423	.0226	.1728
	45	.0392	.2878	.0646	.1178	.1268	.3638
	55	.0628	.3333	.1157	.1207	.0763	.2912
1977	65	.0659	.3281	.1879	.1298	.0475	.2408
	75	.0503	.2805	.2711	.1579	.0325	.2077
	∞	.0217	.1624	.3830	.2415	.0228	.1087

it also decreases with age.

Table 5.A. White Males. Ratios $d_{ki}/d_{\cdot i}$ for Cancers

Age	1968		1969		1970		1971		1972	
	(1)*	(2)**	(1)	(2)	(1)	(2)	(1)	(2)	(1)	(2)
50-55	.0721	.1261	.0741	.1275	.0767	.1285	.0774	.1295	.0812	.1248
55-60	.0785	.1331	.0812	.1370	.0857	.1383	.0862	.1407	.0892	.1441
60-65	.0791	.1375	.0822	.1387	.0841	.1437	.0851	.1431	.0880	.1418
65-70	.0706	.1384	.0735	.1403	.0758	.1415	.0815	.1449	.0816	.1443
85+	.0087	.0800	.0096	.0833	.0108	.0848	.0120	.0860	.0116	.0854

Table 5.A. (Continued)

Age	1973		1974		1975		1976		1977	
	(1)	(2)	(1)	(2)	(1)	(2)	(1)	(2)	(1)	(2)
50-55	.0824	.1293	.0893	.1386	.0933	.1383	.0961	.1431	.1004	.1466
55-60	.0899	.1398	.0960	.1465	.0981	.1484	.1035	.1522	.1058	.1584
60-65	.0918	.1436	.0980	.1528	.1010	.1526	.1035	.1583	.1078	.1648
65-70	.0844	.1450	.0872	.1510	.0923	.1548	.0951	.1591	.0998	.1617
85+	.0126	.0883	.0137	.0903	.0150	.0971	.0228	.0897	.0174	.1024

Table 5.B. White Females. Ratios $d_{ki}/d_{\cdot i}$ for Cancers

Age	1968		1969		1970		1971		1972	
	(1)	(2)	(1)	(2)	(1)	(2)	(1)	(2)	(1)	(2)
50-55	.0394	.3411	.0413	.3389	.0470	.3370	.0512	.3359	.0595	.3344
55-60	.0329	.3135	.0395	.3188	.0422	.3212	.0454	.3213	.0521	.3205
60-65	.0257	.2742	.0292	.2803	.0324	.2783	.0350	.2821	.0384	.2801
65-70	.0191	.2238	.0208	.2320	.0220	.2356	.0255	.2416	.0276	.2400
85+	.0029	.0635	.0029	.0653	.0032	.0673	.0036	.0686	.0035	.0688

Table 5.B. (Continued)

Age	1973		1974		1975		1976		1977	
	(1)	(2)	(1)	(2)	(1)	(2)	(1)	(2)	(1)	(2)
50-55	.0534	.3404	.0579	.3481	.0648	.3517	.0714	.3574	.0771	.3570
55-60	.0525	.3216	.0604	.3333	.0658	.3320	.0691	.3387	.0734	.3480
60-65	.0434	.2856	.0489	.2936	.0550	.2987	.0600	.3034	.0643	.3090
65-70	.0303	.2459	.0346	.2488	.0387	.2536	.0438	.2630	.0472	.2706
85+	.0036	.0668	.0041	.0694	.0046	.0758	.0048	.0757	.0051	.0790

*(1) - Lung Cancer
**(2) - Other Cancers

4.3. Over the calendar period (1968-77), and (roughly) for each *given age group*, there is an increasing trend in mortality (in particular, in CHD). Mortality from Other Causes (including External) decreases, too, over this period, but exhibits variation in some age groups, especially in younger ages. Of course, this category includes a rather large number of diseases of different origins; some of them may be subject to quite substantial seasonal and periodical variation.

ACKNOWLEDGEMENTS: I thank Mr. G. Samsa for preparing the data for the use by the the computer, and to Ms. Y. Tomabechi for writing an appropriate program and computing tables for this paper.

REFERENCES

[1] Doll, R. and Peto, R., The causes of cancer: Quantitative estimates of avoidable risks of cancer in the United States today, J. Nat. Cancer Inst. **66** (6):1193-1308 (1981).

[2] Elandt-Johnson, Regina C., Crude, direct and indirect standardized rates: which (if any) should be used?, Inst. of Statist. Mimeo Series No. 1414, pp. 1-34 (1982).

[3] Elandt-Johnson, Regina C., Analysis of distributional patterns of deaths from different causes, Scand. Actu. J., March (1984).

[4] International Classification of Diseases Adapted (ICDA) for use in the United States, Eighth Revision. U.S. DHEW, Public Health Service Publication No. 1693 (1968).

[5] Proceedings of the Conference on the Decline in Coronary Heart Disease Mortality, Havlik, R.J. and Feinlieb, M. (eds.), U.S. DHEW Public Health Service, NIH Publication No. 79-1610 (1979).

[6] Vital Statistics of the United States, 1968-77, Vol. II. Mortality. Part A. DHEW Publications (1972-1982).

BIOSTATISTICS: Statistics in Biomedical, Public Health
and Environmental Sciences, edited by P.K. Sen

SOME STATISTICAL ASPECTS IN EVALUATING FAMILY PLANNING PROGRAMS

Azza R. Karmous

Department of Biostatistics
University of North Carolina at Chapel Hill

Sara H. Salama

Department of Geography
University of North Carolina at Chapel Hill

During the last twenty years, various findings have been published from international research on family planning program. This research has sought to determine program effects on fertility and to measure its demographic impact.

Our goal is to review and criticize several of the statistical methods that are used in evaluating some of these family planning programs. This includes a review of some matching studies that used areas or individuals as units of study. Examination is also made of other studies which were based on acceptor data, the couple-year of protection technique, and births-averted analysis which views human reproduction as a renewal process. Finally, some of the regression analyses and life table techniques that have been employed in program evaluation are reviewed. A discussion of the advantages and the limitations of each approach is presented along with very brief results.

KEY WORDS: Family planning program, evaluation, couple-year of protection, randomization, matching, Pearl formula, life table technique, births-averted analysis, regression analysis

1. INTRODUCTION

In most instances the term "family planning" refers to the conscious effort on the part of couples or individuals to regulate fertility by delaying or spacing births, thereby limiting their numbers. In other words, it implies taking direct action to avoid conception or a live birth. Throughout the course of human history, family planning has been practiced in one form or another. It was only in the early 1960s, however, that family planning started to gain appreciable acceptance as a policy instrument due to the growing awareness of the dampening effect of rapid population growth on economic development and the availability of new contraceptive methods.

Family planning programs are an organized effort to provide birth control information and services on a voluntary basis to a target population seeking to lower fertility and improve overall community health. A primary aim of family planning programs is to spread effective birth regulation practices, thereby helping to reduce maternal and child deaths, helping to improve child health, and helping to lessen medically and illegally induced abortions. The programs provide information services and supplies that enable people to regulate fertility both in terms of completed family size and the spacing of children and, if possible, to choose among the most effective, safe, and acceptable means of fertility control. In addition, a program may include activities aimed at persuading couples to adopt services or aimed at reducing desired fertility. The typical program includes: an informational component; service points at hospitals, health centers, special clinics, mobile clinics, offices of private physicians, and, occasionally, delivery of supplies by paraprofessional and lay personnel; a variety of contraceptive methods, usually free or at subsidized prices; facilities for male and female sterilization; and sometimes provision for induced abortion.

The aim of this review is to focus on the methods of measurements relevant to different objectives and to mention some of the approaches that have been used in evaluating family planning programs. Two types of matching studies are reviewed: matching of areal units which manifest program effects, and matching of individuals which show that fertility declines among program acceptors are greater than among non-acceptors matched for other characteristics. Other studies that adopt multiple regression analysis indicate that areas with higher acceptance rates tend to have larger decreases in fertility. We briefly mention some of these studies showing their advantages as well as their limitations. One limitation is the set of demands made in terms of data. Some methods, such as births-averted analysis, require measurements on a number of aspects of the reproductive process that are not often available. Areal analysis would benefit from a richer variety of characteristics per area to permit more elaborate model-building, but for many countries even a basic set of variables does not exist. Another problem is the ability to operationalize the key concept in a given approach; e.g., in births-averted analysis, the data required to capture the substitution effect and the potential fertility of acceptors are not easy to specify. Another need is to gain knowledge on the range of precision to be expected from observational studies. Some limitations are temporal, one example being life table analysis of contraceptive use. It is limited by the need to extrapolate from short-run observations to long-run patterns of use. In survey analysis one faces the problem of converting cross-sectional observations into dynamic models. In births-averted analysis, one would like to convert cohort estimates into period measures.

In general, many of the techniques of evaluation are faced with the twin problems of estimating long-range impacts from short-range studies, and of detecting short-range impacts in long-range programs. Improvements in the power of these techniques will come from attention to the limitations, and from cumulative experience in their application. What is needed is the application of different techniques within the same country. Given the limitation of any one technique, stronger statements concerning the demographic impact of a program will be possible when distinct approaches reach the same conclusion. Finally, policymakers can facilitate the evaluation of a family planning program by specifying the magnitude as well as the nature of the impact they are seeking and the time perspective within which the effect is to be realized.

Section Two deals with the objectives of the programs in general, how to measure program performance, and includes a discussion of some common problems in program evaluation. Section Three discusses program performance and progress using acceptor data and indicates how mathematical models can be used in program evaluation. Section Four explains the Couple Year of Protection index as a measure of family planning program output. Section Five discusses the births-averted method as a means of evaluating programs. Section Six compares several matching studies and shows their advantages as well as the problems associated with their applications. Section Seven discusses regression analysis of areal data as a method of evaluation. Finally, Section Eight is concerned with life table techniques used in program evaluation.

2. EVALUATION OF FAMILY PLANNING PROGRAMS

Many investigators view family planning program evaluation as a process that includes measurements of goal achievement, feedback of information for adaptive decision-making, and examination of a wide variety of processes to determine how and why the program was or was not successful. Implicit in this viewpoint is an interest in the evaluation of program activities and tasks as well as the total program, and examination of unexpected and undesirable effects as well as expected and desirable ones. The process of evaluation is an important component of the program and is particularly valuable in advancing new programs.

Since evaluation is done in terms of objectives, the objective of the program must be defined in some way so that success and failure can be measured. Chandrasekaran and Freymann (1964) indicate that many of the objectives of a program can be ordered in an hierarchical fashion, progressing from immediate objectives to the ultimate goal. The immediate objective is the extent to which the population makes use of the services offered by the program. This can be measured by determining the number and the characteristics of acceptors, acceptance trends, and the rate of acceptance. The intermediate objective may be called "use-effectiveness." Evaluation of this factor means assessing the effectiveness of the contraceptive methods used under the program, taking into account the length of time an acceptor utilizes such contraception. The ultimate objective of the program is its demographic impact; that is, measuring the effect of a program on fertility. This can be achieved by two methods. One is by calculation of births averted, which is focused on the biological determinants of fertility. The other is a more broadly-based approach that attempts to relate the macro-demographic aspects of fertility, such as age-specific rates, to program input while taking into account or controlling for the socioeconomic factors which may also affect fertility. The immediate and intermediate objectives as well as the ultimate goal are each subject to evaluation.

Reynolds (1972) was concerned with the type of performance evaluation that relates to the demographic impact of the program. He was concerned with two broad areas, including first of all explanatory examinations of program processes to determine how and why a program was or was not successful. Also, he sought to measure program performance through three factors, including:

*Evaluation of Program Effects. Effects are measures of the outcomes. From an earlier paper (1970) three levels of effects are delineated. These include primary effects, or changes in awareness, knowledge, attitudes, and motivation; behavioral effects, such as trial or adoption of a particular form of behavior; and status effects, or changes in fertility, health, economics, and social status.

*Evaluation of Program Effectiveness. Measures of effectiveness compare achievement to some goal, standard, recognized need, or competitive program. Reynolds (1971) had also shown that many programs measure effectiveness by comparing numbers of acceptors enrolled to some target figure. Often the acceptor figures are simply accumulated and assumed to represent active users. This ignores program and contraceptive dropouts and results in gross overestimates of number of women served.

*Evaluation of Program Efficiency. These are measures of the relative costs of providing services or of achieving effects. Often they are expressed in ratios, such as births averted per dollar, or patients seen per physician-hour. They are usually compared to some goal, standard, and/or competing program.

Ultimately, the success of any family planning program must be demonstrated in terms of demographic changes and the benefits to life that accrue. The approach for measuring demographic effects of family planning programs has been described by Bogue (1970). He recognized the difficulty in demonstrating that a given program has been responsible for a given decline in fertility.

There are other common problems in program evaluation. One is the direct and indirect effect of a program. For example, couples may be influenced directly or indirectly by the program to practice birth control, but may obtain services through private channels. A related problem is that some who have accepted a method under the program may drop out but transfer to private sources for services and supplies and continue to control fertility. Assessment of the indirect effect of the program is further complicated by the non-program effects; i.e., declines of fertility not causally attributable to the family planning program may be due to a number of factors, like changing demographic, economic, and social conditions and/or cultural practices after commencement of the program. There is also the issue of determining fertility change in the absence of the program. This is very difficult to assess. Efforts to cope with such problems have included the use of controlled experiments, matching studies, and multivariate analyses of individual and areal data.

Heed must also be paid to the effects of substitution. This may take any of several forms including enrollment as an acceptor in the program by an individual who has been privately controlling fertility, or the shift of an individual from a less effective to a more effective method within the program. A related problem is assessing the likelihood that new acceptors would have adopted fertility control in the absence of the program. The expected fertility of acceptors is important in evaluating the demographic impact of a program in order to compare the observed and potential fertility of acceptors. How best to estimate this potential is a difficult issue, given that acceptors are not a representative subsample of all couples. They tend to differ with respect to parity, prevalence of sterility, motivation to limit childbearing and experience of contraceptive practice. We need to examine short-term and long-term effects. This involves questioning when a birth has been averted, and whether an assessment is satisfactory while the woman is still subject to the risk of childbearing, or only when she is no longer susceptible.

Other standard problems include such things as conceptual issues, since there is some confusion over the very meaning of the term "evaluation." This is due in part to a lack of agreement about the purpose of program evaluation. Also, there is the sociopolitical problem. Who decides to become involved with a family planning program often becomes a political question. Finally, there are technical problems. The inadequacy of broad theories in relating social action programs to society is troublesome for family planning program evaluation. This makes it hard to fit family planning programs into an overall framework of economic, social, and planned development thereby making evaluation of success difficult.

3. PROGRAM EVALUATION BASED ON ACCEPTOR DATA

Family planning program evaluation deals with several groups of measurements. Some of the major ones are measurements of input, such as money spent, man-hours worked, and materials invested; measurements of output, such as goods and services provided; and measurements of the effects of a program on fertility. Very often, counts of types and quantities of program outputs are the only data that are available. They are useful because they give a picture of the current activity and the initial results of program input, although they reveal nothing about the effects of these results.

There are two concepts crucial to the issue of measurement. "Acceptability" is defined as the extent to which couples favor the use of family planning methods, while "effectiveness" measures the protection it provides against pregnancy. It is very difficult to obtain data on "program acceptor" and "method acceptor" because of the problem of multiple counts; i.e., whether the same person takes advice from two or more clinics served by the program. Information on previous attendance at other clinics run by the program, if obtained routinely on the individual clinic record, can assist in minimizing such duplication in program acceptor counts. The information obtained from the acceptors on age and number of living children is likely to contain errors, just as in the case of other data obtained through inquiries, most notably in developing countries.

Ross et al.(1972) studied the demographic characteristics of acceptors in a number of national family planning programs through the calculation of acceptance rate, acceptance index, and percentage distribution of acceptors. The median can be considered as a useful statistic for comparing different populations by a specific characteristic, but it suffers from the deficiency that it does not control for differences in the base populations belonging to different categories of the characteristic.

Most programs compile monthly data on the number of acceptors by method accepted. The trend in monthly acceptors is usually represented by a line or bar diagram. Representation of the data through tables becomes more expressive when the data are shown by quarter or by full year. Data on characteristics of acceptors, such as age of the mother or living children, are usually compiled less frequently. Regarding this, Ross (1972), following a review of the data obtained in a number of programs, concluded that acceptance rates among younger women appear to rise, and higher proportions of women with few children enroll as program operations continue. Some investigators felt that acceptance data by method might be compiled for each month and trends investigated through time series analysis. In general, only relatively simple analyses of time trends have been carried out thus far in the evaluation of the progress of family planning programs. Even such simple analyses, however, have provided useful guidance for program administration.

Mathematical models can be used to test hypotheses concerning the processes which underlie the trend in the number of acceptors. Chandrasekaran (1969) developed a model assuming a two-stage process by which a person becomes an acceptor:

(i) A certain number (N) of women are "prospective acceptors" at the start of program operation and an additional number (M) become so during each month of the operation of the program.

(ii) There is a time-lag between a woman becoming a "prospective acceptor" and actually accepting a method from the program, the time-lag being determined by the law of consecutive independent trials. The chance of a woman who is a "prospective acceptor" at the beginning of any month becoming an acceptor during that month is a constant, $p < 1$. For purposes of simplicity, he assumed that no woman who becomes a "prospective acceptor" in any month becomes an acceptor in the same month.

Under these assumptions, Chandrasekaran showed that the number of acceptors Y in the t-th month after the start of the program is given by

$$Y_t = M - (M - N.p)\, q^{t-1}, \text{ where } q=1-p.$$

and if $M > N.p$

$$Y_t = K + a\, b^{t-1}.$$

Where $b < 1$, b was estimated using the total number of acceptors per year,

$$a = (\Sigma_2 Y - \Sigma_1 Y)\, [(b-1)/(b_{12}-1)^2] \text{ is negative}$$

$$b_{12} = \frac{\Sigma_3 Y - \Sigma_2 Y}{\Sigma_2 Y - \Sigma_1 Y},$$

and $$K = 1/12\, [\Sigma_1 Y - \{(b_{12}-1)/(b-1)\}.a] \text{ is positive.}$$

$\Sigma_1 Y$ is the number of acceptors for the period 1964-1965 and similar definitions apply to $\Sigma_2 Y$ and $\Sigma_3 Y$.

The model was fitted to the data on the monthly number of IUD acceptors in one of the counties in Taiwan during the three-year period 1964 to 1967. This revealed the need for using different values of the parameter in different periods.

Assessment of program performance and progress using acceptor data is directed to the improvement of administration and management of family planning programs. Acceptance of services and devices from a family planning program is an important step in the utilization of the facilities provided by the program. It may aid in the achievement of individual goals in respect to family formation and/or national goals set by the program. However, the methods described above depend largely on data available through routine records which are not always accurate.

4. PROGRAM EVALUATION BASED ON COUPLE YEAR OF PROTECTION (CYP)

Since programs place different emphases on different contraceptive methods and provide different mixes of contraceptives, simple counts of types and quantities of contraceptives distributed do not give a summary picture of the overall achievement of the program. Therefore, the couple-year of protection index was proposed as a measure of family planning program output. For the Pakistan family planning program the CYP is defined by

$$CYP = C/100 + O/13 + 2.5\ I + 5\ (TL + V),$$

where C = conventional contraceptives distributed; O = cycles of oral pills distributed; I = IUD insertions; TL = tubal ligations performed; and V = vasectomies performed. The equation provides a measure of the joint impact of all program methods distributed each month by the relative protection each affords. It gives an indication of overall prevalence of contraceptive use. The equation states that for every 100 conventionals distributed, or for every 13 cycles of pills, the program will provide a year of complete protection for the couple.

The conversion factor for a method may vary from program to program, and from region to region. The authors used the CYP index in estimating the births averted, where

$$\text{Births averted} = CYP/3.$$

This index has the advantage of summarizing the overall output of a program in terms of time of contraceptive protection provided. Couple-Year of Protection is a composite index that measures the output of family planning programs, including new acceptors and returning clients. It was proposed by Wishik and Chen (1973) as part of their evaluation of the Pakistan family planning program. They used data on "application," including number of IUDs inserted and contraceptive pills distributed during a year. According to Wishik, this index can be used as an achievement index. When a family planning program offers several birth control methods, the CYP index provides a way of comparing and summing the protection afforded by different contraceptive methods by assigning to each method a CYP conversion factor equal to the average length of time one application of that method confers protection upon a couple.

The prevalence of protection during any period includes both the amount of protection resulting from contraceptives distributed during that present period, as well as continuing protection carried over from past distribution. The method usually employed to determine prevalence of contraceptive practice is to conduct a sample survey. The CYP approach provides an alternative based on service statistics and does not require a survey. For example, the current CYP prevalence for IUDs would equal the sum of all IUDs inserted during the same period minus removals and expulsions, as well as an allowance for mortality. The "protection prevalence ratio" relates the CYP prevalence score to the total number of CYP that would be needed to protect all the eligible couples in the population.

$$\text{Protection prevalence ratio} = \frac{\text{CYP prevalence index in a year}}{\text{No. of eligible couples in that year}}$$

CYP achievement is useful for assessing total program output and cost-effectiveness, such as dollars spent per CYP achieved. CYP prevalence is also beneficial in estimating service coverage and for forecasting future service load. Both indexes may be used for comparisons over time, or among different places and programs.

The CYP method is based on routine data that is usually collected during the operation of the program. Like information concerning age, certain social and demographic characteristics of acceptors, or data on length and continuity of users at the peripheral clinic where the supplies pass to the clients, these data should be collected on a regular basis whatever the nature of the program. These will provide the required information for the computation of the CYP conversion factor.

There are numerous advantages of the CYP index. First, it may serve as a measure of achievement. Achievement may be noted for current time periods among geographic areas, for comparison between such units, or for assessment of trends among individual units over time. It may be related to current rate of program inputs (such as money), related to the size of the base population that program aims to serve, or related to the estimated residual of candidates for current service (i.e., the number of couples who have not yet accepted a contraceptive method, but who are still eligible for acceptance). This is important for programs that offer a long-term method, such as IUDs. Achievement may be used as an index in the assessment of new or experimental approaches within a program as an aid to research and evaluation. Second, as a measure of current prevalence, the CYP index provides a means to make comparison over time for long-term methods as well as for renewable or short-term ones. It can be used as a measure of contraceptive practice occurring from past and current efforts, or by geographic areas. It can be related to the span of the past program, or to the total amount of program input, or related to the size of the target population. Prevalence can be related to the total estimated number of residual candidates eligible for program services. And as a measure of future prevalence, the CYP index offers further flexibility to a program administrator by allowing him to estimate the amount of contraceptive practice that will occur over future time periods. It can aid in planning programs, especially in accounting for cost, manpower, and/or new locations. It may have possible use in the projection of population trends in a given area. Again, this method estimates the maximum impact of a certain quantity of contraceptives in a given population. It is built around the assumption that supplies dispensed will not only be used, but used efficiently.

There are also some problems with the CYP method of estimation. Many of the supplies dispensed will be wasted. Also, the method makes no distinction between one woman who is protected for three years and six women who are protected for six months each. This approach estimates expected fertility in terms of the average birth interval for all non-contracepting women, regardless of their age or parity. Finally, the CYP method is suggested as an administrative device, not as a measure of demographic changes or births averted accruing from the activities of a family planning program. Where data is limited, however, the CYP method may still be one of the best possible measures.

5. PROGRAM EVALUATION BASED ON BIRTHS AVERTED

Perrin and Sheps in their basic paper (1964) viewed the phenomenon of human reproduction as a Markov renewal process. According to their formulation, during any month a married female can be assumed to be in one and only one of the following states with respect to reproductivity:

S_0 = fecundable and subject to a constant monthly risk of conception;
S_1 = pregnant;
S_2 = infecundable owing to the temporary anovulation that follows abortion or miscarriage;
S_3 = anovulatory following a stillbirth;
S_4 = anovulatory following a live birth.

The timing of a woman's pregnancy and birth is determined by the sequence of states visited and the length of time spent in each. They assumed that for a given pregnancy outcome, lengths of pregnancy and lengths of the following anovulatory periods are independently distributed. The probabilities that a pregnancy will end in abortion, stillbirth, or live birth sum up to 1. To quantify the reproduction process as a renewal process, the family-building must be formulated in such a way that intervals between consecutive births, or between successive pregnancies, behave as independent, identically distributed random variables. To meet these qualifications they assumed homogeneity among women, homogeneity in time, and sufficiently large reproductive periods.

They showed that

$$P[\text{passage into } S_4 \text{ at } t] = E[N_4(t)] - E[N_4(t-1)];$$

i.e., the probability of a live birth occurring in month t is equal to the expected number of live births in t months less the number expected in (t-1) month. By letting $t \to \infty$, they demonstrated that the asymptotic fertility is equal to the inverse of the mean recurrence time for live births. This can be used as an approximation of the exact fertility rate for homogeneous cohorts which are under observation for a reasonable length of time.

One of the problems with the model is that it does not allow for the probability of conception. Neither is allowance made for the fact that the length of gestation and/or infecundability may depend on such factors as age or birth order in the individual female. Hence, the period for which this model can be assumed to hold for each woman must be restricted to an inverval of at most from 10 to 15 years in the middle of the childbearing age.

Following Perrin and Sheps (1964), Potter (1969, 1970) presented an analytical method that permitted estimates of births averted by one segment of contraception X. A "segment" of contraception is defined as the period of usage of a method between its initiation and its next interruption owing to accidental pregnancy, a switch to another method, or contraception cessation. His approach involves comparisons of pairs of renewal processes representing two cohorts of couples alike in all respects except that during the initial interval one cohort exploits the contraceptive under assessment while the other does not. They are identical in fecundity including an equally long residual reproductive period. Both cohorts start the preperiod (initial pregnancy interval) in the same initial state and during the preperiod (preperiod plus the succeeding pregnancy intervals), both cohorts follow the same repetitive family planning strategy. Therefore, any difference between the two cohorts with respect to time of next conception or cumulative births during the residual reproductive period of T months is due to the presence and/or absence of one segment of X during the preperiod.

Potter showed that the net delay of next conception due to the impact of a segment of X is given by

$$I_{x,f} = U_{01,x}(A) - U_{01,f} ,$$

where $U_{01,x}$ represents the expected delay to next conception in presence of one segment of X; $U_{01,f}$ represents the expected delay to next conception in absence of contraception; A represents the interval from initiation of contraception to end of anovulatory period.

The births averted per segment of X is given by

$$B = I_{x,f}/U_{44,x} ,$$

where $U_{44,x}$ is the mean of the birth interval in the absence of contraception. That is, the births averted per segment of X is given by the difference between the expected births during the residual reproductive period in the absence of contraception and the expected births after the practice of one segment of X during the preperiod.

Potter's model does not include factors such as sterility, mortality, or age-related changes in fecundity. Therefore, it is realistic only for the first half of the childbearing period when sterility is infrequent and changes in fecundity are slow.

6. PROGRAM EVALUATION BASED ON MATCHING

Matching of variables is a very useful method of adjusting for bias in observational (non-controlled) studies in which subjects self-select procedures or treatments. It can also be used to increase the efficiency of controlled trials in which subjects are randomly allocated to treatments, procedures, or programs whose effects are to be determined. Randomization within matching insures that potential bias - whether due to subject or experiment, whether conscious or subconscious - does not influence the assignment of subjects to treatments and, within the limits of probability, balances the influences of other potential disturbing variables by spreading them over all treatment groups. It permits the use of statistical methods for estimation and for the testing of hypotheses about differences between treatments.

Many studies are extant which estimate differences in fertility trends between program service acceptors and non-acceptors. In the following portion of this review, our aim is to compare some of these studies. The comparison serves to illuminate some of the questions which should be considered in designing other studies. A major purpose of the studies was to compare pre- and post-acceptance differences in the fertility of acceptors with differences in fertility during the same period among non-acceptors. In each instance non-acceptors presumably have access to contraceptives from non-program sources. Adjustments were made for potential biases in estimates of fertility differences by matching acceptors and non-acceptors on disturbing variables although the procedures and specific matching variables differed. Matching on disturbing variables was accomplished by comparing the characteristics of acceptors with those of non-acceptors.

A study to evaluate the demographic impact of an IUD program by matching was presented by Chang et al. (1969). The IUD acceptors are matched with the non-acceptors who have characteristics similar to the acceptors at the time of first insertion. Fertility rates of the acceptors before and after the first acceptance can be compared with the corresponding rates of the matches. Four townships in Taiwan were selected for the pilot study. To minimize possible bias in estimating the fertility of the IUD acceptors before this acceptance (this group included more postpartum cases than average) the average fertility over three years before first acceptance is used. The total IUD acceptors were sorted into groups and matched by year of acceptance, age, education, and number of live births. The same grouping was done for the non-acceptors. On the average, the acceptors were selected so that their preinsertion fertility was higher than the corresponding fertility of the married women in the township. The difference in fertility between the IUD acceptors and the married women in general was found to increase with age. Because of the matching procedures, which take the number of previous live births into account, the preinsertion fertility rates of the acceptors and the corresponding fertility rates of the matches were about the same. This solved one of the serious criticisms of some earlier work concerning the demographic impact of family planning programs.

Three indices are used to compare the postinsertion fertility with the corresponding fertility of the matches, including first of all the "percentage decline of fertility before and after acceptance." Second, the "index of fertility decline of the acceptors against the matches" is given by:

$$\text{Index} = \frac{\text{Expected Fertility - Observed Fertility}}{\text{Expected Fertility}} \times 100.$$

The term "expected fertility" assumes that the postinsertion fertility of the acceptors would have declined at the same rate as did the matches had they not accepted the IUD. That is, the index is the percentage decline of the expected fertility caused by the IUD program, which may be regarded as the net effect of the IUD program. Finally, "births averted by insertion of 1000 IUDs" is obtained by the numerator of the formula above. This is the total effect of IUD acceptance, not only the effect of the actual use of IUDs. The IUD program was able to lower the fertility of the acceptors and decrease the number of births averted. The demographic impact of the program correlated negatively with age and parity of the acceptors and was significantly greater among women of lower educational status.

Evaluation of the impact of family planning programs by matching was first attempted by Takeshita (1964). The acceptors of this program were mainly using conventional contraceptive methods. They were matched with the nonacceptors in the area. The criteria of matching was wife's age, marriage duration, and number of living children. Comparison was made of the birth rate for the five years before and after entry into the program. Although their analysis indicates that the program was moderately effective, the fact that the fertility of non-acceptors had also declined caused some concern. This concern seems relevant in the case of an IUD program, where much stronger motivation is needed for acceptance than for conventional methods. Matching methods underestimate the real program impact: the fertility decline of the matches is caused partly by the influence of the program activities. Moreover, it might be that the matches did not accept the IUD because they had accepted some other method; or, on the contrary, the matches may be less motivated so that they did not accept the IUD.

A different type of matching study was done by Rochat et al. (1968). They used data from seven pairs of small rural counties in Georgia. One of each pair had an active family planning program for two or more years prior to the study and the other did not. The counties were matched on the basis of geographical location, racial distribution, and population size. Fertility trends were then compared before and after the program. They examined the fertility by three variables: race, age, and parity. The authors concluded that the rate of fertility decline decreased for whites but increased among Negroes. A marked decrease in high birth-order births and an increase in first births for Negroes in the study area suggested that family planning programs helped accelerate the decline in fertility in the study area. They found that changes in estimated fertility do not correlate precisely with the age distribution of active contraceptors. This may be related to imprecise age-specific population estimates, different intervals of contraceptive service for women of different ages, or the different risks of becoming pregnant without public contraceptive services for women of different ages. This suggests that more frequent and precise measurement of population characteristics, and perhaps more sophisticated analysis of the contraceptive users, is required in order to demonstrate clearly the impact of family planning programs on fertility.

Okada (1969) studied women giving birth in a single hospital and whether they accepted postpartum program contraception within four months of delivery. The respondents in this study are low-income, urban Negro women with high indices of social and personal disorganization. The study population consisted of fecund Negro women who had a live-born infant at a D.C. general hospital in a certain year. Each member of the study group was matched by a nonparticipant who also delivered at the same hospital during the same month, and was within the same age, parity, and marital status categories. The main objectives of the study were to measure the reduction in pregnancies as a result of participation in the birth control program, to determine whether certain demographic characteristics may be related to differential participation in the program, and to measure the use-effectiveness or continuation rate of the program. A comparison of the pregnancy rates between the matched study and control groups indicates that the program was responsible for a substantial reduction in the number of pregnancies among the study group members. Comparisons of pregnancy rates before and after for both groups tend to confirm this conclusion.

Another matching study was reported by Johnson et al.(1973). They assessed family planning program effects on birth rates by determining the differences between pre- and post-program fertility among acceptors and non-acceptors (otherwise similar to acceptors). In order to accomplish this, they measured the births occurring in each group before and after the acceptors entered the program, directly from the birth records. Before matching, the birth records of acceptors were computer-scanned in order to eliminate any duplication. One limitation of their study is that the fertility rates obtained are based on confirmed matches of live births with all acceptors during designated periods, so this may have underestimated fertility. This resulted from the incomplete records. A further limitation is the bias resulting from incorrect numbers.

There are additional limitations to these studies (cf. Wells (1975)). For example, in Chang et al., acceptors and non-acceptors were not matched on their "motivation for family limitation." While all three studies showed large short-term differences in favor of lower fertility for acceptors, long-range studies are required in order to determine the extent to which differences change or persist over the remaining reproductive life span. Johnson et al.(op. cit.) reported that in his study about 40% of those accepting a contraceptive during the last month of a six-month interval had a birth during that interval. This suggests that seasonal fluctuations in births may well influence seasonal acceptance patterns.

Certainly, these patterns of acceptance may in turn influence seasonal fluctuations in births. Such seasonal fluctuations may give investigators clues to the biological and social variables which are operative in affecting seasonal fertility variation. Further, and germane to our discussion here, seasonal analysis has implications for the design and operation of health services, such as family planning programs. For example, by regressing periods of increased birth rates nine months to arrive at times of conception, family planning services could be stepped up during a particular time for maximum effectiveness. Discussion of the best and worst times of the year to give birth is important. A desire to control the season of birth may be an incentive to embark on fertility planning. Again, the seasonal dimension can be of critical importance in determining medical research priorities, and in planning and administering preventive and curative health programs.

Chang et al. used an open interval as one matching variable, but matching was not exact and dampened the effect of variation in birth intervals since open intervals for non-acceptors were based upon time between date of last birth and July 1 of the year of acceptance. It was not clear from his description how he distributed the acceptors and non-acceptors in his matching procedure; nor was it clear what procedure was followed if matches could not be found. No data were shown for non-acceptors, nor were data given on the number of disqualified who were substituted for the non-acceptors. Takeshita et al. (op. cit.) gave few details of the pair-matching procedure employed, except that age matches were defined as agreement within five years. Okada listed the matching variables used but gave no details about the procedures employed in matching tolerances. Migration and loss-to-follow-up were reasons for incomplete linkages of records in some studies.

Greenberg (1968) had discussed other difficulties and suggested various approaches for evaluating social programs where areas or communities are used as experimental units. One of the major problems is establishing a set of comparable experimental areas which can be randomly assigned to different treatment procedures including serving as a control. He suggests testing alternative programs rather than program versus no program. Cross-communication between control and program areas must either be measured or kept to the minimum. For measurement and evaluation of a community program, if the double-blind principle cannot be used, one should try to select observers who were not involved in providing services in that community. In order to solve the problem of number and timing of measurements to be taken after the start of the program, Greenberg suggests several sample designs for taking repeated measurements in the different program areas. This will minimize the possibility of repeated measurements on the same persons. To help deal with the issue of confounding program merits with personnel, Greenberg recommends rotation of staff. Staff members involved in measurement of program results might also bias the results.

The matching techniques mentioned above help to reduce bias in observational as opposed to experimental studies. Observational and experimental studies can be used to evaluate programs. Matching is only one of several statistical procedures for handling biases due to disturbing variables. Other techniques include adjustment by weighted subclass means and by regression methods. However, subgroup means and matching procedures have an advantage over regression adjustment in that specific functional relationships between response and disturbing variables need not be specified, although it is implicitly assumed that they are the same in both treatment comparison groups. Finally, unless motivation to practice family planning is added to future studies of acceptor and non-acceptor fertility, it is doubtful that matching designs are suitable for provision of much additional information on program effects.

7. PROGRAM EVALUATION BASED ON REGRESSION ANALYSIS OF AREAL DATA

The use of areas rather than individuals as the unit of analysis is one approach that has been employed in demographic research. In the earlier period, lack of data on individuals constrained researchers in differential fertility to rely on areal comparisons. A question that always arises in the evaluation of family planning programs is how much of the reduction in fertility is due to the program. A reasonable way to answer this is by regression analysis, which tests the relationship between levels of fertility and program inputs or acceptance rates, after taking into account socioeconomic and demographic factors.

Hermalin (1971) analyzed the relation between fertility and acceptance rates in the Taiwan program. His model postulates that an area's level of socioeconomic development, its demographic status in terms of age at marriage and child mortality, and its level of IUD acceptance before the program influence its level of fertility. He utilizes the socioeconomic characteristics as indicators of the influence of modernization and urbanization on fertility. By means of multiple regression (path analysis), intercorrelations among causal and dependent variables were examined for 331 administrative areas in Taiwan. He proposed the following model:

$$Y_i = \beta_0 + \beta_1 X_{1i} + \beta_2 X_{2i} + \beta_{3i} X_{3i} + \beta_4 X_{4i} + \beta_5 X_{5i} + \varepsilon_i$$

where Y_i respresents fertility of the i-th area; X_{1i} is a measure of child mortality; X_{2i} is the proportion of male labor force in agriculture; X_{3i} is a measure of educational level; X_{4i} is age at marriage; X_{5i} is the rate of acceptance of program contraception; and ε_i is the error term representing all unmeasured influences on fertility.

The analysis showed a consistently significant negative relation between acceptance rates and age-specific marital fertility rates in four age groups. Hermalin's analysis confirms that IUD acceptances in Taiwan have reduced fertility rates. Further results strengthen the likelihood that acceptance of IUDs is not merely a substitution for other forms of contraceptive use.

Schultz (1971) used the areal framework to analyze the effect of regional allocation of program personnel, as well as certain socioeconomic and demographic variables, on regional birth rates in Taiwan. His approach was to observe birth data directly and develop a conceptual and statistical framework within which he could obtain unbiased estimates of the average and marginal effects of program activities on birth rates. His model was based on considerations of the environmental constraints and opportunities that influence the number of surviving children parents want, the level of child mortality, and the family planning activities viewed as facilitating the avoidance of unwanted births. Schultz proposed an analysis of the interregional variation in birth rates as one approach to program evaluation. This took into account both the environemntal changes thought to alter parent's desired fertility, and the activities of the family planning program thought to help avert unwanted births. Multiple linear regression and nonlinear estimation techniques are used to analyze variation in birth rates across 361 small administrative regions in Taiwan from 1964-1978. He proposed the following model:

$$Y_i = \beta_0 + \beta_1 X_{1i} + \beta_2 X_{2i} + \beta_3 X_{3i} + \beta_4 X_{4i} + \varepsilon_i$$

where Y_i, X_{2i}, X_{3i} and ε_i are defined as in Hermalin's model; X_{1i} is the reciprocal of child survival probability; and X_{4i} is worker-months per women aged 15-49.

Through the use of regression analysis, Schultz found an inverse association between regional age-standardized crude birth rates and the regional allocation of field personnel between 1965-1968. The marginal annual effectiveness of the personnel has diminished as the personnel inputs have accumulated over time. There was a statistically significant negative relationship between cumulative field-worker activity and age-specific birth rates for women over age 30.

Hermalin had stated that areal analysis may be a preferred code of testing an hypothesis and not merely a second-best alternative when individual correlations are unavailable. On the other hand, the most well known problem associated with areal analysis is the so-called "ecological fallacy;" i.e., whether ecological correlations can validly be used as substitutes for individual correlation. Robinson (1950) had shown that correlations observed to hold among areas cannot be assumed to hold among individuals. While it is theoretically possible for the two to be equal, the conditions under which this can happen are far from those ordinarily encountered in the data.

8. PROGRAM EVALUATION BASED ON LIFE TABLE TECHNIQUES

Effectiveness of contraception has been conventionally measured by the Pearl pregnancy rate:

$$\text{Pearl rate} = \frac{\text{Total accidental pregnancies}}{\text{Total months of contraceptive exposure}} \times 1200.$$

This index gives the number of pregnancies per 100 years of contraceptive exposure.

The Pearl rate is not a pure measure of contraceptive effectiveness because it is sensitive to the length of observation. The formula assumes a constant pregnancy rate for a given population in all months following the initiation of contraceptive practice. Pregnancy rates vary with duration of contraception use: they tend to decline, sometimes steeply; at other times they increase as couples abandon contraception, then decline gradually. In view of these relationships, Pearl's formula can no longer be considered adequate for the evaluation of contraceptive effectiveness. Therefore, attention has been focused on an alternative measure, the "Cumulative Failure Rate", based on a life table approach.

The life table procedure can be used for the computation of cumulative rates of continuation and discontinuation of contraceptive practices in general or of specific contraceptive methods. Potter (1966) showed that the cumulative pregnancy and continuation rates may be computed as gross rates using a single-decrement life table. His procedure involves three main steps. First, each couple is classified by number of exposure months and terminal status, separately for each pregnancy interval. Next, these data are reorganized so that for each exposure month i, the number of exposed couples n_i and the number of accidental pregnancies a_i are determined. Then, again for each exposure month i, a conception rate $P_i = a_i/n_i$ is calculated. Its complement q_i defines the proportion of couples who survived that month without conceiving. The complement of the cumulative product of these survival probabilities, $1 - p(k)$, yields the cumulative k-month failure rate. The cumulative one-year failure rate can serve as a summary measure for comparison purposes. It is the proportion of women who would be expected to conceive during the first year of exposure if subject to the monthly conception rates observed in a sample, and if there was no dropout for reasons other than accidental pregnancy during that year.

Potter (1967) further computed the cumulative rates as net rates using a multiple-decrement life table which takes into account the competing effects of both pregnancies and other types of discontinuation. According to Potter, the essential idea was developed in collaboration with C. Tietze and refinements relating to standard errors were added later with the help of B.G. Greenberg, N.L. Johnson, and W.K. Poole. Data are taken from one of the Taiwan population studies and estimate the cumulative proportion of women losing their IUD to pregnancy, expulsion, and/or removal within 12 months of insertion. It required that the three specific rates sum to the 12-month rate of device loss for the three reasons combined. Each of the women coded "first segments" (the period of use from adoption of a contraceptive to its first interruption) are assigned a length and a terminal status. The terminal status was classified as pregnancy, expulsion, removal, or continuing user. For each month the following rates are computed:

Rate of device loss from:

$$\text{Pregnancy } (Q_{xp}) = \frac{\text{Number of pregnancies during month } (x,x+1)}{\text{Actual number of women exposed}} .$$

$$\text{Expulsion } (Q_{x\ell}) = \frac{\text{Number of expulsions during month } (x,x+1)}{\text{Actual number of women exposed}} .$$

$$\text{Removal } (Q_{xr}) = \frac{\text{Number of removals during month } (x,x+1)}{\text{Actual number of women exposed}} .$$

$$\text{All causes } (q_x) = \frac{\text{Total observed terminations during month } (x,x+1)}{\text{Actual number of women exposed}} .$$

Each monthly rate is a conditional probability: the probability of losing the device during the next month if retaining it up to the start of the month. The cumulative rates of retention and loss for all reasons can be estimated using the same procedure as in the single-decrement life table. The probability of retention is the complement of q_x. The accumulation of the initial 12 months yields the probabilities of retaining the device through the first year. The complement of this probability is the proportion of women expected to lose the device before the end of the first year. The difference between the proportion retaining it x month or longer and (x+1) month or longer yields the expected fraction of women losing it during the monthly interval (x,x+1). An advantage of calculating proportions of women retaining and losing the device in this manner is that it allows segments having the terminal status of "continuing user" to be utilized, as well as those segments with known types of termination. While this approach enables one to combine terminated segments with the unterminated segments of continuing users, it gives no protection against the biases arising from dropout.

REFERENCES

[1] Bogue, D. (ed.), Family Planning Improvement through Evaluation: A manual of Basic Principles, Chicago, Community and Family Study Center, University of Chicago (1970).

[2] Chandrasekaran, C. and Freymann, M.W., Evaluating community family planning programs, Public Health and Population Changes, Sheps, M.C. and Ridley, J.C. (eds.). University of Pittsburgh Press (1964).

[3] Chandrasekaran, C. (ed.), In Measuring the Effect of Family Planning Programs on Fertility, Chandrasekaran, C. and Hermalin. A.I. (eds.). Liege, Belgium, Ordina Editions (1969).

[4] Chang, M.C., Liu, T.H., and Chow, L.P., Study by matching of the demographic inpact of an IUD program, Milbank Memorial Fund Quarterly **58**:137-157 (1969).

[5] Greenberg, B.G, Evaluation of social programs, In: Readings in evaluation research, Compiled by Francis G. Caro, New York, Russell Sage Found. 155-75 (1971).

[6] Hermalin, A.I., Taiwan: Appraising the effect of family planning program, through areal analysis, Taiwan Population Studies Working Paper No. 14, Ann Arbor, Population Studies Center, University of Michigan (1971).

[7] Johnson, J.T., Ann, T.B., and Corsa, L., Assessment of family planning programme effects on births: Preliminary results obtained through direct matching of birth and programme acceptor records, Pop. Studies **27**:85-96 (1973).

[8] Perrin, E.B. and Sheps, M.C., Human reproduction: A stochastic process, Biometrics **20**:28-45 (1964).

[9] Potter, R.G. Jr., Application of life table techniques to measurement of contraceptive effectiveness, Demography **3**:297-304 (1966).

[10] Potter, R.G. Jr., The multiple decrement life table as an approach to the measure of use-effectiveness and demographic effectiveness of contraception. Contributed paper, Sydney Conference (International Union for the Scientific Study of Population (1967), 869-83 (1967).

[11] Potter, R.G. Jr., Renewal theory and birth averted, International Union for the Scientific Study of Population, General Conference, London (1969), 145-149 (1969).

[12] Potter, R.G. Jr., Births averted by contraception: an approach through renewal theory, Theor. Pop. Biol. **1**:251-272 (1970).

[13] Okada, L.M., Use of matched pairs in evaluation of a birth control program, Public Health Reports **84**:445-450 (1969).

[14] Reynolds, J., A framework for the selection of family planning program evaluation topics. Manuals for evaluation of family planning and population programs, No. 1, New York: Division for Program Development and Evaluation, International Institute for the Study of Human Reproduction, Columbia University (1970).

[15] Reynolds, J., Methods for estimating future caseloads of family planning programs, Family Planning Perspectives **3**:56-61 (1971).

[16] Reynolds, J., Evaluation of family planning program performance: A critical review, Demography 9:69-85 (1972)

[17] Robinson, W.S., Ecological correlations and the behaviour of individuals, Amer. Sociol. Rev. 15:351-357 (1950).

[18] Rochat, R.W., Tyler, C.W. Jr., and Schoenbucher, A.K., The effect of family planning in Georgia on fertility in selected rural counties. In Advances in Planned Parenthood, 6. Proceedings of the American Association of Planned Parenthood Physicians, Boston, 9-10, Excerpta Medica Foundation (1970).

[19] Ross, J.A., Germain, A., Forrest, J.E., and Ginneken, J.V., Findings from family planning research, Reports on Population/Family Planning, No. 12, New York, The Population Council (1972).

[20] Schultz, T.P., Evaluation of population policies: A framework for analysis and its application to Taiwan's family planning program (R-643-AID), Santa Monica, Rand (1971).

[21] Takeshita, J.Y., Peng, J.Y., and Kiu, P.K.C., A study of the effectiveness of the pregnancy health program in Taiwan, Eugenics Quarterly 11:222-233 (1964).

[22] Wells, H.B., Measuring the Effect of Family Planning Program on Fertility. Chandrasekaran, C. and Hermalin, A I. (eds,), Liege, Belgium, Ordina Editions, 215-241 (1967).

[23] Wishik, S.M. and Chen, K.H., Couple year of protection: A measure of family planning program output. Manual of Family Planning and Population Programs, No. 7, International Institute for the Study of Human Reproduction, Columbia University, New York (1973).

BIOSTATISTICS: Statistics in Biomedical, Public Health
and Environmental Sciences, edited by P.K. Sen

MULTISTATE LIFE TABLES: THEORY AND APPLICATION

El-Sayed Nour and C.M. Suchindran

Department of Biostatistics
University of North Carolina

A multistate life table is a generalization of the conventional life table. Members of a cohort are followed over time as they move between various states of life. In this paper, the statistical theory of multistate life tables is reviewed. Computational procedures for use with different types of demographic data are outlined. The model is applied to the study of marital status patterns for black and white females in North Carolina in 1970.

KEY WORDS: Survival analysis, cohort analysis, Markov process, competing risks, heterogeneity, marital status

1. INTRODUCTION

The life table is the central mathematical model in Demography. It is concerned with the description of duration variables which are encountered in various demographic applications. Originally developed as a statistical model for measuring mortality, the life table is easily adapted to the study of other phenomena such as morbidity, nuptiality, birth interval dynamics and migration. The basic version of this model considers situations where a well-defined population is exposed to the risk of a certain force of decrementation over successive time periods. The underlying stochastic process is that of a discontinuous Markov process with a transient state (an individual being alive) and an absorbing state (an individual being dead). The model is used to compute statistics such as survivorship probabilities, the mean residual life time function and the like.

A major extension of the basic life table model is the recently developed multistate increment-decrement life tables. These tables result by enlarging the transient and absorbing sets of the process to allow for more than one state in each. Thus, the tables may be used to study demographic processes where individual members of a population move over a set of discrete states during their lifetime, and where increments (entrants) as well as multiple decrements (exits) from various life statuses are possible. Examples of such processes include marital status, place of residence, labor force participation and educational attainments as well as related differentials in mortality by cause of death.

The probability structure underlying multistate life tables is in most instances a time-inhomogeneous Markov process or, more generally, an age-dependent semi-Markov process (Hennessey (1980), Hoem (1972), Hoem and Jensen (1982), Keyfitz (1980), Krishnamoorthy (1979), Ledent (1980), Mode (1982), Nour and Suchindran (1983b), Rogers (1980), Schoen and Land (1979), Willekens et al. (1982)).

Accordingly, the model has a long-standing foundation in the theory of stochastic processes. Early specifications of the model in the context of bio-social applications include Fix and Neyman (1951), Chiang (1964), Sverdrup (1965) and Hoem (1970). It was not until recently, however, with the advent of high speed computers that the first nontrivial empirical applications of the model to demographic processes appeared (Hoem and Fong (1976), Krishnamoorthy (1979), Rogers (1973), Rogers and Ledent (1976), Schoen (1975), Schoen and Nelson (1974)).

The purpose of this article is to present some methodological developments that are an outgrowth of the application of multistate life table approach to demographic processes. Particular attention will be paid to the following:

(1) specification of computational formulas for the different functions of the model under the various observational plans used for demographic data collection,

(2) identification of appropriate biometric functions which are easily interpreted and which may be needed for subsequent demographic computations (such as those involving population projections), and

(3) the limitations of the Markovian probability structure as a description of the demographic process being considered.

A brief specification of the basic aspects of the multistate model is given in section 2. Possible generalizations of the Markovian model are also outlined. In section 3, computational strategies are discussed for various types of demographic data. The competing risks model which arises in the hypothetical situation where the state space is reduced is also considered. The various methods are illustrated by a practical example.

2. THE MULTISTATE LIFE TABLE MODEL

2.1. THE BASIC MODEL

Consider the entire closed age interval $T=[0,\omega]$ and let $\{X(t);t\varepsilon T\}$ be a non-homogeneous Markovian stochastic process where, for each fixed $t\varepsilon T$, $X(t)$ has a finite state space H. The state space H consists of a set H_1 of ℓ transient states and a set H_2 of n absorbing states. The major assumption underlying this process is that all individuals who are found in a given state at the same age will exhibit the same probabilistic behavior regardless of their previous paths.

Define $q_{ij}(t,s)=P\{X(t)=i|X(s)=j\},i,j\varepsilon H$ and $p_i(t)=P\{X(t)=i\},i\varepsilon H$. It is assumed that the following limits exist and are continuous for all $i,j\varepsilon H$:

$$\lim_{\Delta t \to 0} \frac{1 - q_{ii}(t + \Delta t, t)}{\Delta t} = r_{ii}(t) \; , \; i \epsilon H,$$

$$\lim_{\Delta t \to 0} \frac{q_{ij}(t + \Delta t, t)}{\Delta t} = r_{ij}(t) \; , \; i \neq j \; , \quad i,j \epsilon H. \tag{1}$$

For $i \neq j$, the function $r_{ij}(t)$ is usually called the force of transition from state j to state i at age t. Clearly $r_{jj}(t) = -\sum_{i \neq j} r_{ij}(t)$. Also, $r_{ij}(t) = 0$ for $j \epsilon H_2$.

Define the $(\ell+n)x(\ell+n)$ matrices $\underset{\sim}{Q}(t,s)$ and $\underset{\sim}{R}(t)$ by $\underset{\sim}{Q}(t,s) = (q_{ij}(t,s))$ and $\underset{\sim}{R}(t) = (r_{ij}(t))$ respectively. Also, define the $(\ell+n)x1$ vector $\underset{\sim}{p}(t)$ by $\underset{\sim}{p}(t) = (p_i(t))$. The matrix $\underset{\sim}{Q}(t,s)$ may be represented in the canonical form

$$\underset{\sim}{Q}(t,s) = \left[\begin{array}{c|c} \underset{\sim}{Q}_{11}(t,s) & \underset{\sim}{0} \\ \hline \underset{\sim}{Q}_{21}(t,s) & \underset{\sim}{I} \end{array}\right] \tag{2}$$

where $\underset{\sim}{Q}_{11}(t,s)$ is an $(\ell x \ell)$ matrix describing transitions among transient states in H_1, $\underset{\sim}{Q}_{21}(t,s)$ is an $(nx\ell)$ matrix governing transitions from a transient state $j\epsilon H_1$ to an absorbing state $i\epsilon H_2$, $\underset{\sim}{0}$ is an $(\ell \, x \, n)$ matrix of zeros and $\underset{\sim}{I}$ is an (nxn) identity matrix. Similarly, the matrix $\underset{\sim}{R}(t)$ may be partitioned as

$$\underset{\sim}{R}(t) = \left[\begin{array}{c|c} \underset{\sim}{R}_{11}(t) & \underset{\sim}{0} \\ \hline \underset{\sim}{R}_{21}(t) & \underset{\sim}{0} \end{array}\right] . \tag{3}$$

Since assumption (1) and the Markov property imply the validity of the Kolmogorov forward and backward differential equations,

$$\frac{d\underset{\sim}{Q}(t,s)}{dt} = \underset{\sim}{R}(t).\underset{\sim}{Q}(t,s) \quad , \; \underset{\sim}{Q}(t,t) = \underset{\sim}{I} \quad . \tag{4}$$

For a given $\underset{\sim}{R}(t)$, $t\epsilon T$, there exists a unique solution to (4) satisfying the properties of a probability transition matrix (Feller (1940)). This solution may be represented as

$$\underset{\sim}{Q}(t,s) = \underset{\sim}{I} + \int_s^t \underset{\sim}{R}(\tau) \, \underset{\sim}{Q}(\tau,s) d\tau \; . \tag{5}$$

In addition, the vector of unconditional probabilities p(t) satisfies the recursive relation

$$\underset{\sim}{p}(t) = \underset{\sim}{Q}(t,s) \, \underset{\sim}{p}(s) \; . \tag{6}$$

Let the matrix function $\underset{\sim}{E}(t,s) = (e_{ij}(t,s))$ be such that $e_{ij}(t,s)$ is the expected length of time spent in state i within the age interval (s,t) conditionally for a process in state j at age s. Also, define the matrix function $\underset{\sim}{M}(t,s) = (m_{ij}(t,s))$ where $m_{ij}(t,s)$ is the mean number of visits to state i over the age period (s,t) conditionally for a process in state j at age s. Fix and Neyman (1951) emphasized the importance of computing biometric functions such as those just defined. In demographic applications, these functions provide useful summary measures for the pattern in which the process evolves over time.

It may be shown that (Nour and Suchindran (1983b))

$$\underset{\sim}{E}(t,s) = \int_s^t \underset{\sim}{Q}(\tau,s)d\tau \tag{7}$$

and

$$\underset{\sim}{M}(t,s) = \int_s^t \underset{\sim}{B}(\tau)\underset{\sim}{Q}(\tau,s)d\tau \quad , \tag{8}$$

where

$$\underset{\sim}{B}(t) = (b_{ij}(t)) \text{ such that}$$

$$b_{ij}(t) = \begin{cases} 0 \quad , & i = j \\ r_{ij}(t), & i \neq j \end{cases} .$$

The canonical representations of the matrices $\underset{\sim}{E}(t,s)$ and $\underset{\sim}{M}(t,s)$ are

$$\underset{\sim}{E}(t,s) = \left[\begin{array}{c|c} \underset{\sim}{E}_{11}(t,s) = \int_s^t \underset{\sim}{Q}_{11}(\tau,s)d\tau & \underset{\sim}{0} \\ \hline \underset{\sim}{E}_{21}(t,s) = \int_s^t \underset{\sim}{Q}_{21}(\tau,s)d\tau & (t - s)\underset{\sim}{I} \end{array}\right] \tag{9}$$

and

$$\underset{\sim}{M}(t,s) = \left[\begin{array}{c|c} \underset{\sim}{M}_{11}(t,s) = \int_s^t \underset{\sim}{B}_{11}(\tau)\underset{\sim}{Q}_{11}(\tau,s)d\tau & \underset{\sim}{0} \\ \hline \underset{\sim}{M}_{21}(t,s) = \int_s^t \underset{\sim}{B}_{21}(\tau)\underset{\sim}{Q}_{11}(\tau,s)d\tau & \underset{\sim}{0} \end{array}\right] . \tag{10}$$

Note that $\underset{\sim}{1}\cdot\underset{\sim}{E}_{11}(t,s)$, where $\underset{\sim}{1}$ is an $(1x\ell)$ vector of ones, has the general element $e_j(t,s)$ which is the expected "survival time" (i.e., time spent in the transient set H_1) in the age interval (s,t) conditionally for a process in state j at age s. Similarly, the elements of the matrix $\underset{\sim}{1}\cdot\underset{\sim}{E}_{21}(t,s)$, where $\underset{\sim}{1}$ is an (1xn) vector of ones, represent the total lengths of time expected to be "lost" to the various causes of death under current mortality conditions.

The life table is often used for purposes of population projections by providing estimates for the survivorship ratios in the corresponding stationary population,

$s_{ij}(T_2,T_1)$, $i,j\varepsilon H$. The ratios $s_{ij}(T_2,T_1)$ represent the probability of survival to be in state i and age group T_2 conditionally for a process in state j and age group T_1. Define the matrix $\underset{\sim}{S}(T_2,T_1) = (s_{ij}(T_2,T_1))$. For simplicity, let $T_1 = (x,x+u)$ and $T_2 = (x+u,x+2u)$. The matrix $\underset{\sim}{S}(T_2,T_1)$ is clearly a measure of location for the matrices $\underset{\sim}{Q}(x+u+s,x+s)$ for $0 \leq s \leq u$. A natural location measure may be the weighted mean (Nour and Suchindran (1983b)),

$$\underset{\sim}{S}(T_2,T_1) = \int_0^u \underset{\sim}{Q}(x+u+s,x+s) \quad . \quad \underset{\sim}{W}(x,s)ds \ , \tag{11}$$

where $W(x,s)$, $0<s<u$, is a set of diagonal matrices such that

$$\underset{\sim}{0} \leq \underset{\sim}{W}(x,s) \leq \underset{\sim}{I} \text{ and } \int_0^u \underset{\sim}{W}(x,s)ds = \underset{\sim}{I}$$

2.2. THE COMPETING RISKS MODEL

The stochastic process under study, $\{X(t);t\varepsilon T\}$, has a finite state space H. Consider the related Markovian process $\{X^c(t);t\varepsilon T\}$ which arises in the hypothetical situation where the state space H is restricted to a nonempty set H_c, $H_c \subset H$. For any $i,j\varepsilon H_c$, define the forces of transition $r^c_{ij}(t)$ for the process $\{X^c(t);t\varepsilon T\}$ exactly as $r_{ij}(t)$ are defined for $\{X(t);t\varepsilon T\}$. Assume that

$$r^c_{ij}(t) = r_{ij}(t) \ , \ i \neq j \ , \ i,j\varepsilon H^c \quad . \tag{12}$$

This is the usual assumption employed in the development of the competing risks model (e.g. Gail (1975), Tsiatis (1975), Fleming (1978)). It serves to assure that various life table functions which describe the process $\{X^c(t);t\varepsilon T\}$ are well defined. Consequently, the matrix $\underset{\sim}{R}^c(t) = (r^c_{ij}(t))$ can be obtained from the original matrix $\underset{\sim}{R}(t) = (r_{ij}(t))$ by deleting the rows and columns of $\underset{\sim}{R}(t)$ not corresponding to states in H^c. The diagonal elements of $\underset{\sim}{R}(t)$ are obtained from the condition $\sum_{i\varepsilon H^c} r^c_{ij}(t) = 0$. The matrix $\underset{\sim}{R}^c(t)$ is then used to compute the various life table functions as described in the previous section.

Assumption (12) may also be extended to cover situations where the pattern of movements in the transient set H_1 is altered by changing the values of certain elements of $\underset{\sim}{R}(t)$. These situations are useful for studying the effects of changing transition patterns on the associated mortality process (i.e., movements to the absorbing set H_2). According to (12), such changes do not affect the values of the off-diagonal elements of $\underset{\sim}{R}(t)$. Only the diagonal elements are recomputed. Clearly, assumption (12) must be critically evaluated in the context of each particular application.

2.3. GENERALIZATIONS

The various life table functions derived in the previous section are independent of the past transition history of the process being considered, as a result of the underlying Markovian assumption. This situation is acceptable for ordinary life tables where the transient set H_1 is composed of one state and the process does not extend beyond first passages. For multistate life tables, however, this assumption may be inadequate, since the past history of individual movements among transient states may influence subsequent behavior.

The advantage of the Markovian formulation is that it requires data that are obtained from the usual demographic sources. A reconsideration of its basic assumptions has been motivated, however, by the recent availability of large longitudinal data files on human populations where repeated observations are made on the same individuals over long periods of time. Methods that allow for the use of such data on sample paths have obvious advantages. Mode (1976) and Littman and Mode (1977) propose a formulation in which a life table is presented as an absorbing age-dependent semi-Markov process. Such a process is defined in terms of the state space H and a $(\ell+n)x(\ell+n)$ one step transition matrix $\underset{\sim}{A}(x,t) = (A_{ij}(x,t))$. The function $A_{ij}(x,t)$ is the probability that an individual of age x entering state j at time t=0 makes a transition to state i during the interval (0,t]. Making the assumption that all individuals who enter a given state at the same age, at any time, will have identical probabilistic behavior, the various life table functions are derived in terms of $\underset{\sim}{A}(x,t)$ and are, accordingly, specific to the ages at which the process enters the various transient states.

The structure can be further generalized by considering more characteristics of sample paths. These may include the number of transient states visited by a given age, the ages when entering these states and calendar time variability in the basic functions. This is the customary procedure for accounting for population heterogeneity by grouping the population on variables that are important in the analysis. In general, the grouping variables may be exogenous, or they may be characteristics of individuals' previous life histories.

Population heterogeneity can also be partially accommodated by introducing covariates into the life table model. Let $\underset{\sim}{Z}$ be a vector with $k \geq 1$ covariates (i.e., explanatory variables whose values are known at the individual level). The usual practice is to formulate the forces of transition $r_{ij}(t)$ as a regression equation of $\underset{\sim}{Z}$ and t (e.g. Coleman (1964, 1981); Cox (1975); Kalbfleisch and Prentice (1979); Menken et al. (1981); Hennessey (1983)).

Unobservable or intrinsic variability among individuals may also be taken into account, as failure to do so may lead to biases in measuring and interpreting the various life table functions (e.g. Heckman and Singer (1982); Nour and Suchindran (1983a); Vaupel et al. (1979)). Each individual is assigned an unobservable parameter, say ε_i for individual i, according to a certain probability distribution. The process on the population level is a mixture of individual processes, each of which is treated as a homogeneous process.

As the realism of the Markovian assumption seems to be a problem in multistate life tables, the need arises for methods for validating the model. At present there is no well-tested method available for this purpose. Primitive tests suggested by Singer and Spilerman (1976) include examining whether the product of estimated transition matrices for two successive age groups equals the transition probability matrix estimated from the period that spans the two initial periods. Also, the probability matrix $\underset{\sim}{Q}(t,s)$ estimated from the model may be compared to the actual probability matrix computed directly from the data (on sample paths).

3. COMPUTATIONAL STRATEGIES

3.1. GENERAL

All life table functions described in (2.1) may be derived from a knowledge of the matrix of transition probabilities $\underset{\sim}{Q}$. For a given age group (x,x+h), and by applying the mean value theorem, equation (5) may be written as

$$\underset{\sim}{Q}(x+h,x) = \underset{\sim}{I} + \underset{\sim}{R}^*(x,h)\int_0^h \underset{\sim}{Q}(x+u,x)du \ , \tag{13}$$

where the elements of $\underset{\sim}{R}^*(x,h)$, $r^*_{ij}(x,h)$, are such that

$$\min_{0 \le u \le h} r_{ij}(x+u) \le r^*_{ij}(x,h) \le \max_{0 \le u \le h} r_{ij}(x+u).$$

Rearranging (13),

$$\underset{\sim}{R}^*(x,h) = [\underset{\sim}{Q}(x+h,x) - \underset{\sim}{I}][\int_0^h \underset{\sim}{Q}(x+u,x)du]^{-1} \ . \tag{14}$$

Recalling that $\underset{\sim}{Q}(x+h,x)$ satisfies the Chapman-Kolmogorov equations, namely

$$\underset{\sim}{Q}(x+h,0) = \underset{\sim}{Q}(x+h,x).\underset{\sim}{Q}(x,0), \tag{15}$$

equation (14) becomes

$$\underset{\sim}{R}^*(x,h) = [\underset{\sim}{Q}(x+h,0) - \underset{\sim}{Q}(x,0)][\int_0^h \underset{\sim}{Q}(x+u,0)du]^{-1} \ . \tag{16}$$

The function $\underset{\sim}{R}^*(x,h)$ as defined by (16) is the usual matrix of age-specific life table occurrence/exposure rates of transitions in the state space H (Willekens et al. (1982)). Solutions for $\underset{\sim}{Q}(x+h,x)$ in terms of these rates are given here.

A solution for $\underset{\sim}{Q}$ may be obtained under a linear integration hypothesis (Rogers and Ledent (1976)). According to this hypothesis,

$$\int_0^h \underset{\sim}{Q}(x + u)du = \frac{h}{2}[\underset{\sim}{Q}(x + h,x) + \underset{\sim}{I}] .$$

Substituting in (13), it follows that

$$\underset{\sim}{Q}(x + h,x) = (I - \frac{h}{2}\underset{\sim}{R}^*(x,h))^{-1}(I + \frac{h}{2}\underset{\sim}{R}^*(x,h)). \qquad (17)$$

The application of (17), however, can lead to implausible results such as negative elements of $\underset{\sim}{Q}$ (Hoem and Jensen (1982); Nour and Suchindran (1984a)). These problems arise when h becomes too large.

Alternatively, a solution for Q may be obtained by assuming that

$$\frac{d\underset{\sim}{R}^*(x,u)}{du} = \underset{\sim}{Q} \text{ , } 0 \le u \le h \qquad (18)$$

(Nour and Snchindran (1984a,b)). This assumption states that the rate of transition from state j to state i for all $i,j \varepsilon H$ over the age interval $(x,x+h)$ is the same for all possible subsets of that interval. This is a standard assumption used in the analysis of age dependent demographic phenomena, where the width of the interval h must be selected to minimize the within interval variability in the age-specific transition rates.

Under this assumption, and recalling (2), the solution to (13) becomes

$$\underset{\sim}{Q}_{11}(x + u,x) = \exp\{u\underset{\sim}{R}^*_{11}(x,h)\} \text{ , } 0 \le u \le h \text{ ,}$$

and

$$\underset{\sim}{Q}_{21}(x + u,x) = \underset{\sim}{R}^*_{21}(x,h).\underset{\sim}{R}^{*-1}_{11}(x,h)[\underset{\sim}{Q}_{11}(x + u,x) - \underset{\sim}{I}] \qquad (19)$$

$$0 \le u \le h \text{ ,}$$

provided that $\underset{\sim}{R}^*_{11}(x,h)$ is a non-singular matrix. It follows from (9) that

$$\underset{\sim}{E}_{11}(x + u,x) = \underset{\sim}{R}^{*-1}_{11}(x,h)[\underset{\sim}{Q}_{11}(x + u,x) - \underset{\sim}{I}] \text{ , } 0 \le u \le h \text{ ,}$$

and

$$\underset{\sim}{E}_{21}(x + u,x) = \underset{\sim}{R}^*_{21}(x,h)\underset{\sim}{R}^{*-1}_{11}(x,h)[\underset{\sim}{E}_{11}(x + u,x) - u\underset{\sim}{I}] \text{ ,} \qquad (20)$$

$$0 \le u \le h \text{ .}$$

Similarly, (10) leads to

$$\underset{\sim}{M}_{11}(x+u,x) = \underset{\sim}{B}^{*}_{11}(x,h)\underset{\sim}{E}_{11}(x+t,x) \ , \quad 0 \le u \le h \quad ,$$

and

$$\underset{\sim}{M}_{21}(x+u,x) = \underset{\sim}{Q}_{21}(x+u,x) \ , \quad 0 \le u \le h \ . \qquad (21)$$

The computational formulas (17), (19), (20), and (21) may be extended beyond a particular age group using the Chapman-Kolmogorov equations (Nour and Suchindran (1984b)).

A computational formula for the elements of $R^{*}(x,h)$ under assumption (18) is straightforward. The rates $r^{*}_{ij}(x,h)$ are defined as

$$r^{*}_{ij}(x,h)\Delta t + o(\Delta t) = P[\text{an individual in state } j \text{ at age } x{+}t \text{ will be in state } i \text{ at age } x{+}t{+}\Delta t] \ , \quad i \ne j \ , \quad 0 \le t \le h \ .$$

Let $N_{ij}(x,t)$ be the number of transitions from state j to state i over the interval $(x,x+t)$ and $N_j(x+t)$ be the number of individuals in state j at age x+t, $0 \le t \le h$, for the life table population. Then

$$r^{*}_{ij}(x,h)\Delta t + o(\Delta t) = \frac{N_{ij}(x,t+\Delta t) - N_{ij}(x,t)}{N_j(x+t)} \ , \quad i \ne j \ .$$

This leads to

$$r^{*}_{ij}(x,h)N_j(x+t) = \frac{d}{dt} N_{ij}(x,t) \ .$$

Integrating between 0 and h and noting assumption (18) leads to

$$r^{*}_{ij}(x,h) = \frac{N_{ij}(x,h)}{\int_0^h N_j(x+t)dt} \ , \quad i \ne j \ , \qquad (22)$$

which is the usual definition of the occurrence/exposure rate of transition from state j to state i during the age interval $(x,x+h)$. For i=j, $r^{*}_{ij}(x,h)$ is determined by the condition $\sum_i r^{*}_{ij}(x,h) = 0$.

To use (22) in practice, an additional assumption is introduced: each $r^{*}_{ij}(x,h)$ is adequately approximated by the corresponding rate observed for the population under study, say $R_{ij}(x,h)$. Equating life table rates to observed occurrence/exposure rates implies sectional stationarity, i.e., within each age interval and within each life status of the model the observed population is stationary. Alternative assumptions for the relationship between $r^{*}_{ij}(x,h)$ and $R_{ij}(x,h)$ are found in the literature. These include a sectional stability assumption (Keyfitz (1970); Keyfitz et al. (1972)), as well as other structures. These alternative structures, however, seem to make very little difference in numerical applications (Hoem and Jensen (1982)).

3.2. PERIOD LIFE TABLES

These tables are computed by assuming that period data (i.e., data across age groups observed for a particular calendar period) provide adequate estimates of life table functions to be used in longitudinal analysis. The current occurrence/exposure rates observed in the population are assumed to be equal to the life table rates, across all age groups. The resulting tables do not describe the experience of an actual cohort. Instead, it assumes a synthetic cohort that is subject to the conditions observed in the actual population during the particular period. Caution must be exercised, therefore, in making longitudinal inferences for real cohorts based on these period estimates.

Period life tables are excellent summary measures of prevailing conditions. If data are available, then time trends as well as differences among populations can be identified from a comparison of these tables.

3.3. LIFE TABLES FROM COHORT DATA

The computational formulas discussed in (3.1) can be used with one of two types of cohort data: event-history data, which record the exact time and sequence of various individual transitions, and aggregate change data, which record the number and nature of transitions made by the members of the population in some time interval (Hannan (1982), Tuma et al. (1979)). Although the latter type of data is the only data required for computing the model, event-history data may be needed if the analyst wishes to experiment with different values for the width h of age intervals. In addition, event-history data permit the generalization of the Markovian model along the lines described in (2.3).

To compute (22) from real cohort data, where ascertainments are made at discrete time points, we may assume, for appropriately small values of h, that

$$N_j(x+t) = N_j(x)\exp\{\nu_j t\} \quad , \; 0 \le t \le h$$

where ν_j is the appropriate rate of change in $N_j(x)$, $j \in H$, over $(x,x+h)$. Then (22) becomes

$$r^*_{ij}(x,h) = N_{ij}(x,h)\,\frac{\ell n\, N_j(x+h) - \ell n\, N_j(x)}{h(N_j(x+h) - N_j(x))} \quad , \; i \ne j \; . \tag{23}$$

Alternatively, the model may be determined by first computing the transition probabilities $q_{ij}(x+t,x)$ with t being integer multiples of h, as

$$q_{ij}(x+t,x) = \frac{N_{ij}(x,t)}{N_j(x)} \quad , \ i,j\varepsilon H \ . \tag{24}$$

Appropriate adjustments must be made in (24) to account for loss to follow-up if it occurs. A common procedure is to replace $N_j(x)$ by $N_j(x) - \frac{h}{2}W_j(x)$, where $W_j(x)$ are those lost to follow-up out of state j during the interval $(x,x+h)$(consult, for example, Elandt-Johnson and Johnson (1980)). A study of embeddability may be carried out by comparing the life tables obtained under (23) and (24) respectively.

When data are obtained through a retrospective survey, the method of ascertainment has the consequence of including only life histories for individuals who are alive in a certain life status at the time of interview. (Thus, no transitions to the death states are observed.) Examples include data collected for ever-married women, data collected for individuals with a terminal disease and data collected for women with a given completed family size. The inherent selectivity of such data can lead to biases (Hoem (1969); Aalen et al.(1980); Borgan (1980); Heckman (1979); Sheps and Menken (1973)).

The usual procedure for dealing with the lack of observed data transitions is to delete the rows and columns which correspond to these states in the manner implied by assumption (12). Alternatively, this assumption has also been used to include rates for death transitions estimated from other data sources (Espenshade (1982)). On the other hand, restricting the survey to those in a certain life status at the time of interview is tantamount to observing a purged Markov process. Hoem (1969) demonstrates that if the life status at the time of interview constitutes an absorbing proper subset A of states in H (every state in A need not be absorbing), then life histories after inclusion are not affected by selection biases. In addition, these biases are not present for life histories preceding entry into A provided that the risk of moving into A is the same for all states in H-A. In these situations, the purged and original processes coincide. Otherwise, the results must be interpreted with caution.

4. AN APPLICATION

In this section we describe an application of multistate life tables to study some aspects of marital status patterns for black and white females in North Carolina in 1970. Previous applications of multistate life tables to marital status patterns include Espenshade (1982), Nour and Suchindran (1984b), Schoen and Nelson (1974) and Willekens et al.(1982).

The calculations are based on data across age groups which are observed for

North Carolina females during 1970. The methods employed for constructing these period life tables are Markovian. It is useful, therefore, to assess the adequacy of the Markovian model in describing marital status processes.

A simple example based on data on marital histories of ever-married women in the United States as reported in the June 1980 Current Population Survey is given in Table 1. The transient set H_1 is composed of the four possible marital states: single, married, widowed and divorced. The actual and estimated probabilities $q_{ij}(t,s)$ from these data are compared. Based on these limited results, the Markovian assumption appears to be adequate in following the various individual marital transitions. The differences between actual and estimated probabilities increase as the sample size decreases.

Table 1. Actual and Estimated Matrices $\underset{\sim}{Q}(t,s)$ for Women Aged 40 or Older at Time of Interview; United States Current Population Survey, 1980. (estimated probabilities are given in parentheses)

	Starting State			
	Single	Married	Widowed	Divorced
	$\underset{\sim}{Q}(30,20)$			
	.1214 (.1214)	--- (---)	--- (---)	--- (---)
	.8450 (.8346)	.9245 (.9362)	.6529 (.6506)	.6879 (.7699)
	.0082 (.0108)	.0186 (.0161)	.3160 (.3238)	.0076 (.0107)
	.0254 (.0332)	.0569 (.0477)	.0311 (.0256)	.3045 (.2193)
Sample Size:	17314	10690	75	295
	$\underset{\sim}{Q}(40,20)$			
	.0259 (.0259)	--- (---)	--- (---)	--- (---)
	.9051 (.8922)	.8961 (.9097)	.7256 (.7757)	.7873 (.8625)
	.0216 (.0271)	.0345 (.0286)	.1735 (.1788)	.0141 (.0250)
	.0474 (.0548)	.0694 (.0617)	.1009 (.0455)	.1986 (.1125)
Sample Size:	17314	10690	75	295

Life table functions for North Carolina females are given in subsequent tables. These functions represent the experiences of hypothetical cohorts of individuals who are subject to age-specific marital status transition rates and death rates that have been observed in North Carolina, 1970. Tables 2, 3 and 4 give examples of the matrix functions $\underset{\sim}{Q}$, $\underset{\sim}{E}$ and $\underset{\sim}{M}$ respectively. In Table 5, summary statistics on lifetime experiences of mortality, marriage, marital disruption and remarriage are presented.

The results shown in Tables 2, 3 and 4 highlight some racial differences in marital status patterns in North Carolina in 1970. Blacks have a higher mean age at marriage (as evidenced by the larger proportion remaining single at age 50 for those who are single at age 20). Also, because of higher mortality for blacks, married black females are more likely to experience a widowhood. They are exposed, in addition, to higher rates of divorce. Those black females who are currently widowed or divorced are less likely to change their marital status than their white counterparts. It is also seen that black females have significantly higher mortality than whites and that these mortality differentials affect marital status patterns.

The same racial differences are reflected in the summary measures given in Table 5. These measures are computed from the matrix functions $\underset{\sim}{Q}$, $\underset{\sim}{E}$ and $\underset{\sim}{M}$ after a proper interpretation of their elements in the current context of marital status analysis. Table 5 describes the life time experience of a hypothetical cohort of 15-year-old single women. Similar tables may be constructed for other combinations of age and marital status.

Table 2. Examples of the Matrix $\underset{\sim}{Q}(t,s)$; White and Black Females, North Carolina, 1970

Whites: Starting State					Blacks: Starting State				
Single	Married	Widowed	Divorced	Dead	Single	Married	Widowed	Divorced	Dead
		Q(50,20)					Q(50,20)		
.0735	0	0	0	0	.1510	0	0	0	0
.7490	.8151	.6939	.8133	0	.5539	.6798	.3312	.6601	0
.0405	.0449	.1809	.0445	0	.0727	.0970	.4744	.0895	0
.0845	.0929	.0741	.0932	0	.0763	.0999	.0403	.1174	0
.0525	.0471	.0511	.0490	1	.1461	.1233	.1541	.1330	1
		Q(50,30)					Q(50,30)		
.5650	0	0	0	0	.4418	0	0	0	0
.3215	.8229	.5251	.7693	0	.3241	.7018	.2120	.5750	0
.0145	.0433	.3777	.0374	0	.0320	.0911	.6186	.0619	0
.0306	.0927	.0479	.1473	0	.0343	.0985	.0199	.2263	0
.0684	.0411	.0493	.0460	1	.1678	.1086	.1495	.1368	1
		Q(85,50)					Q(85,50)		
.3330	0	0	0	0	.2887	0	0	0	0
.0010	.0088	.0025	.0037	0	.0003	.0030	.0005	.0012	0
.0190	.2458	.2914	.0856	0	.0168	.1956	.1873	.0650	0
.0010	.0166	.0028	.0544	0	.0051	.0811	.0065	.3140	0
.6460	.7288	.7037	.8563	1	.6891	.7203	.8057	.6198	1

Table 3. Examples of the Matrix $\underset{\sim}{E}(t,s)$; White and Black Females, North Carolina, 1970

	White					Black				
	Starting State					Starting State				
	Single	Married	Widowed	Divorced	Dead	Single	Married	Widowed	Divorced	Dead
$\underset{\sim}{E}(50,20)$	5.58	0	0	0	0	10.32	0	0	0	0
	22.51	27.95	17.11	26.29	0	16.46	26.19	7.87	20.95	0
	.35	.41	11.59	.40	0	.70	1.06	20.03	.89	0
	1.04	1.20	.83	2.82	0	1.02	1.51	.48	6.69	0
	.52	.44	.47	.49	30	1.50	1.24	1.62	1.47	30
$\underset{\sim}{E}(50,30)$	14.20	0	0	0	0	13.25	0	0	0	0
	4.89	18.39	7.33	13.50	0	4.83	17.19	2.66	9.86	0
	.09	.32	11.97	.24	0	.21	.78	15.89	.43	0
	.26	1.00	.38	5.92	0	.29	1.15	.15	8.43	0
	.56	.29	.32	.34	20	1.42	.88	1.30	1.28	20
$\underset{\sim}{E}(85,50)$	25.89	0	0	0	0	21.98	0	0	0	0
	.76	16.34	2.41	3.97	0	.81	13.92	1.02	2.87	0
	.45	7.71	24.08	2.27	0	.48	7.10	20.34	1.74	0
	.12	3.13	.40	19.43	0	.15	3.01	.19	19.82	0
	7.78	7.82	8.11	9.33	35	11.58	10.97	13.45	10.57	35

Table 4. Examples of the Matrix $\underset{\sim}{M}(t,s)$; White and Black Females, North Carolina, 1970

Whites					Blacks				
Starting State					Starting State				
Single	Married	Widowed	Divorced	Dead	Single	Married	Widowed	Divorced	Dead
		$\underset{\sim}{M}(50,20)$					$\underset{\sim}{M}(50,20)$		
0	0	0	0	0	0	0	0	0	0
1.063	.185	.946	1.171	0	.881	.143	.495	1.068	0
.057	.065	.048	.064	0	.091	.124	.047	.113	0
.223	.265	.175	.255	0	.165	.241	.081	.208	0
.053	.047	.051	.049	0	.146	.123	.154	.133	0
		$\underset{\sim}{M}(50,30)$					$\underset{\sim}{M}(50,30)$		
0	0	0	0	0	0	0	0	0	0
.406	.114	.654	.997	0	.445	.090	.280	.815	0
.018	.058	.028	.048	0	.038	.112	.022	.073	0
.056	.199	.084	.152	0	.055	.189	.030	.111	0
.068	.041	.049	.046	0	.168	.109	.150	.137	0
		$\underset{\sim}{M}(85,50)$					$\underset{\sim}{M}(85,50)$		
0	0	0	0	0	0	0	0	0	0
.068	.052	.205	.326	0	.074	.041	.096	.276	0
.044	.633	.132	.209	0	.047	.609	.061	.177	0
.010	.213	.031	.052	0	.010	.184	.013	.035	0
.646	.729	.704	.856	0	.689	.720	.806	.620	0

Table 5. Summary Measures of Life Time Marital Experiences by Race for Hypothetical Cohorts of 15-Year-Old Single Women; North Carolina, 1970

	Measure	Whites	Blacks
(1)	Proportion Ever Marrying	.950	.861
(2)	Expected Age at First Marriage	22.75	30.43
(3)	Mean Age at First Marriage for those Marrying	20.42	24.20
(4)	Expected Number of Marriages Per Woman	1.20	1.00
(5)	Expected Number of Years Spent in the Married State	39.42	27.50
(6)	Proportion of Life Expectancy Spent in the Married State	.64	.50
(7)	Expected Duration of a Marriage	32.85	27.50
(8)	Proportion of Marriages Ending in Divorce	.35	.30
(9)	Proportion of Marriages Ending in Widowhood	.49	.51
(10)	Expected Number of Years Spent in the Divorced State	5.45	4.54
(11)	Proportion of Life Expectancy Spent in the Divorced State	.09	.08
(12)	Expected Number of Years Spent in the Widowed State	8.90	8.11
(13)	Proportion of Life Expectancy Spent in the Widowed State	.14	.15
(14)	Expected Duration of a Divorce	13.10	15.36
(15)	Expected Duration of a Widowhood	15.20	17.40
(16)	Expected Number of Remarriages	.25	.14

ACKNOWLEDGEMENTS: This research was supported by training grant HD07237 from the National Institute of Child Health and Human Development and by research grant R01-AG-04131 from the National Institute of Aging. We thank Larry Sink for the programming assistance.

REFERENCES

[1] Aalen, O.O., Borgan, O., Keiding, N. and Thorman, J., Interaction between life history events. Nonparametric analysis for prospective and retrospective data in the presence of censoring. Scand. J. of Statist. 7 (4): 161-171 (1980).

[2] Borgan, O., Applications of non-homogeneous Markov chains in medical studies. Nonparametric analysis for prospective and retrospective data. University of Oslo, Institute of Mathematics. Statistical Research Report 1980:8 (1980).

[3] Chiang, C.L., A stochastic model of competing risks of illness and competing risks of death. In: Stochastic Models in Medicine and Biology, Gurland, J. (ed.), Madison: University of Wisconsin Press, pp. 323-354 (1964).

[4] Coleman, J.S., Introduction to Mathematical Sociology, New York: Free Press (1964).

[5] Coleman, J.S., Estimating individual-level transition probabilities for multi-state life tables. Paper presented at the Conference on Multi-dimensional Mathematical Demography held at the University of Maryland, College Park, March 23-25 (1981).

[6] Cox, D.R., Regression models and life tables (with discussion), J. Roy. Statist. Soc., Series B 34:187-220 (1972).

[7] Elandt-Johnson, R.C. and Johnson, N.L., Survival Models and Data Analysis, New York: Wiley (1980).

[8] Espenshade, T., Marriage, divorce and remarriage from retrospective data: a multiregional approach. Paper presented at the Annual Meeting of the Population Association of America, San Diego (1982).

[9] Feller, W., On the integrodifferential equations of purely discontinuous Markoff processes, Trans. Amer. Math. Soc. 48:488-515 (1940).

[10] Fix, E. and Neyman, J., A simple stochastic model of recovery, relapse, death and loss of patients, Human Biology 23:205-241 (1951).

[11] Fleming, T.R., Nonparametric estimation for nonhomogeneous Markov processes in the problem of competing risks, Ann. of Statist. 6 (5):1057-1070 (1978).

[12] Gail, M., A review and critique of some models used in competing risks analysis , Biometrics 31:209-222 (1975).

[13] Hannan, M.T., Multistate demography and event history analysis. Working Papers. International Institute for Applied Systems Analysis, Laxenburg, Austria (1982).

[14] Heckman, J.J., Sample selection bias as a specification error, Econometrica 47 (1):153-161 (1979).

[15] Heckman, J.J. and Singer, B., Population heterogeneity in demographic models. In: Multidimensional Mathematical Demography, Land, K.C. and Rogers, A. (eds.), New York: Academic Press, pp. 567-599 (1982).

[16] Hennessey, J.C., An age dependent, absorbing, semi-Markov model of work histories of the disabled, Math. Biosciences 51:283-304 (1980).

[17] Hennessey, J.C., Testing the predictive power of a proportional-hazard semi-Markov population model, Math. Biosciences 67:193-212 (1983).

[18] Hoem, J.M., Purged and partial Markov chains, Scand. Act. J. 52:147-155 (1969).

[19] Hoem, J.M., Probabilistic fertility models of the life table type, Theoretical Pop. Biol. 1:12-38 (1970).

[20] Hoem, J.M., Inhomogeneous semi-Markov processes, select actuarial tables, and duration-dependence in demography. In: Population Dynamics, Greville, T.N.E. (ed.), New York: Academic Press, pp. 251-296 (1972).

[21] Hoem, J.M. and Fong, M.S., A Markov chain model of working life tables. Working Paper No. 2. Laboratory of Actuarial Mathematics, University of Copenhagen, Denmark (1976).

[22] Hoem, J.M. and Jensen, U.F., Multistate life table methodology: a probabilist critique. In: Multidimensional Mathematical Demography, Land, K.C. and Rogers, A. (eds.), New York: Academic Press, pp. 155-264 (1982).

[23] Kalbfleisch, J.D. and Prentice, R.L., The Statistical Analysis of Failure Time Data, New York: Wiley (1980).

[24] Keyfitz, N., Finding probabilities from observed rates or how to make a life table, The Amer. Statist. 24 (1):28-33 (1970).

[25] Keyfitz, N., Multidimensionality in population analysis, Sociological Methodology 1980:191-218 (1980).

[26] Keyfitz, N., Preston. S. and Schoen, R., Inferring probabilities from rates: extension to multiple decrement, Scand. Act. J., 55 (1):1-13 (1972).

[27] Krishnamoorthy, S., Classical approach to increment-decrement life tables: an application to the study of the marital status of United States females, 1970, Math. Biosciences 44:138-154 (1979).

[28] Ledent, J., Multistate life tables: movement versus transition perspectives, Environment and Planning A 12:533-562 (1980).

[29] Littman, G.S. and Mode, C.J., A non-Markovian model of Taichung medical IUD experiment, Math. Biosciences 34:279-302 (1977).

[30] Menken, J., Trussell, J., Stempel, D. and Babokol, O., Proportional hazards life table models: an illustrative analysis of socio-demographic influences on marriage dissolution in the United States, Demography 18:181-200 (1980).

[31] Mode, C.J., On the calculation of the probability of current family-planning status in a cohort of women. Math. Biosciences 31:105-120 (1976).

[32] Mode, C.J., Increment-decrement life tables and semi-Markovian processes from a sample path perspective. In: Multidimensional Mathematical Demography, Land, K.C. and Rogers, A. (eds.), New York: Academic Press, pp. 535-565 (1982).

[33] Nour, E. and Suchindran, C.M., The effect of individual variability on the life table, Communications in Statistics (Theory and Methods) 12 (3):323-339 (1983a).

[34] Nour, E. and Suchindran, C.M., A general formulation of the life table, Math. Biosciences 63:241-252 (1983b).

[35] Nour, E. and Suchindran, C.M., The construction of multi-state life tables: a comment on the article by Willekens et al., Pop. Studies 38 (2) (1984a).

[36] Nour, E. and Suchindran, C.M., Multi-state mortality by cause of death: a life table analysis, J. Roy. Statist. Soc. Series A 147 (1984b).

[37] Rogers, A., The multiregional life table, J. Math. Sociol. 3:127-137 (1973).

[38] Rogers, A., Introduction to multi-state mathematical demography, Environment and Planning A 12:489-498 (1980).

[39] Rogers, A. and Ledent, J., Increment-decrement life tables: a comment, Demography 13:287-290 (1976).

[40] Schoen, R., Constructing increment-decrement life tables, Demography 12: 313-324 (1975).

[41] Schoen, R. and Land, K.C., A general algorithm for estimating a Markov-generated increment-decrement life table for applications to marital status patterns, J. Amer. Statist. Assoc. 74:761-776 (1979).

[42] Schoen, R. and Nelson, V.E., Marriage, divorce and mortality: a life table analysis, Demography 11:267-290 (1974).

[43] Sheps, M.C. and Menken, J.A., Mathematical Models of Conception and Birth, Chicago: University of Chicago Press (1973).

[44] Singer, B. and Spilerman, S., Some methodological issues in the analysis of longitudinal surveys, Ann. of Econ. and Social Measurements 5 (4):447-474 (1976).

[45] Sverdrup, E., Estimates and test procedures in connection with stochastic models for deaths, recoveries and transfers between different states of health, Scand. Act. J. 48:184-211 (1965).

[46] Tsiatis, A., A non-identifiability aspect of the problem of competing risks, Proceedings of the National Academy of Sciences 72:20-22 (1975).

[47] Tuma, N.D., Hannan, M.T. and Groeneveld, L.P., Dynamic analysis of event histories, Amer. J. Sociol. 84:820-854 (1979).

[48] Vaupel, J.W., Manton, K.G. and Stallard, E., The impact of heterogeneity in individual frailty on the dynamics of mortality, Demography 16:439-454 (1979).

[49] Willekens, F.J., Shah, I., Shah, J.M. and Ramachandran, P., Multi-state analysis of marital status life tables: theory and applications, Pop. Studies 36:129-144 (1982).

EPIDEMIOLOGY AND ENVIRONMENTAL BIOSTATISTICS

BIOSTATISTICS: Statistics in Biomedical, Public Health
and Environmental Sciences, edited by P.K. Sen

RISK FACTOR PATTERNS AMONG SUBGROUPS IN CASE-CONTROL STUDIES: AN APPLICATION OF MULTINOMIAL LOGISTIC REGRESSION

Neil Dubin and Bernard S. Pasternack

Institute of Environmental Medicine
New York University Medical Center
550 First Avenue
New York, New York 10016

When more than two disease categories are to be compared by means of logistic regression, a multinomial extension of the usual binary (binomial) regression model is appropriate. Although one may compare the several disease subgroups in pairs by the standard method, advantages of the multinomial approach include simultaneous estimation of the subgroup-specific parameters and direct hypothesis testing involving as many of the subgroups as required. Disadvantages primarily involve lack of necessary computer software and much increased computer costs.

The method is applied to a large case-control study of breast cancer involving three unordered disease categories and both categorical and continuous explanatory variables. Epidemiologic interpretation of multinomial regression outputs is emphasized, particularly with respect to the assessment of differences in risk factor patterns among case subgroups.

KEY WORDS: Multinomial logistic regression, risk factors, case-control studies

INTRODUCTION

Usual binary (binomial) logistic regression concerns the estimation of the probability of response versus non-response (or diseased versus disease-free, etc.) as a function of several explanatory variables. The technique has proven extremely useful in epidemiology, clinical studies of disease, and bioassay. If more than two groups are to be compared, a multinomial logistic model becomes appropriate. This would occur in a case-control study for which there were both hospital-based and population-based controls to compare to cases, or several case subgroups to compare to controls or, conceivably, many types of both cases and controls. Alternatively, one may need to discriminate between several diseases on the basis of symptoms or other factors. Or, response to various agents may be one of several levels of toxicity. One option in the polychotomous-response situation is to perform separate binomial logistic regressions between the various groups two at a time [Breslow and Day (1980), pp. 111-112; Begg and Gray (1984)]. Nevertheless, a unified, multinomial approach reflects the underlying statistical model, permitting simultaneous estimation of parameters and straightforward hypothesis testing.

The theoretical framework for the multinomial logistic model has been developed extensively [Mantel (1966), Cox (1966), Walker and Duncan (1967), Bock (1970), Jones (1975), Prentice and Breslow (1978), Levin and Shrout (1981), Anderson (1982a)]. Widespread application of the technique, however, has been severely limited by the unavailability of computer programs to perform the necessary iterative maximum likelihood calculations. Multinomial logistic analysis with an entirely categorical set of independent variables may be performed with such computer packages as GLIM. The incorporation of continuous variables into the regression has generally required that individual researchers develop their own programs, although SAS may be used in the case of an ordinal response variable.

Our own application of multinomial logistic regression uses data from a recently completed case-control study of breast cancer in a screened population [Dubin, Hutter, Strax, et al (1984); Dubin, Pasternack and Strax (1984)]. As we shall see below, this study involved potentially unordered multiple case subgroups, as well as both continuous and categorical risk factors. We use a computer program appropriate to this data situation, one recently developed by Thomas, Dewar and Goldberg (1983), which we adapted for use on the CDC Cyber.

MINIMAL BREAST CANCER STUDY

The Minimal Breast Cancer Study was motivated by the enrollment of large numbers of women in breast cancer screening programs in the 1970's. Breast cancer detected in a screened population is more likely to be pre-invasive or minimally invasive than breast cancer detected in usual clinical practice. At issue was whether such *minimal* breast cancer is an inherently less invasive form of disease which would remain subclinical if undetected. If so, many women may unnecessarily undergo mastectomy. Different distributions of epidemiologic risk factors between minimal cases and clinically invasive cases would support (although by no means prove) this hypothesis. Such risk factors include age, weight, race, family history of breast cancer, reproductive factors, and religion. An absence of risk factor differences would support the hypothesis that minimal breast cancer detected by screening would otherwise progress to clinical disease. Histopathologic review of original hospital material and reports was used to classify breast cancer as minimal (*in situ* or invasive ≤ 1 cm) or clinical (invasive > 1 cm). Control women without breast cancer were used to establish the validity of the study population, enabling one to compare relative risk estimates to other studies. Concerns regarding validity were substantial, given the numerous potential sources of selection bias inherent in using a screened population [Dubin and Pasternack (1984)]. Sample sizes for the logistic analysis consisted of 85 minimal cases, 618 clinical cases and 1965 controls.

The trichotomous nature of the response variable and the availability of data regarding numerous epidemiologic risk factors make this study suitable for multinomial logistic regression analysis. The particular pertinence of this technique, compared to separate binomial logistic regressions, stems from our objective to assess risk factor differences among the disease subgroups: the model is designed not only to detect a significant association between a risk factor and a disease subgroup, but also whether this relationship differs from one subgroup to another.

Although the clinical manifestation of the case subgroups is ordered (no disease vs. minimal disease vs. clinical disease), a principal goal of the study was to assess the evidence that the two cancer subgroups have different etiologies. If the etiologies were different, the ordered response model would be inappropriate. The unordered model was therefore applied, as it is the more general of the two.

MULTINOMIAL LOGISTIC MODEL

In the general case of d mutually exclusive response or disease categories, following the notation of Jones (1975), we write the probability that an observation is in category k as

$$P_k(x_i,\beta) = \exp(x_i'\beta_k) \;/\; \sum_{m=1}^{d} \exp(x_i'\beta_m) \qquad (1)$$

$$\text{for } k = 1,2,\ldots,d$$

where there are n independent observations on the p independent variables, $x_i'=(1,x_{i1},\ldots,x_{ip})$, $i = 1,2,\ldots,n$, and the $\beta_m' = (\beta_{m0}, \beta_{m1},\ldots,\beta_{mp})$, $m = 1,2,\ldots,d$ are the risk coefficients to be estimated. The explanatory variables comprising the x_i may be either categorical or continuous. Of the d(p+1) parameters in the model (1) only (d-1)(p+1) may be uniquely determined. Parameter estimation is accomplished by Newton-Raphson maximization of the log-likelihood

$$\ell(\beta) = \sum_{i=1}^{n} \sum_{k=1}^{d} z_{ki} \log P_k(x_i,\beta) \qquad (2)$$

with respect to the β, where $z_{ki} = 1$ if the ith observation is in disease category k [Thomas et al (1984)]. The appropriateness of this log-likelihood for case-control studies was established by Anderson (1972) and Prentice and Pyke (1979), provided that the sampling probabilities depend only upon disease status and not upon the explanatory variables. The parameters $\beta_{10}, \beta_{20},\ldots, \beta_{d0}$ are nuisance parameters related to absolute disease incidence and the sampling probabilities for the disease categories [Breslow and Day (1980), pp. 192-205]. Fitting the model requires estimation of d-1 of these nuisance parameters, the remaining one being redundant. Similarly for the jth explanatory variable there is a set of regression coefficients $\beta_{1j}, \beta_{2j}, \ldots, \beta_{dj}$, one for each disease category, d-1 of which may be estimated. Although it has been customary, to avoid redundancy, to set all the coefficients to zero for one disease category and then estimate the remaining disease-specific parameters relative to that category, it is not necessary to do so. Further, it may be advisable to do otherwise in order to keep the number of parameters to a minimum. Suppose we follow custom and constrain $\beta_d' = 0$. If explanatory variable j were a good discriminator of disease category d, but provided no discrimination between the other disease groups, the model would have to fit the d-1 parameters $\beta_{1j}, \ldots, \beta_{d-1,j}$ in order to describe this relationship. If, however, one did not so constrain the model, the single parameter β_{dj} would suffice. The disease-factor coefficients β_{kj} may be interpreted as log relative risks. If we set $\beta_{qj} = 0$ the remaining β_{kj}, $k \neq q$, are relative risks of disease category k versus disease category q. If not all the d-1 parameters are included in the model, the interpretation of an included risk coefficient will be as relative to the disease categories for which coefficients were not included.

APPLICATION OF THE MODEL

For the Minimal Breast Cancer study, a trinomial model (d=3) was appropriate. We considered p = 18 potential risk factors, four of which are continuous variables and the remainder indicator variables. Consequently, there were 3-1 = 2 nuisance parameters to estimate and 3x18 = 54 risk coefficients, of which 2x18 = 36 may be uniquely determined. This includes main effects only; the examination of interaction terms requires additional parameters. The large number of main effect parameters alone suggests that forward selection would be more practical than backward elimination as a stepwise variable-selection procedure. Before presenting the full stepwise approach, however, we illustrate the univariate and bivariate cases in turn.

Consider the risk factor FAMILY, which is an indicator for having a mother or sister with a history of breast cancer. The three possible disease-factor coefficients are MINIMAL-FAMILY, CLINICAL-FAMILY, and CONTROL-FAMILY, but only two may be included in the model at once. Clearly, if family history of breast cancer were not associated with the subject's disease status, none of these terms should be statistically significant in the regression. If, however, family history were related to an individual's risk of breast cancer, and the same relationship held for minimal and clinical disease, then this should be reflected by the parameter CONTROL-FAMILY having the greatest statistical significance of the three. Or, both forms of the disease may be associated with family history, but the relative risks may differ significantly. In this case two of the disease-factor coefficients would be required to adequately describe the relationship. Finally, suppose only one form of the disease were associated with family history, in which case the one pertinent disease-specific coefficient should be significant.

Actual logistic regression results for this risk factor are given in Table 1. Model 1 included only the nuisance parameters, but the score statistic to enter each of the three possible FAMILY coefficients was calculated. CLINICAL-FAMILY had the highest Z-test and was included in Model 2. At that point neither of the other two family history coefficients were significant ($Z = \pm 1.03$), the numerical equivalence of the test reflecting their redundancy in the model. The epidemiologic interpretation of this analysis would be that family history of breast cancer was associated with clinical, but not minimal, disease. The relative risk may be calculated as $\exp(\hat{\beta}) = \exp(0.339) = 1.40$. Formally, this is the relative odds of clinical breast cancer versus minimal or no breast cancer among women with a family history compared to women with no family history. If one wished to obtain a relative risk for clinical cases strictly versus controls, this could be easily accomplished by including the term MINIMAL-FAMILY in the regression model in spite of its statistical non-significance. The coefficient of CLINICAL-FAMILY thus obtained (model not shown) gave the relative risk comparing cases to controls, $\exp(0.320) = 1.38$, a negligible quantitative difference. In general for a multinomial model, one must pay extra attention to the identification of the implicit reference group, which depends upon which disease-factor parameters are included in the model and which may be risk-factor-specific.

The score statistics were also calculated for the disease-factor coefficients of another risk factor, LNAGE, the natural logarithm of the subject's age at last screening (Table 1, Model 2). The score statistic was highly significant for all three; the most significant, CONTROL-LNAGE ($Z = -9.98$) was entered into Model 3, along with CLINICAL-FAMILY. The negative coefficient for CONTROL-LNAGE ($\hat{\beta} = -2.40$) indicates a negative relationship between being disease-free and increasing age; in other words, increasing age is a risk factor for breast cancer. The score test indicated no additional contribution would be made by the addition of another LNAGE coefficient ($Z = \pm 0.232$), so that the relationship with age could not be judged different for minimal and clinical disease. The case-control relative risk associated with a ten-year increase in age from 40 to 50 years was $\exp[2.40(\ell n\ 50 - \ell n\ 40)] = 1.71$.

Investigation of possible interactions between risk factors must also reflect the trinomial structure of the model. Given Model 3 (Table 1), which included

CLINICAL-FAMILY and CONTROL-LNAGE, one might wish to examine interaction between age at last screen and family history (AGXFAM = [LNAGE-4][FAMILY], as we defined it). As with main effects, there are three potential disease-interaction coefficients. To assess CLINICAL-AGXFAM, one must have included both main effects CLINICAL-LNAGE and CLINICAL-FAMILY. Only the latter was included in Model 3. Thus, in order to calculate score statistics for entering CLINICAL-AGXFAM and CONTROL-AGXFAM, one must first have obtained Model 4, which included all the necessary main effects. CONTROL-AGXFAM (Z = 2.21) was entered into Model 5, at which point the score statistic for entering CLINICAL-AGXFAM was not significant (Z = 0.821). We removed this latter interaction term's forced main effect, CLINICAL-LNAGE, to obtain Model 6, the log-likelihood for which changed negligibly from Model 5 ($\chi^2_1 = 0.20$). To summarize our findings as represented by Model 6, the minimal and clinical case subgroups had distinct relationships with respect to the family history main effect, but not the age main effect of the interaction between the two. To illustrate the calculation of relative risks, the risk of clinical breast cancer, relative to controls, associated with a family history of breast cancer was, at age 35 years, $\exp\{.881 - .580 - 1.87 \times [(\ell n\ 35)-4]\} = 3.10$, and at age 65 years, 0.98. The analogous relative risks for minimal breast cancer, relative to controls, were 1.29 and 0.40, respectively. The relative risks for clinical disease followed the expected interactive relationship between age and family history [Dubin, Pasternack and Strax (1984)]. However, it does not seem reasonable to conclude that family history had a protective effect against minimal breast cancer in older women. Rather it seems more plausible to suggest that our results were evidence of a shift in diagnosis, that women with a family history tended to be diagnosed at earlier ages and with more invasive disease. One could also formulate an alternative, Model 7 (Table 1), which included a slightly different set of disease-factor coefficients in which only clinical disease was modelled to have an interactive relationship between age and family history. The relative risks of clinical disease associated with family history change only a little from Model 6, being estimated as 2.8 at 35 years of age and 1.05 at 65 years of age. For minimal breast cancer, the relative risk associated with a family history is unity at all ages. This model may be easier to interpret, but there was a decrease in the likelihood compared to Model 6.

For the purpose of selecting a "best" model from among several, one may use

$$\min\{-\ell(\hat{\beta}) + \nu\},$$

where $\hat{\beta}$ is the maximum likelihood estimate and ν is the number of parameters [Jones (1975)]. Using this criterion one would select Model 6 as best from among those in Table 1.

In epidemiologic studies, numerous risk factors are the rule. We performed a stepwise forward selection procedure from among all 54 potential disease-factor coefficients. At each step the parameter with the greatest score statistic was entered into the model, stopping when the significance level for each remaining parameter was greater than 0.10. The final model of ten main effects (plus two nuisance parameters) resulting from this stepwise procedure is given in Table 2. Parameters are listed in the order in which they were entered, along with their definitions. This single stepwise trinomial analysis reproduced the important findings of the Minimal Breast Cancer Study, which had previously been analyzed by conventional methods (Mantel-Haenszel analysis and binomial logistic regression) [Dubin, Hutter, Strax, et al (1984); Dubin, Pasternack and Strax (1984)]. Family history, weight, and late age at first birth were confirmed as usual risk factors for clinical breast cancer, but appeared unrelated to minimal breast cancer. Black race and lactation history, on the other hand, were associated with minimal, but not clinical, disease. Increasing age was, as expected, a risk factor for all breast cancer. An additional association, that of minimal breast cancer with early age at first birth, previously found to be significant by conventional analysis, was absent from the final model. However, it would have been the next association to enter the model had it not just exceeded ($p = 0.103$) the cut-off significance level of 0.10. The remaining significant risk factors in Table 2 pertain to breast symptomatology, number of screenings and year of first screening. These represent important sources of selection bias, requiring regression adjustment as confounders to ensure the validity of relative risk estimates [Dubin, Pasternack and Strax (1984); Dubin and Pasternack (1984)]. A past history of breast lump or cyst was significantly associated with minimal disease only; otherwise the confounding effects of these factors are the same for both breast cancer subcategories.

With respect to the original hypothesis of the study, substantial risk factor differences were found between minimal and clinical cases. Even so, risk factor patterns alone cannot suffice to establish minimal and clinical breast cancer as separate disease entities. Other considerations, such as histopathologic differences and the implausibility of alternative explanations would have to be established [Dubin, Hutter, Strax, et al (1984)]. One such alternative hypothesis explains risk factor differences as being related to enhancement or inhibition of breast cancer. Under this hypothesis we would expect an ordered response relationship to apply: if MEDIUM and HIGH are indicator variables describing level of exposure to a risk factor, then MINIMAL-MEDIUM and CLINICAL-HIGH should be simultaneously significant in the model. Such effects were apparent only for weight (suitably categorized), but not the other variables in the analysis (results not shown). However, the sample size of minimal cases (85) may not have been adequate to detect such subtleties.

Addition of interactive effects to the full main effects model would be required for a complete epidemiologic analysis. Doing this poses no special problems beyond ensuring that all necessary main effects are forced in, regardless of statistical significance, prior to the assessment of interaction. The computing expense, however, increases substantially with each additional parameter.

Such costs will also apply if one needs to calculate standard relative risks using a uniform reference group (usually controls) for all risk factors. This requires, for d distinct disease categories, fitting a model with d-1 parameters for each pertinent risk factor. If many nonsignificant parameters are added, increased computing costs could be accompanied by appreciably increased variability (but not bias) of estimated coefficients [Day, Byar and Green (1980)], even though there should still be an improvement over separate binomial regressions [Begg and Gray (1984)]. Consider the model in Table 2, for which the estimated regression coefficients are labelled $\hat{\beta}$. We fitted a subsequent model which included not only these main effects, but also MINIMAL-WEIGHT, CLINICAL-BLACK, CLINICAL-LUMPPAST, MINIMAL-FAMILY, CLINICAL-NUR2UP and MINIMAL-AGBIRHI, which enables one to calculate risk relative to controls for all independent variables in the model. The resultant estimates of parameters of interest are labelled $\hat{\beta}^*$ in the table. These were generally similar to the $\hat{\beta}$, although there was a greater than 10% change in magnitude for MINIMAL-LUMPPAST and CLINICAL-AGBIRHI. The ratio of the variances, $\hat{Var}(\hat{\beta}^*) / \hat{Var}(\hat{\beta})$, for the main effects of interest were 1.08, 1.05, 1.01, 1.01, 1.06, 1.11, 1.04, 1.03, 1.16 and 1.09, respectively, by order of entry as in Table 2. Of course, use of the $\hat{\beta}^*$ would only be required for variables with entry order 5 through 10; the others were already in case-control format. As an example, consider MINIMAL-BLACK, for which $\exp(\hat{\beta}) = \exp(0.873) = 2.39$ (approximate 95% confidence interval: 1.95, 2.95) and $\exp(\hat{\beta}^*) = \exp(0.898) = 2.45$ (approximate 95% confidence interval: 1.95, 3.09); the latter confidence limits are a little wider. The resulting increase in variance, which could be greater than here illustrated, represents a statistical cost associated with using a standard reference group for relative risk estimates.

OTHER CONSIDERATIONS

In most data situations there will be inevitable tradeoffs between simultaneous parameter estimation by means of the multinomial approach and separate parameter estimation by means of individual binomial regressions. As we have shown, multinomial regression can concisely identify, in a one-step modelling process, the key associations inherent in the data. Hypothesis testing and relative risk estimation are fully flexible. However, unless one has unlimited computer time one will be *practically* constrained with respect to the number of risk factors and interactions that can be examined simultaneously. For this reason it was not possible in our application to use a backward-elimination variable-selection procedure, even one of main effects alone, to choose among the 54 possible disease-factor associations. Even if we had screened variables by means of univariate score statistics, using a cut-off of p = 0.20, a 29-parameter initial model would have been required. The principal drawbacks of performing separate binomial regressions, in which one may easily handle the comparatively fewer parameters, are that it requires the fitting of several regression models and allows only informal assessment of parameter differences among them; joint tests of parameters from different regressions are not available. Begg and Gray (1984) have proposed a hybrid procedure which uses parameter estimates and individual variances from separate regressions, and requires, for joint hypothesis testing across regressions, the calculation of the joint covariance matrix of all parameters. They found that, in general, asymptotic relative efficiencies of their method are high, but decrease as the number of parameters and diagnostic categories increase.

In certain special cases one may be able to use a single binomial model to compare associations between explanatory variables and several disease categories, provided the latter can be appropriately defined by indicator variables [Breslow and Day (1980), pp. 111-112]. For example, in the Minimal Breast Cancer Study, one may wish to compare risk factors among women screened only once to those among women who underwent multiple screenings [Dubin and Pasternack (1982)]. Another example is given by Whittemore and McMillan (1982) for the analysis of matched case-control data involving several histologic types of lung cancer, as related to radiation and cigarette smoking.

One may specify an ordered polychotomous relationship among the disease subgroups [Walker and Duncan (1967), McCullagh (1980), Anderson and Philips (1981), Anderson (1982a)]. Anderson (1982b) considers a family of logistic regression models which includes both ordered and unordered response models as special cases, the likelihood ratio test being used to choose between them. Such a procedure could be applied to the Minimal Breast Cancer Study. On the other hand, Mantel (1966) recommends use of the unordered response model, relying upon the adequacy of the data to correctly associate intensity of response with intensity of dosage or exposure.

ACKNOWLEDGEMENTS: The authors wish to express their gratitude to Dr. Duncan Thomas for making available the NOMINAL computer program. This work was supported by Center Grants CA-16087 and CA-13343 from the National Cancer Institute, and Center Grant ES-00260 from the National Institute of Environmental Health Sciences.

REFERENCES

[1] Anderson, J.A., Separate sample logistic discrimination. Biometrika 59: 19-35 (1972).

[2] Anderson, J.A., Logistic regression methods in risk assessment. In: Prentice, R.L. and Whittemore, A.S. (eds). Environmental Epidemiology: Risk Assessment. Philadelphia: Society for Industrial and Applied Mathematics: 205-215 (1982a).

[3] Anderson, J.A., Regression modelling for ordered categorical variables. Tech Report No. 52, Department of Biostatistics, University of Washington (1982b).

[4] Anderson, J.A. and Philips, P.R., Regression, discrimination and measurement models for ordered categorical variables. Appl. Statist. 30:22-31 (1981).

[5] Begg, C.B. and Gray, R., Calculation of polychotomous logistic regression parameters using individualized regressions. Biometrika 71:11-18 (1984).

[6] Bock, R.D., Estimating multinomial response relations. In: Bose, R.C., Chakravarti, I.M., Mahalanobis, P.C., Rao, C.R. and Smith, K.J.C. (eds). Essays in Probability and Statistics. Chapel Hill: University of North Carolina Press:111-132 (1970).

[7] Breslow, N.E. and Day, N.E., Statistical Methods in Cancer Research. Volume 1. The Analysis of Case-Control Studies. IARC Sci. Publ. 32, 338 p. (1980).

[8] Cox, D.R., Some procedures connected with the logistic qualitative response curve. In: David, F.N. (ed). Research Papers in Statistics: Essays in Honor of J. Neyman's 70th Birthday. London: Wiley:55-71 (1966).

[9] Day, N.E., Byar, D.P. and Green, S.D., Overadjustment in case-control studies. Am. J. Epidemiol. 112:696-706 (1980).

[10] Dubin, N., Hutter, R.V.P., Strax, P., Fazzini, E.P., Schinella, R.A., Batang, E.S., Pasternack, B.S., Epidemiology of minimal breast cancer among women screened in New York City. JNCI (in press, 1984).

[11] Dubin, N. and Pasternack, B.S., Case-control analysis of breast cancer in a screened population: Implications for the assessment of environmental exposure risk factors. In: Prentice, R.L. and Whittemore, A.S. (eds). Environmental Epidemiology: Risk Assessment. Philadelphia: Society for Industrial and Applied Mathematics:154-172 (1982).

[12] Dubin, N. and Pasternack, B.S., Breast cancer screening data in case-control studies. Am. J. Epidemiol. 120:8-16 (1984).

[13] Dubin, N., Pasternack, B.S. and Strax, P., Epidemiology of breast cancer in a screened population. Cancer Detect. Prev. 7:87-102 (1984).

[14] Jones, R.H., Probability estimation using a multinomial logistic function. J. Statist. Comput. Simul. 3:315-329 (1975).

[15] Levin, B. and Shrout, P.E., On extending Bock's model of logistic regression in the analysis of categorical data. Commun. Statist-Theor. Meth. A10: 125-147 (1981).

[16] Mantel, N., Models for complex contingency tables and polychotomous dosage response curves. Biometrics 22:83-95 (1966).

[17] McCullagh, P., Regression models for ordinal data (with discussion). J.R. Statist. Soc. B42:109-142 (1980).

[18] Prentice, R.L. and Breslow, N.E., Retrospective studies and failure time models. Biometrika 65:153-158 (1978).

[19] Prentice, R.L. and Pyke, R., Logistic disease incidence models and case-control studies. Biometrika 66:403-411 (1979).

[20] Thomas, D.C., Dewar, R.A.D., and Goldberg, M.S., NOMINAL: Generalized logistic regression program for nominal outcome data. FORTRAN computer program (1983).

[21] Thomas, D.C., Siemiatycki, J., Goldberg, M. and Dewar, R., Discovery of risks: Statistical methods for relating several exposure factors to several diseases in case-heterogeneity studies. Unpublished manuscript (1984).

[22] Walker, S.H. and Duncan, D.B., Estimation of the probability of an event as a function of several independent variables. Biometrika 54:167-179 (1967).

[23] Whittemore, A.S. and McMillan, A., Analyzing occupational cohort data: Application to US uranium miners. In: Prentice, R.L. and Whittemore, A.S. (eds). Environmental Epidemiology: Risk Assessment. Philadelphia: Society for Industrial and Applied Mathematics:65-81 (1982).

Table 1. Trinomial logistic regression analysis of the Minimal Breast Cancer Study: Family history of breast cancer and age at last screen

Model	No. of parameters[a]	Log-likelihood	Disease - factor parameters	$\hat{\beta}$	Score statistic[b]
1	2	-1797.80	CONTROL - FAMILY		-1.68
			MINIMAL - FAMILY		-1.23
			CLINICAL- FAMILY		2.34
2	3	-1795.01	CONTROL - FAMILY		1.03
			MINIMAL - FAMILY		-1.03
			CLINICAL- FAMILY	0.339	
			CONTROL - LNAGE		-9.98
			MINIMAL - LNAGE		3.28
			CLINICAL- LNAGE		8.94
3	4	-1742.52	CLINICAL- FAMILY	0.283	
			CONTROL - LNAGE	-2.40	
			MINIMAL - LNAGE		0.232
			CLINICAL- LNAGE		-0.232
4	6	-1741.29	CONTROL - FAMILY	0.631	
			CLINICAL- FAMILY	0.890	
			CONTROL - LNAGE	-2.74	
			CLINICAL- LNAGE	-0.381	
			CONTROL - AGXFAM		2.21
			CLINICAL- AGXFAM		-1.84
5	7	-1738.09	CONTROL - FAMILY	0.576	
			CLINICAL- FAMILY	0.875	
			CONTROL - LNAGE	-2.88	
			CLINICAL- LNAGE	-0.284	
			CONTROL - AGXFAM	1.85	
			CLINICAL- AGXFAM		0.821
6	6	-1738.19	CONTROL - FAMILY	0.580	
			CLINICAL- FAMILY	0.881	
			CONTROL - LNAGE	-2.64	
			CONTROL - AGXFAM	1.87	
7	6	-1740.16	CLINICAL- FAMILY	0.322	
			CONTROL - LNAGE	-2.71	
			CLINICAL- LNAGE	-0.150	
			CLINICAL- AGXFAM	-1.57	

[a]Includes two nuisance parameters.
[b]Z-test.

Table 2. Trinomial logistic regression analysis of the Minimal Breast Cancer Study: Final model from stepwise forward selection of main effects

No. of parameters[a]: 12
Log-likelihood: -1545.57

Entry order	Disease - factor parameters	Definition of risk factor	Score statistic when entered[d]	$\hat{\beta}^b$	$\hat{\beta}^{*c}$
1	CONTROL - LUMPNOW	Indicates breast lump or cyst at initial screen	-12.08	-1.30	-1.39
2	CONTROL - LNAGE	Natural logarithm of age at last screen	-11.54	-3.05	-3.14
3	CONTROL - PRE1973	Indicates first screen prior to 1973	-8.98	-1.77	-1.78
4	CONTROL - LNNSCR	Natural logarithm of number of screens	10.18	0.827	0.842
5	CLINICAL - WEIGHT	Weight in pounds minus 135	3.22	0.00681	0.00704
6	MINIMAL - BLACK	Indicates black race	3.10	0.873	0.898
7	MINIMAL - LUMPPAST	Indicates breast lump or cyst prior to initial screen	2.80	0.789	0.954
8	CLINICAL - FAMILY	Indicates mother or sister with breast cancer history	2.61	0.414	0.384
9	MINIMAL - NUR2UP	Indicates ≥ 2 children nursed ≥ 2 months	2.18	0.642	0.596
10	CLINICAL - AGBIRHI	Indicates 1st birth ≥ 30 years of age or nulliparity	1.96	0.206	0.174

[a] Includes two nuisance parameters.

[b] Regression coefficients when all ten significant main effects are included in the model.

[c] Regression coefficients when all ten significant main effects, plus MINIMAL-WEIGHT, CLINICAL-BLACK, CLINICAL-LUMPPAST, MINIMAL-FAMILY, CLINICAL-NUR2UP and MINIMAL-AGBIRHI, are included in the model.

[d] Z-test.

BIOSTATISTICS: Statistics in Biomedical, Public Health
and Environmental Sciences, edited by P.K. Sen

ON STATISTICAL INFERENCES ABOUT COVARIATE-ADJUSTED PROPORTIONS

David G. Kleinbaum, Lawrence L. Kupper,
Chung-Yi Suen, and Sherman A. James

University of North Carolina at Chapel Hill

The comparison of groups using covariate-adjusted proportions based on fitting logistic and linear models is discussed with regard to two alternative procedures for computing such adjusted proportions following maximum likelihood estimation of model coefficients. Situations when tests about linear functions of adjusted proportions are not equivalent to tests about linear functions of adjusted logits are examined, and procedures are developed for testing statistical hypotheses in these situations. Methods are presented for the coding of dummy variables distinguishing comparison groups in order to facilitate the employment of appropriate statistical testing procedures. A detailed numerical example is provided to illustrate the key concepts.

KEY WORDS: Covariate-adjusted proportions; logistic models; linear models

1. INTRODUCTION

An analogue of classical analysis of covariance (ANOCOVA) for dichotomous outcomes using logistic modeling has been considered by Lee (1981). Here, we propose an alternative procedure for covariate-adjustment using the logistic model. We also describe other hypothesis testing procedures and strategies not previously addressed by Lee, some of which require modification of standard large sample likelihood ratio methods (Kleinbaum, Kupper, and Morgenstern, 1982). In particular, we focus on hypotheses about linear functions of covariate-adjusted proportions, and show that such hypotheses are not necessarily equivalent to linear hypotheses involving the logistic model parameters. For this situation, we then provide alternative large sample testing procedures involving Taylor-series approximations for the variances and covariances of adjusted sample proportions. We also discuss some issues concerning the coding of dummy variables used to distinguish comparison groups, and concerning the choice of a linear versus a logistic model for carrying out covariate-adjustment.

Two examples of the dichotomous-outcome ANOCOVA situation arising from our recent consulting experiences derive from a hypertension control intervention program (The Edgecombe County Project) conducted in a rural North Carolina county by investigators from the Department of Epidemiology during 1981-83 (Wagner et al., 1984). A baseline survey of this county was conducted in 1981 to learn about demographic, environmental, and psycho-social factors that might explain the extent to which existing hypertensives were aware of, under treatment for, or in control of their problem. The 539 hypertensives surveyed were divided into seven subgroups reflecting successive stages in the control of hypertension based on their awareness, treatment, and control status. Among analyses conducted, logistic modeling was used to derive age and race adjusted proportions for the seven hypertension subgroups for binary outcome variables such as employment and marital status. One contrast of interest involved comparing adjusted proportions for three hypertensive subgroups combined into an "unaware" category with the adjusted proportions for the remaining four subgroups designated as an "aware" category. A review by these authors of the statistical testing procedure typically used to make this comparison revealed that the null hypothesis actually tested involved a linear contrast of the logits of the proportions rather than of the proportions themselves. The fact that these two null hypotheses are not generally equivalent motivated the development of the statistical methodology to be described below.

Another illustration of this problem comes from a 1983 re-survey of Edgecombe County designed to evaluate the effectiveness of an intervention effort involving three "experimental" townships and two "control" townships. An important research goal was to assess whether the proportion of hypertensives under treatment at time of re-survey was higher for the combined experimental (i.e., intervened upon) townships than for the combined control townships, taking into account important covariates such as age, race, and sex. Methods for making statistical inferences about this type of comparison of adjusted proportions will be discussed in this paper.

2. THE ANOCOVA LOGISTIC MODEL

We consider the following logistic model for the analysis of covariance situation:

$$P = \frac{1}{1 + \exp\left[-(\alpha + \sum_{i=1}^{R} \gamma_i V_i + \sum_{j=1}^{S-1} \beta_j G_j)\right]} , \tag{1}$$

where V_i, i=1,..., R, denote covariates of interest (not restricted as to measurement scale), and G_j, j=1, 2,..., S-1, denote dummy variables that distinguish S comparison groups of interest. Note that the G_j may be indicators of either a single multi-level nominally treated variable or may identify a cross-classification of several distinct nominally treated variables. (A discussion of some issues regarding the coding of the G_j will appear later in the paper.) Also, no product terms of the form $(V_i \times G_j)$ appear in model (1), which is in line with the usual no interaction assumption for classical ANOCOVA (Kleinbaum and Kupper, 1978). Some assessment regarding the validity of this no interaction assumption should constitute part of any analysis plan in order to check whether or not model (1) is appropriate.

The logit form of model (1) is given as follows:

$$\text{logit}(P) = \ell n \left(\frac{P}{1 - P}\right) = \alpha + \sum_{i=1}^{R} \gamma_i V_i + \sum_{j=1}^{S-1} \beta_j G_j \ . \qquad (2)$$

Thus, as is well known, the parameters in the logistic model contribute <u>linearly</u> to the logit of individual risk, rather than to the risk itself. Consequently, as we discuss below, tests about linear functions of these parameters are not, in general, equivalent to tests about linear functions of P.

3. HYPOTHESES OF INTEREST

The general null hypothesis of interest concerns linear functions of covariate-adjusted proportions for the various groups being compared. This necessitates our specifying the following group-specific version of model (2):

$$\text{logit}(P_g) = \alpha_g + \sum_{i=1}^{R} \gamma_i V_i \ , \qquad (3)$$

where g = 1, 2,..., S; also,

$$\alpha_g = \alpha + \sum_{j=1}^{S-1} \beta_j G_j(g) \ ,$$

where $G_j(g)$ denotes the value of G_j for persons in group g. Based on fitting model (3), let $\hat{P}_g$ denote a covariate-adjusted proportion for group g calculated using one of two alternative averaging schemes to be described in the next section. Then, with P_g being the population analogue of $\hat{P}_g$, the general null hypothesis of interest may be expressed in matrix notation as

$$H_0: \underset{\sim}{C}\,\underset{\sim}{P} = \underset{\sim}{0} \ , \qquad (4)$$

where

$$\underset{\sim}{P} = \begin{bmatrix} P_1 \\ \vdots \\ P_S \end{bmatrix}$$

is a column vector with S elements, and $\underset{\sim}{C}$ is a T x S hypothesis matrix with $\text{rank}(\underset{\sim}{C}) = T < S$.

Three $\underset{\sim}{C}$ matrices of interest will now be discussed.

(a)

$$\underset{\sim}{C} = \begin{bmatrix} 1 & -1 & 0 & . & . & . & 0 \\ 1 & 0 & -1 & . & . & . & 0 \\ 1 & 0 & 0 & . & . & . & -1 \end{bmatrix}, \text{ where } T = (S - 1) .$$

For this $\underset{\sim}{C}$ matrix, $\underset{\sim}{C}\ \underset{\sim}{P} = \underset{\sim}{0}$ can be equivalently expressed as $P_1 = P_2 = \ldots = P_S$, from which it follows that $\text{logit}(P_1) = \text{logit}(P_2) = \ldots = \text{logit}(P_S)$. Further, from (2), equality among these S logits is equivalent to $\beta_1 = \ldots = \beta_{S-1} = 0$. Thus, for this particular choice for $\underset{\sim}{C}$, H_0: $\underset{\sim}{C}\ \underset{\sim}{P} = \underset{\sim}{0}$ can be equivalently tested using a standard multiple-partial likelihood ratio statistic obtained by fitting two (full and reduced) versions of the logistic model (1). For large samples, this statistic will have an approximate χ^2 distribution under H_0, with (S - 1) degress of freedom.

(b) $\underset{\sim}{C}$ is a (1 x S) row vector with one element equal to 1, another equal to -1, and the remaining elements all equal to 0; for example, we could have

$$\underset{\sim}{C} = (1 \quad -1 \quad 0 \ldots 0).$$

Given this choice for $\underset{\sim}{C}$, it follows that $\underset{\sim}{C}\ \underset{\sim}{P} = \underset{\sim}{0}$ can equivalently be expressed as $P_{g_1} = P_{g_2}$, where g_1 and g_2 denote the two groups corresponding to the two non-zero elements of $\underset{\sim}{C}$. It can be shown (an example is provided in Section 6) that, in general, it is possible to choose an appropriate coding scheme for the G_j variables so that the null hypothesis of the equality of a subset of the P_g's is equivalent to the null hypothesis that one or more β's in model (1) are equal to zero. Thus, a test of H_0: $P_{g_1} = P_{g_2}$ (or, more generally, a test of the equality of two or more P_g's) can be carried out via a likelihood ratio test using appropriately specified full and reduced models based on (1).

(c) $\underset{\sim}{C}$ is a general (1 x S) row vector representing any contrast of interest; i.e., $\underset{\sim}{C}$ has the general structure

$$\underset{\sim}{C} = (c_1, c_2, \ldots, c_S), \text{ with } \sum_{j=1}^{S} c_j = 0 .$$

An example for S = 5 is given by

$$\underset{\sim}{C} = (1/3 \quad 1/3 \quad 1/3 \quad -1/2 \quad -1/2),$$

so that

$$\underset{\sim}{C}\ \underset{\sim}{P} = \frac{(P_1 + P_2 + P_3)}{3} - \frac{(P_4 + P_5)}{2} .$$

In general, any null hypothesis <u>not</u> directly expressible as an equality among a subset of the P_g cannot be equivalently expressed as a null hypothesis involving the logits of the P_g. For example, the null hypothesis that $\underset{\sim}{C}\,\underset{\sim}{P}$ above is equal to zero is not equivalent to

$$H_0: \frac{\text{logit}(P_1) + \text{logit}(P_2) + \text{logit}(P_3)}{3} - \frac{\text{logit}(P_4) + \text{logit}\ (P_5)}{2} = 0\ ,$$

even though the latter null hypothesis can be expressed as a linear function of the β_j, the particular form of which depends on the coding scheme used to define the G_j variables. Consequently, the usual approach for carrying out large sample likelihood ratio tests about linear functions of the β_j will not generally be applicable here, and so some modification of standard statistical methodology is required.

4. COVARIATE-ADJUSTED PROPORTIONS

Lee (1981) proposed the following formula for obtaining covariate-adjusted propostions:

$$\hat{P}_g^* = \frac{1}{N} \sum_{k=1}^{N} \frac{1}{1 + \exp\left[-(\hat{\alpha}_g + \sum_{i=1}^{R} \hat{\gamma}_i V_{ik})\right]}\ , \tag{5}$$

where

$$\hat{\alpha}_g = \hat{\alpha} + \sum_{j=1}^{S-1} \hat{\beta}_j G_j(g)\ ,$$

$\hat{\alpha}$, $\hat{\gamma}$, and $\hat{\beta}_j$ denote maximum likelihood estimates of their respective parameters, and V_{ik} is the value of V_i for the k-th individual in the total sample, k = 1, 2, ..., N.

Lee's formula provides covariate-adjusted proportions which treat each of the comparison groups as having an identical joint distribution of covariates, namely, that distribution based on the entire combined sample of N persons. This procedure thus has the same rationale as classical ANOCOVA, which adjusts for disparities in covariate distributions over groups by artificially assuming a common covariate distribution based on the combined sample.

Although conceptually reasonable, Lee's formula is computationally cumbersome for two reasons: (1) it requires the covariate-adjusted proportion for each group to be computed as an average over all N individuals in the study; and, (2) it necessitates the use of very cumbersome formulae (described below) for the estimated variances and covariances of such adjusted proportions, which are then needed for inference-making involving linear contrasts of such proportions.

An alternative formula, which has conceptual merit on its own, which is computationally simpler, and which provides a very good approximation to expression (5), is given as follows:

$$\hat{P}_g^{\dagger} = \frac{1}{1 + e^{-(\hat{\alpha}_g + \sum_{i=1}^{R} \hat{\gamma}_i \bar{V}_i)}} ; \qquad (6)$$

here, $\bar{V}_i$ denotes the arithmetic mean for covariate V_i based on averaging over the combined sample of N persons.

In contrast to Lee's formula (5), which is an average of estimated proportions, formula (6) is an estimated proportion computed at the overall average of each covariate. In general, expressions (5) and (6) are not equivalent; this is in contrast with the classical ANOCOVA situation where the model is linear, and not logistic, so that the analogous formulae are identical (Lee, 1981).

Formula (6) has conceptual justification in that it corrects for unequal covariate distributions in the various groups by artificially restricting all groups to have the same set of mean covariate values. Although this restriction does not imply that the joint distribution of covariates will be the same in each group, this latter requirement is generally unnecessary for adequate covariate-adjustment.

We will also illustrate below via an example that formulae (5) and (6) can be expected to produce similar numerical results. Furthermore, as we show in the next section, formula (6) is considerably easier to work with for certain inference-making situations.

5. HYPOTHESIS TESTING PROCEDURES

To test the null hypothesis

$$H_0: \underset{\sim}{C}\,\underset{\sim}{P} = \underset{\sim}{0}$$

as defined by (4), the following large sample chi-square statistic may be used:

$$\chi^2 = (\underset{\sim}{C}\,\hat{\underset{\sim}{P}})' \, [\underset{\sim}{C}\, \hat{Var}(\hat{\underset{\sim}{P}})\, \underset{\sim}{C}']^{-1} \, (\underset{\sim}{C}\,\hat{\underset{\sim}{P}}) , \qquad (7)$$

where $\hat{\underset{\sim}{P}}$ denotes a vector of covariate-adjusted proportions and $\hat{Var}(\hat{\underset{\sim}{P}})$ denotes the estimated variance-covariance matrix of $\hat{\underset{\sim}{P}}$. Under H_0, χ^2 will have, for large samples, approximately a central chi-square distribution with T degrees of freedom.

We have already mentioned that tests about the equality of subsets of a group of covariate-adjusted proprotions can equivalently be carried out (with proper coding of the G_j variables) using likelihood ratio tests about appropriate β_j's in the logistic model (1). Consequently, our interest here concerns the form that (7) takes when the hypothesis matrix $\underset{\sim}{C}$ represents a general linear contrast of the P_g's. For this situation, the null hypothesis H_0: $\underset{\sim}{C}\,\underset{\sim}{P} = \underset{\sim}{0}$ can be expressed in the simple form

$$H_0: \sum_{g=1}^{S} s_g P_g = 0 \ ,$$

where $\sum_{g=1}^{S} c_g = 0$; in this case, the test statistic (7) can be alternatively expressed as the standard normal variable

$$Z = \frac{\sum_{g=1}^{S} c_g \hat{P}_g}{\sqrt{\mathrm{Var}\left(\sum_{g=1}^{S} c_g \hat{P}_g\right)}} \ , \tag{8}$$

where

$$\mathrm{Var}\left(\sum_{g=1}^{S} c_g \hat{P}_g\right) = \sum_{g=1}^{S} c_g^2 \mathrm{Var}(\hat{P}_g) + 2 \sum_{g<g'}\sum c_g c_{g'} \mathrm{Cov}(\hat{P}_g, \hat{P}_{g'}) \ . \tag{9}$$

To employ (8) requires deriving a computational expression for its denominator. The structure of this expression will depend on which form of covariate-adjusted proportion is used [either (5) or (6) above] and hence on the variance-covariance matrix attendant with that choice of estimate. Because the covariate-adjusted proportions (5) and (6) are nonlinear functions of random variables (i.e., the estimated logistic regression function coefficients), their variance-covariance matrices cannot be exactly determined as in classical linear model theory. Consequently, we have derived formulae for variances and covariances based on first-order Taylor-series approximations involving the logistic function. Here, we present such formulae for the two computational procedures (5) and (6) for obtaining covariate-adjusted proportions.

5.a. VARIANCE-COVARIANCE FORMULAE FOR $\hat{P}_g^*$

From (5), and with

$$\hat{Y}_{kg} = \hat{\alpha}_g + \sum_{i=1}^{R} \hat{\gamma}_i V_{ik}$$

and

$$\hat{P}_{kg} = \frac{1}{1 + e^{-\hat{Y}_{kg}}} \ ,$$

where $k = 1,\ldots, N$ and $g = 1,\ldots, S$, we find using a first order Taylor-series approximation that:

$$N^2\hat{Var}(\hat{P}_g^*) \approx \sum_{k=1}^{N} \sum_{\ell=1}^{N} \hat{P}_{kg}\hat{P}_{\ell g}(1 - \hat{P}_{kg})(1 - \hat{P}_{\ell g})\hat{Cov}(\hat{Y}_{kg}, \hat{Y}_{\ell g}) \; ; \qquad (10)$$

and, for $g < g'$,

$$N^2\hat{Cov}(\hat{P}_g^*, \hat{P}_{g'}^*) \approx \sum_{k=1}^{N} \sum_{\ell=1}^{N} \hat{P}_{kg}\hat{P}_{\ell g'}(1 - \hat{P}_{kg})(1 - \hat{P}_{\ell g'})\hat{Cov}(\hat{Y}_{kg}, \hat{Y}_{\ell g'}) \; , \qquad (11)$$

where

$$\hat{Cov}(\hat{Y}_{kg}, \hat{Y}_{\ell g}) = \hat{Var}(\hat{\alpha}_g) + \sum_{i=1}^{R} V_{ik}\hat{Cov}(\hat{\alpha}_g, \hat{\gamma}_i) + \sum_{i=1}^{R} V_{i\ell}\hat{Cov}(\hat{\alpha}_g, \hat{\gamma}_i)$$

$$+ \sum_{i=1}^{R} V_{ik}V_{i\ell}\hat{Var}(\hat{\gamma}_i) + 2\sum_{i<j}\sum V_{ik}V_{j\ell}\hat{Cov}(\hat{\gamma}_i, \hat{\gamma}_j),$$

and

$$\hat{Cov}(Y_{kg}, Y_{\ell g'}) = \hat{Cov}(\hat{\alpha}_g, \hat{\alpha}_{g'}) + \sum_{i=1}^{R} V_{ik}\hat{Cov}(\hat{\alpha}_g, \hat{\gamma}_i)$$

$$+ \sum_{i=1}^{R} V_{i\ell}\,\hat{Cov}\,(\hat{\alpha}_{g'}, \hat{\gamma}_i) + \sum_{i=1}^{R} V_{ik}V_{i\ell}\hat{Var}(\hat{\gamma}_i)$$

$$+ 2\sum_{i<j}\sum V_{ik}V_{j\ell}\hat{Cov}(\hat{\gamma}_i, \hat{\gamma}_j) \; .$$

Using (10) and (11) in conjunction with (9), we can now compute the denominator for the Z-statistic (8).

5.b. VARIANCE-COVARIANCE FORMULAE FOR $\hat{P}_g^\dagger$

Based on the alternative formula (6) for covariate adjusted proportions, we obtain the following Taylor-series approximations for the estimates of the needed variances and covariances:

$$\hat{Var}(\hat{P}_g^\dagger) \approx (\hat{P}_g^\dagger)^2(1 - \hat{P}_g^\dagger)^2\hat{Var}(\hat{Y}_g) \; ; \qquad (12)$$

and (for $g < g'$),

$$\hat{Cov}(\hat{P}_g^\dagger, \hat{P}_{g'}^\dagger) \approx \hat{P}_g^\dagger \hat{P}_{g'}^\dagger(1 - \hat{P}_g^\dagger)(1 - \hat{P}_{g'}^\dagger)\hat{Cov}(\hat{Y}_g, \hat{Y}_{g'}), \qquad (13)$$

where

$$\hat{Y}_g = \hat{\alpha}_g + \sum_{i=1}^{R} \hat{\gamma}_i\overline{V}_i,$$

$$\hat{Var}(\hat{Y}_g) = \hat{Var}(\hat{\alpha}_g) + 2\sum_{i=1}^{R} \overline{V}_i\hat{Cov}(\hat{\alpha}_g, \hat{\gamma}_i)$$

$$+ \sum_{i=1}^{R} \bar{V}_i^2 \, \hat{Var}(\hat{\gamma}_i) + 2\sum_{i<j}\sum \bar{V}_i \bar{V}_j \hat{Cov}(\hat{\gamma}_i, \hat{\gamma}_j) \ ,$$

and, for g < g',

$$\hat{Cov}(\hat{Y}_g, \hat{Y}_{g'}) = \hat{Cov}(\hat{\alpha}_g, \hat{\alpha}_{g'}) + \sum_{i=1}^{R} \bar{V}_i \left[\hat{Cov}(\hat{\alpha}_g, \hat{\gamma}_i) + \hat{Cov}(\hat{\alpha}_{g'}, \hat{\gamma}_i) \right]$$

$$+ \sum_{i=1}^{R} \bar{V}_i^2 \hat{Var}(\hat{\gamma}_i) + 2\sum_{i<j}\sum \bar{V}_i \bar{V}_j \hat{Cov}(\hat{\gamma}_i, \hat{\gamma}_j) \ .$$

Regarding the use of formulae (10) and (11) as opposed to (12) and (13), the former pair involve, for each comparison group, the computation of the adjusted proportion for each individual in the entire study sample, whereas the latter pair of formulae only require the computation of adjusted proportions as a function of overall mean covariate values. In either case, the values for the estimated variances and covariances of the $\hat{\alpha}_g$ and $\hat{\gamma}_i$ can be obtained from the printout of standard packaged programs for logistic regression.

6. EFFECT OF CODING

It is well known regarding classical ANOVA and ANOCOVA that statistical inferences about a set of orthogonal contrasts involving group means can be conveniently made by fitting a single regression model and using judicious coding of the group indicator variables in the model. In particular, a coding scheme can be defined so that each contrast corresponds to one particular group indicator variable in the model, and so that the collection of such indicator variables constitutes a pairwise orthogonal set. Then, a test of significance concerning a given contrast can be equivalently carried out by making a partial F test to assess whether the corresponding regression coefficient is significantly different from zero. To illustrate why such orthogonal coding can be useful when making inferences about contrasts among covariate-adjusted proportions, we will consider the following situation motivated by the Edgecombe Study mentioned earlier.

Suppose that we wish to compare covariate-adjusted proportions for S = 5 townships, and that we have a single (i.e., R = 1) covariate AGE. In particular, let us focus on that contrast which compares the first three ("experimental") groups with the last two ("control") groups. So, this contrast is of the form

$$L = \frac{(P_1 + P_2 + P_3)}{3} - \frac{(P_4 + P_5)}{2} \ , \qquad (14)$$

and the model being considered is

$$\text{logit}(P) = \alpha + \gamma(\text{AGE}) + \sum_{j=1}^{4} \beta_j G_j \ . \tag{15}$$

Consider now the coding scheme for the G_j given in Table 1.

Table 1. Coding Scheme for Edgecombe Example

	G_j \ g	Indicator Variable (G_j) G_1	G_2	G_3	G_4
Group Number (g)	1	2	2	0	0
	2	2	-1	1	0
	3	2	-1	-1	0
	4	-3	0	0	1
	5	-3	0	0	-1

The columns in Table 1 are clearly pairwise orthogonal. Furthermore, it can be shown that the regression coefficients β_j can each be expressed in terms of a unique contrast involving the logits of the P_g, as follows:

$$5\beta_1 = \frac{\text{logit}(P_1) + \text{logit}(P_2) + \text{logit}(P_3)}{3} - \frac{\text{logit}(P_4) + \text{logit}(P_5)}{2} \tag{16}$$

$$3\beta_2 = \text{logit}(P_1) - \frac{\text{logit}(P_2) + \text{logit}(P_3)}{2} \ , \tag{17}$$

$$2\beta_3 = \text{logit}(P_2) - \text{logit}(P_3) \ , \tag{18}$$

$$2\beta_4 = \text{logit}(P_4) - \text{logit}(P_5) \ . \tag{19}$$

As discussed earlier, the null hypothesis H_0: L = 0, where L is given by (14), is not generally equivalent to testing H_0: $\beta_1 = 0$ based on fitting (15) under the coding scheme in Table 1 and the attendant relationships (16) - (19). Only in the situation when $\text{logit}(P_1) = \text{logit}(P_2) = \text{logit}(P_3)$, so that $P_1 = P_2 = P_3$, and when $\text{logit}(P_4) = \text{logit}(P_5)$, so that $P_4 = P_5$, would the above null hypotheses be equivalent. If at least one of these two sets of equalities does not hold, then one is forced to use the testing procedure discussed in Section 5 to test H_0: L = 0 directly.

However, the cumbersome methodology of Section 5 can be avoided under the following circumstances. Suppose one fits model (15) using the coding scheme in

Table 1. Based on (17) - (19), the (3 degrees of freedom) likelihood ratio test of H_0: $\beta_2 = \beta_3 = \beta_4 = 0$ provides an assessment of the reasonableness of the assumption of within-"experimental" group and within-"control" group homogeneity of covariate-adjusted proportions. If the test statistic is significant, then one must test H_0: $L = 0$ directly as described in Section 5. On the other hand, if this test for heterogeneity of within-group proportions is not close to significance, then it would be reasonable to test (without fitting another model to the data) the null hypothesis H_0: $\beta_1 = 0$ and to assume that this likelihood ratio test is essentially equivalent to testing H_0: $L = 0$, where L is given by (14). Of course, if one concludes that $\beta_2 = \beta_3 = \beta_4 = 0$, a test of H_0: $\beta_1 = 0$ based on the reduced model $\text{logit}(P) = \alpha + \gamma(\text{AGE}) + \beta_1 G_1$ could be expected to be slightly more statistically efficient. A numerical example illustrating the use of the coding scheme in Table 1 will be discussed in Section 8.

7. CHOICE OF MODEL FORM: LINEAR OR LOGISTIC

Although the logistic function is a very popular choice of model form for binary response data, it is certainly not the only type of model available, another reasonable choice being a linear model. In particular, suppose a linear model of the form

$$P = \alpha + \sum_{i=1}^{R} \gamma_i V_i + \sum_{j=1}^{S-1} \beta_j G_j \tag{20}$$

is considered instead of (1), with group-specific form given by

$$P_g = \alpha_g + \sum_{i=1}^{R} \gamma_i V_i \ , \tag{21}$$

where α_g is defined as in (3). It then follows that covariate-adjusted proportions defined analogously to (5) and (6) are identical; in particular,

$$\hat{P}'_g = \frac{1}{N}\sum_{k=1}^{N} (\hat{\alpha}_g + \sum_{i=1}^{R} \hat{\gamma}_i V_{ik}) = \hat{\alpha}_g + \sum_{i=1}^{R} \hat{\gamma}_i \bar{V}_i \tag{22}$$

where $\hat{\alpha}_g$ and $\hat{\gamma}_i$ are estimated by maximum likelihood methods. It is also clear that any test of a linear hypothesis of the form (4) can be equivalently expressed in terms of a linear hypothesis involving the β_j's. Thus, when compared with logistic modeling, the use of a linear model is somewhat computationally advantageous.

Nevertheless, the question still remains as to which choice of model form is more appropriate from a more general perspective which considers the underlying process being modeled, the types of parameters of interest, the fit of the model, and the precision of estimators and the power of tests of hypotheses. Kleinbaum, Kupper and Morgenstern (1982) have pointed out that a linear model may yield predicted probabilities outside the permissible (0, 1) range, although this may be more of an indictment of the method of estimation than of the appropriateness of the linear model. They also state that a linear model is possibly a better choice when the primary parameter of interest is a difference of proportions, whereas the logistic model is to be preferred when the parameter of interest is a ratio of proportions. In addition, they point out, with regard to interaction assessment, that a linear model allows direct quantification of deviations from additivity, whereas a logistic model can directly measure deviations from a multiplicative state of no interaction. Finally, they point out that the choice of model should be based on a combination of theoretical factors (e.g., biologic plausibility) and data-based results (e.g., the fit of the model, the appropriateness of the signs and magnitudes of the estimated regression coefficients relative to a priori expectations based on the process under study, etc.).

In practice, it is usually the case that neither the logistic nor the linear model can be given a blanket recommendation over the other. Considering all perspectives, we recommend that the choice of model, whether linear, logistic or other form, reflect a combination of theoretical and data-based considerations.

8. A NUMERICAL EXAMPLE WITH DISCUSSION

The following numerical example is based on the Edgecombe County re-survey data described earlier, for which comparisons among age-adjusted proportions for five townships are of interest. We consider both the logistic model (15) as well as its linear model counterpart of the general form (20). The (orthogonal) coding scheme used for the G_j variables in each of these two models is given in Table 1. Estimates of crude and age-adjusted proportions are presented in Table 2, with adjustments under model (15) obtained using both formulae (5) and (6), and adjustments under the analogous linear model obtained using (22). Unconditional maximum likelihood estimation was employed to fit both the logistic model (using SAS's LOGIST procedure) and the linear model [using the MAXLIK package of Kaplan and Elston (1972)].

Inspection of the entries in Table 2 leads to the following general impressions. For the proportions in column A (which are based on the actual Edgecombe County re-survey data), there does not appear to be much difference among the five townships. For the proportions in columns B and C (which are based on artificial manipulation of the data in the two control groups for illustrative purposes), there is clearly variation in the proportions from township to township. More specifically, the column B proportions manifest between, but not within, experimental and control groups variation. In contrast, the column C proportions exhibit within control group (but not within experimental group) differences; however, the average proportions fro the column C experimental and control groups do not differ.

The hypothesis testing results are presented in Table 3, and they support the impressions gleaned from our examination of Table 2. The first null hypothesis of no differences among the five townships [namely, H_0: $\beta_1 = \beta_2 = \beta_3 = \beta_4$, or equivalently, $P_1 = P_2 = P_3 = P_4 = P_5$] is, for all four testing procedures, not rejected for the column A proportions, but is strongly rejected for the column B and column C proportions. Thus, further testing is required to determine where the significant differences lie among the proportions in columns B and C. Note that the test statistics $(\chi^2)^*$ and $(\chi^2)^\dagger$, based on (5) and (6) respectively, have essentially the same values for all situations considered in Table 3. This suggests that the less statistically cumbersome covariate-adjusted proportion formula (6) would be sufficiently accurate for practical purposes.

As expected, the null hypothesis of within-group homogeneity (namely, H_0: $\beta_2 = \beta_3 = \beta_4 = 0$, or equivalently, $P_1 = P_2 = P_3$ and $P_4 = P_5$) is not rejected for the column A and column B proportions, but is strongly rejected for the proportions in column C. Based on these results and on our earlier discussions concerning the connection between tests about β's and tests about P's, it is not at all surprising that the χ^2_{LR} test of H_0: $\beta_1 = 0$ gives about the same test statistic value as the $(\chi^2)^*$ and $(\chi^2)^\dagger$ tests of H_0: L = 0 for the column B proportions; and, in addition, χ^2_{LR} is much larger than $(\chi^2)^*$ and $(\chi^2)^\dagger$, as well as being highly significant, for the column C proportions.

Our conclusion regarding the column B proportions is that there is no within-group differences among the adjusted proportions (and hence among the adjusted logits), but that there is a significant difference between the experimental and control groups with regard to their average proportions (or, equivalently, their average logits).

Because of the large difference between the two proportions in the control group for the column C proportions, it turns out, even though $\hat{L} = \frac{1}{3}\sum_{g=1}^{3}\hat{P}_g - \frac{1}{2}\sum_{g=4}^{5}\hat{P}_g$ is close to zero in value (which can be seen by checking the $\hat{P}^*_g$ and $\hat{P}^\dagger_g$ values under column C in Table 2), β_1 is significantly different from zero, a fact that can be appreciated by computing the value of $\frac{1}{3}\sum_{g=1}^{3}\text{logit}(\hat{P}_g) - \frac{1}{2}\sum_{g=4}^{5}\text{logit}(\hat{P}_g)$ using the column C values for $\hat{P}^*_g$ and $\hat{P}^\dagger_g$ in Table 2. Our conclusion regarding the column C proportions is that there is significant within- (control) group heterogeneity of adjusted proportions; also, the experimental and control groups do not differ with respect to their average adjusted proportions, but do differ significantly with respect to their average adjusted logits. The column C data clearly demonstrate that tests about covariate-adjusted proportions are not necessarily equivalent to tests about covariate-adjusted logits.

Finally, the $(\chi^2_{LR})'$ results based on the use of the linear model generally led to the same conclusions as those based on the logistic model. The only difference was that the test of H_0: $\beta_1 = 0$ under column C was not significant, the reason being that, for the linear model, this is actually a test about adjusted proportions, and not about adjusted logits.

ACKNOWLEDGEMENTS: This research study was partially supported by NHLBI Grant No. HL 24003, and by NIEHS Grant No. ES 07018. Dr. James is supported by a NHLBI research career development award, Grant No. 5K04-HL01011.

REFERENCES

[1] Kaplan, E.G. and Elston, R.C., A subroutine package for maximum likelihood estimation. Institute of Statistics Mimeo Series No. 823, University of North Carolina at Chapel Hill (1972).

[2] Kleinbaum, D.G. and Kupper, L.L., Applied Regression Analysis and Other Multivariable Methods, Duxbury Press: North Scituate, Massachusetts (1978).

[3] Kleinbaum, D.G., Kupper, L.L. and Morgenstern, H., Epidemiologic Research: Principles and Quantitative Methods, Lifetime Learning Publications (1982).

[4] Lee, J., Covariance adjustment of rates based on the multiple logistic regression model, J. Chron. Dis. 34:415-426 (1981).

[5] Wagner, E. H., James, S.A., Reresford, S.A.A., Storgatz, D.S., Grimson, R.C., Kleinbaum, D.G., Williams, C.A., Critchin, L.M., and Ibrahim, M.A., The Edgecombe County high blood pressure control program: I. Correlates of uncontrolled hypertension at baseline, Amer. J. Pub. Health, in press (1984).

Table 2. Estimated Proportions for Edgecombe County Re-Survey Data**

Township	Intervention Status	Sample Size	Crude Proportions A	B	C	Crude Logits A	B	C	Values of $\hat{P}^*_g$ A	B	C	Values of $\hat{P}^\dagger_g$ A	B	C	***Values of $\hat{P}'_g$ A	B	C
1	Exper.	99	.61	.61	.61	.45	.45	.45	.54	.55	.54	.55	.55	.55	.54	.55	.56
2	Exper.	45	.64	.64	.64	.58	.58	.58	.64	64	.64	.67	.66	.67	.61	.63	.63
3	Exper.	88	.58	.58	.58	.32	.32	.32	.57	.57	.57	.58	.58	.58	.57	.58	.58
4	Control	104	.57	.30	.96	.28	-.85	3.18	.61	.33	.97	.64	.30	.98	.56	.35	.93
5	Control	89	.53	.34	.25	.12	-.66	-1.10	.56	.36	.27	.57	.34	.23	.55	.36	.29

**The proportions in column A are based on the actual Edgecombe County re-survey data; those in columns B and C derive from perturbations of the actual control groups data made for illustrative purposes.

***The maximum likelihood estimates for the linear model (20) were based on the use of the following function:

$$P = \hat{\alpha} + \hat{\gamma}(\text{AGE}) + \sum_{j=1}^{4} \hat{\beta}_j G_j = \begin{cases} .999 \text{ if } \hat{P} > .999 \,, \\ .001 \text{ if } \hat{P} < .001 \,, \\ \hat{P} \text{ otherwise.} \end{cases}$$

Table 3. Hypothesis Testing Results for Edgecombe County Re-Survey Data**

Null Hypothesis	χ^2_{LR}			$(\chi^2)^*$			$(\chi^2)^\dagger$			$(\chi^2_{LR})'$		
	A	B	C	A	B	C	A	B	C	A	B	C
1. $\beta_1 = \beta_2 = \beta_3 = \beta_4 = 0$ ($P_1 = P_2 = P_3 = P_4 = P_5$)	2.44	27.20	147.35	2.51	28.16	331.03	2.50	29.86	330.64	1.14	27.35	137.40
2. $\beta_2 = \beta_3 = \beta_4 = 0$ ($P_1 = P_2 = P_3$ and $P_4 = P_5$)	2.27	1.58	144.41	2.33	1.63	223.20	2.35	1.64	237.71	1.08	0.91	135.92
3. $\beta_1 = 0$	0.01	25.48	8.69	--	--	--	--	--	--	0.27	26.77	1.66
4. L = 0	--	--	--	0.01	27.48	1.03	0.01	29.05	0.03	--	--	--

** χ^2_{LR} is a likelihood ratio statistic based on the logistic model (15) and the coding scheme in Table 1; $(\chi^2)^*$ involves the use of expressions (7), (10) and (11) under model (15); $(\chi^2)^\dagger$ involves the use of expressions (7), (12) and (13) under model (15); $(\chi^2_{LR})'$ is a likelihood ratio statistic for the linear model analogue (20) of the logistic model (15). The parameter L is defined by expression (14).

BIOSTATISTICS: Statistics in Biomedical, Public Health
and Environmental Sciences, edited by P.K. Sen

LIMITED OCCUPATIONAL MORTALITY DATA AND STATISTICS

Michael J. Symons, Donald L. Doerfler, and Yang C. Yuan

Department of Biostatistics and Occupational Health Studies Group
School of Public Health
University of North Carolina at Chapel Hill

From principles of survival analysis statistical approaches are identified for a major problem with occupational mortality studies when few deaths, as with cause specific mortality, and no exposure data on the individuals are available. Specifically, this limitation is the general inadequacy of a geographic population as a comparison group for employed populations. Qualitative aspects of the so-called healthy worker effect are briefly presented, while the essential features of statistical strategies for part of this issue are sketched in detail. Although the occupational epidemiology literature focuses on selection bias as the source of the problems with the standardized mortality ratio, the potential inadequacy of the proportional hazards model for occupational mortality evaluations is emphasized here. A practical, non-proportional hazards model, patterned after the slope-ratio design in bioassay, may provide a more sensitive approach.

KEY WORDS: Survival analysis, proportional hazards, standardized mortality ratio, non-proportional models, slope-ratio design, bioassay, occupational epidemiology

1. INTRODUCTION AND SUMMARY

In the evaluation of the cause specific mortality experience of an occupational cohort, statistical analysis options are limited with only a few observed deaths and when no exposure evaluations on individuals are available. The standardized mortality ratio (SMR) is the historical summary statistic for such studies. It is the maximum likelihood estimator for a presumed constant ratio of the employee's rates, at every age band, to those of the population of an appropriate geographic area which is composed of employable and unemployable individuals. The so-called healthy worker effect represents a major criticism of the SMR as a summary statistic, since employees are typically healthier than members of the general population. The unavailability of mortality rates for employees in the United States that may reasonably represent an unexposed group is a very fundamental limitation.

Two alternative single parameter models are suggested that circumvent some of the criticisms of the SMR, but require mortality rates of employable individuals at the ages of hire. The first is an SMR construction presuming the appropriate standard rates are those of a general population multiplied at each age band by the ratio of the rate for employables at the ages of hire to the corresponding rate for a general population. This synthetic schedule of rates then is proportional to that of the general population at all ages as determined by mortality rates for an employed group at the ages of hire.

A second schedule of synthetic rates is suggested presuming a regression from the mortality rates at the ages of hire for employables to those at the ages of retirees in the general population. The rationale being that the mortality rates at the older ages are by definition those of the survivors of all younger ages, and is compatible with a dying off of those at greater risk in the general population during the younger ages. An SMR construction, presuming a proportional effect on mortality from exposure in the workplace, is a convenient choice. However, the non-proportional character of this synthetic standard may warrant a parameter reflecting this understructure. It appears that such a parameter may be more relevant to quantify an excess risk of the employed group than the proportional one estimated by the SMR based upon any standard schedule of rates. Further, it appears to accommodate the lack of fit frequently exhibited by the SMR, namely the tendency for SMRs to increase with age or cumulative exposure.

Another selection force operating on those employed is that of leaving the industry or moving from the jobs of hazardous exposures by the most sensitive individuals. Statistical aspects of competing risks modelling may be helpful in this regard, but this approach requires reliable data on the reason for an exit from an area or the reason for quitting the industry and better follow-up of short-term workers. Analysis experience with even limited data on these points may be the best guide to addressing statistically these complex issues.

Each of these qualitative criticisms of occupational mortality studies could be addressed statistically given the required data. It is in the quality and extent of data, rather than in development of statistical techniques, where the greater need is identified. For example, the adversary climate of government and industry seems to inhibit a good faith collection and retention of data by companies. Presently, there is no protection afforded to a company having data that incriminates itself even if prudent industrial hygiene practices were maintained in past times by its management. Further, similar conflicts of interest induce workers to understate personal exposures, for example, smoking level, thereby limiting the scientific apportionment of injury to occupational and non-occupational factors. Perhaps incentives could be devised to assure the collection and retention of quality data by industry and the cooperation of employees.

Attention is directed to a one parameter, non-proportional hazards model as an alternative to the constant rate ratio estimated by the SMR. Support in two areas for this alternative summary statistic is developed. First, this model is patterned after a slope-ratio design from bioassay where age is a proxy for dose, which in occupational mortality applications can be viewed as highly correlated with cumulative exposure. Second, criticisms of the SMR can be seen as evidence for lack of fit of the proportional model, for example, in its tendency to increase when computed for older age bands, longer periods of hire, or greater accumulations of exposure. The problems exhibited by the SMR seemed to be described generally as selection bias in the occupational epidemiology literature. Although there is reason for this concern, it may be a lesser one than simply an inadequate proportional hazards model.

Brief discussion of the potential implications from the inadequacy of a proportional hazards model for the unknown carcinogenic process is also offered.

2. A PARTITION OF THE HEALTHY WORKER EFFECT

Selection biases pose difficulties for any analysis strategy of occupational mortality studies. As a total, McMichael, Haynes, and Tyroler (1975) termed them the so-called "healthy worker effect." Meaning, in order to be and to stay employed one must physically go to work. As the population of any geographic area contains some who cannot go to work, mortality rates for the United States, or any single state, are thought to be greater than for the employed population of the geographic area. However, quantification of this idea has not been clearly set out, probably due to the absence of an operational definition. As a step in this direction, consider the healthy worker effect as composed of two selective forces:

(i) Selection at Hire: Some individuals in the general population do not appear before, or get past, the industrial employment offices at the usual ages of hire. Self-imposed or real physical/medical limitations are obstacles to employment. This selection force should be considered stronger for some industries or companies.

(ii) Selection by the Workplace Environment: For those initially employed, the most sensitive individuals leave the industry due to physical or mental intolerance for the environment of the workplace. Exposures, as well as the stresses of holding a job, individually or in combination, may result in the exit of some individuals from the employment roles of some companies. A lesser effect, job changes to positions of less exposure or stress, may also be operative.

In the context of these two presumed selective forces, the required data are generally not available. Even so, there are statistical methods that are applicable. For the selection of those employed, cause specific mortality rates at the ages of hire for these workers form the basis of two strategies for constructing synthetic schedules of more appropriate mortality rates. For the selection out of the industry of those most sensitive to a workplace environment, reliable data on the reason for exit and better follow-up of these individuals would permit the application of competing risks to assist in the evaluation of the mortality experience of an employed group. Only for selection at hire are the statistical alternatives considered in further detail.

3. PARSIMONIOUS MODELS OF MORTALITY FOR SELECTION AT HIRE

As frequently the interest of an occupational mortality study is the risk from a specific cause of death, there may be few deaths upon which to estimate the hazard. Recall that with the constant hazard, θ, of an exponential model, the observed information available in the sample is proportional to the observed number of failures. Specifically,

$$\ell n L(\theta) = \sum_{i=1}^{n} \{\delta_i \ell n\theta - \theta t_i\} \tag{1}$$

is the log-likelihood for n exponential observations t_i, where δ_i indicates a failure (δ_i=1) or a censored observation (δ_i=0) at t_i. The negative of the second partial of the log-likelihood is the observed information; visually,

$$-\frac{\partial^2 \ell n L(\theta)}{(\partial\theta)^2} = \sum_{i=1}^{n} \delta_i/\theta^2 \ . \tag{2}$$

Consequently when the number of failures is small, only the simplest models should be considered since the observed information is proportional to the number of failures in the sample. Applications requiring an evaluation of a specific occupational mortality cause of death will be of special interest.

A. Standardized Mortality Ratio

Let t_i be the follow-up time for the ith member of an occupational cohort, whose hazard can be written as

$$\lambda_i(t) = \theta R_i(t), \tag{3}$$

where $R_i(t)$ is a known cause specific death rate appropriate for the ith worker at time t. For $0 \le u \le t_i$, $R_i(u)$ is the pattern of standard rates relevant for the pertinent demographic factors, e.g., sex, race and calendar year of each achieved age. The above multiplicative form is termed a proportional hazards model, following Cox (1972). While at risk to a specific cause of death, the survival function for a worker surviving to time t_i, i.e., being censored by the end of follow-up, being lost during study, or by dying of another cause, is given by

$$S[t_i;\ \theta, R_i(u)] = \exp\{-\int_0^{t_i} \theta R_i(u)\ du\}. \tag{4}$$

The log-likelihood for θ then is given by

$$\ell nL(\theta) = \sum_{i=1}^{n} \{\delta_i \ell n[\theta R_i(t_i)] - \theta\int_0^{t_i} R_i(u)du\}. \tag{5}$$

By direct calculation,

$$\frac{\partial \ell nL(\theta)}{\partial\theta} = \sum_{i=1}^{n} \{\delta_i/\theta - \int_0^{t_i} R_i(u)du\} \tag{6}$$

and then

$$\hat{\theta} = \sum_{i=1}^{n} \delta_i / \sum_{i=1}^{n} \int_0^{t_i} R_i(u)du = O/E, \tag{7}$$

the ratio of the total observed number of deaths from the specific cause to the total "expected" deaths under the schedules of rates $R_i(\cdot)$, for $i=1,2,\ldots,n$. These are expected, in that as Breslow (1978) has shown,

$$E(\delta_i) = E\{\int_0^{t_i} R_i(u)du\}, \tag{8}$$

where the length of follow-up is infinite. The estimator (7), SMR = O/E, was presented by Kilpatrick (1962) as the maximum likelihood estimator of the unkonwn constant ratio of the study to standard population specific rates. He used a Poisson model and a person-years summary of these same data. Breslow (1975, 1976, 1978) exhibited the SMR as the maximum likelihood estimator for the above survival analysis presentation of these data.

However, in actual cohort mortality studies the equality (8) holds only approximately, since in finite length studies follow-up losses are certain. Consequently, for common causes of death with even a short follow-up period or for more rare causes but with a long follow-up period, the inequality with equation (8) can be substantial. The approximation of the survival function (4) by the cumulative of the hazard (3) is essence of the difficulty. The Appendix contains elaboration on this point, as well as illustration of the differences in two alternative calculations of expected numbers, represented by each side of equation (8). Each has been used in practice. The result is to stress the use (7) for estimation of the common age specific rate ratio, although the sum of the expected values of the δ_i perhaps is more properly called the expected number of deaths. Further, this is a good example where following likelihood principles results in clarification of statistical practice.

For estimation of θ and testing of the null hypothesis that θ is one, the observed information in the data is

$$-\frac{\partial^2 \ell nL(\theta)}{(\partial\theta)^2} = \sum_{i=1}^{n} \delta_i/\theta \; . \tag{9}$$

Noting (9) and (8), the expected information reflects the observed and expected number of deaths, respectively. The expected information in the sample increases directly with the number of individuals n and with the follow-up time t_i for each individual in the cohort. The integral in (8) can be approximated as follows:

$$\int_0^{t_i} R_i(u)du \doteqdot \sum_{j=1}^{J_i} R_i(u_j) + f_i(J_i+1)R_i(u_{J_i}+1), \tag{10}$$

where $R_i(u_j)$ is the constant, cause specific death rate in the standard population for follow-up year u_j and restricted to the relevant demographic characteristics of individual i, who is followed for J_i complete years and for fraction $f_i(J_i+1)$ of the last year. Fractions $f_i(j)$ for each year could be introduced if only portions of years, or none of some, were considered at risk. A common practice is to let $R_i(\cdot)$ be constant for five year intervals rather than single years and take $f_i(J_i+1)$ equal to one-half. These details and variations are reviewed by Symons and Taulbee (1984).

Breslow (1975) showed that Rao's efficient score test for H_0: $\ell n\theta=0$ is given by $(O-E)^2/E$. Likelihood ratio or Wald's criteria might also be used, but they are not so simple computationally. For hand calculations, the score statistic is convenient and can be expected to have power comparable to that of likelihood ratio and Wald test statistics, as shown by Lee, Harrell, Tolley and Rosati (1983) for proportional hazard models with covariables.

B. Two Synthetic Standard Constructions

Within the context of a proportional effect on the occupation mortality cause with biological relevance, an SMR type estimator of this effect is one approach for using any synthetic standard. Each of the two standards presented attempts to address the concern for a reduced risk of death at a typical age of hire by using a presumed available estimate of the age-at-hire death rate among employees for the cause of interest. One standard is proportional to the general population at all ages in the same proportion as the mortality rate at the age of hire for those employed to that for those in the general population. The second is a regression of the mortality rates at the age of hire for those employed to that of the general population at an age of retirement. As the latter synthetic standard is a non-proportional pattern, another summary statistic is also presented that may be more sensitive than the SMR, especially if the effect of the workplace environment is progressively greater at older ages, as would be consistent with a mortality rate that increased with cumulative exposure. The age at which death occurs is incorporated into this estimator, circumventing one criticism of the SMR; see Gaffey (1976). Use of the age pattern of deaths may be appropriate for some applications, for example with long, low level exposures, which may provide a more powerful test statistic for this bioassay inspired alternative to the SMR.

1. Synthetic Standard Proportional to Rates of Hirees

Let $R_i^*(x_0)$ be the cause specific rate for employees at the ages of hire in the study cohort, or other comparable group of employed individuals, perhaps obtained from Social Security files as suggested by Goldsmith (1975). These may be available from the occupational mortality literature, but unfortunately such reporting is not customary; Pell, O'Berg and Karrh (1978) is an illustrative exception. The age band 15-25 years of age may be required to provide a reliable estimate for an individual with comparable demography, for example, having the same sex and race, and using $x_0 = 20$ as a representative age of hire.

The calculation of an SMR estimate (7), using an approximation following (10), would be accomplished by replacing the general population schedule of rates $R_i(x_j)$ with

$$R_i^P(x_j) = [R_i^*(x_0)/R_i(x_0)]R_i(x_j) \ , \qquad (11)$$

where x_j is a representative age for the jth age band. In Figure 1, the schedule of $\ln\{R_i^P(x_j)\}$ is presented approximately by the longer dashed lines at the logarithm at the midpoints of the 10 year age bands from age 15 through age 75. Although the pattern of death rates in this age range is largely exponential in nature, observed rates at the upper age bands are usually slightly less than those predicted by an exponential trend. The fit was improved by plotting against log-age, i.e., fitting a Weibull rather than Gompertz hazard as preferred by Cook, Doll and Fellingham (1969). Also, fitting an additive constant to a general exponential trend with age, as described by Makeham (1860), improves the representation of general population rates, but considerably complicates the determination of the synthetic standard.

With the availability of a mortality rate for the employed segment of the population at an age of hire, a more suitable standard schedule of rates may be prepared. The first synthetic schedule is motivated by the very essence of the SMR itself, namely, that the effect of the workplace on the mortality rates of its employees is in constant proportion at all ages. Visually this is portrayed in Figure 1 as a constant difference between the Smoothed U.S. Standard and the Proportional Synthetic Standard when the corresponding logarithms of the rates are plotted against log-age.

2. Synthetic Standard Regressing to U.S. Rates of Retirees

At the ages of hire a general population may have mortality rates in excess of those in a demographically similar employed group. (However, the previous synthetic standard, nor the one to be described, requires this difference necessarily to be positive.) The thinking is that the unemployable portion in the general population is subject to greater death rates. With aging the tendency would be for these individuals to be depleted from the population more quickly than those initially employed and working in reasonably safe conditions. Consequently, the death rates of those survivors at the ages of retirement in the general population would then be more like those initially employed and surviving to the same ages of retirement than at the ages of hire. If every workplace environment posed serious threats to the longevity of its employees, this line of argument fails. In such harsh workplace environments, the survivors to retirement age would represent a very hardy group of individuals.

The second synthetic standard is a schedule of rates that progresses smoothly from reduced rates at the ages of hire for employed individuals to rates at a specified age of retirement for the general population survivors. Such a regression strategy is not new. Yates (1934) suggested a similar regression-to-the-mean phenomena in examining the pulmonary function of workers employed in dusty environments as a means to address the presence of workers at hire with the superior lung function and tolerance to irritant dust.

One simple way of trending the rates from those of the employed at the ages of hire to those of the general population survivors is to use a linear pattern in the death rates over log-age as depicted by the shorter dashed line in Figure 1 for the Non-Proportional Synthetic Standard. This construction implies greater rates than the U.S. Standard at ages past retirement, but seems useful at least as a first approximation. Therefore, a slightly less steep slope is probably more appropriate for the unexposed pattern of rates, but then the slope described should be somewhat conservative. Also, due to the decrease in reliability of death certificate coding of cause of death at the older ages, an SMR calculation would be terminated at this age band of retirement where the regression ends.

Under proportional hazards the observed over expected ratio (7) is a desirable estimator, utilizing the exponential of the logarithm of the rates in Figure 1 for the Non-proportional Synthetic Standard to obtain the expected number of deaths. However, by this latter construction a parameter reflecting a non-proportional effect of a detrimental workplace environment may be more relevant than the proportional one given in (3). For example, if the effect increases with cumulative time in the workplace environment, a descriptive one parameter model is a slope. An estimate is obtained by fitting a line to the logarithm of the observed age specific death rates, but required to go through the logarithm of the death rate at the log-age of hire in the employed group. There is an analogy to the bioassay design with the power dose metameter in a slope-ratio design; see Finney (1964). In this application, advancing age is a proxy for cumulative exposure. The null hypothesis is that the slope in the logarithm of death rates across log-age is given in Figure 1 by

$$\beta_o = \{\ell n[R_i(70)] - \ell n[R_i^*(20)]\}/[\ell n(70) - \ell n(20)]. \tag{12}$$

In the presence of detrimental effects to mortality from the workplace environment, alternative values of the slope parameter are greater than β_o.

Following Frome (1983), maximum likelihood estimation of the slope descriptive of the experience of the cohort could be obtained from the Poisson regression likelihood

$$L(\beta) = \prod_{j=o}^{J} \{e^{-\lambda_j(\beta)P_j}[\lambda_j(\beta)P_j]^{d_j}\} , \tag{13}$$

where d_j is the observed number of deaths in the jth age band; P_j is the number of person-years of experience accrued in age band j; and

$$\lambda_j(\beta) = R^*(e^{a_o}) \exp\{\beta(a_j-a_o)\}, \tag{14}$$

with a_j being the logarithm of the midpoint age of the jth age band, i.e., $x_j = \exp(a_j)$. The subscript i has been suppressed, indicating that only one race-sex group is being considered. Differentiating the logarithm of $L(\beta)$ with respect to β yields the following equation that would require iterative methods to obtain the maximum likelihood estimate, $\hat{\beta}$, satisfying:

$$R^*(e^{a_o}) \sum_{j=1}^{J} (a_j-a_o)\ e^{\hat{\beta}(a_j-a_o)}P_j = \sum_{j=1}^{J} (a_j-a_o)d_j. \tag{15}$$

The null hypothesis value β_o in (12) is a natural initial estimate and convergence to $\hat{\beta}$ should not be difficult by a Newton-Ralphson algorithm.

A significance test of the null hypothesis $H_o:\beta=\beta_o$ versus the alternative hypothesis $H_a:\beta>\beta_o$ can be approximated by

$$z_s = (\hat{\beta} - \beta_o)/\sqrt{\hat{V}(\hat{\beta})} , \tag{16}$$

where the variance of the maximum likelihood estimator for β can be approximated by the inverse of the negative of the second partial of the log-likelihood evaluated at $\hat{\beta}$, specifically

$$\hat{V}(\hat{\beta}) \doteq \{R^*(e^{a_o}) \sum_{j=1}^{J} (a_j-a_o)^2 e^{\hat{\beta}(a_j-a_o)}P_j\}^{-1} . \tag{17}$$

A value of the test statistic exceeding the $(1-\alpha)100$ percentile of the standard normal distribution supports rejecting $\beta=\beta_0$, which would be appropriate when the workplace conditions are detrimental. A score statistic will not have its usual simplifications since β_0 is not zero, but it or the likelihood ratio statistic could be computed in place of the Wald statistic (16).

4. EXAMPLE

The slope-ratio model and variations of a proportional hazards model are illustrated with data originally in Dement (1980) and updated in Dement, et al. (1983). For white males working one or more months between 1940 and 1975 with chrysotile asbestos, the numbers of deaths, d_j, for lung cancers (ICDA 162-163) and the person-years, P_j, of work experience accrued were summarized by 10 year age bands: 15-24, 25-34, ..., 65-74, and 75-84. Indexing these age bands by j, the a_j's are the natural logarithm of the midpoints of these age bands, namely, $a_0 = \ell n(20)$, $a_1 = \ell n(30), \ldots$, and $a_6 = \ell n(80)$. [With such a long study period, the numbers of deaths and person years could be tabulated by age bands for shorter calendar year periods. General population age specific lung cancer death rates would then be employed as described by Monson (1974)]. In addition, the lung cancer death rate for an employed group at the age band of hire, denoted by $R^*(20)$, was approximated by the data in Pell, O'Berg and Karrh (1978).

These data are exhibited in Table 1 and the various standards are graphically displayed in Figure 1. The logarithm of the observed age specific lung cancer rates are also plotted as open squares in Figure 1. The proportional models can be projected onto Figure 1 as a line parallel to one of the standards or as points at a constant distance above (for an excess) the points determining the straight line smoothings in Figure 1. The amount of separation from the standard is determined by the observed rates. The slope-ratio model parameter is the slope of a line with intercept at $[\ell n(20), \ell n R^*(20)]$ and whose magnitude is determined by the observed rates.

Test statistics were computed for each of three models using data in Table 1. Two variations on proportional models were included. The slope-ratio model is determined by the age specific hazard (14). Each model has one parameter in the generic Poisson log-likelihood

$$\ell n L(\theta) = \sum_{j=o}^{J} \{-\lambda_j(\theta)P_j + d_j\, \ell n[\lambda_j(\theta)P_j]\} \ . \tag{18}$$

The three models can be specified by their use of one parameter in the age specific hazard as follows:

$$\lambda_j(\theta=\gamma) = \exp(\gamma)\ R(e^{a_j}) \tag{19}$$

Table 1. Lung Cancer Deaths and Person-Years of Experience by Ten Year Age Bands for White Male Chrysotile Asbestos Textile Workers, 1940-1975, and 1960 U.S. General Population for White Males

Age Band	Person Years	Lung Cancer Deaths ICDA 162-163	Observed Rate	1960 U.S. Male Standard†		
				Population	Deaths ICDA 162-163	Rate
15-24	4023.1	0	(0.000 000 928)*	10,500,224	13	0.000 001 238
25-34	9039.2	0	0.000 000	9,904,838	132	0.000 013 326
35-44	9645.6	0	0.000 000	10,524,984	1029	0.000 097 767
45-54	7009.8	16	0.002 283	9,147,704	4480	0.000 489 740
55-64	2743.9	11	0.004 009	6,883,578	9536	0.001 385 326
65-74	606.5	8	0.013 189	4,606,007	9760	0.002 118 972
75-84	71.7	0	0.000 000	1,804,478	3195	0.001 770 595

*Approximated as 75% of the corresponding U.S. lung cancer rate, following the age specific mortality ratios in Table 3 and lung, bronchus and trachea rates in Figure 3 of Pell, O'Berg and Karrh (1978).

†Population counts are from 1960 Census of Population, Characteristics of the Population. Deaths are from 1960 Vital Statistics of the United States, Volume II: Mortality.

Source: Dr. John M. Dement kindly provided the person-years and numbers of lung cancer deaths for this purpose.

is the first model (Model I), the familiar SMR model described earlier with equation (3). The second model (Model II) is

$$\lambda_j(\theta=\alpha) = \exp(\alpha)\ R^*(e^{a_0})\ \exp[B(a_j-a_0)], \tag{20}$$

where the slope B was determined by a least squares fit of a line to the points $[a_j,\ \ell nR(e^{a_j})]$. A graphical fit should be sufficient, however, as can be seen in Figure 1. It uses the smoothed U.S. Standard rates and is required to pass through the point $[a_0,\ \ell n\, R^*(e^{a_0})]$. This is the Proportional Synthetic Standard in Figure 1 and follows from a parallel line bioassay design. The third model (Model III) is given by

$$\lambda_j(\theta=\beta) = R^*(e^{a_0})\ \exp[\beta(a_j-a_0)], \tag{21}$$

the slope-ratio bioassay alternative. Two other proportional models not included are variants of (19). The first replaces $R(\cdot)$ with the proportional standard given in (11) and the second replaces $R(\cdot)$ by (21) with β specified as

$$\beta_0 = \{\ell nR(e^{a_5}) - \ell nR^*(e^{a_0})\}/(a_5-a_0). \tag{22}$$

Notice for this application, the data in 75-84 year age band were not used.

The Rao-Score, Wald, and likelihood ratio strategies each determines approximately a one degree of freedom chi-square test statistic. Each is described generically as follows using θ to represent the single parameter of each model as in equation (18) for $\ell nL(\theta)$. Let

$$U(\theta) = \frac{\partial \ell nL(\theta)}{\partial\theta} \tag{23}$$

denote the first partial of the log-likelihood with respect to the parameter of the model and

$$I(\theta) = -\frac{\partial U(\theta)}{\partial\theta} = -\frac{\partial^2 \ell nL(\theta)}{(\partial\theta)^2} \tag{24}$$

Now the three test statistics can be specified as follows:

(i) The Rao-Score (RS) one degree of freedom chi-square is

$$\chi^2_{RS} = [U(\theta_0)]^2/I(\theta_0)\ , \tag{25}$$

where θ_0 is the null hypothesis value of the model parameter. For the first and second models, γ_0 and α_0 are zero, respectively, and for the slope-ratio model the null hypothesis value of β is approximated by β_0 in (22).

(ii) The Wald Statistics (WS) one degree of freedom chi-square is

$$\chi^2_{WS} = (\hat\theta-\theta_0)^2/\hat V(\hat\theta)\ , \tag{26}$$

where $\hat{\theta}$ is the maximum likelihood estimate of θ. Specifically, $\hat{\theta}$ is the value of the parameter θ at which $\ell nL(\theta)$ is maximized. This requires that $U(\theta) = 0$ and $I(\hat{\theta})$ be non-negative. Also, $V(\hat{\theta}) = [I(\hat{\theta})]^{-1}$.

(iii) The Likelihood Ratio (LR) statistic is also a one degree of freedom chi-square and

$$\chi^2_{LR} = -2[\ell nL(\theta_0) - \ell nL(\hat{\theta})]. \tag{27}$$

The test statistics, parameter estimates, and their estimated standard errors are given in Table 2 for the chrysotile asbestos cohort data in Table 1. Clearly the test statistics for each model reject the null hypothesis value for its corresponding model parameter. There is very strong evidence for an excess of lung cancer mortality among these workers. The nature of the patterns of this excess is the important issue. These results are suggestive of two points worthy of further study. First, directly incorporating information about the perceived lower rate of lung cancer among employees than among the general population at the

Table 2. Analysis Results from Three Models for the Chrysotile Asbestos Data

		Parameter			Test Statistics		
Model	Description	θ_0	$\hat{\theta}$	Standard Error	Rao-Score	Wald Statistic	Likelihood Ratio
I	Constant rate ratio	$\gamma_0=0$	1.295	0.1690	67.4	58.7	39.8
II	Intercept	$\alpha_0=0$	1.706	0.1690	129.1	101.9	62.1
III	Slope	$\beta_0 \doteq 6.17$	7.816	0.1575	140.6	108.8	64.5

ages of hire can increase the statistical leverage of the proportional hazards model. Is the pattern of excess being measured from too high a standard? The test statistics for Model II are about double those for Model I. Second, the slope-ratio may be more sensitive than a proprotional hazards approach. The test statistics for Model III are also about double those for Model I. Is the excess lung cancer proportional at all ages, or progressively greater at older ages? The issue identified is a choice between models. A resolution is difficult without biologic understanding of the cancer mechanism; other alternatives to proportional hazards should also be sought.

5. A BIOASSAY PERSPECTIVE

The essential difficulty with occupational mortality studies can be viewed from a bioassay perspective. The discussions of the synthetic standard options in the previous subsections attempt to identify a pattern over age for the unexposed but employed individuals. The synthetic standards correspond to the responses for the standard preparation in bioassay, with the dose metameter corresponding to age in the analysis of occupational mortality studies. However, there are usually no data for the employed but unexposed group. In the absence of any data, analysis choices are largely subjective and consequently may be difficulty to defend.

The two proposed synthetic standards depend primarily on the addition of one data point, namely, the cause specific mortality rate for employed individuals at the age of hire, hopefully providing a more appropriate schedule of rates for the unexposed employees than that by using the general population specific rates. That such a schedule of rates is appropriate for unexposed employees makes sense only with an assumption that the effects of exposure on mortality are not immediate. This is probably reasonable for many causes of death, especially those resulting from chronic exposure, but the assumption is not always reasonable, for example with accidental deaths.

To summarize, Table 3 contains a correspondence between features of an occupational mortality evaluation with those of a bioassay, where statistical models and procedures are long-standing. The main observation is that data on the comparison group are generally not available from occupational mortality studies. The synthetic standards were devised to partially remedy this shortcoming. With each synthetic standard, a proportional model and the corresponding observed over expected ratio estimator (3) is one analysis strategy. However, the analogue to the slope-ratio model from bioassay, suggests a reasonable alternative with biologic support, namely, a detrimental effect that increases with cumulative exposure and is measured in proxy by age.

6. DISCUSSION AND CONCLUSION

An association between the way of constructing the synthetic standards and the summary statistic for the underlying bioassay model raises at least two questions. First, is a proportional hazards model appropriate? If a non-proportional hazard seems to fit occupational mortality data better than a proportional one, are there any implications about the basic carcinogenic process? Second, are there other possibilities for the rates of the unexposed employees?

Table 3. Comparison of Bioassays and Occupational Mortality Evaluations

Aspect	Type of Study: Bioassay	Occupational Mortality
1. Nature of Study	Experimental	Observational
2. Groups	Test/Standard Preparations	Exposed/Unexposed Employee Groups
3. Data on	Both Groups	Exposed Only*
4. Response Trend on	Dose	Time Employed or Age
5. Parallel Model	Parallel Line Assay	Proportional Rates
a. Group Differences	Estimated	Relative to Known Pattern
b. Common Slope	Estimated	Assumed Known
c. Linearity	Testable	Not Strictly Needed
6. Non-Proportional Model	Slope-Ratio	Alternative Slope
7. a. Group Difference	Estimated	Relative to Known Slope
b. Common Intercept	Estimated	Assumed Known
c. Linearity	Testable	Assumed

*That there are no data on the employed, strictly unexposed, group is the main limitation of occupational mortality studies.

Regarding the relevance of the proportional hazards model, most of the references to the SMR statistic are very critical; see especially Gaffey (1976) and Wong (1970). Symons and Taulbee (1981) review several of the criticisms of the SMR and provide suggestions for improving its appropriateness. Unfortunately, their suggestion to restrict the age range to improve its applicability requires medical input on the natural history of the disease. Breslow (1975, 1976, 1977) presents the SMR as the maximum likelihood estimator for a proportional hazards model, but he does not consider goodness of fit. However, he mentions using direct rates rather than the indirect adjustment of the SMR approach. The proportional and non-proportional approaches can each be viewed as fitting the observed rates and then testing for a common intercept or a common slope, respectively. Breslow (1983) provides a recent comprehensive review, including Poisson regression of which likelihood (18) is an example. However, his paper is oriented towards proportional hazards and applications where data on the accumulated exposure are available. Again, the common situation where data are very limited is of primary concern here.

Empirically the support for the proportional hazards model is scattered. Amandus (1982) reports on the analysis of the mortality experience of a cohort of coal mine workers and notes the non-parallelism in some cumulative hazard plots. However, he does not include a quantitative assessment. Dement (1980, 1982) assembled and reported the results of the analysis of the lung cancer mortality of a cohort of chrysotile workers. Dement (1983) contains the updated data used for the example. Gaynor (1983) provides further modelling and analysis of the same data. He presents non-parametric cumulative hazard curves that suggest deviation from the proportional hazards model, but with so few deaths beyond age 65, a clear violation is not reported. Visually, the cumulative hazard plots in Figure 4.1 of Gaynor (1983) suggest that the pattern of increased risk for all exposed groups would not be inconsistent with the non-proportional hazard in (14). For an application of evaluating fertility in an occupational setting, Boyle (1983) used an external standard for birth rates, as in (3), and found that this proportional model was inadequate.

In addition, there is considerable evidence for lack of fit to occupational mortality by the proportional hazards model in the critical comments surrounding the SMR. Although usually attributed to unestimable biases as part of the healthy worker effect, another interpretation is that the ratio of the study to standard age specific rates simply tends not to be constant. For example, McMichael (1976) notes the tendency of the SMR to increase when restricted to older ages. Gilbert (1982) reviews the occupational epidemiology literature in this regard and notes that the SMR tends to increase with longer exposure, longer employment and older age. The non-proportional model (14) could accommodate these phenomena observed with the SMR. When exposure data are not available and using age as a proxy for cumulative exposure, model (14) can be viewed as a slope-ratio design from bioassay when the pattern in the unexposed group is considered known.

Brief comments are made concerning the implications of non-proportionality for the underlying cancer risk. Breslow (1977) remarks that the multiplicative effect on the hazard due to membership in the occupational cohort is implied by multistage theories of carcinogenesis in which the last stage is affected by the occupation exposure. Whittemore (1977) establishes this relationship. If a non-proportional model seems more appropriate for occupational mortality data of site specific cancers, then either the multistage theory is inappropriate or other stages than the last are affected by the occupational exposure. The fit of proportional hazards model has implications to the underlying carcinogenic theory and, perhaps, for undue concern about selection bias in the occupational epidemiological literature. Critical evaluation of its fit is needed.

Regarding other schedules of standard mortality rates suggested by Figure 1, there are two in addition to the pair of synthetic schedules described above.

They are the smoothed U.S. general population death rates and another synthetic one with slope β_0 in (12) but originating at the logarithm of the general population death rate at log-age a_0. The first is usual choice for standard rates and was used by (19). The latter is more of a logical possibility, however.

Finally, it is helpful to note that if data on exposure for each employee are available, even an indication of categorically "exposed" or "not exposed," internal comparisons are then possible. The Cochran-Mantel-Haenszel procedure was set out by Cochran (1954) and was refined by Mantel and Haenszel (1959) to accommodate very slim two-by-two tables. It is applicable to many problems, particularly relevant is the survival application described by Mantel (1966). Gilbert (1979), 1982) and Gilbert and Marks (1979) demonstrated its applicatility to occupational mortality with their analysis of the mortality experience of workers exposed to radiation. Breslow (1975, 1983) indicated and also endorsed the use of internal standards, where the pooled experience of the exposure groups becomes the standard schedule of rates under the null hypothesis. Also, Woolson (1981) showed that the indirect adjustment procedure, of which the SMR is an example, can be viewed as a limiting case of the Cochran-Mantel-Haenszel procedure when the unexposed group becomes very large relative to the exposed group.

The point of mentioning the Cochran-Mantel-Haenszel procedure is to illustrate that statistically powerful methods are available for the evaluation of the mortality experience of an occupational cohort with appropriate data. With even more data on the level of exposure than just whether an individual was exposed or not, the time dependent covariates sketched by Breslow (1983) and utilized by Gaynor (1983) can be incorporated into the principles of survival analysis to utilize exposure by calendar year worked, for example. However, critical evaluation of the proportional hazards assumption in those cases still seems prudent.

ACKNOWLEDGEMENTS: This work was started in 1980 while MJS was on leave with the Health Effects Research Laboratory, U.S. Environmental Protection Agency, Research Triangle Park, North Carolina. Various subsequent fragments of support have come from the Epi-Stat Unit of the Cancer Research Center, and from the Occupational Health Studies Group, School of Public Health, University of North Carolina at Chapel Hill. This research support is gratefully acknowledged.

REFERENCES:

[1] Amandus, H., A comparison of analytical techniques for occupational mortality studies with an empirical example. Ph.D. Dissertation in Epidemiology. University of North Carolina at Chapel Hill (1982).

[2] Bittinger, C.S., A comparison of two methods of calculating the expected number of deaths for the standardized mortality ratio. Report for Master of Science in Public Health Degree in Biostatistics. University of North Carolina at Chapel Hill (1982).

[3] Boyle, K.E., Survival model for fertility evaluation. Ph.D. Dissertation in Biostatistics. University of North Carolina at Chapel Hill (1983).

[4] Breslow, N.E., Analysis of survival data under the proportional hazards model, International Statist. Rev. **43**:45-58 (1975).

[5] Breslow, N.E., Some statistical models useful in the study of occupational mortality. Environmental Health: Quantitative Methods. Proceedings of a Conference on Environmental Health, Whittemore, A. (ed.). Philadelphia, SIAM: 88-103 (1977).

[6] Breslow, N.E., The proportional hazards model: applications in epidemiology, Comm. in Statist.:Theory and Methods A7(4):315-332 (1978).

[7] Breslow, N.E., Lubin, J.H., Marek, P., and Laugholz, B., Multiplicative models and cohort analysis, J. Amer. Statist. Assoc. **78**:1-12 (1983).

[8] Chiang, C.L., On constructing current life tables, J. Amer. Statist. Assoc. **67**:538-541 (1972).

[9] Cochran, W.R., Some methods for strengthening the common chi-square tests Biometrics **10**:417-450 (1954).

[10] Cook, P.J., Doll, R., and Fellingham, S.A., A mathematical model for the age distribution of cancer in man, International J. of Cancer **4**:93-112 (1969).

[11] Cox, D.R., Regression models and life tables (with discussion), J. Roy. Statist. Soc. Series B, **34**:187-220 (1972).

[12] Dement, J.M., Estimation of dose and evaluation of dose-response in a retrospective cohort mortality study of chrysotile asbestos textile workers. Ph.D. Dissertation in Environmental Sciences and Engineering. University of North Carolina at Chapel Hill (1980).

[13] Dement, J.M., Harris, Jr. R.L., Symons, M.J., and Shy, C.M., Estimates of dose-response for respiratory cancer among chrysotile asbestos textile workers, Ann. of Occup. Hygiene **26**:869-887 (1982).

[14] Dement, J.M., Harris, R.L., Symons, M.L., and Shy, C.M., Exposures and mortality among chrysotile asbestos workers. Part II: Mortality, Amer. J. of Industrial Med. **4**:421-433 (1983).

[15] Enterline, P.E., Not uniformly true for each cause of death, J. of Occupational Med. **17**:127-128 (1975).

[16] Frome, E.L., The analysis of rates using Poisson regression models, Biometrics **39**:665-674 (1983).

[17] Gaffey, W.R., A critique of the standardized mortality ratio, J. of Occupational Med. **18**:157-160 (1976).

[18] Gaynor, J.J., A framework that incorporates repeated measurements into the hazard. Ph.D. Dissertation in Biostatistics. University of North Carolina at Chapel Hill (1983).

[19] Gilbert, E.S., The assessment of risks from occupational exposures to ionizing radiation. Energy and Health, Breslow, N.E. and Whittemore, A. (eds.), Society for Industrial and Applied Mathematics, Philadelphia: 209-225 (1979).

[20] Gilbert, E.S., Some confounding factors in the study of mortality and occupational exposures, Amer. J. Epidemiol. 116:177-188 (1982).

[21] Gilbert, E.S. and Marks, S., An analysis of the mortality of workers in a nuclear facility, Radiation Research 79:122-148 (1979).

[22] Goldsmith, J., What we expect from an occupational cohort? J. Occupational Med. 17:126-217 (1975).

[23] Goldsmith, J.R. and Hirshberg, M.S., Mortality and industrial employment, J. Occupational Med. 18:161-164 (1976).

[24] Hill, I.D., Computing man years at risk, Brit. J. Preventive and Social Med. 26:132-134 (1972).

[25] Kilpatrick, S.J., Occupational mortality indices, Population Studies 16: 175-182 (1962).

[26] Lee, K.L., Harrell, F.E., Jr., Tolley, H.D., and Rosati, R.A., A comparison of test statistics for assessing the effect of concomitant variables in survival analysis, Biometrics 39:341-350 (1983).

[27] Levine, R.M., Symons, M.J., Balogh, S.A., Arndt, D.M., Kawsandik, N.T. and Gentile, J.W., A method to monitor the fertility of workers. 1. Method and pilot studies, J. of Occupational Med. 22:781-791 (1980).

[28] Makeham, W.M., On the law of mortality and the construction of annuity tables, J. Instit. of Actuaries 8 (1960).

[29] Mantel, N., Evaluation of survival data and two new rank order statistics arising in its consideration, Cancer Chemotherapy Rep. 50:163-170 (1966).

[30] Mantel, N. and Haenszel, N., Statistical aspects of the analysis of data for retrospective studies of disease, J. National Cancer Instit. 22:719-748 (1959).

[31] Marsh, G.M., OCMAP: a user oriented occupational cohort mortality analysis program, Amer. Statistician 34:245 (1980).

[32] McMichael, A.J., Standardized mortality ratios and the healthy worker effect: scratching beneath the surface, J. Occupational Med. 18:165-168 (1976).

[33] McMichael, A.J., Haynes, S.G., and Tyroler, H.A., Observations on the evaluation of occupational mortality data, J. Occupational Med. 17:128-131 (1975).

[34] Monson, R.R., Analysis of relative survival and proportional mortality, Computers and Biomedical Res. 1:325-332 (1974).

[35] Ortmeyer, C.E., Costello, J., Morgan, W.K.C., Swecker, S., and Peterson, M., The mortality of Appalachian coal miners, 1963 to 1971, Arch. of Environ. Health 29:67-72 (1974).

[36] Pell, S., O'Berg, M.T. and Karrh, B.W., Cancer epidemiology surveillance in the Du Pont Company, J. Occupational Med. 20:725-740 (1978).

[37] Scott, L.M., Bahler, K.W., DeLaGarza, A., and Lincoln, T.A., Mortality of uranium and non-uranium workers, Health Physics 23:555-557 (1972).

[38] Shindell, S., Weisberg, F.J., and Geifer, E.E., The "healthy worker effect"-Fact or artifact? J. Occupational Med. **20**:807-811 (1978).

[39] Silcock, H., The comparison of occupational mortality rates, Population Studies **13**:183-192 (1959).

[40] Symons, M.J. and Taulbee, J.D., Practical considerations for approximating relative risk by the standardized mortality ratio, J. Occupational Med. **23**: 413-416 (1981).

[41] Symons, M.J. and Taulbee, J.D., Chapter 2: Statistical evaluation of the risk of cancer mortality among industrial populations. Statistical Methods for Cancer Studies, Cornell, R.G. (ed.). Marcel Dekker, Inc., New York (1984).

[42] Whittemore, A., Epidemiological implications of the multistage theory of carcinogenesis. Environmental Health: Quantitative Methods. Proceedings of a Conference on Environmental Health, Whittemore, A. (ed.). Philadelphia, SIAM: 72-87 (1977).

[43] Wong, O., Further criticisms on epidemiological methodology in occupational studies, J. Occupational Med. **19**:220-222 (1977).

[44] Wong, O., Utidjian, M.D., and Karten, V.S., Retrospective evaluation of reproductive performance of workers exposed to ethylene dibromide, J. Occupational Med. **21**:98-102 (1979).

[45] Woolson, R.F., Rank tests and a one-sample log rank test for comparing observed survival data to a standard population, Biometrics **37**:687-698 (1981).

[46] Yule, G.U., On some points relating to vital statistics, more especially statistics of occupational mortality (with discussion), J. Roy. Statist. Soc. **94**:1-84 (1934).

APPENDIX: Alternative Calculations of Expected Number of Deaths

Breslow (1978) shows that the probability of failure under study is equal to the expected value of the cumulative hazard. Specifically,

$$E(\delta_i) = E\{\int_0^{t_i} \lambda_i(u)du\}. \tag{A1}$$

For standardized mortality calculations, each side of the equation (A1) has been utilized for computing the expected number of deaths.

In applications, inequality rather than equality exists between the probability of failure and the expected cumulative hazard. The critical step in the derivation of the equality of (A1) requires an interchange in the order of integration. Although not required to be known, the hazard of being lost to follow-up must be continuous for the order of integration to be interchanged. But for any finite length study the hazard of being censored has a discontinuity at the end of study, when all those alive are censored indicating an instantaneous, infinite increase in risk.

The two approaches used in calculating the number of expected deaths in a cohort mortality study are now described. Both methods require age-specific death rates for the cause of death of interest. The Modified Life Table Methods has been used almost exclusively in occupational cohort mortality studies. The description by Hill (1972) and the computer program implementations by Monson (1974) and Marsh (1980) are commonly used for applications to cohort mortality studies. Mechanically, the person-years of experience accumulated by the cohort of workers over the observation period in each age band multiplied by the corresponding age-specific death rate determines the number of expected deaths for the age band. Summing over all the age bands gives the total number of expected deaths.

The other approach, the Life Table Method, requires the use of life table conditional probabilities, but it has been less frequently applied to occupational mortality study data. For two such applications, see Scott, et al. (1972) and Ortmeyer, et al. (1974). Also in fertility studies the approach has recently been used by Wong, et al. (1979) and modified by Levine, et al. (1980). Essentially, the age-specific death rates are transformed into conditional probabilities of dying for the corresponding age band by one of the standard methods from current life table construction. Chiang (1972) presents the various conversions of rates to conditional probabilities of death that are utilized in current life table construction. The probability of dying during the observation period of the study can then be calculated for each individual. The sum of these probabilities over all individuals is taken as the total number of expected deaths.

The number of expected deaths calculated from the Modified Life Table Method is always larger than that calculated by the Life Table Method. The magnitude of the discrepancy in the two methods of calculating the number of expected deaths depends upon the cause of death being studied, the age distribution of the cohort, and the length of the study period. For a rare cause of death, such as a site specific cancer, the results from the two approaches are the same for all practical purposes. However, the calculation of number of expected deaths for all causes of death can depend dramatically on the method utilized. The older the age distribution of the cohort members and the longer the period of observation, the greater will be the discrepancy in the number of expected deaths computed by the two methods.

Algebraic demonstration of these results is straightforward. By examining series expansions of current life table conversions of specific rates into conditional probabilities of death, the numerical value for the probability of dying is readily seen to be exceeded by that of the corresponding rate. For example, the linear conversion of the age specific rate, R_x, to the conditional probability of death, q_x, during the age interval, indexed by x, is

$$q_x = R_x[1 + \tfrac{1}{2}R_x]^{-1} \quad \text{(A2)}$$

Using the binomial to expand $(1 + \tfrac{1}{2}R_x)^{-1}$ in a series yields

$$q_x = R_x[1 - \tfrac{1}{2}R_x + (\tfrac{1}{2}R_x)^2 - (\tfrac{1}{2}R_x)^3 + \ldots]$$

and then

$$q_x = R_x - \tfrac{1}{2}R_x^2 + 1/4R_x^3 - 1/8R_x^4 + \ldots, \quad \text{(A3)}$$

the series converging for R_x less than one. Since the age specific rates are positive,

$$q_x < R_x, \quad \text{(A4)}$$

strictly, since $\tfrac{1}{2}R_x^2$ is positive. Consequently, for any age band the expected value of the right side of (A1) exceeds that of the left side of (A1). Then the division by these expected numbers in an SMR ratio of observed to expected and will produce an inequality in the opposite direction. The inequality also holds for two or more age bands, and hence the SMRs by the Life Table Method will exceed those by the Modified Life Table Method.

However, when the rate R_x is small so that R_x^2 and higher power terms are negligible, then

$$q_x \doteqdot R_x. \quad \text{(A5)}$$

Notice that this is a numerical approximation; the probability of dying and the corresponding rate are two distinct concepts.

Several cohorts of employed populations were used by Bittinger (1982) that illustrate the practical differences in these two alternatives for computing the expected number of deaths. Table A1 contains the results for one example that shows a 20% relative discrepancy in the standardized mortality ratio for all causes of death computed using the expected numbers of death by the two methods. The magnitude of the differences possible depends upon the length of follow-up, the age distribution of the workforce, the size of the workforce and the cause of death under study. For more common causes of death, the death rates are larger, which is the single most important factor affecting the size of the discrepancy. Longer follow-up, older workers, and larger workforces also contribute to greater differences in these two computational strategies.

Table A1. Standardized Mortality Ratios by Two Methods of Calculating the Expected Numbers for Large Cohort

Cause of Death (number of deaths)	Modified Life Table Method SMR	Life Table Method SMR'	Ratio SMR/SMR'
Leukemia (25)	1.382	1.385	0.998
All Neoplasms (463)	0.976	1.018	0.959
Ischemic Heart Disease (1017)	0.954	1.056	0.904
All Causes (2373)	0.941	1.180	0.800

Adapted from Table 5 of Bittinger (1982).

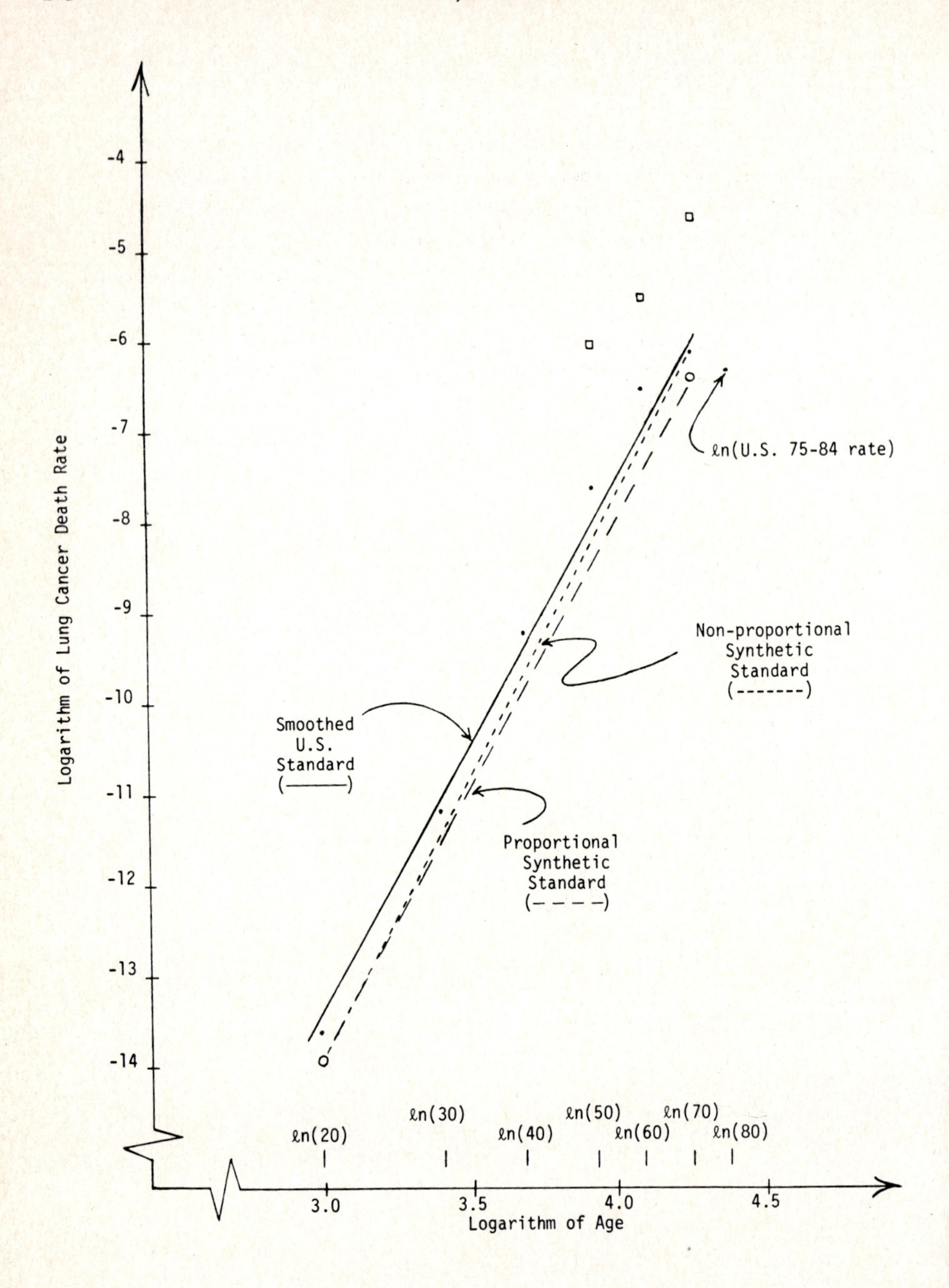

Figure 1. Smoothed U.S. Lung Cancer Rates, Synthetic Standards, and Observed Rates (□)

GENERAL METHODOLOGY

BIOSTATISTICS: Statistics in Biomedical, Public Health
and Environmental Sciences, edited by P.K. Sen

NONPARAMETRIC ESTIMATION OF INTERCEPTS AFTER A PRELIMINARY TEST ON PARALLELISM OF SEVERAL REGRESSION LINES

M.G. Akritas, A.K. Md.E. Saleh and P.K. Sen
Univ. of Rochester, Carleton University and Univ. of North Carolina

For several simple linear models, the problem of estimating the intercepts after a preliminary test on parallelism of the lines is considered. Nonparametric estimators of intercepts are proposed for the problem and their asymptotic behavior is studied. Comparisons with the conventional estimation procedures are made and allied (asymptotic) efficiency results are studied.

KEY WORDS AND PHRASES: Preliminary test estimators, nonparametric tests, tests for parallelism, rank estimators, asymptotic relative efficiency

1. INTRODUCTION

For each $\ell = 1,\ldots,p$, $p \geq 2$, let $X_{\ell j}$, $j = 1,\ldots,n_\ell$ be independent random variables with distribution functions (d.f.'s)

$$F_\ell(x) = P(X_{\ell j} \leq x) = F(x - \theta_\ell - \beta_\ell C_\ell), \quad -\infty< <\infty \tag{1.1}$$

where $\underset{\sim}{C}_n^{(\ell)} = (C_{\ell 1},\ldots,C_{\ell n\ell})'$ is a vector of known constants (not all equal) and $\underset{\sim}{\theta} = (\theta_1,\ldots,\theta_p)'$, $\underset{\sim}{\beta} = (\beta_1,\ldots\beta_p)'$ are unknown parameters; the form of F may or may not be known but we will assume that F belongs in the class $\mathcal{F}$ of all symmetric absolutely continuous d.f.'s with absolutely continuous probability density functions (p.d.f.'s) f having finite Fisher information

$$I(f) = \int_{-\infty}^{\infty} \left(\frac{f'(x)}{f(x)}\right)^2 f(x)\, dx < \infty \tag{1.2}$$

We are primarily interested in the estimation of the vector $\underset{\sim}{\theta}$. In many problems it may be a priori known that $\beta_1 = \beta_2 = \cdots = \beta_p \equiv \beta$, say, in which case the estimate of the intercept vector $\underset{\sim}{\theta}$ will involve the common estimate of β; such an estimator will be denoted by $\hat{\underset{\sim}{\theta}}$. On the other hand, when the slopes are known to be unequal an estimate of $\underset{\sim}{\theta}$ will depend on the estimates of the components of $\underset{\sim}{\beta}$, i.e., $\beta_1,\ldots,\beta_p$; such an estimator will be denoted by $\tilde{\underset{\sim}{\theta}}_n$. When the truth about the slopes $\beta_1,\ldots,\beta_p$ is not well supported but it is suspected that they are equal, a preliminary test of significance concerning the parallelism can be made and if H_0: $\beta_1 = \cdots = \beta_p$ is tenable, the estimate $\hat{\underset{\sim}{\theta}}_n = (\hat{\theta}_{n_1}^{(1)},\ldots,\hat{\theta}_{n_p}^{(p)})'$ is used, while $\tilde{\underset{\sim}{\theta}}_n = (\tilde{\theta}_{n_1}^{(1)},\ldots,\tilde{\theta}_{n_p}^{(p)})'$ is used when H_0 is not tenable. Such an estimator after a preliminary test on parallelism will be denoted by $\bar{\underset{\sim}{\theta}}_n = (\bar{\theta}_n^{(1)},\ldots,\bar{\theta}_n^{(p)})'$.

The object of this paper is to investigate the asymptotic properties of $\bar{\underset{\sim}{\theta}}_n$.

The need for making a preliminary test for parallelism arises quite naturally in simple analysis of covariance models $E(X_{\ell j}) = \mu + \alpha_\ell + \beta_\ell c_{\ell j}$ and in particular in some bioassay problems (see also Sen (1971, 1972)).

The effects of preliminary tests of significance upon estimation (viz. bias and mean square error) have been studied in various special cases by Bancroft (1944), Han and Bancroft (1968) and Mosteller (1948) among others. Ahsanullah and Saleh (1972) considered the model (1.1) for p = 1 and studied these effects for classical least squares estimators under normality assumptions. Recently, Saleh and Sen (1978) and Sen and Saleh (1979) considered the nonparametric formulation to this class of problems. In this paper we consider the problem of estimating the intercepts $\underset{\sim}{\theta}$ after a preliminary test on parallelism. Along with some notations, the proposed estimators are introduced in Section 2. The asymptotic distributions are derived in Section 3 while the resulting expressions for the asymptotic bias and asymptotic dispersion matrix are given in Section 4. The asymptotic relative efficiency (ARE) results are presented in Section 5.

2. THE PROPOSED ESTIMATORS

In this section we will define explicitly the estimators we will consider and will introduce the necessary notation and assumptions.

Let n_ℓ, $c_{\ell j}$, $\ell = 1,\ldots,p$, $j = 1,\ldots,n_\ell$ be as defined in connection with relation (1.1). We will assume that

$$c_{\ell j} \text{ are all bounded} \tag{2.1}$$

$$n^{-1}n_\ell \to \lambda_\ell,\ 0 < \lambda_\ell < 1,\ \ell = 1,\ldots,p \tag{2.2}$$

where $n = \sum_{\ell=1}^{p} n_\ell$, and

$$\bar{c}_{n_\ell}^{(\ell)} = n_\ell^{-1}\sum_{j=1}^{n_\ell} c_{\ell j} \to \bar{c}_\ell\ ,\ n_\ell^{-1}\sum_{j=1}^{n_\ell}(c_{\ell j} - \bar{c}_{n_\ell}^{(\ell)})^2 \to Q_\ell \tag{2.3}$$

with $|\bar{c}_\ell| < \infty$ and $0 < Q_\ell < \infty$, $\ell = 1,\ldots,p$, so that

$$n^{-1}\sum_{\ell=1}^{p}\sum_{j=1}^{n_\ell}(c_{\ell j} - \bar{c}_{n_\ell}^{(\ell)})^2 \to \sum_{\ell=1}^{p}\lambda_\ell Q_\ell \equiv Q$$

The score function corresponding to $F \in \mathcal{F}$ will be denoted by

$$\psi(u) = -\frac{f'(F^{-1}(u))}{f(f^{-1}(u))} \tag{2.4}$$

Next, let $\phi(u)$, $u\varepsilon(0,1)$ be a nondecreasing, skew symmetric (i.e., $\phi(u) + \phi(1-u) = 0$, $\forall\ u\varepsilon(0,1)$) and square integrable score function, let $\phi^+(u) = \phi(\frac{1}{2} + \frac{1}{2}u)$, $u\varepsilon(0,1)$ and set

$$a_n(i) = E\ \phi(U_{n,i}) \quad \text{or} \quad \phi(\tfrac{i}{n+1}) \tag{2.5}$$

$$a_n^+(i) = E\ \phi^+(U_{n,i}) \quad \text{or} \quad \phi^+(\tfrac{i}{n+1}) \tag{2.6}$$

for $i=1,\ldots,n$ where $U_{n,1} < \cdots < U_{n,n}$ are the order statistics of a sample size n from $U(0,1)$. Finally, for every real constant a,b let $R^{(\ell)}_{n_\ell,j}(a,b) \equiv R_{n_\ell,j}(b)$ $(R^+_{n_\ell,j}(a,b))$ be the rank of $X_{\ell j} - a - bc_{\ell j}$ ($|X_{\ell j} - a - bc_{\ell j}$) among $X_{\ell 1} - a - bc_{\ell 1},\ldots,X_{\ell n_\ell} - a - bc_{\ell n_\ell}$ $(|X_{\ell 1} - a - bc_{\ell 1}|,\ldots,|X_{\ell n \ell} - a - bc_{\ell n_\ell}|)$, for $\ell = 1,\ldots,p$ and set

$$T^{(\ell)}_{n\ell}(a,b) = \sum_{j=1}^{n_\ell} \text{sign}(X_{\ell j} - a - bc_{\ell j})a_n^+(R^{(\ell)}_{n_\ell,j}) \tag{2.7}$$

$$S^{(\ell)}_{n_\ell}(b) = \sum_{j=1}^{n_\ell} (c_{\ell j} - c^{(\ell)}_{n_\ell})a_{n_\ell}(R^{(\ell)}_{n_\ell,j}(b)) \tag{2.8}$$

$\ell = 1,\ldots,p$, and

$$S_n(\underset{\sim}{b}) = \sum_{\ell=1}^{p} S^{(\ell)}_{n_\ell}(b_\ell) \tag{2.9}$$

For notational convenience, $R^{(\ell)}_{n_\ell,j}$, $T^{(\ell)}_{n_\ell}$, $S^{(\ell)}_{n_\ell}$, S_n will stand for $R^{(\ell)}_{n_\ell,j}(0,0)$, $T^{(\ell)}_{n_\ell}(0,0)$, $S^{(\ell)}_{n_\ell}(0)$, $S_n(\underset{\sim}{0})$, respectively.

Note that, given $\underset{\sim}{X}^{(\ell)}_{n_\ell} = (X_{\ell 1},\ldots,X_{\ell n_\ell})'$, $S^{(\ell)}_{n_\ell}(b)$ is $\downarrow$ in b, for $b\varepsilon(-\infty,\infty)$. Also, under the model (1.1) with $\theta_\ell = \beta_\ell = 0$, $T^{(\ell)}_{N_\ell}$ and $S^{(\ell)}_{n_\ell}$ both have distributions symmetric about 0.

The three types of estimators we will consider are given by (2.10), (2.11), and (2.16), respectively.

$$\hat{\theta}^{(\ell)}_n = \tfrac{1}{2}[\sup\{a:T^{(\ell)}_{n_\ell}(a,\hat{\beta}_n) > 0\} + \inf\{a:T^{(\ell)}_{n_\ell}(a,\hat{\beta}_n) < 0\}] \tag{2.10}$$

where $\hat{\beta}_n = \frac{1}{2}[\sup\{b:S_n(b\underset{\sim}{1}) > 0\} + \inf\{b:S_n(b\underset{\sim}{1}) < 0\}]$ is an Adichie (1967)-type estimate of the slope under $H_0:\beta_1 = \cdots = \beta_p = \beta$,

$$\tilde{\theta}^{(\ell)}_{n_\ell} = \tfrac{1}{2}[\sup\{a:T^{(\ell)}_{n_\ell}(a,\tilde{\beta}^{(\ell)}_{n_\ell}) > 0\} + \inf\{a:T^{(\ell)}_{n_\ell}(a,\tilde{\beta}^{(\ell)}_{n_\ell}) < 0\}] \tag{2.11}$$

where $\tilde{\beta}^{(\ell)}_{n_\ell} = \frac{1}{2}[\sup\{b:S^{(\ell)}_{n_\ell}(b) > 0\} + \inf\{b:S^{(\ell)}_{n_\ell}(b) < 0\}]$ is an Adichie-type estimate of the slope β_ℓ. Note that under $H_0:\beta_1 = \cdots = \beta_p$, $\underset{\sim}{\hat{\theta}}_n$ is a translation invariant, robust and consistent estimator of $\underset{\sim}{\theta}$; $\tilde{\underset{\sim}{\theta}}_n$ enjoys the same properties under the general model in (1.1).

For the preliminary test on the parallelism of the regression lines, we use the nonparametric test due to Sen (1969) which is based on the statistic

$$L_n = \sum_{\ell=1} [S^{(\ell)}_{n_\ell}(\hat{\beta}_n)]^2/(n_\ell A^2_\phi Q_\ell) \tag{2.12}$$

where

$$A^2_\phi = \int_0^1 \phi^2(u)du - (\int_0^1 \phi(u)du)^2 \tag{2.13}$$

It is shown in Sen (1969) (see also relation (3.10) below) that under

$$K_n: \underset{\sim}{\theta}_n = \underset{\sim}{\theta},\ \underset{\sim}{\beta}_n = \beta\underset{\sim}{1} + n^{-\frac{1}{2}}t \tag{2.14}$$

where $\underset{\sim}{t} = (t_1,\dots,t_p)$ satisfies $\sum_{\ell=1}^{p} t_\ell\lambda_\ell Q_\ell = 0$, the asymptotic distribution of L_n is χ^2 with p-1 degrees of freedom and noncentrality parameter

$$\Delta = \gamma^2 A_\phi^{-2} \sum_{\ell=1}^{p} t^2_\ell \lambda_\ell Q_\ell \tag{2.15}$$

where $\gamma \equiv \gamma(\phi,\psi) = \int_0^1 \phi(u)\psi(u)du$ with ψ defined in (2.4) and ϕ in connection with (2.5). Thus, the preliminary test accepts or rejects H_0 at level α according as $L_n < \chi^2_{p-1,\alpha}$ or $L_n \geq \chi^2_{p-1,\alpha}$ where $\chi^2_{p-1,\alpha}$ is the upper $100(1-\alpha)$ percentile of the chi-square distribution, and the preliminary test estimator $\bar{\underset{\sim}{\theta}}_n$ will be defined by

$$\bar{\underset{\sim}{\theta}}_n = \begin{cases} \hat{\underset{\sim}{\theta}}_n & \text{if } L_n \leq \chi^2_{p-1,\alpha} \\ \tilde{\underset{\sim}{\theta}}_n & \text{if } L_n > \chi^2_{p-1,\alpha} \end{cases} \tag{2.16}$$

3. THE ASYMPTOTIC DISTRIBUTION OF $\bar{\underset{\sim}{\theta}}_n$.

The proof of the main result of this section will be given in Theorem 3.3 and its proof will be based on Theorems 3.1 and 3.2. We begin by stating some known results for convenient reference.

It follows from the basic theorems of Chapter 4 of Hájek and Šidák (1967) that for each $\ell = 1,\dots,p$

$$L(n^{-\frac{1}{2}}(T^{(\ell)}_{n_\ell}, S^{(\ell)}_{n_\ell})' \mid \theta_\ell = \beta_\ell = 0] \to N(\binom{0}{0},\ A^2_\phi\lambda_\ell \begin{pmatrix}1 & 0\\ 0 & Q_\ell\end{pmatrix})). \tag{3.1}$$

Next, Theorem 3.1 of Jurečková (1969) implies that under $\beta = 0$ and for any finite positive constant k,

$$\sup\{n^{-\frac{1}{2}}|S_{n_\ell}^{(\ell)}(n^{-\frac{1}{2}}b) - S_{n_\ell}^{(\ell)} + \lambda_\ell n^{\frac{1}{2}} bQ_\ell\gamma| : |b| \leq k\} \xrightarrow{P} 0, \qquad (3.2)$$

so that

$$\sup\{n^{-\frac{1}{2}}|S_n(n^{-\frac{1}{2}}\underset{\sim}{b}) - S_n + n^{\frac{1}{2}}\sum_{\ell=1}^{p}\lambda_\ell b_\ell Q_\ell\gamma|;\ |\underset{\sim}{b}| < k\} \xrightarrow{P} 0, \qquad (3.3)$$

where $|\underset{\sim}{b}| = \max\{|b_1|,\ldots,|b_p|\}$. Finally, Theorem 7.2 of Kraft and van Eeden (1972) implies that under $\theta_\ell = \beta_\ell = 0$,

$$\sup\{n^{-\frac{1}{2}}|T_{n_\ell}^{(\ell)}(n^{-\frac{1}{2}}(a,b)) - T_{n_\ell}^{(\ell)} + \lambda_\ell n^{\frac{1}{2}}(a+bc_{n_\ell}^{(\ell)})\gamma|;\ |a| \leq k, |b| \leq k\} \to 0. \qquad (3.4)$$

Note that relations (3.2), (3.3), and (3.4) are also true under $\{K_n\}$ (see (2.14) by contiguity.

Theorem 3.1. Under K_n, the distribution of

$$\underset{\sim}{V}_n = (n^{\frac{1}{2}}(\hat{\underset{\sim}{\theta}}_n - \underset{\sim}{\theta}), n^{-\frac{1}{2}}(\hat{S}_{n_1}^{(1)},\ldots,\hat{S}_{n_p}^{(p)})'$$

where $\hat{S}_{n_\ell}^{(\ell)} \equiv S_{n_\ell}^{(\ell)}(\hat{\beta}_n)$, $\ell = 1,\ldots,p$, converges to a 2p-variate normal distribution with mean

$$(t_1\bar{c}_1,\ldots,t_p\bar{c}_p,\ \gamma\lambda_1 t_1 Q_1,\ldots,\gamma\lambda_p t_p Q_p)'$$

and covariance matrix $\Sigma = (\Sigma_{rs})$, r,s = 1,2, where $\Sigma_{12} = \Sigma_{21} = 0$,

$$\Sigma_{11} = \frac{A_\phi^2}{\gamma^2 Q}\underset{\sim}{c}\,\underset{\sim}{c}' + \frac{A_\phi^2}{\gamma^2}\,\mathrm{Diag}(\frac{1}{\lambda_\ell}),\ \Sigma_{22} = A_\phi^2\ \mathrm{Diag}(\lambda_\ell)B\ \mathrm{Diag}(\lambda)$$

where $\underset{\sim}{c}' = (\bar{c}_1,\ldots,\bar{c}_p)$ and $B = \mathrm{Diag}(\frac{Q_\ell}{\lambda_\ell}) - (\frac{Q_i Q_i}{Q})_{i,j=1}^{p}$.

Proof. Assume without loss of generality that $\underset{\sim}{\theta} = \underset{\sim}{0}$. From the relative compactness of $\sqrt{n}\,\hat{\beta}_n$ (Sen (1969)), the definition of $\hat{\underset{\sim}{\theta}}_n$ and relations (3.2) - (3.4) it follows that

$$n^{\frac{1}{2}}\hat{\theta}_{n_\ell}^{(\ell)} - (n^{-\frac{1}{2}}T_{n_\ell}^{(\ell)}\gamma^{-1}\lambda_\ell^{-1} - n^{\frac{1}{2}}\hat{\beta}_n\bar{c}_{n_\ell}^{(\ell)}) \xrightarrow{P} 0\ , \qquad (3.5)$$

$$n^{\frac{1}{2}}\hat{\beta}_n - n^{-\frac{1}{2}}S_n\gamma^{-1}Q^{-1} \xrightarrow{P} 0\ , \qquad (3.6)$$

$$n^{-\frac{1}{2}}S_{n_\ell}^{(\ell)} - (n^{-\frac{1}{2}}S_{n_\ell}^{(\ell)} - n^{\frac{1}{2}}\hat{\beta}_n\lambda_\ell Q_\ell\gamma) \xrightarrow{P} 0\ . \qquad (3.7)$$

Relations (3.5) and (3.6) imply

$$n^{\frac{1}{2}}\hat{\theta}_{n_\ell}^{(\ell)} - (n^{-\frac{1}{2}}T_{n_\ell}^{(\ell)}\gamma^{-1}\lambda_\ell^{-1} - n^{-\frac{1}{2}}S_n\gamma^{-1}Q^{-1}\bar{c}_{n_\ell}^{(\ell)}) \overset{P}{\to} 0, \tag{3.8}$$

while relations (3.6) and (3.7) imply

$$n^{-\frac{1}{2}}\hat{S}_{n_\ell}^{(\ell)} - (n^{-\frac{1}{2}}S_{n_\ell}^{(\ell)} - n^{-\frac{1}{2}}S_nQ^{-1}\lambda_\ell Q_\ell) \overset{P}{\to} 0 \tag{3.9}$$

Relations (3.8) and (3.9) imply that $\underset{\sim}{V}_n$ is, under K_n, asymptotically equivalent to

$$n^{-\frac{1}{2}}(T_{n_\ell}^{(\ell)}\gamma^{-1}\lambda_\ell^{-1} - S_n\gamma^{-1}Q^{-1}\bar{c}_{n_\ell}^{(\ell)},\ \ell=1,\dots,p,\ S_{n_\ell}^{(\ell)} - S_nQ^{-1}\lambda_\ell Q_\ell,\ \ell=1\dots,p)'$$

where $(a_\ell,\ \ell=1,\dots,p,\ b_\ell,\ \ell=1,\dots,p)$ denotes the vector $(a_1,\dots,a_p,\ b_1,\dots,b_p)$. But the distribution of the above under K_n is the same as the distribution of

$$n^{-\frac{1}{2}}(T_{n_\ell}^{(\ell)}(0,-n^{-\frac{1}{2}}t_\ell)\gamma^{-1}\lambda_\ell^{-1} - S_n(-n^{-\frac{1}{2}}\underset{\sim}{t})\gamma^{-1}Q^{-1}\bar{c}_{n_\ell}^{(\ell)},\ \ell=1,\dots,p,$$

$$S_{n_\ell}^{(\ell)}(-n^{-\frac{1}{2}}t_\ell) - S_n(-n^{-\frac{1}{2}}\underset{\sim}{t})Q^{-1}\lambda_\ell Q_\ell,\ \ell=1,\dots,p)$$

under H*: $\underset{\sim}{\theta} = \underset{\sim}{0}$, $\underset{\sim}{\beta} = \underset{\sim}{0}$ (by the fact that the distribution of $T_{n_\ell}^{(\ell)}(a,b)$ under $\theta_\ell = a_\ell$, $\beta_\ell = b_\ell$ is the same as that of $T_{n_\ell}^{(\ell)}(a - a_\ell,\ b - b_\ell)$ under $\theta_\ell = 0$, $\beta_\ell = 0$, and similarly for $S_{n_\ell}^{(\ell)}$). Using again the results in (3.2) - (3.4) the above has, under H*, the same distribution as

$$n^{-\frac{1}{2}}(T_{n_\ell}^{(\ell)}\gamma^{-1}\lambda^{-1} - S_n\gamma^{-1}Q^{-1}\bar{c}_{n_\ell}^{(\ell)},\ \ell=1,\dots,p,\ S_{n_\ell}^{(\ell)} - S_nQ^{-1}\lambda_\ell Q_\ell,\ \ell=1,\dots,p)$$

$$+\ (t_\ell\bar{c}_{n_\ell}^{(\ell)} - (\sum_{m=1}^{p}\lambda_m t_m Q_m)Q^{-1}\bar{c}_{n_\ell}^{(\ell)},\ \ell=1,\dots,p,\ \lambda_\ell t_\ell Q_\ell\gamma - (\sum_{m=1}^{p}\lambda_m t_m Q_m)\gamma Q^{-1}\lambda_\ell Q_\ell,\ \ell=1,\dots,p)$$

However, by assumption in the definition of K_n, $\sum_{m=1}^{p}\lambda_m t_m Q_m = 0$ and thus it follows immediately that the constant term converges to $(t_\ell\bar{c}_\ell,\ \ell=1,\dots,p,\gamma\lambda_\ell t_\ell Q_\ell,\ \ell=1,\dots,p)$ as claimed in the theorem. Using now the result in (3.1) it follows after straightforward computations that the asymptotic covariance of the above vector is indeed as claimed in the theorem.

<u>Corollary 3.1</u>. The random variables $n^{\frac{1}{2}}(\underset{\sim}{\hat{\theta}}_n - \underset{\sim}{\theta})$ and $L_n = \sum_{\ell=1}^{p}\frac{n_\ell^{-1}}{A_\ell^2 Q_\ell}(S_{n_\ell}^{(\ell)})^2$ are asymptotically independent.

<u>Corollary 3.2</u>. Under $\{K_n\}$ we have that L_n is asymptotically equivalent to

$$n^{-\frac{1}{2}}(S_{n_\ell}^{(\ell)},\ \ell=1,\dots,p)\ \frac{1}{A_\phi^2}\ [\mathrm{Diag}(\frac{1}{\lambda_\ell Q_\ell}) - Q^{-1}\underset{\sim}{1}\ \underset{\sim}{1}']n^{-\frac{1}{2}}(S_{n_\ell}^{(\ell)},\ \ell=1,\dots,p)'. \quad (3.10)$$

Proof. Relation (3.9) implies that $n^{-\frac{1}{2}}(S_{n_\ell}^{(\ell)},\ \ell=1,\dots,p)'$ is asymptotically equivalent to

$$[I_p - (\frac{\lambda_\ell Q_\ell}{Q},\ \ell=1,\dots,p)'\underset{\sim}{1}_p']n^{-\frac{1}{2}}(S_{n_\ell}^{(\ell)},\ \ell=1,\dots,p)'$$

above I_p is the pxp identify matrix and $\underset{\sim}{1}_p$ is the p-dimensional column vector with unit entries. Also,

$$L_n = n^{-\frac{1}{2}}(\hat{S}_{n_\ell}^{(\ell)},\ \ell=1,\dots,p)\ \mathrm{Diag}(\frac{1}{A_\phi^2\lambda_\ell Q_\ell})\ n^{-\frac{1}{2}}(\hat{S}_{n_\ell}^{(\ell)},\ \ell=1,\dots,p)'$$

The corollary follows now by some straightforward matrix operations.

Theorem 3.2. Under K_n, the distribution of

$$(n^{\frac{1}{2}}(\underset{\sim}{\tilde{\theta}}_n - \underset{\sim}{\theta}),\ n^{-\frac{1}{2}}(S_n^{(1)},\dots,S_{n_p}^{(p)}))'$$

converges to a normal distribution with mean

$$(0,\dots,0,\ \lambda_1 t_1 Q_1\gamma,\dots,\lambda_p t_p Q_p\gamma)'$$

and covariance matrix $\tilde{\Sigma} = (\tilde{\Sigma}_{rs}),\ r,s = 1,2$ where

$$\tilde{\Sigma}_{11} = \frac{A_\phi^2}{\gamma^2}\ \mathrm{Diag}(\frac{1}{\lambda_\ell} + \frac{\bar{c}_\ell^2}{\lambda_\ell Q_\ell}),\quad \tilde{\Sigma}_{22} = A_\phi^2\ \mathrm{Diag}(\lambda_\ell Q_\ell),\ \tilde{\Sigma}_{12} = -\frac{A_\phi^2}{\gamma}\ \mathrm{Diag}(\bar{c}_\ell).$$

Proof. Assume without loss of generality that $\underset{\sim}{\theta} = \underset{\sim}{0}$. Relation (3.4) and the definition of $\underset{\sim}{\tilde{\theta}}_n$ imply that under K_n, $n^{\frac{1}{2}}\underset{\sim}{\tilde{\theta}}_n$ is asymptotically equivalent to

$$(n^{-\frac{1}{2}}T_{n_\ell}^{(\ell)}\lambda_\ell^{-1}\gamma^{-1} - n^{\frac{1}{2}}\tilde{\beta}_{n_\ell}^{(\ell)}c_{n_\ell}^{(\ell)},\ \ell=1,\dots,p)$$

and using relation (3.2) and the definition of $\tilde{\beta}_{n_\ell}^{(\ell)}$ this is, under K_n again, equivalent to

$$(n^{-\frac{1}{2}}T_{n_\ell}^{(\ell)}\lambda_\ell^{-1}\gamma^{-1} - n^{-\frac{1}{2}}S_{n_\ell}^{(\ell)}\lambda_\ell^{-1}\gamma^{-1}Q_\ell^{-1}c_{n_\ell}^{(\ell)},\ \ell=1,\dots,p)$$

which is the same as

$$(n^{-\frac{1}{2}}T_{n_\ell}^{(\ell)}(0,-n^{-\frac{1}{2}}t_\ell)\lambda_\ell^{-1}\gamma^{-1} - n^{-\frac{1}{2}}S_{n_\ell}^{(\ell)}(-n^{-\frac{1}{2}}t_\ell)\lambda_\ell^{-1}\gamma^{-1}Q_\ell^{-1}\bar{c}_{n_\ell}^{(\ell)},\ \ell=1,\dots,p)$$

under H^*: $\underset{\sim}{\theta} = \underset{\sim}{0}$, $\underset{\sim}{\beta} = \underset{\sim}{0}$. Using relations (3.2), (3.4) the above is, under H^*

asymptotically equivalent to

$$(n^{-\frac{1}{2}}T_{n_\ell}^{(\ell)}\lambda_\ell^{-1}\gamma^{-1} - n^{-\frac{1}{2}}S_{n_\ell}^{(\ell)}\lambda_\ell^{-1}\gamma^{-1}Q_\ell^{-1}\bar{c}_{n_\ell}^{(\ell)}, \quad \ell=1,\ldots,p)$$

since the constants cancel each other. Thus, we have shown that the last expression above has the same asymptotic distribution under both K_n and H. Since the asymptotic mean of $n^{\frac{1}{2}}S_{n_\ell}^{(\ell)}$ is $\lambda_\ell t_\ell Q_\ell \gamma$, as was shown in the proof of Theorem 3.1, Theorem 3.2 follows from relation (3.1).

Theorem 3.3. Under K_n and with the notation introduced in Theorems 3.1 and 3.2, the ditribution of $n^{\frac{1}{2}}(\bar{\underset{\sim}{\theta}}_n - \underset{\sim}{\theta})$ converges to

$$\bar{G}_p(\underset{\sim}{x};\underset{\sim}{t}) = G_p(\underset{\sim}{x} - \mathrm{Diag}(\bar{c}_\ell)\underset{\sim}{t};\ \underset{\sim}{0},\ \Sigma_{11})H_{p-1}(L_\alpha;\Delta)$$

$$+\int_{E(\underset{\sim}{t})} G_p(\underset{\sim}{x} + \mathrm{Diag}(\frac{\bar{c}_\ell}{\gamma\lambda_\ell Q_\ell})\underset{\sim}{z};\ 0,\ \mathrm{Diag}(\frac{A\phi^2}{\gamma^2\lambda_\ell}))\ dG_p(\underset{\sim}{z};\ \underset{\sim}{0},\ \tilde{\Sigma}_{22})$$

where $G_p(\underset{\sim}{a};\underset{\sim}{0},\underset{\sim}{B})$ denotes the p-dimensional normal cdf with mean zero and covariance matrix B, $H_{p-1}(y;\Delta)$ denotes the cdf of a non-central chi-square variable with p-1 degrees of freedom and noncentrality parameter Δ and,

$$E(\underset{\sim}{t}) = \{\underset{\sim}{z}:(\underset{\sim}{z}+\gamma\,\mathrm{Diag}(\lambda_\ell Q_\ell)\underset{\sim}{t})'\ W(\underset{\sim}{z}+\gamma\,\mathrm{Diag}(\gamma_\ell Q_\ell)\underset{\sim}{t}) > L_\alpha\}$$

where W is the matrix of the quadratic form in (3.10).

Proof. By the definition of $\bar{\underset{\sim}{\theta}}_n$ in (2.16) it follows that

$$P_{K_n}[n^{\frac{1}{2}}(\bar{\underset{\sim}{\theta}}_n - \underset{\sim}{\theta}) \le \underset{\sim}{x}] = P_{K_n}[n^{\frac{1}{2}}(\bar{\underset{\sim}{\theta}}_n - \underset{\sim}{\theta}) \le x,\ L_n \le L_\alpha]$$

$$+ P_{K_n}[n^{\frac{1}{2}}(\tilde{\underset{\sim}{\theta}}_n - \underset{\sim}{\theta}) \le \underset{\sim}{x},\ L_n > L_\alpha] \tag{3.11}$$

where $L_\alpha = \chi^2_{p-1,\alpha}$. By Corollary 3.1, Theorem 3.1 and the convergence in distribution of L_n stated in connection with (2.14), (2.15) we have

$$\lim P_{K_n}[\sqrt{n}(\hat{\underset{\sim}{\theta}}_n - \underset{\sim}{\theta}) \le x,\ L_n \le L_\alpha] = \lim P_{K_n}[\sqrt{n}(\hat{\underset{\sim}{\theta}}_n - \underset{\sim}{\theta}) \le \underset{\sim}{x}]\ \lim P(L_n \le L_\alpha)$$

$$= G_p(\underset{\sim}{x} - \mathrm{Diag}(\bar{c}_\ell)\underset{\sim}{t};\ \underset{\sim}{0},\ \Sigma_{11})H_{p-1}(L_\alpha;\Delta).$$

Next, it follows from Theorem 3.2 that the asymptotic conditional distribution of $n^{\frac{1}{2}}(\tilde{\underset{\sim}{\theta}}_n - \underset{\sim}{\theta})$ given that $n^{\frac{1}{2}}(S_{n_1}^{(1)},\ldots,S_{n_p}^{(p)} = \underset{\sim}{z}$ is normal with mean

$\mathrm{Diag}(\frac{\bar{c}_\ell}{\gamma\lambda_\ell Q_\ell})\ (\underset{\sim}{z} - \gamma\ \mathrm{Diag}(\lambda_\ell Q_\ell)\underset{\sim}{t})$ and covariance matrix $\tilde{\Sigma}_{11} - \tilde{\Sigma}_{12}\Sigma_{22}^{-1}\Sigma_{21} = \frac{A_\phi^2}{\gamma^2}\ \mathrm{Diag}(\frac{1}{\lambda_\ell})$. Thus, noting the asymptotically equivalent form of L_n given in (3.10), it follows that the second term on the right-hand side of (3.11) equals the second term in the expression for $\bar{G}_p(\underset{\sim}{x};\underset{\sim}{t})$ and this completes the proof.

4. ASYMPTOTIC BIAS AND MEAN SQUARED ERROR OF THE ESTIMATOR

In this section we will define expressions for the asymptotic bias and dispersion matrices of the estimators which we will use in the next section for making comparisons.

Let $\bar{\xi}(\underset{\sim}{t})$ and $\bar{\Gamma}(\underset{\sim}{t})$ stand for the asymptotic bias and dispersion matrix of $\bar{\underset{\sim}{\theta}}_n$ under K_n; thus, in the notation of Theorem 3.3,

$$\bar{\underset{\sim}{\xi}}(\underset{\sim}{t}) = \int_{E^p} \underset{\sim}{x}\ d\bar{G}_p(x;t) \tag{4.1}$$

$$\bar{\Gamma}(t) = \int_{E^p} \underset{\sim}{x}\underset{\sim}{x}'dG_p(x;t) \tag{4.2}$$

In order to calculate these quantities we will use the relation

$$\int_{[\underset{\sim}{z}'W\underset{\sim}{z}>c]} dG_p(\underset{\sim}{z};\underset{\sim}{a},\tilde{\Sigma}_{22}) = \exp(-\tfrac{1}{2}\delta)\sum_{r=0}^{\infty}\frac{1}{r!}\ (\tfrac{1}{2}\delta)^r[1 - H_{p-1+24}(c;0)]$$

where c is any positive constant, W is given in the definition of $E(\underset{\sim}{t})$ and $\delta = \underset{\sim}{a}'W\underset{\sim}{a}$, and we will differentiate both sides with respect to $\underset{\sim}{a}$. Differentiating once we get

$$\int_{[\underset{\sim}{z}'W\underset{\sim}{z}>c]} \tilde{\Sigma}_{22}^{-1}(\underset{\sim}{z} - \underset{\sim}{a})\ dG_p(\underset{\sim}{z};\underset{\sim}{a},\tilde{\Sigma}_{22}) = W\underset{\sim}{a}[H_{p-1}(c;\delta) - H_{p+1}(c;\delta)]$$

or

$$\int_{[(\underset{\sim}{z}+\underset{\sim}{a})'W(\underset{\sim}{z}+\underset{\sim}{a})>c]} \underset{\sim}{z}\ dP_p(\underset{\sim}{z};\underset{\sim}{0},\ \tilde{\Sigma}_{22}) = \tilde{\Sigma}_{22}W\underset{\sim}{a}[H_{p-1}(c;\delta) - H_{p+1}(c;\delta)] \tag{4.3}$$

Differentiating twice we get

$$\int_{[\underset{\sim}{z}'W\underset{\sim}{z}>c]} (-\tilde{\Sigma}_{22}^{-1} + \tilde{\Sigma}_{22}^{-1}(\underset{\sim}{z} - \underset{\sim}{a})(\underset{\sim}{z} - \underset{\sim}{a})'\tilde{\Sigma}_{22}^{-1})\ dG_p(\underset{\sim}{z};\underset{\sim}{a},\tilde{\Sigma}_{22})$$

$$= W[H_{p-1}(c,\delta) - H_{p+1}(c;\delta)] - W\ \underset{\sim}{a}\ \underset{\sim}{a}'W[H_{p-1}(c;\delta) - 2H_{p+1}(c;\delta) + H_{p+3}(c;\delta)]$$

so that

$$\int_{[(\underset{\sim}{z}+\underset{\sim}{a})'W(\underset{\sim}{z}+\underset{\sim}{a})>c]} \underset{\sim}{z}\underset{\sim}{z}'\, dG_p(\underset{\sim}{z};\underset{\sim}{0},\tilde{\Sigma}_{22}) = [1-H_{p-1}(c;\delta)]\tilde{\Sigma}_{22}+\tilde{\Sigma}_{22}W\tilde{\Sigma}_{22}[H_{p-1}(c;\delta) - H_{p+1}(c;\delta)]$$

$$- \tilde{\Sigma}_{22}W\,\underset{\sim}{a}\,\underset{\sim}{a}'W\tilde{\Sigma}_{22}(H_{p-1}(c;\delta) - 2H_{p+1}(c;\delta) + H_{p+3}(c;\delta)] \qquad (4.4)$$

From (4.1) we have

$$\underset{\sim}{\bar{\xi}}(\underset{\sim}{t}) = \text{Diag}(\bar{c}_t)\underset{\sim}{t}\, H_{p-1}(L_\alpha;\Delta) - \int_{E(\underset{\sim}{t})} \text{Diag}(\frac{\bar{c}_\ell}{\gamma\lambda_\ell Q_\ell})\underset{\sim}{z}\, dG_p(\underset{\sim}{z};\underset{\sim}{0},\tilde{\Sigma}_{22})$$

so that, writing H_r instead of $H_r(L_\alpha;\Delta)$, (4.3) implies

$$\underset{\sim}{\bar{\xi}}(\underset{\sim}{t}) = \text{Diag}(\bar{c}_\ell)\underset{\sim}{t}\, H_{p-1} - \text{Diag}(\bar{c}_\ell)\underset{\sim}{t}\, [H_{p-1} - H_{p+1}]$$

$$= \text{Diag}(\bar{c}_\ell)\underset{\sim}{t}\, H_{p+1} \qquad (4.5)$$

Next, with $\Gamma_1(\underset{\sim}{t})$ given in (4.7), (4.2) and (4.4) imply,

$$\bar{\Gamma}(\underset{\sim}{t}) = \Gamma_1(\underset{\sim}{t})\, H_{p-1} + \text{Diag}(\frac{A_\phi^2}{\gamma^2\lambda_\ell})[1 - H_{p-1}]$$

$$+ \text{Diag}(\frac{\bar{c}_\ell}{\gamma\lambda_\ell Q_\ell}) \int_{E(\underset{\sim}{t})} \underset{\sim}{z}\underset{\sim}{z}'\, dG_p(\underset{\sim}{z};\underset{\sim}{0},\tilde{\Sigma}_{22})\, \text{Diag}(\frac{\bar{c}_\ell}{\gamma\lambda_\ell Q_\ell})$$

$$= \Gamma_1(\underset{\sim}{t})H_{p-1} + \text{Diag}(\frac{A_\phi^2}{\gamma^2\lambda_\ell})(1 - H_{p-1}) + \text{Diag}(\frac{\bar{c}_\ell}{\gamma\lambda_\ell Q_\ell})\tilde{\Sigma}_{22}\, \text{Diag}(\frac{\bar{c}_\ell}{\gamma\lambda_\ell Q_\ell})(1 - H_{p-1})$$

$$+ \text{Diag}(\frac{\bar{c}_\ell}{\gamma\lambda_\ell Q_\ell})\tilde{\Sigma}_{22}\, W\tilde{\Sigma}_{22}\, \text{Diag}(\frac{\bar{c}_\ell}{\gamma\lambda_\ell Q_\ell})(H_{p-1} - H_{p+1})$$

$$- \text{Diag}(\frac{\bar{c}_\ell}{\gamma\lambda_\ell Q_\ell})\tilde{\Sigma}_{22}\, W\, \gamma^2\, \text{Diag}(\gamma_\ell Q_\ell)\, \underset{\sim}{t}\, \underset{\sim}{t}'\, \text{Diag}(\gamma_\ell Q_\ell)W\, \tilde{\Sigma}_{22}\text{Diag}(\frac{\bar{c}_\ell}{\gamma\lambda_\ell Q_\ell})(H_{p-1}$$

$$- 2H_{p+1} + H_{p+3})$$

Recall now that $\text{Diag}(\frac{A_\phi^2}{\gamma^2\lambda_\ell}) = \tilde{\Sigma}_{11} - \tilde{\Sigma}_{12}\tilde{\Sigma}_{22}^{-1}\tilde{\Sigma}_{21}$ and $\text{Diag}(\frac{\bar{c}_\ell}{\gamma\lambda_\ell Q_\ell}) = \tilde{\Sigma}_{12}\tilde{\Sigma}_{22}^{-1}$, so that the second and third term on the right-hand side of the above relation is $\tilde{\Sigma}_{11}(1 - H_{p-1})$; also the matrix of the fourth term is $\tilde{\Sigma}_{11} - \frac{A_\phi^2}{\gamma^2}[\text{Diag}(\frac{1}{\lambda_\ell}) + Q^{-1}\,\underset{\sim}{c}\,\underset{\sim}{c}']$ and finally the matrix in the last term is $\text{Diag}(\bar{c}_\ell)\underset{\sim}{t}\,\underset{\sim}{t}'\text{Diag}(\bar{c}_\ell)$. Thus, after some calculations, we have

$$\bar{\Gamma}(\underset{\sim}{t}) = \Gamma_1(\underset{\sim}{t})H_{p-1} + \tilde{\Sigma}_{11}(1 - H_{p+1}) - \frac{A_\phi^2}{\gamma^2}[\text{Diag}(\frac{1}{\lambda_\ell}) + Q^{-1}\,\underset{\sim}{c}\,\underset{\sim}{c}'](H_{p-1} - H_{p+1})$$

$$- \text{Diag}(\bar{c}_\ell)\underset{\sim}{t}\,\underset{\sim}{t}' \,\text{Diag}(\bar{c}_\ell)(H_{p-1} - 2H_{p+1} + H_{p+3})$$

$$= \Gamma_1(\underset{\sim}{t})H_{p+1} + \tilde{\Sigma}_{11}(1 - H_{p+1})$$

$$+ \text{Diag}(\bar{c}_\ell)\underset{\sim}{t}\,\underset{\sim}{t}' \,\text{Diag}(\bar{c}_\ell)(H_{p+1} - H_{p+3}) \tag{4.6}$$

where

$$\Gamma_1(\underset{\sim}{t}) = \text{asymptotic disperson matrix of } n^{\frac{1}{2}}(\hat{\underset{\sim}{\theta}}_n - \underset{\sim}{\theta})$$

$$= \int \underset{\sim}{x}\,\underset{\sim}{x}' \,dG_p(\underset{\sim}{x} - \text{Diag}(\bar{c}_\ell)\underset{\sim}{t};\underset{\sim}{0},\Sigma_{11}) = \Sigma_{11} + \text{Diag}(\bar{c}_\ell)\underset{\sim}{t}\,\underset{\sim}{t}' \,\text{Diag}(\bar{c}_\ell). \tag{4.7}$$

We close this section by introducing some more notation

$$\Gamma_2(\underset{\sim}{t}) = \text{asymptotic dispersion matrix of } n^{\frac{1}{2}}(\tilde{\underset{\sim}{\theta}}_n - \underset{\sim}{\theta}) = \tilde{\Sigma}_{11} \tag{4.8}$$

$$\xi_1(\underset{\sim}{t}) = \text{asymptotic bias of } n^{\frac{1}{2}}(\hat{\theta}_n - \underset{\sim}{\theta}) = \text{Diag}(\bar{c}_\ell)\,\underset{\sim}{t} \tag{4.9}$$

$$\xi_2(\underset{\sim}{t}) = \text{asymptotic bias of } n^{\frac{1}{2}}(\tilde{\underset{\sim}{\theta}}_n - \underset{\sim}{\theta}) = \underset{\sim}{0} \tag{4.10}$$

5. ASYMPTOTIC COMPARISONS OF ESTIMATORS

First note that when $\bar{c}_\ell = 0$, $\ell=1,\dots,p$ we have that all estimators have the same asymptotic distribution with no bias. From now on we will assume that not all $\bar{c}_\ell$ are zero.

Theorem 5.1. Under $H_0: \beta_1 = \cdots = \beta_p$ (i.e., $\underset{\sim}{t} = \underset{\sim}{0}$ in K_n) it follows that

$$\Gamma_2(\underset{\sim}{0}) - \Gamma_1(\underset{\sim}{0}),\ \Gamma_2(\underset{\sim}{0}) - \bar{\Gamma}(\underset{\sim}{0}) \text{ and } \bar{\Gamma}(\underset{\sim}{0}) - \Gamma_1(\underset{\sim}{0})$$

are all positive definite whenever $\text{Diag}(\frac{\bar{c}_\ell^2}{\gamma_\ell Q_\ell}) - \frac{1}{Q}\underset{\sim}{c}\underset{\sim}{c}'$ is so where $\Gamma_1(\underset{\sim}{t})$, $\Gamma_2(\underset{\sim}{t})$ and $\Gamma(\underset{\sim}{t})$ are given by (4.7), (4.8) and (4.6) respectively.

Proof. We have

$$\Gamma_2(\underset{\sim}{0}) - \Gamma_1(\underset{\sim}{0}) = \tilde{\Sigma}_{11} - \Sigma_{11}$$

$$= \frac{A_\phi^2}{\gamma^2}\left[\text{Diag}(\frac{\bar{c}_\ell^2}{\lambda_\ell Q_\ell}) - \frac{1}{Q}\underset{\sim}{c}\,\underset{\sim}{c}'\right],$$

next,

$$\Gamma_2(\underset{\sim}{0}) - \bar{\Gamma}(\underset{\sim}{0}) = \tilde{\Sigma}_{11} - \Sigma_{11}(1 - H_{p+1}) - \Sigma_{11}\,H_{p+1} = H_{p+1}(\tilde{\Sigma}_{11} - \Sigma_{11})$$

as before and

$$\bar{\Gamma}(0) - \Gamma_1(0) = \tilde{\ddagger}_{11}\,(1 - H_{p+1}) + \ddagger_{11}\,H_{p+1} - \ddagger_{11}$$

$$= 1 - H_{p+1})\,(\tilde{\ddagger}_{11} - \ddagger_{11})$$

as before again, and thus the result follows.

We define the asymptotic relative efficiency (ARE) in terms of the ratio of the p-th root of the a.d.m. as in Sen and Saleh (1979). Then, for $\underset{\sim}{t} = \underset{\sim}{0}$, the ARE of $\bar{\theta}_{\sim n}$ relative to $\hat{\theta}_{\sim n}$ is given by

$$e(\bar{\theta}_{\sim n} : \hat{\theta}_{\sim n} | \underset{\sim}{0}) = \{|\bar{\Gamma}_1(\underset{\sim}{0})| / |\Gamma(\underset{\sim}{0})|\}^{1/p}$$

$$= \frac{|I_p + Q^{-1}\underset{\sim}{c}\,\underset{\sim}{c}'\,\mathrm{Diag}(\lambda_\ell)|^{1/p}}{|I_p + Q^{-1}\underset{\sim}{C}\,\underset{\sim}{C}'\,\mathrm{Diag}(\lambda_\ell) + (1 - H_{p+1}(L_\alpha;0)\,[\mathrm{Diag}(\frac{\bar{c}_\ell^2}{\lambda_\ell Q_\ell}) - Q^{-1}\underset{\sim}{c}\,\underset{\sim}{c}']\,\mathrm{Diag}(\lambda_\ell)|^{1/p}} \tag{5.1}$$

where

$$\mathrm{Diag}(\frac{\bar{c}_\ell^2}{\lambda_\ell Q_\ell}) - Q^{-1}\underset{\sim}{c}\,\underset{\sim}{c}' \text{ is positive semi-definite} \tag{5.2}$$

Hence (5.1) cannot exceed 1. The larger the characteristic roots of (5.2), the smaller the value of (5.1). Since $H_{p+1}(L_\alpha;0) < 1 - \alpha = H_{p-1}(L_\alpha;0)$, (5.1) also depends on α, p and $\mathrm{Diag}(\lambda_\ell)$. However, since $1-H_{p+1}(L_\alpha;0)$ is usually small, (5.1) is generally not very different from 1.

$$e(\bar{\theta}_{\sim n} : \tilde{\theta}_{\sim n} | \underset{\sim}{0}) = \{|\Gamma_2(\underset{\sim}{0})| / |\Gamma(\underset{\sim}{0})|\}^{1/p}$$

$$= \frac{|I + Q^{-1}\underset{\sim}{c}\,\underset{\sim}{c}'\,\mathrm{Diag}(\lambda_\ell) + \mathrm{Diag}(\frac{\bar{c}_\ell^2}{\lambda_\ell Q_\ell}) - Q^{-1}\underset{\sim}{c}\,\underset{\sim}{c}']\,\mathrm{Diag}(\lambda_\ell)|^{1/p}}{|I + Q^{-1}\underset{\sim}{c}\,\underset{\sim}{c}'\,\mathrm{Diag}(\lambda_\ell) + (1 - H_{p-1}(L_\alpha;0))\,[\mathrm{Diag}(\frac{\bar{c}_\ell^2}{\lambda_\ell\,\ell}) - Q^{-1}\underset{\sim}{c}\,\underset{\sim}{c}']\,\mathrm{Diag}(\lambda_\ell)|^{1/p}} \tag{5.3}$$

so that by (5.2) and the above discussion

$$e(\bar{\theta}_{\sim n} : \hat{\theta}_{\sim n} | 0) \le 1 \le e(\bar{\theta}_{\sim n} : \tilde{\theta}_{\sim n} | \underset{\sim}{0}) \ , \ \forall\, \underset{\sim}{c} \tag{5.4}$$

Next we consider the case that $t \neq 0$. First note that we have

$$e(\hat{\theta}_{\sim n} : \tilde{\theta}_{\sim n} | \underset{\sim}{t}) = \{|\Gamma_2(\underset{\sim}{t})| / \Gamma_1(\underset{\sim}{t})|\}^{1/p}$$

$$= \frac{|\mathrm{Diag}(\frac{1}{\lambda_\ell} + \frac{\bar{c}_\ell^2}{\lambda_\ell Q_\ell})|^{1/p}}{|\mathrm{Diag}(\frac{1}{\lambda_\ell}) + Q^{-1}\underset{\sim}{c}\,\underset{\sim}{c}' + \frac{\gamma^2}{A_\phi^2}\mathrm{Diag}(\bar{c}_\ell)\underset{\sim}{t}\,\underset{\sim}{t}'\,\mathrm{Diag}(\bar{c}_\ell)|^{1/p}}$$

$$= \frac{|\mathrm{Diag}(\frac{1}{\lambda_\ell} + \frac{\bar{c}_\ell^2}{\lambda_\ell Q_\ell})|}{|\mathrm{Diag}(\frac{1}{\lambda_\ell} + \frac{\bar{c}_\ell^2}{\lambda_\ell Q_\ell}) - \mathrm{Diag}(\frac{\bar{c}_\ell^2}{\lambda_\ell Q_\ell}) + Q^{-1}\underset{\sim}{c}\,\underset{\sim}{c}' + \frac{\lambda^2}{A_\phi^2}\mathrm{Diag}(\bar{c}\ )\underset{\sim}{t}\,\underset{\sim}{t}'\,\mathrm{Diag}(\bar{c}_\ell)|^{1/p}}$$

$$\frac{|\mathrm{Diag}(\frac{1}{\lambda_\ell} + \frac{\bar{c}_\ell^2}{\lambda_\ell Q_\ell})|^{1/p}}{|\mathrm{Diag}(\frac{1}{\lambda_\ell} + \frac{\bar{c}_\ell^2}{\lambda_\ell Q_\ell}) - \mathrm{Diag}(\frac{\bar{c}_\ell}{\lambda_\ell Q_\ell})\,[G - \frac{\gamma^2}{A_\phi^2}\mathrm{Diag}(\lambda_\ell Q_\ell)\underset{\sim}{t}\,\underset{\sim}{t}']\,\mathrm{Diag}(\bar{c}_\ell)|^{1/p}} \tag{5.5}$$

where $G = I_p - Q^{-1}\underset{\sim}{1}_p\underset{\sim}{1}_p'\,\mathrm{Diag}(\lambda_\ell Q_\ell)$ since $\mathrm{Diag}(\frac{\bar{c}_\ell^2}{\lambda_\ell Q_\ell}) - Q^{-1}\underset{\sim}{C}\,\underset{\sim}{C}' = \mathrm{Diag}(\frac{\bar{c}_\ell}{\lambda_\ell Q_\ell})$. $G\,\mathrm{Diag}(\bar{c}_\ell)$. Note that G is p.s.d. of rank p-1, while $\frac{\sigma^2}{A_\phi^2}\mathrm{Diag}(\lambda_\ell Q_\ell)\underset{\sim}{t}\,\underset{\sim}{t}'$ is p.s.d. of rank 1, its only characteristic root is equal to Δ (see (2.15)) and all others are equal to zero. Since $\mathrm{Diag}(\frac{1}{\lambda_\ell Q_\ell})$ is p.d. we conclude that whenever $G - \frac{\lambda^2}{A_\phi^2}\mathrm{Diag}(\lambda_\ell Q_\ell)\underset{\sim}{t}\,\underset{\sim}{t}'$ is positive definite, (5.5) exceeds one, i.e., whenever Δ is small (5.5) exceeds one, and the opposite conclusion holds when Δ is large. In fact, as $\Delta \to \infty$ (5.5) converges to zero indicating that $\hat{\underset{\sim}{\theta}}_n$ loses its efficiency as $\underset{\sim}{t}$ moves away from the origin.

Let us next consider the ARE of $\bar{\underset{\sim}{\theta}}_n$ relative to $\hat{\underset{\sim}{\theta}}_n$. Now with $H_r \equiv H_r(\chi^2_{p-1,\alpha};\Delta)$,

$$e(\bar{\underset{\sim}{\theta}}_n : \hat{\underset{\sim}{\theta}}_n | \underset{\sim}{t}) = \{|\Gamma(\underset{\sim}{t})|/|\bar{\Gamma}(\underset{\sim}{t})|\}^{1/p}$$

$$= \{|\Gamma_1(\underset{\sim}{t})|/|H_{p+1}\Gamma_1(\underset{\sim}{t}) + (1 - H_{p+1})\Gamma_2(\underset{\sim}{t}) + \mathrm{Diag}(\bar{c}_\ell)\underset{\sim}{t}\,\underset{\sim}{t}'\,\mathrm{Diag}(\bar{c}_\ell)(H_{p+1} - H_{p+3})|\}^{1/p}$$

$$= |H_{p+1}I_p + (1 - H_{p+1})\Gamma_1^{-1}(\underset{\sim}{t})\Gamma_2(\underset{\sim}{t}) + \Gamma_1^{-1}(\underset{\sim}{t})\,\mathrm{Diag}(\bar{c}_\ell)\underset{\sim}{t}\,\underset{\sim}{t}'\,\mathrm{Diag}(\bar{c}_\ell)(H_{p+1} - H_{p+3})|^{1/p}$$

where for $\Gamma_1^{-1}(\underset{\sim}{t})\,\Gamma_2(t)$, the bounds in (5.5) apply. Since $H_{p+1} - H_{p+3} \geq 0$ we conclude from (5.5) and (5.6) that for small $\underset{\sim}{t}$, $\Gamma_1^{-1}(\underset{\sim}{t})\,\Gamma_2(\underset{\sim}{t})$ has roots greater than 1, so that (5.6) is smaller than 1. On the other hand, as Δ increases, i.e., $\underset{\sim}{t}$ moves away from $\underset{\sim}{0}$, the roots of $\Gamma_1^{-1}(\underset{\sim}{t})\,\Gamma_2(\underset{\sim}{t})$ become small while $H_{p+1} - H_{p+3}$ converges to zero as $\Delta \to \infty$. Hence, as $\Delta \to \infty$ (5.6) increases indicating that $\bar{\underset{\sim}{\theta}}_n$

becomes more efficient than $\hat{\theta}_n$; this is mainly due to the bias of $\hat{\theta}_n$ which increases linearly with t (see (4.9)) while (4.5) is a bell shaped function which is equal to zero at both $\Delta = 0$ and $\Delta = +\infty$. Similarly,

$$e(\bar{\theta}_n : \tilde{\theta}_n | t) = \{|\Gamma(t)|/|\bar{\Gamma}(t)|\}^{1/p}$$

$$= \{(1 - H_{p+1}) I_p + H_{p+1} \Gamma_2^{-1}(t) \Gamma_1(t) + \Gamma_2^{-1}(t) \operatorname{Diag}(\bar{c}_\ell)\, tt' \operatorname{Diag}(\bar{c}_\ell)(H_{p+1} - H_{p+3})\}^{1/p} \tag{5.7}$$

so that for small t, i.e., Δ close to 0, (5.7) will be greater than 1 while as $\Delta \to \infty$ (5.7) converges to 1 since $H_r \to 0$ for $r = p+1,\ p+3$. Hence, $\bar{\theta}_n$ has an edge over the unrestricted estimator $\tilde{\theta}_n$ when $C \neq 0$ and t is not equal to zero. From the above we conclude that the PTE $\bar{\theta}_n$ possesses good robust-efficiency properties against t away from 0.

ACKNOWLEDGEMENT: This research has been supported by NSERC grant no. A3088 ((Canada), the National Heart, Lung and Blood Institute, Contract NIH-NHLBI-71-2243-L from the National Institute of Health (U.S.A.) and NSF grant no. 7905536.

REFERENCES

[1] Adichie, J.N., Estimates of regression parameters based on ranks, Ann. Math. Statist. 38:894-904 (1967).

[2] Ahsanullah, M. and Saleh, A.K.Md.E., Estimation of intercepts in a linear regression model with one dependent variable after a preliminary test of significance, Rev. Inter. Statist. Inst. 40:139-145 (1972).

[3] Bancroft, I.A., On biases in estimation due to use of preliminary test of significance, Ann. Math. Statist. 15:190-204 (1944).

[4] Hájek, J. and Šidák, Z., Theory of Rank Tests, Academic Press, New York (1967).

[5] Han, C.P. and Bancroft, T.A., On pooling means when variance is unknown, J. Amer. Statist. Assoc. 63:1333-1342 (1968).

[6] Jurečková, J., Asymptotic linearity of rank statistic in regression parameter, Ann. Math. Statist. 40:1889-1900 (1969).

[7] Kraft, C.H. and Van Eeden, C., Linearited rank estimates and signed rank estimates for the general linear hypothesis, Ann. Math. Statist. 43:42-57 (1972).

[8] Mosteller, F., On pooling data, J. Amer. Statist. Assoc. 43:231-242 (1948).

[9] Saleh, A.K.Md.E. and Sen, P.K., Nonparametric estimation of location parameter after a preliminary test on regression, Ann. Statist. 6:154-168 (1978).

[10] Sen, P.K., On a class of rank order tests for the parallelism of several regression lines, Ann. Math. Statist. 40:1668-1683 (1969).

[11] Sen, P.K., Robust statistical procedures in problems of linear regression with special reference to quantitative bioassays, I., Rev. Inter. Statist. Inst. 39:21-38 (1971).

[12] Sen, P.K., Robust statistical procedures in problems in linear regression with special reference to quantitative bioassays, II., Rev. Inter. Statist. Inst. 40:161-172 (1972).

[13] Sen, P.K. and Saleh, A.K.Md.E., Nonparametric estimation of location parameter after a preliminary test on regression in the multivariate case, J. Multivariate Analysis 9:322-331 (1979).

BIOSTATISTICS: Statistics in Biomedical, Public Health
and Environmental Sciences, edited by P.K. Sen

COVARIANCE ANALYSIS WITH LOG-LINEAR AND LOGIT MODELS

V.P. Bhapkar and K.W. Teoh

Department of Statistics
University of Kentucky
Lexington, KY

The relationship between the log-linear models and logit models have been discussed in the categorical data literature. However, there is considerable divergence of views as regards the relative advantages and disadvantages of these models. In this article, we show how 'generalized logit' models contain just as much relevant information as the log-linear models. Further, we demonstrate the flexibility and ease of interpretation that is afforded by such generalized logit models especially in situations where there are random concomitant variables.

KEY WORDS: Categorical data; log-linear, logit, generalized logit and nested models; intrinsic treatment effect.

1. INTRODUCTION

For studying relationships among several dimensions of a contingency table log-linear models are used extensively in categorical data with Poisson or multinomial distributional assumptions. Similarly, logit models have been found to be very useful for studying the dependence of the primary response on one or several design factors.

The relationships between the log-linear models and logit models have been discussed in the categorical data literature. However, there is considerable divergence of views as regards the relative advantages and disadvantages of these models.

Thus, at one end of the spectrum, we have the remarks from Bishop (1969) that essentially claim that the log-linear models are more general than the logit models. While discussing the log-linear models, as given by Method I in her terminology, and logit models as given by some other method, she states (Bishop, 1969, ff 384): "We discuss the advantages and disadvantages of each method and demonstrate that some models that can be fitted by Method I cannot be fitted by either of the other methods." She goes on to state later (Bishop, 1969, ff 390-391): "If further dichotomies were available the number of available logit models would be increased (if rates were based successively on different variables) but the complement of models would still be less than that available with the contingency-table method." Although these remarks have been made in the context of dichotomous variables for which logits are obtained, the remarks essentially apply to the general case as well with polytomous variables, using multivariate logits rather than only simple binary logits.

At the other end of the spectrum, we have views that mainly emphasize the advantages of using logit models. Thus, for example, Levin and Shrout (1981, ff 1) state: "One is that the larger number of parameters necessary to specify a given model is often an embarrassment to the statistician in his attempt to justify his models to a client who is, after all, mainly interested in the dependence of the response structure on the explanatory variables." These authors are referring to the difficulty in handling the larger number of parameters with log-linear models.

The purpose of this article is two-fold. First of all, we show how 'generalized logit' models contain just as much relevant information as the log-linear models. By generalized logits we mean not only the binary or multivariate logits with a given single response, but also some additional logits with respect to other responses when necessary. The main advantage of such generalized models is that they describe the dependence of one response (using logits) or several responses (using generalized logits) on the combinations of one or more design factors. This fact renders such generalized logit models easier to interpret than the log-linear models. The side advantage of the generalized logit models is that they are more parsimonious than the log-linear models in that the former do not contain the nuisance parameters (μ's in the notation of (2.1), (2.10), (2.15) etc.) which have to satisfy essential constraints in view of the underlying probability distributional structure.

The second purpose of the article is to describe the flexibility that is afforded by logit models (or generalized logit models) especially in situations where we have more than one random response such that one dimension corresponds to the primary response while the remaining random dimensions describe the levels of concomitant variables.

Section 2 presents the discussion concerning log-linear models and generalized logit models while Section 3 considers the situation in the presence of random covariates. Section 4 considers some numerical examples.

2. LOG-LINEAR AND GENERALIZED LOGIT MODELS

For log-linear models we use a notation which is slightly different from the 'u-notation' in Bishop, Fienberg and Holland (1975). The difference in the notation is important especially in the case where one or more dimensions correspond to categories of some design factors, rather than those of random responses, in the terminology of Bhapkar and Koch (1968). As in Bishop, et al. (1975) we shall deal only with hierarchical models where the presence of any interaction term involving specific dimensions requires inclusion of all lower order interaction terms involving those dimensions.

In order to point out the main concepts we shall confine our discussions mostly to three dimensional contingency tables. Consider thus an $I \times J \times K$ table $\underset{\sim}{N}$ of observed counts $\{n_{ijk},\ i = 1,\ldots,I,\ j = 1,\ldots,J,\ k = 1,\ldots,K\}$. We use the standard summation convention $n_{ij+} = \sum_k n_{ijk}$, etc., $n_{+j+} = \sum_i \sum_k n_{ijk}$, etc.

We now consider the different underlying probability distributions separately.

2.1 One Response, Two Factors

Suppose now that only one dimension represents levels of the random response. Assume that the third dimension level k represents the random response, and that the counts n_{ijk}, $k = 1,\ldots,K$ have the multinomial distribution with fixed sample size n_{ij+} and probabilities π_{ijk}, $k = 1,\ldots,K$, so that $\pi_{ij+} \equiv \sum_k \pi_{ijk} = 1$. We have such multinomial vectors for all factorial combinations (ij), $i = 1,\ldots,I$, $j = 1,\ldots,J$.

We use the log-linear representation

$$R:\ \ln \pi_{ijk} = \mu + \mu_i^{(1)} + \mu_j^{(2)} + \alpha_k^{(3)} + \mu_{ij}^{(1,2)} + \alpha_{ik}^{(1,3)} + \alpha_{jk}^{(2,3)} + \alpha_{ijk}^{(1,2,3)}. \quad (2.1)$$

Here the parameters satisfy the usual side constraints $\sum_i \mu_i^{(1)} = 0$, etc., $\sum_k \alpha_k^{(3)} = 0$, $\sum_i \mu_{ij}^{(1,2)} = \sum_j \mu_{ij}^{(1,2)} = 0$, $\sum_i \alpha_{ik}^{(1,3)} = 0,\ldots,\sum_k \alpha_{ijk}^{(1,2,3)} = 0$, etc.

The main difference from the u-notation in Bishop, et al, (1975) is that the μ-parameters (e.g. μ, $\mu_i^{(1)}$, $\mu_{ij}^{(1,2)}$ etc.) are written differently from the α-parameters in order to stress that μ's are not genuinely free parameters, but that these are determined uniquely by the α-parameters in view of the basic constraints

$$1 = \sum_{k=1}^{K} \pi_{ijk} = \exp\{\mu + \mu_i^{(1)} + \mu_j^{(2)} + \mu_{ij}^{(1,2)}\} \sum_k \exp\{\alpha_k^{(3)} + \alpha_{ik}^{(1,3)} + \alpha_{jk}^{(2,3)} + \alpha_{ijk}^{(1,2,3)}\}. \quad (2.2)$$

The other minor difference from the u-notation is that Bishop, et al. (1975) use the log-linear representation for the expected count $m_{ijk} = n_{ij+}\pi_{ijk}$, while (2.1) uses the log-linear representation for the intrinsic parameter π_{ijk} without combining it with extraneous sample size n_{ij+}.

We shall denote an independent set of $(I-1)(J-1)(K-1)$ parameters among $\{\alpha_{ijk}^{(1,2,3)}\}$, say by choosing $i = 1,\ldots,I-1$, $j = 1,\ldots,J-1$; $k = 1,\ldots,K-1$, by the vector $\underset{\sim}{\alpha}^{(1,2,3)}$. Similarly, $\underset{\sim}{\alpha}^{(1,3)}$, $\alpha^{(3)}$ denote vectors of $(I-1)(K-1)$ and $K-1$ independent parameters, respectively.

The non-trivial log-linear models are then obtained from (2.1) by considering linear models for highest order non-zero interaction terms arising from representation (2.1); e.g. we have the model

$$\underset{\sim}{\alpha}^{(1,2,3)} = \underset{\sim}{X}\underset{\sim}{\beta}. \tag{2.3}$$

Here $\underset{\sim}{X}$ is a known $(I-1)(J-1)(K-1) \times t$ matrix of rank $t < (I-1)(J-1)(K-1)$ and $\underset{\sim}{\beta}$ is a vector of t unknown parameters. As one model we have

$$\underset{\sim}{\alpha}^{(1,2,3)} = \underset{\sim}{0}. \tag{2.4}$$

Other hierarchical log-linear models would follow by taking (2.4) in conjunction with linear models for $\underset{\sim}{\alpha}^{(1,3)}$ and/or $\underset{\sim}{\alpha}^{(2,3)}$. In general, a log-linear model would be alternatively written as the expression (2.1) incorporating additional restrictions of the type (2.3), (2.4) or further such hierarchical structures; for example the log-linear model

$$M: \ell n\, \pi_{ijk} = \mu + \mu_i^{(1)} + \mu_j^{(2)} + \alpha_k^{(3)} + \mu_{ij}^{(1,2)} + \alpha_{ik}^{(1,3)} + \alpha_{jk}^{(2,3)} \tag{2.5}$$

states that the three-dimensional interaction is absent.

Suppose now we consider the multivariate logits

$$\lambda_{ijk}^{(3)} = \ell n \frac{\pi_{ijk}}{\pi_{ijK}}, \quad k = 1,\ldots,K-1 \tag{2.6}$$

and its unique representation

$$R': \lambda_{ijk}^{(3)} = \theta_k^{(3)} + \theta_{ik}^{(1,3)} + \theta_{jk}^{(2,3)} + \theta_{ijk}^{(1,2,3)}, \tag{2.7}$$

with the side-constraints $\sum_i \theta_{ik}^{(1,3)} = 0$, $\sum_j \theta_{jk}^{(2,3)} = 0$, $\sum_i \theta_{ijk}^{(1,2,3)} = \sum_j \theta_{ijk}^{(1,2,3)} = 0$ for every $k = 1,\ldots,K-1$. The relationship between the α and θ-parameters is seen to be

$$\theta_k^{(3)} = \alpha_k^{(3)} - \alpha_K^{(3)}, \quad \theta_{ik}^{(1,3)} = \alpha_{ik}^{(1,3)} - \alpha_{iK}^{(1,3)}$$

$$\theta_{jk}^{(2,3)} = \alpha_{jk}^{(2,3)} - \alpha_{jK}^{(2,3)}, \quad \theta_{ijk}^{(1,2,3)} = \alpha_{ijk}^{(1,2,3)} - \alpha_{ijK}^{(1,2,3)},$$

in view of (2.1) and (2.7). Thus, in this case, the logit linear models obtained from (2.7) are seen to be as informative as the corresponding log-linear models. For example, corresponding to the log-linear model (2.3) we have the logit model

$$\underset{\sim}{\theta}^{(1,2,3)} = \underset{\sim}{Y}\underset{\sim}{\beta}; \tag{2.8}$$

here $\underset{\sim}{\theta}^{(1,2,3)}$ is the vector of (I-1)(J-1)(K-1) independent $\theta_{ijk}^{(1,2,3)}$ parameters and $\underset{\sim}{Y}$ is a matrix of rank t. Furthermore $\underset{\sim}{Y}$ is determined uniquely from $\underset{\sim}{X}$ and the converse is also true. Similarly, corresponding to the log-linear model (2.5), we have the logit model

$$M': \lambda_{ijk}^{(3)} = \theta_k^{(3)} + \theta_{ik}^{(1,3)} + \theta_{jk}^{(2,3)}. \tag{2.9}$$

A similar correspondence holds for all hierarchical log-linear models. Thus, in this probability structure, the logit models obtained from (2.7) give precisely the same information as that given by the corresponding log-linear models obtained from (2.1). Also, the logit models are simpler to interpret since these are not cluttered up with the nuisance μ-parameters.

2.2 Two Responses, One Factor

Consider now the case where two dimensions represent random levels, say j and k respectively, while i denotes the level of the design factor. Here the counts $\{n_{ijk}, j = 1,\ldots,J, k = 1,\ldots,K\}$ have the multinomial distribution with fixed sample size n_{i++} and probabilities π_{ijk} such that $\pi_{i++} \equiv \sum_j\sum_k\pi_{ijk} = 1$. We have such multinomial vectors for i = 1,...,I.

Here the log-linear representation takes the form

$$R: \ln \pi_{ijk} = \mu + \mu_i^{(1)} + \alpha_j^{(2)} + \alpha_k^{(3)} + \alpha_{ij}^{(1,2)} + \alpha_{ik}^{(1,3)} + \alpha_{jk}^{(2,3)} + \alpha_{ijk}^{(1,2,3)}, \tag{2.10}$$

with the usual side constraints; the μ-parameters are such that they satisfy the basic constrants

$$\exp\text{-}[\mu + \mu_i^{(1)}] = \sum_j \sum_k \exp[\alpha_j^{(2)} + \alpha_k^{(3)} + \alpha_{ij}^{(1,2)} + \alpha_{ik}^{(1,3)} + \alpha_{jk}^{(2,3)} + \alpha_{ijk}^{(1,2,3)}].$$

Suppose now that we consider the logits $\lambda_{ijk}^{(3)}$, defined as in 2.1 by

$$\lambda_{ijk}^{(3)} = \ln \frac{\pi_{ijk}^{(3)}}{\pi_{ijK}^{(3)}} = \ln \frac{\pi_{ijk}}{\pi_{ijK}}, \quad k = 1,\ldots,K-1; \tag{2.11}$$

here $\pi_{ijk}^{(3)}$ is the <u>conditional</u> probability of k, <u>given</u> j for the factor level i, i.e.

$$\pi_{ijk}^{(3)} = \pi_{ijk}/\pi_{ij+}.$$

Then, as in (2.1), we have the representation R' given by (2.7) for $\lambda_{ijk}^{(3)}$. Although the information regarding the parameters $\underset{\sim}{\alpha}^{(3)}$, $\underset{\sim}{\alpha}^{(1,3)}$, $\underset{\sim}{\alpha}^{(2,3)}$ and $\alpha^{(1,2,3)}$ in the log-linear representation R is retained in R', the information regarding

$\underset{\sim}{\alpha}^{(1,2)}$, and then $\underset{\sim}{\alpha}^{(2)}$, is lost as pointed out by Bishop (1969). A similar assertion holds for linear hierarchical models obtained from R and R' respectively.

The information concerning $\underset{\sim}{\alpha}^{(1,2)}$ and $\underset{\sim}{\alpha}^{(2)}$ could be retrieved by considering the logits for the other random response, viz.,

$$\lambda_{ijk}^{(2)} = \ell n \frac{\pi_{ijk}^{(2)}}{\pi_{iJk}^{(2)}} = \ell n \frac{\pi_{ijk}}{\pi_{iJk}}, \quad j = 1,\ldots,J-1. \tag{2.12}$$

Here $\pi_{ijk}^{(2)}$ is the conditional probability of j, given the third dimension level k for factor level i; thus $\pi_{ijk}^{(2)} = \pi_{ijk}/\pi_{i+k}$.

Thus the second remark (Bishop, et al. 1969, ff 390-9) of Bishop is not quite accurate. The information regarding $\underset{\sim}{\alpha}^{(1,2)}$, $\underset{\sim}{\alpha}^{(2)}$ is lost if we confine our attention only to $\underset{\sim}{\lambda}^{(3)}$. However, this lost information can be retrieved by using $\underset{\sim}{\lambda}^{(2)}$ along with $\underset{\sim}{\lambda}^{(3)}$.

But this approach of using $\underset{\sim}{\lambda}^{(3)}$ and $\underset{\sim}{\lambda}^{(2)}$ jointly is not quite satisfactory. There is some redundancy, since there are IJ(K-1) independent elements $\lambda_{ijk}^{(3)}$ in the vector $\underset{\sim}{\lambda}^{(3)}$ and, similarly, I(J-1)K independent elements $\lambda_{ijk}^{(2)}$ in $\underset{\sim}{\lambda}^{(2)}$. However, jointly these are not independent since there are only I(JK-1) independent parameters π_{ijk}.

Hence we use the set of independent I(J-1) logits $\lambda_{ijK}^{(2)}$ in conjunction with $\underset{\sim}{\lambda}^{(3)}$. Note that the logits $\lambda_{ijK}^{(2)}$ are functionally independent of logits in $\underset{\sim}{\lambda}^{(3)}$. Thus, jointly, these two sets give a set of IJ(K-1) + I(J-1) = I(JK-1) logits, which contain all the relevant information regarding the α-parameters in the log-linear representation (2.10) for R. The representation of $\lambda_{ijK}^{(2)}$ is seen to be

$$R'': \lambda_{ijK}^{(2)} = \theta_j^{(2)} + \theta_{ij}^{(1,2)} + \theta_{jK}^{(2,3)} + \theta_{ijK}^{(1,2,3)}, \tag{2.13}$$

where

$$\theta_j^{(2)} = \alpha_j^{(2)} - \alpha_J^{(2)}, \quad \theta_{ij}^{(1,2)} = \alpha_{ij}^{(1,2)} - \alpha_{iJ}^{(1,2)}, \quad \theta_{jK}^{(2,3)} = \alpha_{jK}^{(2,3)} - \alpha_{JK}^{(2,3)}$$

and

$$\theta_{ijK}^{(1,2,3)} = \alpha_{ijK}^{(1,2,3)} - \alpha_{iJK}^{(1,2,3)}.$$

Finally abosrbing the third and fourth terms into the first and second terms on the right-hand side of (2.13), we may write

$$R'': \lambda_{ijK}^{(2)} = \phi_j^{(2)} + \phi_{ij}^{(1,2)}. \tag{2.14}$$

Observe that $\theta^{(2,3)}_{jK}$ is determined by $\theta^{(2,3)}_{jk}$, $k = 1,\dots,K-1$, and the full information on these is available from $\underset{\sim}{\lambda}^{(3)}$. The same is true for $\theta^{(1,2,3)}_{ijK}$.

Whatever we have said earlier applies also to hierarchical models obtained from the representation (2.10) for R and the corresponding logit models obtained from R' and R" given by (2.7) and (2.1), respectively.

Denoting the elements $\lambda^{(2)}_{ijK}$ by $\underset{\sim}{\lambda}^{(2)}_K$, we shall refer to the set of elements in the vectors $(\underset{\sim}{\lambda}^{(3)}, \underset{\sim}{\lambda}_K{}^{(2)})$ as the set of generalized logits for this probability distributional structure. We have thus shown that the generalized logit models are as informative as log-linear models; furthermore, they are simpler to interpret due to absence of μ's.

2.3 Three Responses

Here the counts $\{n_{ijk},\ i = 1,\dots,I,\ j = 1,\dots,J,\ k = 1,\dots,K\}$ have the multinomial distribution with fixed sample size n_{+++} and probabilities π_{ijk} such that $\pi_{+++} \equiv \sum_i\sum_j\sum_k\pi_{ijk} = 1$. The log-linear representation here takes the form

$$R:\ \ell n\ \pi_{ijk} = \mu + \alpha^{(1)}_i + \alpha^{(2)}_j + \alpha^{(3)}_k + \alpha^{(1,2)}_{ij} + \alpha^{(1,3)}_{ik} + \alpha^{(2,3)}_{jk} + \alpha^{(1,2,3)}_{ijk}. \quad (2.15)$$

Here, as before, the logits $\lambda^{(3)}_{ijk}$ are defined as

$$\lambda^{(3)}_{ijk} = \ell n\ \frac{\pi^{(3)}_{ijk}}{\pi^{(3)}_{ijK}} = \ell n\ \frac{\pi_{ijk}}{\pi_{ijK}},\quad k = 1,\dots,K-1,$$

where $\pi^{(3)}_{ijk}$ is the conditional probability of k, given the categories i,j of the first two responses, i.e. $\pi^{(3)}_{ijk} = \pi_{ijk}/\pi_{ij+}$.

Similarly, we define logits for dimension 2 as

$$\lambda^{(2)}_{ijK} = \ell n\ \frac{\pi^{(2)}_{ijK}}{\pi^{(2)}_{iJK}} = \ell n\ \frac{\pi_{ijK}}{\pi_{iJK}},\quad j = 1,\dots,J-1$$

where $\pi^{(2)}_{ijK} = \pi_{ijK}/\pi_{i+K}$, for dimension 1 as

$$\lambda^{(1)}_{iJK} = \ell n\ \frac{\pi^{(1)}_{iJK}}{\pi^{(1)}_{IJK}} = \ell n\ \frac{\pi_{iJK}}{\pi_{IJK}},\quad i = 1,\dots,I-1,$$

where $\pi^{(1)}_{iJK} = \pi_{iJK}/\pi_{+JK}$.

Now it is straightforward to check that information regarding $\underset{\sim}{\alpha}^{(3)}$, $\underset{\sim}{\alpha}^{(1,3)}$, $\underset{\sim}{\alpha}^{(2,3)}$ and $\underset{\sim}{\alpha}^{(1,2,3)}$ is provided by the IJ(K-1) independent elements $\lambda^{(3)}_{ijk}$ of $\underset{\sim}{\lambda}^{(3)}$, as in cases 2.1 and 2.2. Next, information concerning $\underset{\sim}{\alpha}^{(2)}$ and $\underset{\sim}{\alpha}^{(1,2)}$ is provided by the I(J-1) independent elements $\lambda^{(2)}_{ijK}$ of $\underset{\sim}{\lambda}^{(2)}_{K}$, as in (2.2). Finally, information concerning $\underset{\sim}{\alpha}^{(1)}$ is provided by the I-1 independent elements $\lambda^{(1)}_{iJK}$ of $\underset{\sim}{\lambda}^{(1)}_{JK}$. Furthermore, the elements of $\underset{\sim}{\lambda}^{(3)}$, $\underset{\sim}{\lambda}^{(2)}_{K}$ and $\underset{\sim}{\lambda}^{(1)}_{JK}$ are functionally independent. Thus, here, the set of generalized logits is provided by elements of $\{\underset{\sim}{\lambda}^{(3)}, \underset{\sim}{\lambda}^{(2)}_{K}, \underset{\sim}{\lambda}^{(1)}_{JK}\}$. The number of independent elements in this set is IJ(K-1) + I(J-1) + (I-1) = IJK-1, which is precisely the number of independent π_{ijk}'s for this probability structure.

Thus, again, although the logit models concerning $\underset{\sim}{\lambda}^{(3)}$ alone are not as general as the log-linear models, as shown by Bishop (1969), her second remark quoted earlier (ff 390-91) is not quite accurate. The logit models obtained from $\underset{\sim}{\lambda}^{(3)}$, $\underset{\sim}{\lambda}^{(2)}_{K}$ and $\underset{\sim}{\lambda}^{(1)}_{JK}$ are as informative as the log-linear models; furthermore, these are simpler to interpret.

3. OTHER MODELS

Now that we have shown in Section 2 the essential equivalence of logit and log-linear models, we will examine in this section several other equally informative logit models.

For the sake of simplicity, let us consider a 2-response, 1-factor situation where one of the response variables (indexed by k) is binary while the other variable (indexed by j) has 3 categories. For clarity, we shall refer to the former as the primary response variable while the latter will be referred to as the covariate.

Let $\lambda_{ij} = \lambda^{(3)}_{ij1} = \ell n(\pi_{ij1}/\pi_{ij2})$, as defined in (2.11), and

$$\Delta_{ij} = \log \frac{\pi_{ij1}\pi_{i32}}{\pi_{ij2}\pi_{i31}} .$$

Δ_{ij} are association measures between the covariate and the response variable for treatment i using the third covariate level as the pivotal category. The following are some possible representations for this problem:

(A) $\lambda_{ij} = \mu + T_i + C_{j(i)}$, $i = 1,\ldots,I$, $j = 1,2,3$;

(B) $\Delta_{ij} = \mu + T_i + C_j + \delta_{ij}$, $i = 1,\ldots,I$, $j = 1,2$;

(C) $\Delta_{ij} = \mu + T_i + C_{j(i)}$, $i = I,\ldots,I$, $j = 1,2$.

All the above three representations are subject to the side constraint $\sum_{i=1}^{I} T_i = 0$. In addition, representation (A) is subject to $\sum_{j=1}^{3} C_{j(i)} = 0$, $i = 1,\ldots,($ and representation (C) is subject to $\sum_{j=1}^{2} C_{j(i)} = 0$, $i = I,\ldots,I$. Also, in (B), $\sum C_j = 0$, $\sum_i \delta_{ij} = \sum_j \delta_{ij} = 0$. Using collapsed tables additional representations can be considered for $\lambda_{ij+} = \ell n(\pi_{ij+}/\pi_{i3+})$:

(D) $\lambda_{ij+} = \mu + T_i + C_j + \delta_{ij}$, $i = 1,\ldots,I$, $j = 1,2$;

(E) $\lambda_{ij+} = \mu + T_j + C_{j(i)}$, $i = 1,\ldots,I$, $j = 1,2$;

with constraints for parameters in (D) and (E) as those for (B) and (A), respectively.

The utility of using such representations and the models therefrom is their interpretability. Representation (B) postulates interactive treatment and covariate effects on the primary response variable. Representations (A) and (C) are subtle variations of (B) which, because of the nested structure, allow for more specific hypotheses to be tested. While representations (D) and (E) may look similar to (A)-(C), they are used primarily to examine the effect of the treatments on the covariate, i.e., in testing the hypothesis H: $T_i = 0$, $i = 1,\ldots,I$.

Some interesting hypotheses that can be tested using the above representations are:

(1) H: $C_{j(i)} = 0$, $i = 1,\ldots,I$

in representation (A) or (C). One can interpret this hypothesis as testing for the effect of the covariate on the response adjusted for the treatment differences. In other words, even if the distribution of the covariate were to be affected by the treatment, this test will reveal the additional effect of the covariate, if any.

(2) H: $C_{j(1)} = \cdots = C_{j(I)}$,

in representation (A) or (C). This provides a check on the interactive effect of treatment and covariate on the response variable. Another interesting model to consider is

(3) H: $T_i = 0$, $i = 1,\ldots,I$

in (A) or (C). This is a valuable hypothesis to test as it reveals the 'intrinsic' effect of the treatments separate from their interactive effects with the covariate. This is the sole treatment effect if the covariate were completely absent.

In implementing the above models, one can test the hypothesis separately or in conjunction with one another. We suggest using the combination of representations (A) and (E), and testing the hypotheses of type (1), (2) and/or (3) above. This will provide answers to most questions regarding the effects of the treatments on the covariate and the response variable, intrinsic and otherwise, as well as the effect of the covariate on the response variable.

Representations (B) and (C) are useful for exploring the effect of treatment on the response-covariate association measures. This can be done, for example, by testing hypotheses of type (3) in conjunction with either (1) or (2) above in (C). The representations (B) and (D) may be used either in their saturated form for testing interactive effects, or in their restricted form for testing simpler models.

4. NUMERICAL EXAMPLES

Consider the data taken form Table 4.3 of Haberman (1978), and for simplicity, let us restrict our analysis and discussion to the 3-dimensional table (Table 1) made up of the 1975 portion of the original data. The log-linear representation is given by

$$\log \pi_{ijk} = \mu + \alpha_k^R + \mu_i^S + \alpha_j^E + \alpha_{ik}^{SR} + \alpha_{ij}^{ER} + \alpha_{ij}^{SE} + \alpha_{ijk}^{SER} \qquad (4.1)$$

$i,k = 1,2$, $j = 1,2,3$ and the α's represent the response, sex, education and interaction effects, and the cell probability

$$\pi_{ijk} = P(\text{Response} = k, \text{ Education} = j \mid \text{Sex} = i.$$

The corresponding logit representation is

$$\lambda_{ij} = \gamma + \gamma_i^S + \gamma_j^E + \gamma_{ij}^{SE} . \qquad (4.2)$$

We have used the μ and α-notation from 2.2 in (4.1) to stress that μ's are dependent parameters. The usual side constraints are assumed to hold for (4.1) and (4.2). The following analyses are carried out using the FUNCAT procedure in SAS (for logistic models) and the BMDP4F procedure in BMDP (for log-linear models).

Table 1

Sex	Education in Years	Response	
		Agree	Disagree
Male	≤8	72	47
	9-12	110	196
	≤13	44	179
Female	≤8	86	38
	9-12	173	283
	≤13	28	187

Table 2 presents the results of log-linear model fitting using BMDP4F. It is clear that the best-fitting log-linear model is given by

$$\log \pi_{ijk} = \mu + \mu_i^S + \alpha_j^E + \alpha_k^R + \alpha_{ij}^{SE} + \alpha_{jk}^{ER} \tag{4.3}$$

In addition, the best-fitting logit interaction model seems to be given by

$$\lambda_{ij} = \gamma + \gamma_j^E. \tag{4.4}$$

Table 2. Log-Linear Interaction Model

Effect	d.f.	Pearson's χ^2	p-value
R	1	121.72	.0000
E	2	286.02	.0000
RxE	2	165.58	.0000
RxS	1	0.04	.8449
ExS	2	14.76	.0006
RxExS	2	5.91	.0521
Saturated	0	0.00	1.0000

NOTE: Each effect refers to the model which includes all effects of similar or lower order except for the effect.

Table 3. Logit Interaction Model

Source	d.f.	χ^2	p-value
Intercept	1	58.99	.0001
Sex	1	0.00	.9647
Education	2	144.48	.0001
Sex * Education	2	5.82	.0545
Residual	0	0.00	1.0000

[We cannot rule out the possibility of using the saturated model itself since the SEX * EDUC effect has a p-value that is very close to .05.]

Alternatively, one can also fit a model to a logit nested representation of the form

$$\lambda_{ij} = \beta + \beta_i^S + \beta_{j(i)}^{E(S)}, \quad \sum_{i=1}^{2} \beta_i^S = 0, \quad \sum_{j=1}^{3} \beta_{j(i)}^{E(S)} = 0 \tag{4.5}$$

The result of fitting this model is shown in Table 4. Again, the saturated model is required. We note here that, unfortunately, the BMDP4F program does not allow fitting models other than the log-linear interaction model.

So how do the analyses differ? In using a log-linear interaction model, one winds up with (4.3) as the best-fitting model. A reasonable interpretation of this model is that responses are directly affected by education level. However, the same model also shows a link between education levels and sex. Thus, indirectly, sex also plays a role in determining the responses through its association with education. It is, perhaps, convenient here to think of education as a random co-variate that is affected by the treatment sex. It is pertinent to note here that in interpreting the log-linear model (4.3), we conclude that education is affected by sex and not the converse because of biological reasons. But if sex is instead some other variable such as family income level, some ambiguity in interpretation will be unavoidable.

Table 4. Logit Nested Model

Source	d.f.	Pearson's χ^2	p-value
Intercept	1	58.99	0.0001
Sex	1	0.00	0.9647
Educ (Sex)	4	145.59	0.0001
Residual	0	0.00	1.0000

Compared to the log-linear model, the logit interaction model (4.2) is not fully informative; this is seen by comparing the unsaturated logit model (4.4) with the log-linear model (4.3) and noting that the sex effect need not be included in the logit model. This is in accordance with the discussion in 2.2 that the information about $\underset{\sim}{\alpha}^{SE}$ is lost if we consider only the conditional logits $\lambda_{ij} \equiv \lambda_{ij1}^{(3)}$. But, even here, the logit nested model (4.5) needs to include the factor sex due to the significant nested effect Education (Sex). It should be pointed out here that if we had used α=.10, all three saturated models (4.1), (4.3) and (4.5) would be acceptable. This would diminish the interpretive ability of model (4.1) beyond the simple conclusion that responses are associated with sex and education levels. In such a case new representations and models should prove useful. To recall the new representation are

$$\text{(M1)} \quad \lambda_{ij}^{(C)} = \alpha^{(C)} + S_i^{(C)} + E_{j(i)}^{(C)}, \quad \sum_{i=1}^{2} S_i^{(C)} = 0, \quad \sum_{j=1}^{2} E_{j(i)} = 0$$

(M2) $$\lambda^R_{ij} = \alpha^R + S^R_i + E^R_{j(i)}, \quad \sum_{i=1}^{2} S^R_i = 0, \quad \sum_{j=1}^{3} E^R_{j(i)} = 0,$$

where

$$\lambda^C_{ij} = \ell n \frac{\pi_{ij+}}{\pi_{i3+}}, \quad i,j = 1,2, \quad \pi_{ij+} = \sum_{k=1}^{2} \pi_{ijk}$$

and

$$\lambda^R_{ij} = \ell n \frac{\pi_{ij1}}{\pi_{ij2}}, \quad i = 1,2, \text{ and } j = 1,2,3.$$

We note that the hypothesis

$$H_1: S^C_i = 0, \quad i = 1,2$$

tests for the effect of sex on the covariate education; the hypothesis

$$H_2: S^R_i = 0, \quad i = 1,2$$

tests for the "intrinsic" effect of sex on the response; the hypothesis

$$H_3: E^R_{j(1)} = E^R_{j(2)}, \quad j = 1,2,3$$

tests for the interactive effect of sex and education; and finally, the hypothesis

$$H_4: E^R_{1(i)} = E^R_{2(i)} = E^R_{3(i)} = 0, \quad i = 1,2$$

tests for the differences due to the covariate Education within each sex level. We note here that all these tests could be carried out singly or simultaneously.

Using weighted least squares (with estimated weights), the value of statistic for testing H_2 in the representation M2 (intrinsic sex effect) has a value of .00257 which is not significant for 1 degree of freedom (d.f.). The value of statistic for simultaneously testing H_2 and H_4 is 147.718 on 5 d.f. Thus, conditional on no sex effect on the response variable, there is a very significant education effect within each level of sex (the difference of χ^2 statistic = 147.715 on 4 d.f.). Further, the statistic for the interactive effect of sex and education has the value 5.867 which is not significant for 2 d.f. These tests show clearly that sex has no influence on the response variable either on its own or in conjunction with education. They also show that education has a strong influence on the response variable even after adjusting for variation in the sex distribution. A final test of the hypothesis H_1 yields the statistic of 8.937 which is significant on 1 d.f. Thus, although sex does not seem to have influenced the response variable, it does affect the distribution of the covariate education. We wish to emphasize here that this is completely different from merely asserting that sex 'interact' with education to influence the response variable. In fact, all the above findings point towards a causative model of the form

$$\text{SEX} \xrightarrow{\text{influences}} \text{EDUCATION} \xrightarrow{\text{influences}} \text{RESPONSES}$$

Remark. Although the above models cannot be fitted using either FUNCAT or BMDP4F, only a simple computer program (of less than 100 lines) is needed to compute the statistics.

REFERENCES:

[1] Bishop, Y.M.M., Full contingency tables, logits and split contingency tables, Biometrics **25**:119-128 (1969).

[2] Bishop, Y.M.M., Fienberg, S.E., and Holland, P.W., Discrete Multivariate Analysis - Theory and Practice. The MIT Press, Cambridge, Massachusetts (1975).

[3] Haberman, S.J., Analysis of Qualitative Data (Vol. 1). Academic Press, New York (1978).

[4] Levin, B. and Shrout, P.E., On extending Bock's model of logistic regression in the analysis of categorical data. Communication in Statistics - Theory and Method A 10:125-147 (1981).

BIOSTATISTICS: Statistics in Biomedical, Public Health and Environmental Sciences, edited by P.K. Sen

STATISTICAL ANALYSIS OF LONGITUDINAL REPEATED MEASURES DESIGNS

Edward Bryant
Ortho Pharmaceuticals
Raritan, NJ

Dennis Gillings
University of North Carolina at Chapel Hill

The longitudinal repeated measures design often is utilized to evaluate treatment effectiveness when response to treatment manifests itself over time. This paper reviews a variety of statistical methods that are used to analyze this type of data and discusses their application to the pharmaceutical clinical trial setting. The use of area-under-the-curve (AUC) analysis to assess treatment main effects is discussed and related to experimental and statistical concerns. A piecewise growth curve analysis approach is introduced with which to model data whose functional form changes during the period of observation.

KEY WORDS: Repeated measures designs, clinical trials, statistical models, piecewise growth curves, area-under-the-curve

1. INTRODUCTION

In many experimental situations, interest lies in examining an intervention where the response manifests itself over time. Pharmaceutical clinical trials to evaluate drug effectiveness often involve the assessment of response data collected over a time following initiation of drug therapy. Decisions regarding the efficacy of one treatment relative to another are based on comparing response along a time continuum.

A common research design that has been used to examine response-over-time phenomena is the longitudinal repeated measures design. The longitudinal study is characterized by measurements taken on the same unit of observation (subject) over a set of occasions. Typically, subjects are assigned on a random basis to one of g treatment groups. Responses to treatment are observed at p pre-specified time points. For the ith subject from the jth treatment group, treatment responses can be denoted by the vector $\underset{\sim}{y}'_{ij} = (y_{ij1}, y_{ij2}, \ldots, y_{ijp})$, $i=1,\ldots,n_j$; $j=1,\ldots,g$. Associated with the response vector $\underset{\sim}{y}_{ij}$ is another vector $\underset{\sim}{t}_{ij}$, where $\underset{\sim}{t}'_{ij} = (t_{ij1}, t_{ij2}, \ldots, t_{ijp})$, which denotes the time of observation. Often these times are the same for all N subjects so that $\underset{\sim}{t}'_{ij} = (t_1, t_2, \ldots, t_p)$ for all i,j. The response data for N subjects from a treatment-by-occasions trial can be expressed in an N x p array as shown in Figure 1.

Selection of an appropriate statistical approach to analyze the repeated measures

FIGURE 1. DATA ARRAY FROM THE REPEATED MEASURES TREATMENT-BY-OCCASIONS DESIGN.

Treatment Group	Subject	Occasion 1	2	...	k	...	p	All Occasions
1	1	y_{111}	y_{112}	...	y_{11k}	...	y_{11p}	
	.	.	.	.	.	.	.	
	.	.	.	.	.	.	.	
	.	.	.	.	.	.	.	
	n_1	y_{n_111}	y_{n_112}	...	y_{n_11k}	...	y_{n_11p}	
Group 1 Means		$\bar{y}_{.11}$	$\bar{y}_{.12}$	...	$\bar{y}_{.1k}$	...	$\bar{y}_{.1p}$	$\bar{y}_{.1.}$
.								
.								
.								
j	1	y_{1j1}	y_{1j2}	...	y_{1jk}	...	y_{1jp}	
	.	.	.	.	.	.	.	
	.	.	.	.	.	.	.	
	.	.	.	.	.	.	.	
	i	y_{ij1}	y_{ij2}	...	y_{ijk}	...	y_{ijp}	
	.	.	.	.	.	.	.	
	.	.	.	.	.	.	.	
	.	.	.	.	.	.	.	
	n_j	y_{n_jj1}	y_{n_jj2}	...	y_{n_jjk}	...	y_{n_jjp}	
Group j means		$\bar{y}_{.j1}$	$\bar{y}_{.j2}$	...	$\bar{y}_{.jk}$	...	$\bar{y}_{.jp}$	$\bar{y}_{.j.}$
.								
.								
.								
g	1	y_{1g1}	y_{1g2}	...	y_{1gk}	...	y_{1gp}	
	.	.	.	.	.	.	.	
	.	.	.	.	.	.	.	
	.	.	.	.	.	.	.	
	n_g	y_{n_ggl}	y_{n_gg2}	...	y_{n_ggk}	...	y_{n_ggp}	
Group g Means		$\bar{y}_{.g1}$	$\bar{y}_{.g2}$	...	$\bar{y}_{.gk}$	...	$\bar{y}_{.gp}$	$\bar{y}_{.g.}$
Combined Group Means		$\bar{y}_{..1}$	$\bar{y}_{..2}$	...	$\bar{y}_{..k}$	...	$\bar{y}_{..p}$	$\bar{y}_{...}$

y_{ijk} = k^{th} response of ith subject from j^{th} group;

$i = 1,\ldots,n_j$;

$j = 1,\ldots,g$;

$k = 1,\ldots,p$.

$\underset{\sim}{y}_{ij}' = (y_{ij1}, \ldots, y_{ijp})$.

data of Figure 1 is determined by the experimental goals, the nature of the phenomena under study, validity of statistical model assumptions for the observed data, and characteristics of the completed experiment such as the occurrence of dropouts and missing data.

This paper reviews statistical methods that have been used to analyze repeated measures data and discusses their application to the clinical trial setting. Experimental and statistical issues related to the choice of analysis method are discussed. Attention is focused on applications to pharmaceutical trials where the form of the time-response curve is unknown and where a main concern is assessing the efficacy of an investigational treatment relative to some reference treatment.

The analysis approaches that are reviewed in the next section include univariate methods such as separate analyses of variance at each occasion and endpoint analysis, and approaches depending on multivariate assumptions such as mixed model analysis of variance and multivariate analysis of variance.

Two additional analysis strategies are introduced. The first is the use of a piecewise growth curve analysis approach with which to model data whose functional form changes during the period of observation. The second is an intuitively appealing univariate approach to assessing treatment main effects using an area-under-the-time-effect curve (AUC) approach.

Data from a clinical trial of an investigational drug for the treatment of atypical depression are analyzed to illustrate the reviewed methods of analysis.

2. REVIEW OF ANALYSIS METHODS

Early treatment of the repeated measures analysis problem often consisted of endpoint analysis, i.e., looking at the first and last observation from a subject's vector of responses and ignoring the rest (Wishart, 1938). Another strategy was to analyze the data using classical regression techniques, thus ignoring the correlated structure of the response variables and treating the data as a repeated cross-sectional design. Tests based on this procedure have been shown (McGregor, 1960; Siddiqui, 1958; Watson, 1955, Elston and Grizzle, 1962) to be too liberal, resulting in too many rejections of the null hypothesis. Wishart (1938) was the first to develop repeated measures statistical techniques that incorporated the correlated structure of the data and utilized all the information on subject ij contained in $\underset{\sim}{y}_{ij}$. His method of fitting polynomial curves to the data was essentially the basis for what is now called growth curve analysis. In this section the statistical literature on approaches to the analysis of longitudinal data is briefly reviewed.

Separate Univariate Analyses

For treatment responses measured at p different observation times, separate univariate analyses can be performed for each observation time to assess significant treatment differences. From the perspective of assessing the overall superiority of one treatment relative to other treatments, how to combine the results of the individual tests is problematic.

For the case when the separate tests are not independent, Brown (1975) presented an approximate test for one-sided alternatives under the assumption of multivariate normality. His method consists of calculating the first two sample moments of $-2 \Sigma \log p$ and equating these to that of a chi-square distribution to derive an approximate distribution. The limitation to one-sided tests restricts used of the procedure to the case of two treatment groups. Alternatively, the Bonferroni inequality (e.g., see Neter and Wasserman, 1974) may be used to combine non-independent tests of significance.

Endpoint Analysis

The repeated measure design is often used in pharmaceutical clinical research to collect safety data (e.g., blood chemistry and hematology, vital signs, incidence of side effects) in order to monitor and assess drug toxicity. While efficacy data are usually collected at each visit, primary interest with respect to treatment efficacy may be directed at patient response at the end of the trial. That is, for efficacy response vector $\underset{\sim}{y}_{ij}' = (y_{ij1},\ldots,y_{ijp})$, it is observation y_{ijp} that is of primary interest. For example, in the treatment of depression, the success of drug therapy may be assessed adequately only after sufficient time has passed to establish tolerable drug levels. The beneficial effects of the drug might be expected to be monotonically increasing until time t_p.

When there are no early dropouts and interest is directed at patient response at the end of the observation period, the problem of assessing treatment effects for the repeated measures trial is reduced to a single univariate analysis of y_{ijp}. The occurrence of early dropouts complicates the picture in that y_{ijp} is not observed for some subjects. When dropouts occur, a univariate analysis of the last recorded value is sometimes performed. This is called endpoint analysis (Gould, 1980).

The use of endpoint analysis utilizes information from each subject randomized to treatment and avoids the problem of non-comparable samples. An underlying assumption of endpoint analysis is that the last available response of a patient who withdrew from the study is representative of what would have been observed if that patient had stayed in the trial. The validity of this assumption is difficult if not impossible to test. Gould (1980) proposed an approach to using endpoint analysis which incorporates reason-for-early termination in deriving a nonparametric endpoint score. Patients who drop out early due to side effects or lack of drug efficacy are assigned the lowest score at time t_p while patients who withdraw cured are assigned the highest score at time t_p.

Another approach using endpoint analysis examines the effect of early dropouts on response. In this approach, a dummy variable which identifies whether or not the endpoint score is an early termination score is included in the ANOVA model. Also included is a treatment-by-dummy variable interaction term. The presence of significant interaction identifies a different direction of association of response level and early termination across the treatment groups. For example, the response of early dropouts in the placebo group may be lower than the average total-group response level while the response of early dropouts in the active treatment group may be higher (reflecting a "cure" and no further need for treatment). Using this approach, the nature of the early dropout phenomena can be examined and adjusted estimates of treatment group differences can be obtained.

Mixed Model Analysis of Variance

a. Traditional Procedure

A three factor mixed model analysis of variance can be constructed for the data of Figure 1 with treatments and occasions as completely crossed fixed factors and subjects as a random factor nested within treatment gruops. Letting y_{ijk} represent the kth measurement from the ith subject in the jth treatment group, the mixed model ANOVA can be written (Winer, 1971)

$$y_{ijk} = \mu + \alpha_j + \beta_k + \pi_{i(j)} + \alpha\beta_{jk} + \pi\beta_{ik(j)} + e_{ijk} \qquad (1)$$

with side conditions $\sum_j \alpha_j = \sum_k \beta_k = \sum_j \alpha\beta_{jk} = \sum_k \alpha\beta_{jk} = \sum_k \pi\beta_{ik(j)} = 0$; where μ, α_j, and β_k are the overall population mean, j^{th} treatment effect, and k^{th} occasion effect, respectively; $\pi_{i(j)}$ is a random effect associated with the ith subject in the jth treatment group; $\alpha\beta_{jk}$ is the interaction effect for the jth treatment and the kth occasion; $\pi\beta_{ik(j)}$ is the interaction effect of the ith subject and k^{th} occasion within the jth treatment group; and e_{ijk} is a random error term.

The random error components can be grouped and expressed as

$$m_{ijk} = \pi_{i(j)} + \pi\beta_{ik(j)} + e_{ijk} \; . \qquad (2)$$

Traditional assumptions about the m_{ijk} (Eisenhart, 1947; Scheffé, 1959; Greenhouse and Geisser, 1959) are that the error terms m_{ijk} have a multivariate normal distribution with expectation zero and the following compound symmetric covariance structure

$$\underset{p\times p}{\Sigma_j} \sim \sigma^2 \begin{bmatrix} 1 & \rho & \dots & \rho \\ \rho & 1 & \dots & \rho \\ \cdot & \cdot & & \cdot \\ \cdot & \cdot & & \cdot \\ \cdot & \cdot & & \cdot \\ \rho & \rho & & 1 \end{bmatrix} = \sigma^2[(1-\rho)I_p + \rho \underset{\sim}{i}\underset{\sim}{i}'] \qquad (3)$$

where $\underset{\sim}{i}$ is a p x 1 vector of 1's

for all treatment groups j; j=1,...,g.

When considering response-over-time data, the compound symmetric covariance structure may not occur in many "real life" situations. This is because observations made on the same individual at two closely spaced time points tend to be more highly correlated than two observations made at longer intervals. For example, blood pressure readings made on the same individual one minute apart are apt to have closer values than readings made one day apart. For longitudinal data with an arbitrary covariance structure the question then is whether some other univariate analysis of variance approach to analysis is appropriate.

b. ε-Corrected Approach

Box (1954) derived the theoretical background for approximate univariate ANOVA procedures for data for one group with a general covariance structure. Geisser and Greenhouse (1958) extended results to the situation of several treatment groups. The test of no treatment-by-occasion interaction in the situation of general covariance structure was found to be the same as the usual F-test for the uniform situation, except with modified degrees of freedom $F((p-1)(g-1)\varepsilon,(p-1)(N-g)\varepsilon)$ where ε is defined by

$$\varepsilon = (\mathrm{tr}\ U'\Sigma U)^2/t\cdot \mathrm{tr}(U'\Sigma U)^2 . \qquad (4)$$

U is a p x t orthonormal contrast matrix of interest of rank t and 'tr' is the trace operator (Rogan et al., 1979). In similar fashion, degrees of freedom for an approximate F-test of no occasion main effects are $F((p-1)\varepsilon,(p-1)(N-g)\varepsilon)$. Box (1954) showed ε is such that $(p-1)^{-1} \le \varepsilon \le 1$. When the population covariance matrix is compound symmetric, then ε equals unity. The F-test of the hypothesis of no treatment effects is distributed exactly as $F(g-1,N-g)$, assuming normality and equal among-group variances.

Greenhouse and Geisser (1959) discussed the use of an approximate test whereby the degrees of freedom for testing within-subject factors are reduced by the correction factor ε. A conservative test consists of equating ε to its minimum possible value of $(p-1)^{-1}$. The authors suggested a less conservative three-stage approach. First, use the F-test with full unreduced degrees of freedom. If the

F-test is significant, test for significance with the reduced degrees of freedom (conservative test). If the conservative test is non-significant, estimate ε from the sample covariance matrix S and use the approximate test. Keselman and Keselman (1984) discuss the use of this procedure in conjunction with a Bonferroni multiple comparisons procedure.

Another strategy is to test directly for within-subject factors using F-ratios based on adjusted degrees of freedom where ε is estimated from the sample covariance matrix S. This estimate, $\hat{\varepsilon}$, is the maximum likelihood estimate for ε (Anderson, 1958) when the population is multivariate normal. Collier et al. (1967) and Stoloff (1970) investigated the effect of using $\hat{\varepsilon}$ for the population value ε. In simulation studies, they found $\hat{\varepsilon}$ to be negatively skewed for values of ε close to unity and positively skewed for very low values of ε, especially when sample size is small. In particular, when the group sample size is less than twice the number of occasions, $\hat{\varepsilon}$ may be seriously biased if ε is 0.75 or above, thus resulting in an overly conservative test.

Huynh and Feldt (1976) found the expected value of the numerator and denominator of equation (4) and developed an estimator that is less biased than the maximum likelihood estimator $\hat{\varepsilon}$, when ε is moderately large, say $\varepsilon > 0.75$. The authors found in a review of the educational research literature that values of $\varepsilon > 0.75$ were very common. Their ratio statistic, $\tilde{\varepsilon}$, is given in terms of $\hat{\varepsilon}$ by:

$$\tilde{\varepsilon} = [N(p-1)\hat{\varepsilon} - 2]/(p-1)[N - g - (p-1)\hat{\varepsilon}]. \tag{5}$$

The advantage of $\tilde{\varepsilon}$ over $\hat{\varepsilon}$ with respect to power is greatest when N is small and ε close to unity. When ε is 0.5 or less, the maximum likelihood estimate $\hat{\varepsilon}$ was found to be less biased than $\tilde{\varepsilon}$.

While compound symmetry and equality of covariances were long regarded in the statistical literature as necessary and sufficient conditions for exact within-subject F-ratios for the mixed model (Winer, 1962; Greenhouse and Geisser, 1959), Huynh and Feldt (1970) and Rouanet and Lepine (1970) independently showed that these assumptions were sufficient but not necessary. Assuming multivariate normality, the necessary and sufficient conditions for within-subject F-ratios to be distributed exactly as an F distribution are

$$U'\textstyle\sum_j U = \lambda\, I_{p-1} \qquad j = 1,2,\dots,g\ , \tag{6}$$

where U is a p x (p-1) orthonormal contrast matrix and I_{p-1} is the identity matrix of rank p-1. Equation (6) states that the F-tests of treatment-by-occasion interaction and occasion main effects are valid if and only if the p-1 contrasts represented by U are independent and equally variable. The condition of compound symmetry is a special case of condition (6); Rouanet and Lepine label (6) as the condition of circularity.

The circularity assumption is less restrictive than the assumption of compound symmetry. However, the likelihood that the circularity assumption is met in the response-over-time design is generally problematic. In fact, when conditions have equal variances, (6) implies that all covariances must be equal, which is just the condition of compound symmetry.

The mixed model assumptions include multivariate normality, circularity, and equal among-group covariance structure. Ito (1969) has shown that for large samples, the effect of violation of the multivariate normality assumption is slight on testing hypotheses about the mean vectors but can be serious for testing covariance matrices (this is parallel to the univariate ANOVA situation).

A sequential procedure has been used to assess circularity and equal among-group covariance matrix assumptions. Box's (1949) multivariate analogue of Bartlett's M-test (1937) for homogeneity of variance is first used to test H_{01}: $U'\sum_1 U = \ldots = U'\sum_g U$, under the assumption of multivariate normality. If H_{01} is not rejected, the circularity hypothesis H_{02}: $U'\sum U = \lambda I$, where $\sum$ is the pooled population covariance matrix across groups, is tested using Mauchly's (1940) sphericity criterion W. Assuming normality, if both H_{01} and H_{02} are not rejected, the data are judged to have the circularity property and the mixed model within-subject F-test results are exact.

Rogan et al. (1979) and Keselman et al. (1980) studied the usefulness of this two-step procedure as a decision rule in applying the ε-correction. They found that "for all but the most minute departures from the validity condition, the rejection of the circularity hypothesis is a *fait accompli.*" Another drawback to the use of these tests is their extreme sensitivity to violations of the assumption of multivariate normality.

Multivariate Analysis of Variance Procedures

A natural extension of the univariate general linear model that has been employed for repeated measures studies is multivariate analysis of variance (MANOVA). As before for the mixed model, MANOVA assumes that the vector of responses of the ith subject in the jth treatment group has a multivariate normal distribution with mean vector $\underset{\sim}{u}_j' = (\mu_{j1},\ldots,\mu_{jp})$ and covariance matrix $\sum$ which is common to the g treatment groups. Response vectors of different individuals are assumed to be independent. In contrast to the mixed model, no assumptions regarding the structure of the covariance matrix are necessary for MANOVA. In this section several approaches to repeated measures analysis using MANOVA are discussed.

a. Profile Analysis

Profile analysis is essentially a MANOVA procedure specifically aimed at handling independent variables that are qualitative rather than quantitative. Thus profile analysis applied to longitudinal designs does not consider the ordered relationship of the times of observation $\underset{\sim}{t}' = (t_1,\ldots,t_p)$. However, a discussion of the procedure is included here for two reasons. First, profile analysis has been often used to analyze longitudinal data, perhaps due to the unavailability of a better procedure. Second, since the test for treatment-by-occasion interaction using traditional mixed model ANOVA is inappropriate for longitudinal data when $U'\Sigma U$ is not compound symmetric (Koch et al., 1980), the profile analysis test offers a better way to assess interaction as a preliminary step to testing for treatment main effects.

The MANOVA model for the data of Figure 1 is

$$Y = XB + \varepsilon. \tag{7}$$

X is an N x g design matrix, B is the g x p matrix of unknown parameters, and ε is an N x p matrix of residuals where each row of ε has an independent multivariate normal distribution with zero expectation and common covariance matrix Σ.

Under the multivariate general linear model defined by (7) the maximum likelihood and least squares estimate of the population parameter matrix B is

$$\hat{B} = (X'X)^{-1}X'Y \text{ and } \underset{\sim}{\hat{\beta}}_k = (X'X)^{-1}X'\underset{\sim}{Y}_k \tag{8}$$

where $B_{gxp} = (\underset{\sim}{\beta}_1,\ldots,\underset{\sim}{\beta}_p)$ is the kth column of the data matrix Y, and $k = 1,\ldots,p$. Thus the parameter estimates for the p different conditions are the same as if the data were considered a separate collection of univariate data for each condition.

Tests concerning parameters of B are constructed using

$$H_0\colon CBU = 0 \tag{9}$$

where C is a c x g contrast matrix of rank c referring to hypotheses on the element within columns of B (comparisons across treatments) and U is a p x u contrast referring to hypotheses on the elements within rows of B (comparisons across conditions).

b. Growth Curve Analysis

When responses are measured along an ordered dimension such as time, interest may be directed toward specifying the form of the response curve as a function of time. The form of the response-over-time curve may be suggested in a particular application by theoretical considerations. However, in many areas of study the knowledge of the natural laws of processes is absent and empirical functions which provide an adequate fit to the data are utilized. Growth curve analysis is a multivariate statistical technique of fitting polynomial or other functions linear in their parameters to longitudinal data. While the technique can be applied to any type of repeated measures design with an ordered dimension for occasions, the name "growth curve analysis" entered the statistical literature because much of the early development of methodologies was motivated by problems concerning growth and maturation.

Wishart (1938) first reported the use of orthogonal polynomials in an attempt to utilize all of the data available in a longitudinal study. Rao (1959) developed procedures for parameter estimation and estimation of confidence bands for the response curve. For the treatment-by-occasions design the growth curve model is described by

$$Y = XBT + \varepsilon \tag{10}$$

where Y is an $N \times p$ data matrix representing the data from Figure 1, T is the $q \times p$ $(q \leq p)$ design matrix within individuals, X is the $N \times g$ design matrix across individuals, B is a $g \times q$ matrix of unknown parameters to be estimated, and ε is an $N \times p$ matrix of random errors. It is assumed that the rows of ε are independently distributed as a p-variate normal distribution with zero expectation and common covariance matrix Σ.

The within-individuals design matrix, T, specifies polynomial powers of the times of observation. For the p-observation case, T is of the form

$$T = \begin{bmatrix} t_1^{\,0} & t_2^{\,0} & \cdots & t_p^{\,0} \\ t_1^{\,1} & t_2^{\,1} & \cdots & t_p^{\,1} \\ t_1^{\,2} & t_2^{\,2} & \cdots & t_p^{\,2} \\ . & . & \cdots & . \\ . & . & \cdots & . \\ t_1^{\,p-1} & t_2^{\,p-1} & \cdots & t_p^{\,p-1} \end{bmatrix} \quad \text{where } t_k = \text{time } k;\ k=1,\ldots,p. \tag{11}$$

Implicit in this construction of T is the assumption that all subjects are observed at the same intervals. A polynomial of order $q-1<p-1$ may provide a sufficient fit to the data. In this case, $T_{p\times p}$ is reduced to $T_{q\times p}$ by eliminating the last p-q rows of (11).

The polynomial parameters of B in (10) are correlated. Using a Cholesky decomposition approach, the design matrix T expressed in natural polynomial coefficients in (11) can be transformed into an orthonormal design matrix T_0. The growth curve model in (10) can then be expressed in orthonormal polynomials as

$$E(Y) = XB_0T_0. \tag{12}$$

Assuming a $q-1^{th}$ order polynomial describes the response curve adequately, the matrix B_0 is a $g \times q$ matrix of orthonormal parameters. Using this framework, independent tests of parameters in B_0 can be performed.

Potthoff and Roy (1964) first considered the model in (12) and derived a weighted least squares estimate of B_0 which involves an arbitrary matrix of weights B^{-1}. They showed that the variance of B_0 increases as G^{-1} departs from Σ^{-1}. The maximum likelihood (and weighted least squares estimate) of B_0 was shown by Khatri (1966) to be

$$\hat{B}_0 = (X'X)^{-1}X'YS^{-1}T_0'(T_0S^{-1}T_0')^{-1} \tag{13}$$

where Σ is estimated by,

$$S = Y'[I_N - X(X'X)^{-1}X']Y.$$

This is identical to Potthoff and Roy's estimate of B_0 when G is set equal to S.

Rao (1965, 1966, 1967) developed a somewhat different approach to growth curve analysis based on an analysis of covariance framework. He showed that unless the weight matrix G^{-1} equals Σ^{-1}, sample information for estimating B_0 is lost. He suggested that the remaining (p-q) polynomials not included in the model, or some subset, might be used as covariables to provide a better estimate of B_0. When all p-q variables are used as covariables, this is equivalent to using $G^{-1} = S^{-1}$ as weights. When no covariable is used, this is equivalent to the unweighted analysis discussed by Potthoff and Roy (1964), i.e., $G^{-1} = I$.

c. Piecewise Growth Curve Models

An alternate strategy within the polynomial curve fitting framework is to fit separate polynomial curves over different subintervals of the period of observation $[t_1, t_p]$. This modeling approach is appropriate when response to treatment follows one pattern for a period of time and another for a subsequent time period. An example is the basal body temperature as a function of the day of a woman's menstrual cycle (Teeter, 1982). Prior to ovulation one (linear) submodel for basal body temperature fits and a second (linear) submodel fits after ovulation. In the context of a pharmaceutical clinical trial, the time period following initiation of treatment may involve a placebo effect. In a subsequent time period, the placebo effect may disappear and give way to a different pattern of response. Figure 2 displays a hypothetical mean time-response curve that suggests the fitting of a piecewise model may be appropriate.

FIGURE 2

HYPOTHETICAL MEAN TIME-RESPONSE CURVES FOR PLACEBO AND ACTIVE TREATMENTS

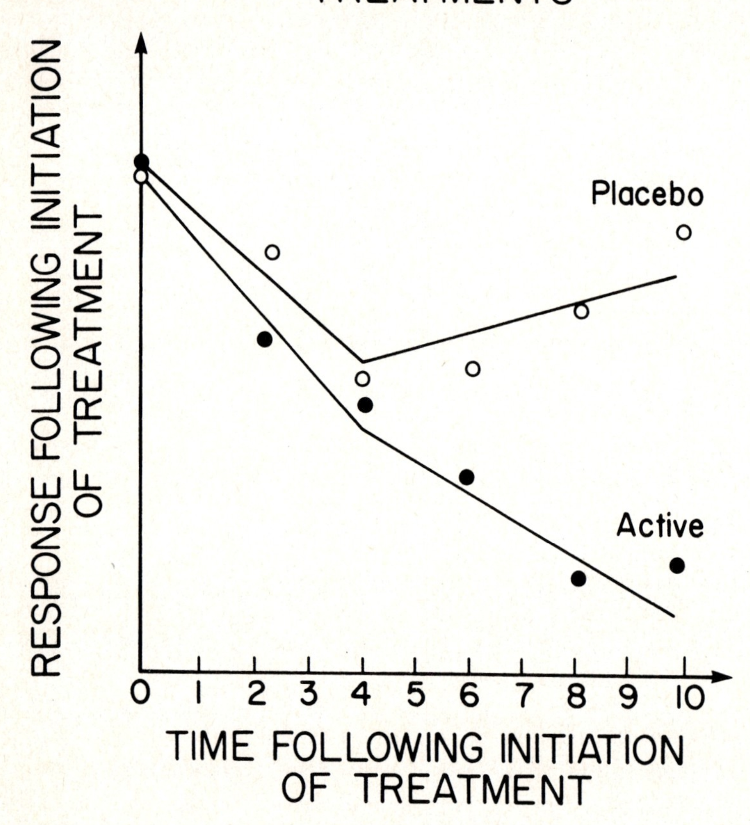

The intersection points of the submodels are referred to as join points and reflect the constraint that $f_i(T) = f_{i+1}(T)$, $t_i \le T \le t_{i+1}$, where T is the join point, and t_i and t_{i+1} are fixed times of observation. Spline function theory adds further constraints that all derivatives of the submodels agree at the join points, which are termed knot points (Smith, 1979). In addition, spline theory selects the knot points for analytical convenience, i.e., they may have no physical interpretation (Feder, 1975). Join points can be assumed to be either known or unknown. Teeter (1982) discusses the use of piecewise linear regression models to estimate the join point when measurement error is present in the independent variables. Situations where the interval containing the join point is assumed either to be known or unknown are considered. A two-segment model is of the form

$$f(t_i) = \begin{cases} \beta_{10} + \beta_{11}t_i + e_{1i} & t_i \le T \\ \beta_{20} + \beta_{21}t_i + e_{2i} & t_i \ge T \end{cases} \tag{14}$$

subject to the constraint $\beta_{10} + \beta_{11}T = \beta_{20} + \beta_{21}T$. Assuming the join point T is known, the unknown parameters β_{j0} and β_{j1} can be estimated with the use of indicator variables. Teeter (1982) applied this model to the cross-sectional ovulation study mentioned above.

In the case of repeated measures data where the observations at the different times of observation are not independent, a piecewise model with continuity constraints within the growth curve model suggests itself. Assuming the join points to be known, a full within-subject piecewise model can be constructed that fits the observed data perfectly. This is analogous to the situation in ordinary growth curve analysis that a $(p-1)^{th}$ order polynomial curve can be constructed to perfectly fit p observation points.

The within-subject polynomial growth curve model for subject ij can be written as

$$E(\underset{\sim}{y}_{ij}^*)' = \underset{\sim}{\beta}'_{ij}\, T^* \qquad q \le p \tag{15}$$

where $\underset{\sim}{y}_{ij}^*$ is the $p \times 1$ vector of responses for subject ij, T^* is a $q \times p$ matrix of powers of observation time values and $\underset{\sim}{\beta}_{ij}$ is a $q \times 1$ matrix of polynomial coefficients to be estimated. When q equals p, the model is a full model which fits the data perfectly. The same data can be fitted exactly through the use of piecewise models with specified join points. Assuming the join points are known and coincide with observed time points, the observations of the p-dimensional observation vector $\underset{\sim}{y}_{ij}^*$ can be grouped into subsets to correspond to the intervals defined by the join points. Within each interval containing $r \ge 2$ data points an $(r-1)^{th}$ order polynomial can be determined which fits the data perfectly. The collection of such submodels then fits exactly the data over $[t_1, t_p]$.

To provide clarity in the development of the piecewise growth curve model, the case of one join point will be considered. The specification of models with two or more join points follows directly. Consider the data presented in Figure 2. Assuming the join point occurs at time 4, a quadratic submodel for interval [0,4] and a cubic submodel for interval [4,10] combine to pass through all observed response values $y_1,\ldots,y_6$. The full submodels can be specified as

$$E(\underset{\sim}{y}_{ij}[t_1,t_3]') = \underset{\sim}{\beta}_{ij1}'T_1 \qquad (16)$$
$$E(\underset{\sim}{y}_{ij}[t_3,t_6]') = \underset{\sim}{\beta}_{ij2}'T_2$$

where T_1 and T_2 are square matrices of powers of observation time values and have order one less than the number of observations in their respective subintervals. The use of linear interpolation to estimate f(t) over $[t_1,t_p]$ can be viewed as a special case of piecewise regression. Here each subinterval is defined by adjacent observations and a first-order (linear interpolated) curve constitutes a full model for the subinterval. In contrast to other piecewise models, however, the linear interpolation method is motivated as an approximation technique with parameters that, in general, may have no direct interpretation. The overall piecewise model can be specified by

$$\begin{aligned} E(\underset{\sim}{y}_{ij}[t_1,t_6]') &= E(\underset{\sim}{y}_{ij}[t_1,t_3]',\underset{\sim}{y}_{ij}[t_3,t_6]') \\ &= [\underset{\sim}{\beta}_{ij1}' \ \underset{\sim}{\beta}_{ij2}'] \begin{bmatrix} T_1 & 0 \\ 0 & T_2 \end{bmatrix} = \underset{\sim}{\beta}_{ij}' \ T. \end{aligned} \qquad (17)$$

Here $\underset{\sim}{y}_{ij}[t_1,t_6]$ is a $(p+1)^{th}$ order vector which contains the response value associated with the join point twice. This reflects the constraint that the submodel endpoints meet at the join point. Using the six-data-point example displayed in Figure 2, this model can be written for the ij^{th} subject in transposed form as

$$\begin{bmatrix} y_{ij1} \\ y_{ij2} \\ y_{ij3} \\ y_{ij3} \\ y_{ij4} \\ y_{ij5} \\ y_{ij6} \end{bmatrix} = \begin{bmatrix} t_1^0 & t_1^1 & t_1^2 & 0 & 0 & 0 & 0 \\ t_2^0 & t_2^1 & t_2^2 & 0 & 0 & 0 & 0 \\ t_3^0 & t_3^1 & t_3^2 & 0 & 0 & 0 & 0 \\ 0 & 0 & 0 & t_3^0 & t_3^1 & t_3^2 & t_3^3 \\ 0 & 0 & 0 & t_4^0 & t_4^1 & t_4^2 & t_4^3 \\ 0 & 0 & 0 & t_5^0 & t_5^1 & t_5^2 & t_5^3 \\ 0 & 0 & 0 & t_6^0 & t_6^1 & t_6^2 & t_6^3 \end{bmatrix} \begin{bmatrix} \beta_0 \\ \beta_1 \\ \beta_2 \\ \beta_3 \\ \beta_4 \\ \beta_5 \\ \beta_6 \end{bmatrix} \begin{matrix} \text{Interval 1 Intercept} \\ \text{Linear} \\ \text{Quadratic} \\ \text{Interval 2 Intercept} \\ \text{Linear} \\ \text{Quadratic} \\ \text{Cubic} \end{matrix}$$

Summary Univariate Procedures

a. Area-Under-the-Curve Analysis

A potentially useful approach to the analysis of treatment main effects in response-over-time studies is area-under-the-curve (AUC) analysis. Response to treatment j might be plotted as a function of time as shown in Figure 3. For this hypothetical curve, one can note a rapid initial response followed by a more gradual reduction in response after a peak is achieved. The response curve $f_j(t)$, when integrated over $[t_1,t_p]$ provides a measure of AUC. That is,

$$AUC_j = \int_{t_1}^{t_p} f_j(t)\, dt. \tag{18}$$

AUC represents the shaded area under the response curve $f_j(t)$ of Figure 3. This parameter, measured in response units multiplied by time units, provides a univariate measure of cumulative response to treatment and can be viewed as a summary of response over the period of observation.

To illustrate further the nature of AUC, (18) can be restated as

$$AUC_j = \int_{t_1}^{t_p} f_j(t)dt = f_u(t_p-t_1) \text{ for some } f_u \tag{19}$$

$$\text{such that } f_{min} \le f_u \le f_{max}.$$

Here f_u represents an average response to treatment exhibited during the period of observation $[t_1,t_p]$. This representation of AUC is shown in Figure 4. Thus the AUC parameter can be viewed as the average response level during the observation period multiplied by the length of observation. In identical manner to one-way ANOVA, the total corrected AUC sum of squares can be partitioned into between- and within-subject sums of squares.

In practice, the form of the response function $f_j(t)$ is unknown. However, the AUC response for each study subject i from treatment j, AUC_{ij}, can be estimated based on the observed response vector y_{ij}. A convenient estimation method is to use linear interpolation between consecutive response coordinates and to estimate AUC as the sum of areas of trapezoids defined by the observed data points.

The trapezoidal-rule AUC estimate for response vector $y' = (y_1,y_2,\ldots,y_p)$ is

$$\widehat{AUC}_{TR} = \sum_{k=1}^{p} h_k(y_k+y_{k+1})/2 = \sum_{k=1}^{p} h_k\, \hat{f}_\xi(k) \tag{20}$$

where h_k is the interval length between y_k and y_{k+1} and $f_\xi(k)$ is an estimate of the average interval response on $[t_k,t_{k+1}]$. So the AUC estimate is a weighted sum of average interval responses, where the weights are interval lengths h_k.

FIGURE 3

HYPOTHETICAL AVERAGE RESPONSE TO TREATMENT j

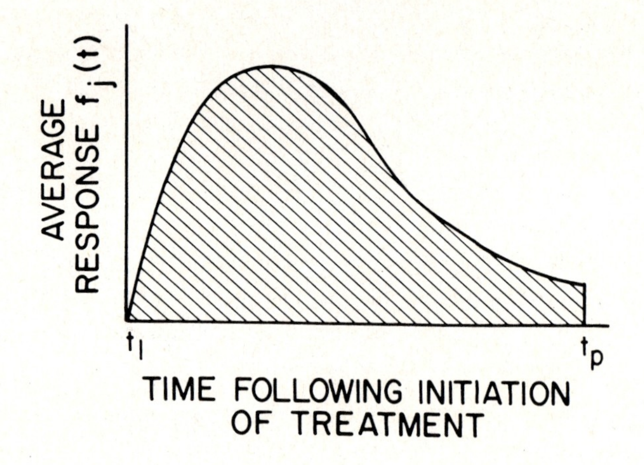

FIGURE 4

TIME-RESPONSE CURVE OF FIGURE 3 SHOWING AVERAGE RESPONSE f_μ

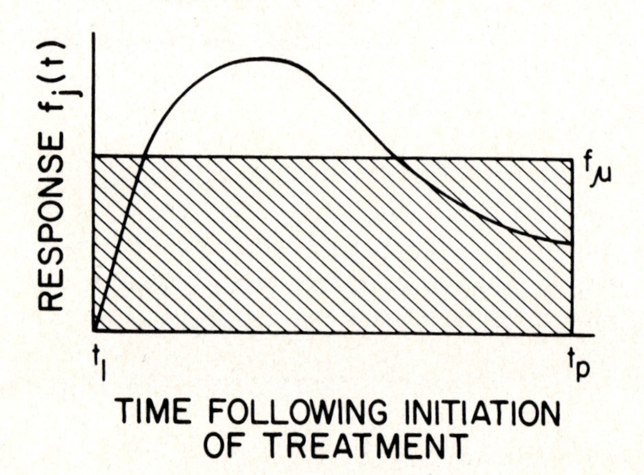

The AUC estimate of (20) also can be rewritten as

$$\hat{AUC}_{TR} = \sum_{k=1}^{p} \frac{(h_{k-1}+h_k)}{2} y_k \quad \text{where } h_0 = h_p = 0. \tag{21}$$

In this construction, $\hat{AUC}$ can be viewed as a weighted sum of the observed responses where the weights are the average interval length of the two intervals before and after the observation. Since no observations are made before t_1 or after t_p, h_o and h_p are defined to be zero. When the intervals are equally spaced, $\hat{AUC}_{TR}$ can be seen to be proportional to the sum of the observed responses where the first and last observations are given a reduced weight of one-half. Specific weighting schemes for experimental situations can also be developed so that the weighting scheme is tailored to reflect a particular objective of the investigator (Bryant, 1983).

The definition of the linear interpolation estimate of AUC in (21) suggests that the AUC analysis of between-subject effects is similar to the mixed model ANOVA main effects test. The ANOVA tables for AUC and mixed model ANOVA are presented in Table 1, which shows that the F-statistics of the two procedures differ in the linear combination of members of observation vector $\underset{\sim}{y}_{ij}$. The mixed model disregards the time interval lengths in $\underset{\sim}{t}' = [t_1,t_2,\ldots,t_p]$ associated with $\underset{\sim}{y}_{ij}$ while the AUC test incorporates the time points of measurement into the test of treatment effects.

b. Zerbe-Walker Test

Zerbe and Walker (1977) introduced a summary measure somewhat similar to the AUC parameter that can be used to assess treatment differences with respect to response curve shape. The Zerbe-Walker parameter d(x,y) is based on cumulating distances between response curves x(t) and y(t) over the interval $[t_1,t_p]$. The distance between two curves x(t) and y(t) over $[t_1,t_p]$ is defined as

$$d(x,y) = \{\int_{t_1}^{t_p} [x(t) - y(t)]^2 \, dt\}^{\frac{1}{2}} \quad . \tag{22}$$

Using (22), the total sum of squared distances from the grand mean curve can be partitioned into sums of squares between groups and within groups.

The similarity of the Zerbe-Walker statistic and the AUC statistic is apparent from review of equations (18) and (22). While the AUC ANOVA approach detects differences in area-under-the-time-response curve, the Zerbe-Walker statistic involves integration of squared differences between response curves.

3. ANALYSIS CONSIDERATIONS

While the general motivation for many repeated measures clinical trials is to determine whether the test treatment is more effective than some comparison treatment(s), the choice of analysis technique involves consideration of the study objectives, knowledge concerning the response phenomena under study, properties of the prospective statistical technique such as model assumptions and power, and characteristics of the observed data such as the presence of missing values.

Study Objectives

An obvious consideration in formulating an analysis plan for repeated measures data is the stated purpose of the study. For example, in pharmaceutical clinical trials safety data are collected at each visit to monitor drug toxicity. While efficacy data may also be collected at each patient visit, the primary interest with respect to treatment efficacy is often patient response at the end of the treatment period. In this case, particularly when missing data are prevalent, univariate analysis of the last scheduled patient visit may be the primary data analytic strategy rather than an analysis approach which considers the entire time vector of responses. However, other analyses to support the primary approach would be necessary.

Expected or desired treatment properties and aspects of disease processes under study should also be considered in the statement of experimental goals and in the subsequent selection of statistical analyses. For example, in the case of assessing analgesics for the relief of severe post-operative pain, rapid onset of drug effect appears desirable. A statistical analysis of experimental data for this situation would use methods that assess time and level of peak effect. In another experimental setting, it may be unimportant whether response is rapid or delayed; for example, a study to compare the efficacy of different anti-hypertensive agents to control high blood pressure is likely to be interested in the overall performance of test drug over a specified period of time.

The repeated measures analysis methods discussed in this paper address differently the general question of whether test treatment is more effective than some comparison treatment over period of observation $[t_1,t_p]$. The Bonferroni procedure null hypothesis is that there are no treatment differences at any time point versus the alternate hypothesis of a significant difference at one or more time points. The mixed model essentially treats time as a class variable and tests for treatment main effects by testing for differences with respect to the unweighted sum of the members of the observation vector y_{ij}. AUC analysis addresses the question of treatment differences over $[t_1,t_p]$ by testing for differences in cumulative (or equivalently, mean) response. Thus, for the anti-hypertensive drug study cited above, AUC analysis may be the method of choice. Multivariate polynomial growth curve analysis approximates response function f(t) over $[t_1, t_p]$ by a polynomial function p(t) and identifies significant differences in curve shape parameters. The Zerbe-Walker test is a univariate test of differences in response curve shape.

Knowledge of the Response Phenomena

The formulation of an analysis plan should consider the theory behind the response phenomena under study. In the case of an exponential process, nonlinear regression techniques can be used to solve for model parameters. For example, human growth from zero to six years has been modeled as an exponential process (Jens and Bayley, 1937). In the field of pharmacokinetics, the uptake and elimination of many types of drugs in the body has been seen to be fit by a double exponential model (Greenblatt and Koch-Weser, 1975). In many areas, little is known about the response phenomena under study, and theoretically-based statistical models cannot be constructed except on an exploratory basis. While reasonable models for the pharmacokinetics of certain drugs are sometimes available, time-response models for the clinical response to the same drugs do not exist. In this case empirical models, constructed to adequately describe the observed data and to address experimental goals, must be used.

Appropriateness of the Statistical Model

The methods of analysis that have been discussed are characterized by different distributional assumptions. The univariate ANOVA, Z-W test and AUC analysis techniques make univariate normality assumptions while the mixed model and MANOVA approaches assume error terms with multivariate normal distributions. The traditional mixed ANOVA model also assumes circularity for tests of within-subject factors.

Growth curve analysis further assumes that all subjects are measured at similar time points and that the response of all subjects are fit by polynomials of identical degree. These assumptions are not strictly necessary for AUC and Z-W analysis, although observations that are poorly spaced and are made at widely different subjects are likely to result in poor, possibly biased response parameter estimates.

An important consideration in the selection of an analysis technique is statistical power. For the test of within-subject factors, a review of the literature indicates that mixed model analysis is at least as powerful as MANOVA when circularity conditions hold. When the circularity assumption is not valid, results by

Rogan et al. (1979) discussed previously indicate that when ε is greater than .75, the ε-adjusted procedures provide the most powerful and robust tests of within-subject factors; as ε departs greatly from unity the multivariate tests are consistently more powerful. With respect to between-subject factors, such as treatment main effects, growth curve analysis generally provides the powerful test, in the sense of detecting any differences in response curve shape. However, an overly sensitive analysis may result when a high-order growth curve model is fitted to irregularly-shaped response curves. Note, though, that while statistical power is an important analysis consideration, selection of an analysis approach based strictly on power is inappropriate. As mentioned previously, the various methods of analysis may be directed at different questions. When the interest of analysis is in detecting significant differences among treatments in average or cumulative response over the interval of observation $[t_1, t_p]$, AUC analysis is a natural choice. In addition, AUC analysis (and Z-W analysis) can be utilized when interest is in examining for treatment differences on subintervals of $[t_1, t_p]$; in general, growth curve analysis is not directed at this interest, but piecewise growth curve analysis does explore subinterval differences.

Characteristics of the Observed Data

Characteristics of the observed study data can affect the choice of method of analysis. In many clinical trial situations, complete data are not obtained for each study subject. Some subjects may miss appointments or drop out of the study early, and omissions in data recording may occur. The existence of missing data can cause serious problems for the analysis and interpretation of repeated measures study results. Exclusion of subject data when there are missing observations in the subject's data vector, in addition to being subject to criticism for loss of information and possible bias, can result in a reduction in sample size and subsequent loss of power. No repeated measures analysis approach is immune to the problems caused by missing data.

When the amount of missing data is not too large, missing data estimation procedures based on least-squares or likelihood approaches (e.g., see Afifi and Elashoff, 1968; Dempster et al., 1976) can be used under the assumption that the cause of the missing data to be estimated is not associated with treatment. Other approaches akin to censoring that do not directly estimate the missing data are also available (Stanish et al., 1978). When dropouts occur and are possibly related to treatment, Gould's approach of assigning scores based on reason-for-withdrawal might be used. When "large" amounts of missing data exist, serious questions about the validity of study results arise. How large is "large" is difficult to define. As a rule of thumb, less than 10% missing data may be ig-

Table 1. ANOVA Table Comparing AUC (Linear Interpolation Estimate) and Mixed Model ANOVA for the Test of Treatment Main Effects

Source		AUC	Mixed Model
Treatment	SSTX	$\sum_j n_j (C\tilde{h})'(\tilde{y}_{.j}-\tilde{y}_{..})(\tilde{y}_{.j}-\tilde{y}_{..})'(C\tilde{h})$	$1/p \sum_j n_j \tilde{j}'(\tilde{y}_{.j}-\tilde{y}_{..})(\tilde{y}_{.j}-\tilde{y}_{..})'\tilde{j}$
	df	$g - 1$	$g - 1$
Error	SSE	$\sum_j \sum_i (C\tilde{h})'(\tilde{y}_{ij}-\tilde{y}_{.j})(\tilde{y}_{ij}-\tilde{y}_{.j})'(C\tilde{h})$	$1/p \sum_j \sum_i \tilde{j}'(\tilde{y}_{ij}-\tilde{y}_{.j})(\tilde{y}_{ij}-\tilde{y}_{.j})'\tilde{j}$
	df	$N - g$	$N - g$
Total	SST	$\sum_j \sum_i (C\tilde{h})'(\tilde{y}_{ij}-\tilde{y}_{..})(\tilde{y}_{ij}-\tilde{y}_{..})'(C\tilde{h})$	$\sum_j \sum_i \tilde{j}'(\tilde{y}_{ij}-\tilde{y}_{..})(\tilde{y}_{ij}-\tilde{y}_{..})'\tilde{j}$
	df	$N - 1$	$N - 1$
F-Test for Test of Treatment Main Effect		$\dfrac{(N-g)\ \sum_j n_j (C\tilde{h})'(\tilde{y}_{.j}-\tilde{y}_{..})(\tilde{y}_{.j}-\tilde{y}_{..})'(C\tilde{h})}{(g-1)\ \sum_j \sum_i (C\tilde{h})'(\tilde{y}_{ij}-\tilde{y}_{..})(\tilde{y}_{ij}-\tilde{y}_{..})'(C\tilde{h})}$	$\dfrac{(N-g)\ \sum_j n_j \tilde{j}'(\tilde{y}_{.j}-\tilde{y}_{..})(\tilde{y}_{.j}-\tilde{y}_{..})'\tilde{j}}{(g-1)\ \sum_j \sum_i \tilde{j}'(\tilde{y}_{ij}-\tilde{y}_{..})(\tilde{y}_{ij}-\tilde{y}_{..})'\tilde{j}}$

$$\underset{p\times(p-1)}{C} = \frac{1}{2}\begin{bmatrix} 1 & 0 & 0 & \dots & 0 & 0 & 0 \\ 1 & 1 & 0 & \dots & 0 & 0 & 0 \\ 0 & 1 & 1 & \dots & 0 & 0 & 0 \\ \cdot & \cdot & \cdot & \dots & \cdot & \cdot & \cdot \\ 0 & 0 & 0 & \dots & 0 & 1 & 1 \\ 0 & 0 & 0 & \dots & 0 & 0 & 1 \end{bmatrix} \quad \underset{(p-1)\times 1}{\tilde{h}} = \begin{bmatrix} h_1 \\ h_2 \\ \cdot \\ \cdot \\ \cdot \\ h_{p-1} \end{bmatrix} \quad \underset{p\times 1}{\tilde{y}_{ij}} = \begin{bmatrix} y_{ij1} \\ y_{ij2} \\ \cdot \\ \cdot \\ \cdot \\ y_{ijp} \end{bmatrix} \quad \underset{p\times 1}{\tilde{j}} = \begin{bmatrix} 1 \\ 1 \\ \cdot \\ \cdot \\ \cdot \\ 1 \end{bmatrix}$$

norable in most cases, but it is necessary to justify each situation on its own merits. The use of sensitivity analyses, where missing data are estimated under various schemes ranging from extreme conservatism with respect to the null hypothesis of no treatment differences (e.g., give missing data of subjects in test treatment group the least favorable observed values) to more liberal schemes (e.g., give missing values the average score of observed values of the appropriate treatment group), is another approach to analysis in the missing data situation. If analyses based on the various assignment schemes yield generally similar conclusions regarding treatment effects, the validity of study conclusions can be judged to be reasonable despite the presence of missing data.

Other data characteristics, such as the level of measurement, form of the distribution of responses, and sample size impact on the choice of analysis. For longitudinal data, the ordered dimension of the times of response measurement should be considered in the selection of statistical technique. Mixed model ANOVA and profile analysis MANOVA both ignore the metric associated with the vector $\underset{\sim}{t}' = (t_1,\ldots,t_p)$ which identifies the p times of observation. AUC analysis focuses the mixed model and growth curve analysis focuses the MANOVA model by considering the ordered nature of the members of $\underset{\sim}{t}$.

4. ANALYSIS OF AN EXAMPLE DATA SET

The illustrative analyses in this section utilize data from a randomized placebo-controlled parallel-group double-blind clinical trial to assess the efficacy of a test drug in the treatment of atypical depression (Davidson et al., 1981). The test drug is believed to stimulate mental activity in depressed patients by inhibiting the neural enzyme, monoamine oxidase (MAO). Outpatients were selected who had persistent depression and who failed to respond to other psychotropic drugs. Drug dosage was adjusted on an individual basis to achieve 90 percent MAO inhibition. A total of 29 patients were randomized to either placebo or test-drug treatment groups and followed for a six-week observation period. At baseline and Weeks 1, 2, 3, 4, and 6, patient depression levels were assessed by physician rater using a modification of the Hamilton Depression Scale (Davidson et al., 1982). The response parameter utilized in the following analyses is the Total Hamilton Score (HAMTOT) obtained by summing the scale items. Due to early dropouts, only 19 subjects completed the full six weeks of treatment.

The average total Hamilton Depression Scale (HAMTOT) scores for placebo and test-drug groups are presented in Table 2 for each time point. Results are displayed for all subjects observed and for subjects completing the six-week period. One can note that the six-week completer group tended to exhibit lower HAMTOT scores over the course of treatment than the all-subjects group. One can also note that while test drug symptomatology steadily decreased over time, the placebo group mean HAMTOT scores initially decreased, but then began to return to baseline levels. A graphical display of HAMTOT means is presented in Figure 5 for the six-week completers.

Table 2. Average HAMTOT Scores by Treatment Group All Subjects and Six-Week Completers

		Weeks Following Initiation of Treatment						
I. All Subjects		0	1	2	3	4	6	End-point
Placebo	n	14	14	14	14	12	8	14
	mean	32.86	21.07	16.71	16.21	16.92	15.00	19.21
Test Drug	n	15	15	15	15	13	11	15
	mean	31.20	19.40	11.60	8.80	5.69	4.55	6.73
	p-value	.5281	.6366	.1886	.0373	.0050	.0165	.0019
II. 6-Week Completers		0	1	2	3	4	6	
Placebo (n=8)		30.25	20.50	11.00	10.62	13.12	15.00	
Test Drug (n=11)		32.36	18.18	12.00	7.91	4.82	4.55	
	p-value	.2055	.1172	.6105	.1275	<.0001	.0165	

Using a Bonferroni multiple comparisons procedure approach, significance is reached when a p-value is less than $\alpha/5$.

Results from separate univariate analysis at each time point are shown in Table 2. Significant treatment differences are present for Weeks 3, 4, and 6 for all subjects and Weeks 4 and 6 for the completers. Using the Bonferroni procedure and an α-value of 0.05, the Week 4 difference is statistically significant for both samples, indicating that the null hypothesis of no treatment differences at any of the time points is rejected. Endpoint analysis of each patient's last response to treatment showed significantly lower symptomatology in the test drug group ($p = 0.002$).

Mixed model analysis of variance results are presented in Table 3 for the first three weeks of treatment and the entire six-week treatment period. Test results of within-subject factors are presented based on no correction factor, based on $\hat{\varepsilon}$, the maximum likelihood estimate of the degrees-of-freedom correction factor ε, and based on the Huynh and Feldt estimator $\tilde{\varepsilon}$. Review of Table 3 indicates a statistically significant treatment-by-time interaction for Weeks 0-6 regardless

FIGURE 5

STUDY 3 PLOT OF HAMTOT MEANS BY TREATMENT GROUP SUBJECTS WITH COMPLETE DATA

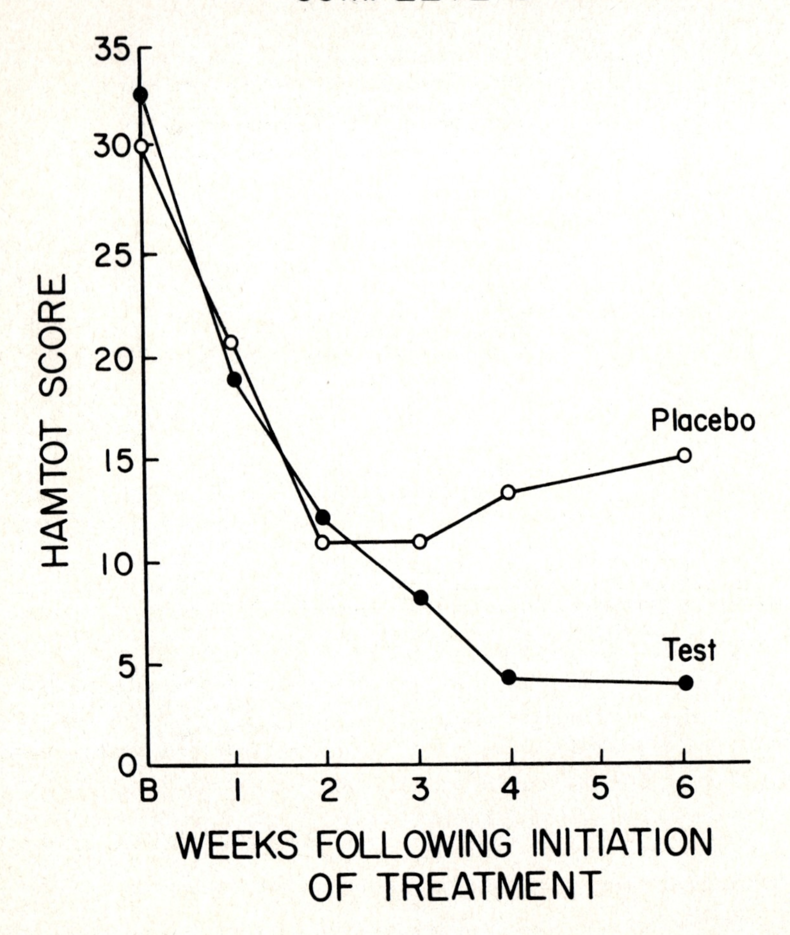

of which correction factor is considered. A test of treatment main effects for Weeks 0-6 is nonsignificant (p = 0.274).

Table 3. Mixed Model Analysis Results ε-Correction Factor Estimates and P-Values for Unadjusted and ε-Adjusted Tests

Estimates of ε	0 - 3 Weeks Only	Entire 0 - 6 Weeks
$\hat{\varepsilon}$	0.914	0.681
$\tilde{\varepsilon}$	1.000	0.923
P-values		
Treatment x time interaction:		
unadjusted	0.239	0.003
$\hat{\varepsilon}$-adjusted	0.241	0.010
$\tilde{\varepsilon}$-adjusted	0.239	0.004
Time:	<0.001*	<0.001*
Treatment:	0.147	0.274

*This level of significance found for unadjusted and ε-adjusted tests.

Though the sample size was small, for illustrative purposes, a growth curve analysis was performed for subjects with complete six-week data (n=19). Initial analyses showed that a polynomial model of order three fit the data adequately (goodness-of-fit p-value = 0.332). Estimates of natural polynomial coefficients for both treatment groups and analysis results are given in Table 4. Review of Table 4 shows significant treatment differences with respect to the linear component (p = 0.004) indicating that the linear component of response to treatment of the test-drug group was significantly larger than that of the placebo group over the six-week observation period.

AUC analysis results for these data, using linear interpolation estimation of response function f(t), are summarized in Table 5. Treatment differences become apparent for the intervals following the Week 2 observation time. Considering cumulative response to treatment over the entire six-week period, differences between the treatment groups in AUC were not found significant (p = 0.180). Included in Table 5 are results from the Zerbe-Walker test for differences in treatment group curve shape for selected subintervals of the treatment period. Estimation of the Z-W parameter was based on using linear interpolation estimation of response function f(t) within each subinterval. Review of Table 5 indicates that the Z-W results are similar to those obtained using AUC analysis.

Table 4. Estimates of Natural Polynomial Coefficients for Third-Order Model with P-Values for Tests of Treatment Differences

	Intercept	Linear	Quadratic	Cubic	
Paremeter Estimate					
Placebo	30.89	-16.13	4.10	-.308	
Test Drug	31.99	-15.64	3.08	-.207	
P-Value for H_0: parameter=0					
Placebo	<0.001	<0.001	<0.001	0.038	
Test Drug	<0.001	<0.001	<0.001	0.093	
Treatment Difference					Overall test of treatment difference
p-value	0.274	0.004	0.748	0.580	p = 0.066

Table 5. AUC and Zerbe-Walker Test Results: Two-Sided P-Values for Test of No Treatment Differences Using Subjects with Complete Data

	Time Period (Weeks)						
	0-1	1-2	2-3	3-4	4-6	0-3	0-6
AUC (Linear Interpolation)	0.537	0.291	0.067	0.013	0.016	0.186	0.180
Zerbe-Walker Test on Selected Intervals	0.558	0.297	0.074	---	---	0.190	---

In summary, these results suggest that the effect of the test-drug on HAMTOT scores began to manifest itself following the Week 2 visit. When considering the entire six-week observation period, no significant differences between treatment groups in cumulative response were noted (AUC p-value = 0.180). A third-order polynomial model was found to provide an adequate fit to the six-week data, and polynomial components up to the cubic term were found to be significantly different from zero for both treatment groups (the p-value of the cubic component for active-drug reached near-significance, p = 0.093). Over this period a significant difference between treatments with respect to the linear component of response was noted (p = 0.004).

The suggestion that drug differences develop beyond Week 2 motivates the use of a piecewise growth curve model. It is assumed that the data (see Figure 5) reflect one form of response for Weeks 0-2 (Time Period 1) and another form of response in the subsequent period of observation (Time Period 2). For example, it might be conjectured that the first two weeks following initiation of treatment represent primarily a period of placebo-effect response.

Based on this framework, a quadratic piecewise model fitted for Weeks 0-2 and a cubic piecewise model fitted for Weeks 2-6 can be constructed so as to fit the data perfectly. A review of response-over-time plots for each of the 19 subjects showed that the response of most subjects in a given treatment group was similar in shape to the average group curve displayed in Figure 5, thus providing some evidence that a segmented model is appropriate.

A reduced quadratic/quadratic model was found to provide an adequate fit to the data (goodness-of-fit p-value = 0.771). Results from analysis using this model are presented in Table 6. Statistically significant differences between treatment groups were found for the linear component during Time Period 2. Differences with respect to the quadratic component for Time Period 1 approached statistical significance ($p = 0.076$).

Several observations concerning a "placebo effect" are suggested from the piecewise model analysis results. From review of the tests of parameters for the placebo group in Time Period 1, the placebo effect for this population can be described as linear in nature. While the quadratic component for placebo is nonsignificant ($p = 0.937$), the linear component is highly significant ($p < 0.001$). Assuming a placebo effect is in operation for both groups during Time Period 1 and that it is linear in nature, the "true" effect of test drug during the first two weeks of treatment can be viewed as occurring through its quadratic component ($p = 0.008$).

Table 6. Piecewise (Two-Segment) Growth Curve Analysis Results. Natural Polynomial Coefficient Estimates for Quadratic/Quadratic Model with P-Values for Tests of Treatment Differences

	Polynomial Parameter Estimate					
	T1INT	T1LIN	T1QUAD	T2INT	T2LIN	T2QUAD
Placebo	30.25	-9.88	0.13	8.89	0.722	0.05
Test Drug	32.36	-18.18	4.00	26.02	-8.642	0.84
p-value for test of H_0: parameter=0						
Placebo	<0.001	<0.001	0.937	---	0.104	0.895
Test Drug	<0.001	<0.001	0.008	---	0.049	0.019
Treatment Difference p-value	0.953	0.989	0.076	---	0.016	0.131

TABLE 7
SUMMARY COMPARISON OF AUC AND OTHER METHODS OF ANALYSIS FOR REPEATED MEASURES DATA

ANALYSIS CONSIDERATION	UNIVARIATE TESTS WITH BONFERRONI MULTIPLE COMPARISON PROCEDURE	MIXED MODEL ANOVA	POLYNOMIAL GROWTH CURVE MODEL	AUC	ZERBE-WALKER TEST
A. Study Objectives	Tests for any differences in treatment group means over $[t_1, t_p]$. No test for treatment-by-time interaction or time main effect.	Tests for consistent differences between treatments over $[t_1, t_p]$.	Tests for any differences between treatment group curve shapes over $[t_1, t_p]$.	Tests for treatment differences w.r.t. mean or cumulative response on $[t_1, t_p]$. Can be used for comparisons on subintervals.	Tests for any differences in treatment group curve shapes on $[t_1, t_p]$. Can be used for comparisons on subintervals.
B. Knowledge of Response Phenomena Under Study	Form of response curve ignored.	Form of response curve ignored.	Form of response considered. When true form of response is not a polynomial, problems of interpretation. When form of response unknown, useful to detect differences in treatment-curve shapes.	Form of response curve ignored. Can be included within meaningful modeling framework. When form of response unknown, can still be estimated.	Form of response ignored. Can be included within meaningful modeling framework. When form of response unknown, can still be estimated. Has less intuitive meaning than AUC parameter.
C. Statistical Model Properties	Univariate normality assumptions. Low power when p is large or when consistent (moderate) differences between treatments.	Multivariate normality assumptions. Assumes circularity for within-subject tests. Overly liberal tests for within-subject tests when no circularity.	Multivariate normality assumptions. Low power for within-subject tests when n is small. Measurement times assumed the same for all subjects. No assumptions regarding $\underset{\sim}{\Sigma}$.	Univariate normality assumptions. Low power to detect treatment differences in response curve shapes. Measurement times need not be the same for all subjects.	Univariate normality assumptions. Power greater than AUC test when treatment x time interaction. Measurement times may not be the same for all subjects.
D. Characteristics of Observed Data	Ignores metric of time vector t. Missing data causes noncomparability of samples at different time points.	Ignores metric of time vector t. Missing data causes noncomparability of samples at different time points.	Considers metric of t. Missing data causes noncomparability of samples at different time points.	Considers metric of t. Missing data causes noncomparability of samples at different time points.	Considers metric of t. Missing data causes noncomparability of samples at different time points.
Comments	"Quick and dirty" method.	Appropriate for profile (as opposed to longitudinal) data. High computation costs if n and/or p is large.	Poor fit possible for irregularly shaped pharmaceutical clinical trial data as p gets large.	Using linear interpolation, Z-W test results similar to AUC test results. AUC parameter is easier to calculate.	

Commentary

A single statistical approach to the analysis of a repeated measures design is unlikely to be universally agreed on because no technique is uniformly superior on every issue. Table 7 summarizes the salient features of methods discussed according to study objectives, knowledge of response phenomena, statistical model properties, data characteristics, and other relevant comments. The depression clinical trial illustrates results of employing the methods discussed and would likely, as a bottom line, emphasize the easy to apply endpoint analysis. Its simplicity in summarizing the effects of a course of treatment is very attractive. Moreover, the presence of dropouts tends to raise questions about any approach that uses the entire time profile of data. However, on a more exploratory level the piecewise growth curve analysis may generate hypotheses regarding the response dynamics of depressed patients in therapy to be tested through further research. In addition, the AUC, mixed model, and Zerbe-Walker analyses provide useful and theoretically based overall summaries that support the endpoint analysis. In practice then, since experimental research interests are many faceted, a combination of techniques is likely to be the most appropriate course in the development of a statistical analysis plan for a repeated measures clinical study.

ACKNOWLEDGEMENTS: Support for this work was provided by Hoechst-Roussel Pharmaceuticals, Inc., and the Center for Epidemiologic Studies of the National Institute of Mental Health. Appreciation is expressed to Dr. Jonathan Davidson for making available the depression data set.

REFERENCES

[1] Afifi, A., and Elashoff, R.M., Missing values in multivariate statistics--I. Review of the literature, J. Amer. Statist. Assoc. **61**:595-604 (1966).

[2] Anderson, T.W., An Introduction to Multivariate Statistical Analysis, New York, Wiley (1958).

[3] Bartlett, M.S., Some examples of statistical methods of research in agriculture, J. Roy. Statist. Soc. Suppl. **4**:137-183 (1937).

[4] Box, G.E.P., A general distribution theory for a class of likelihood criteria, Biometrika **36**:317-346 (1949).

[5] Box, G.E.P., Some theorems on quadratic forms applied to the study of analysis of variance problems, Ann. Math. Statist. **25**:290-302, 484-498 (1954).

[6] Brown, M.B., A method for combining non-independent, one-sided tests of significance, Biometrics **31**:987-992 (1975).

[7] Bryant, E.C., Area-under-the-curve analysis and other analysis strategies for repeated measures clinical trials. Ph.D. dissertation, Department of Biostatistics, University of North Carolina at Chapel Hill (1983).

[8] Collier, R.O., Baker, F.B., Mandeville, G.K., and Hayes, T.F., Estimates of test size for several test procedures based on conventional variance ratios in the repeated measures design, Psychometrika **32**:339-353 (1967).

[9] Davidson, J., Weiss, J., Sullivan, J., Turnbull, C.D., and Linnoila, M., A placebo-controlled evaluation of isocarboxazid in outpatients, Monamine Oxidase Inhibitors: The State of the Art, Youdim and Paykel (eds.), New York, Wiley and Sons, pp. 115-123 (1981).

[10] Davidson, J. and Turnbull, C.D., Loss of appetite and weight associated with isocarboxazid in depression, J. Clin. Psychopharmacology **2**(4):263-266 (1982).

[11] Dempster, A.P., Rubin, D.B., and Hughes, N., Maximum likelihood from incomplete data vis the EM algorithm, Howard University Research Report 5-38, NS-320 (1976).

[12] Eisenhart, C., The assumptions underlying the analysis of variance, Biometrics **3**:1-21 (1947).

[13] Elston, R.C. and Grizzle, J.E., Estimation of time-response curves and their confidence bands, Biometrics **18**:148-159 (1962).

[14] Feder, P.I., On the asymptotic distribution theory in segmented regression problems--identified case, Ann. Statist. **3**(1):49-83 (1975).

[15] Geisser, S. and Greenhouse, S.W., An extension of Box's results on the use of the F distribution in multivariate analysis, Ann. Math. Statist. **29**:885-891 (1958).

[16] Gould, A.L., A new approach to the analysis of clinical drug trials with withdrawals, Biometrics **36**:721-727 (1980).

[17] Greenblatt, D.J. and Koch-Weser, J., Clinical pharmacokinetics, N. Engl. J. Med. **293**:702-704 (1975).

[18] Greenhouse, S.W. and Geisser, S., On methods in the analysis of profile data, Psychometrika **24**:95-112 (1959).

[19] Huynh, H. and Feldt, L.S., Conditions under which mean square ratios in repeated measurements designs have exact F-distributions, J. Amer. Statist. Assoc. **65**:1582-1589 (1970).

[20] Huynh, H. and Feldt, L.S., Estimation of the Box correction for degrees of freedom from sample data in randomized block and split-plot designs, J. Ed. Statist. **1**:69-82 (1976).

[21] Ito, K., On the effect of heteroscedasticity and nonnormality upon some multivariate test procedures. In Krishnaiah, P.R. (ed.). Multivariate Analysis Vol. 2, New York, Academic Press (1969).

[22] Jenns, R.M. and Bayley, N., A mathematical method for studying growth in children, Human Biology **9**:556-563 (1937).

[23] Keselman, H.J., Hogan, J.C., Mendoza, J.L., and Breen, L.J., Testing the validity conditions of repeated measures of F tests, Psych. Bull. **87**:479-481 (1980).

[24] Keselman, H.J. and Keselman, J.C., The analysis of repeated measures designs in medical research, Statist. in Med. 3:185-195 (1984).

[25] Khatri, C.G., A note on a MANOVA model applied to problems in growth curves, Ann. Inst. Statist. Math. 18:75-86 (1966).

[26] Koch, G.G., Amara, I.A., Stokes, M.E., and Gillings, D.B., Some views on parametric and non-parametric analysis for repeated measurements and selected bibliography, Int. Statist. Rev. 4:249-265 (1980).

[27] Mauchly, J.W., Significance test for sphericity of a normal n-variate distribution, Ann. Math. Statist. 29:204-209 (1940).

[28] McGregor, J.R., An approximate test for serial correlation in polynomial regression, Biometrika 47:111-119 (1960).

[29] Neter, J. and Wasserman, W., Applied Linear Statistical Models, Irwin, R.D. (ed.), Homewood, IL (1974).

[30] Potthoff, R.F. and Roy, S.N., A generalized multivatiate analysis of variance model useful especially for growth curve problems, Biometrika 51:313-326 (1964).

[31] Rao, C.R., Some problems involving linear hypotheses in multivariate analysis, Biometrika 46:49-58 (1959).

[32] Rao, C.R., The theory of least squares when the parameters are stochastic and its application to the analysis of growth curves, Biometrika 52:447-458 (1965).

[33] Rao, C.R., Covariance adjustment and related problems in multivariate analysis. In Multivariate Analysis, New York, Academic Press, pp. 87-103 (1966).

[34] Rao, C.R., Least squares theory using an estimated dispersion matrix and its application to measurement of signals, Proceedings of Fifth Berkeley Symposium on Mathematical Statistics and Probability, Vol. I, 355-372 (1967).

[35] Rogan, J.C., Keselman, H.J., and Mendoza, J.L., Analysis of repeated measurements, Brit. J. Math. Statist. Psych. 32:269-286 (1979).

[36] Rouanet, H. and Lepine, D., Comparison between treatments in a repeated-measures design: ANOVA and multivariate methods, Brit. J. Math. Statist. Psych. 23:147-163 (1970).

[37] Siddiqui, M.M., Covariance of least-squares estimates when residuals are correlated, Ann. Math. Statist. 29:1251-1256 (1958).

[38] Smith, P.L., Splines as a useful and convenient statistical tool, Amer. Statistician 33(2):57-62 (1979).

[39] Stanish, W.M., Gillings, D.B., and Koch, G.G., An application of multivariate ratio methods to a longitudinal clinical trial with missing data, Biometrics 34:305-317 (1978).

[40] Stoloff, P.H., Correcting for heterogeneity of covariance for repeated measures designs of the analysis of variance, Educ. and Psych. Measurements 30:909-924 (1970).

[41] Teeter, R.A., Effects of measurement error in piecewise regression models, Ph.D. dissertation, Department of Biostatistics, UNC at Chapel Hill (1982).

[42] Watson, G.S., Serial correlation in regression analysis, Biometrika **42**:327 (1955).

[43] Winer, B.J., Statistical Principles in Experimental Design, New York, McGraw-Hill, (1962), 2nd ed. (1971).

[44] Wishart, J., Growth rate determinations in nutrition studies with the bacon pig and their analysis, Biometrika **30**:16-28 (1938).

[45] Zerbe, G.O. and Walker, J.H., A randomization test for comparison of groups of growth curves with different polynomial design matrices, Biometrics **33**:653-657 (1977).

BIOSTATISTICS: Statistics in Biomedical, Public Health
and Environmental Sciences, edited by P.K. Sen

DESIGNS FOR HITTING THE BULL'S EYE OF A RESPONSE SURFACE

Shoutir Kishore Chatterjee and Nripes Kumar Mandal

Calcutta University

Given a concave (or convex) quadratic response surface, the problem considered is that of estimating its 'bull's eye', i.e., the optimal point in the factor space where the response attains its maximum (or minimum). Assuming that the optimal point is subject to a prior with known first and second moments, it is shown how the design of experimental points should be chosen so as to ensure maximum efficiency in estimating the optimal point. Solutions are obtained with reference to two measures of efficiency. The first is the extended D-optimality criterion which is equivalent to a certain ϕ-optimality criterion. The second (which requires that the shape paremeters also are subject to a prior) is a particular linear optimality criterion represented by the deficiency of the estimated optimal point.

KEY WORDS AND PHRASES: Response surface, quadratic regression, optimum design, extended D-optimality, minimum deficiency, ϕ-optimality, composite design

1. INTRODUCTION

The basic problem of response surface designs can be stated like this: In an experiment there are k quantitative factors whose levels $x_1,\ldots,x_k$ can be accurately controlled. For any choice of these levels there would be a response which can only be observed subject to a random error. The true response is functionally dependent on the levels; however, the functional relation between the true response and the levels (the true response function or response surface) is not completely known. The object is to obtain information about aspects of this relationship on the basis of responses observed at some chosen combination of levels.

Within the above general framework here we are concerned with the following specific problem: Let us assume that the true response function is quadratic and concave (convex) with a finite maximum (minimum) and the variance of the response variable Y is constant, i.e.,

$$\begin{aligned} E(Y|x_1,\ldots,x_k) &= \alpha + \sum_{i=1}^{k} \tau_i x_i - \sum_{i\leq j} \beta_{ij}\, x_i\, x_j \\ &= \alpha + \underline{x}'\underline{\tau} - \underline{x}'\, B\, \underline{x}\ , \qquad (1.1) \end{aligned}$$

$$V(Y|x_1,\ldots,x_k) = \sigma^2$$

where $\underline{x} = (x_1,\ldots,x_k)'$, $\underline{\tau} = (\tau_1,\ldots,\tau_k)'$, $B = (\frac{1}{2}(1+\delta_{ij})\beta_{ij})$, δ_{ij}'s being kronecker deltas. The matrix B is known to be positive definite (p.d.) or negative definite (n.d.) but otherwise the constants α, $\underline{\tau}$, B are unknown. We are concerned with the estimation of the point

$$\underline{\nu} = \tfrac{1}{2}\cdot B^{-1}\underline{\tau} \tag{1.2}$$

where the true response attains its maximum (if B is p.d.) or minimum (if B is n.d.) on the basis of responses observed at N experimental points $(\underline{x}_1,\ldots,\underline{x}_N) = X$. The point $\underline{\nu}$ can be considered as the 'bull's eye' for the response surface. Our main problem is that of choosing the design X so as to get maximum accuracy in hitting the bull's eye, i.e., in estimating $\underline{\nu}$. The problem is of great practical importance because the experimenter, in most cases, primarily aims at finding the best operating conditions for a process which maximizes some features like the yield or purity of the product or minimizes some feature like cost. A great deal of work on response surface designs in general and rotatable designs in particular have been done since the pioneering work of Box and Wilson (1951) and Box and Hunter (1957). However, it cannot be said that the specific problem stated above has received much attention.

The specific problem is generally more difficult than the standard problem of fitting the response surface since estimation of non-linear functions of the parameters in a linear model is involved. In this case any reasonable measure of accuracy of the estimate would, in general, depend on the unknown parameters themselves. The simplest but certainly not a very satisfactory way to avoid this difficulty is to proceed by assigning some prior value to the unknown parameters (local optimality criterion - Fedorov, 1972, p. 187). An alternative approach would be to invoke Bayes principle assuming prior distribution of the parameters themselves. Draper and Hunter (1966, 1967) while considering such problems in a general setup advocated the maximization of the maximal posterior density for selecting the design. But apart from requiring strong assumptions, such an approach seldom gives the solution in an explicit form in particular situations. Here we review some solutions which can be obtained by assuming much less. Specifically, we would assume that the unknown parameters are subject to a prior distribution with some of its moments known. No assumption about the form of the prior would be made.

2. FORMULATION OF THE PROBLEM

For the sake of concreteness we hereafter assume that in (1) B is p.d. so that

$\underline{\nu}$ corresponds to the peak of the response surface. We can then write (1.1) in the form

$$E(Y|\underline{x}) = \alpha_0 - (\underline{x}-\underline{\nu})'\,B(\underline{x}-\underline{\nu}) \tag{2.1}$$

where $\alpha_0 = \alpha + \underline{\nu}'\,B\underline{\nu}$ is the maximal response and ν is given by (1.2).

We assume that ν, B are aubject to a prior under which

$$\mathcal{E}(\underline{\nu}|B) = \underline{c},\ \mathcal{D}(\underline{\nu}|B) = V \tag{2.2}$$

and

$$\mathcal{E}(B^{-1}) = W\ . \tag{2.3}$$

(Note that $\mathcal{E}$ and $\mathcal{D}$ denote mean value and dispersion taken with respect to the prior.) The parameters $\underline{c}$, V and W are supposed to be known. (However, the assumption about the prior distribution of B and relation (2.3) can be dispensed with if we are interested only in the extended D-optimality criterion discussed below.) In any design problem the design points must belong to some bounded experimental region in R^k. Here we assume that the experimental region is given by

$$(\underline{x} - \underline{c})'\,V^{-1}\,(\underline{x} - c) \le \ell^2 \tag{2.4}$$

where $\underline{c}$ and V are as in (2.2) and ℓ^2 is a known number. This is reasonable since, in practice, $\underline{c}$ being known we would naturally want our region to be centered at $\underline{c}$. ℓ^2 can be adjusted to make the ellipsoid (2.4) approximate the actual experimental region closely from inside, whatever its form.

To compare designs we use two optimality criteria which seem to be relevant to our problem. Let $\hat{\underline{\nu}}$ be an estimate of $\underline{\nu}$. The unconditional mean square and product matrix of $\hat{\underline{\nu}} - \underline{\nu}$ is given by $\mathcal{E}\{E[(\hat{\underline{\nu}} - \underline{\nu})(\hat{\underline{\nu}} - \underline{\nu})'|\underline{\nu}, B]\}$ and invocation of <u>D-optimality criterion in an extended sense</u> would then amount to the minimization of the determinant

$$|\mathcal{E}\{E[(\hat{\underline{\nu}}-\underline{\nu})'\,(\hat{\underline{\nu}}-\underline{\nu})'|\nu,\ B]\}| \tag{2.5}$$

by choosing the design X. To develop a second optimality criterion, consider the shortfall in true response at the estimated point from the maximum response α_0:

$$(\hat{\underline{\nu}}-\underline{\nu})'B(\hat{\underline{\nu}}-\underline{\nu}) = \mathrm{tr}\ B(\hat{\underline{\nu}}-\underline{\nu})(\hat{\underline{\nu}}-\underline{\nu})'\ .$$

We shall take N/σ^2 times the unconditional mean shortfall

$$(N/\sigma^2)\mathrm{tr}\ \mathcal{E}\{B\ E[(\hat{\underline{\nu}}-\underline{\nu})(\hat{\underline{\nu}}-\underline{\nu})'|\underline{\nu},\ B]\} \tag{2.6}$$

to measure the deficiency of the design. Our problem is to choose the design X in (2.4) so that (2.5) or (2.6) is minimized. A design which minimizes (2.5) will be called an extended D-optimum design and one which minimizes (2.6) will be called a minimum deficiency design.

We first note that by a nonsingular transformation of the factor space, we can make $\underline{c} = 0$, $V = \ell^{-2} \cdot I_k$, $W = \Lambda = \text{diag}(\lambda_1, \ldots, \lambda_k)$, $\lambda_i > 0$, $i=1,\ldots,k$. For this we have to replace $\underline{x}$, $\underline{\nu}$ and B respectively by

$$\underline{x}^* = \ell^{-1} L(\underline{x}-\underline{c}), \; \underline{\nu}^* = \ell^{-1} L(\underline{\nu}-\underline{c}), \; B^* = \ell^2 L'^{-1} B L^{-1} \tag{2.7}$$

where L is a nonsingular matrix satisfying

$$LVL' = I_k, \; \ell^{-2} LWL' = \Lambda = \text{diag}(\lambda_1, \ldots, \lambda_k) \; . \tag{2.8}$$

For extended D-optimality criterion we would require L to satisfy only the first condition of (2.8). Thus, dropping the asterisks our problem reduces to that of estimating $\underline{\nu}$ in

$$E(Y|\underline{x}) = \alpha_o - (\underline{x}-\underline{\nu})'B(\underline{x}-\underline{\nu}) \tag{2.9}$$

$$= \alpha_o - \underline{x}' B \underline{x} + \underline{\tau}'\underline{x} \tag{2.10}$$

by choosing N design points $\underline{x}_1, \ldots, \underline{x}_N$ in the unit ball

$$\underline{x}' \underline{x} \le 1 \tag{2.11}$$

given that for the extended D-optimality criterion $\underline{X}$ has an a priori distribution with

$$E(\nu|B) = 0, \; \mathcal{D}(\nu|B) = \ell^{-2} \cdot I_k \tag{2.12}$$

and for the minimum deficiency criterion, in addition, B has an a priori distribution with

$$E(B^{-1}) = \Lambda \; . \tag{2.13}$$

Our problem is to minimize (2.5) or (2.6) depending on the criterion chosen by the choice of the design $(\underline{x}_1, \ldots, \underline{x}_N) = X$ in (2.11).

Using the design X we can estimate the parameters α, $\underline{\tau}$, B in (2.10) by least squares method and then estimate $\underline{\nu}$ from $\underline{\nu} = \frac{1}{2} B^{-1} \underline{\tau}$ or

$$\underline{\tau} = 2 \, B \, \underline{\nu} \quad . \tag{2.14}$$

Let us write $\underline{\beta} = (\beta_{11}, \ldots, \beta_{kk}, \beta_{12}, \ldots, \beta_{k-1,k})'$ and use the following notations

$$(1/N)\sum_u x_{iu} = [i], (1/N)\sum_u y_u = [y], (1/N)\sum_u x_{iu}y_u = [iy]$$

$$(1/N)\sum_u x_{iu}x_{ju} = [ij], (1/N)\sum_u x_{iu}x_{ju}y_{ju} = [ijy]$$

$$(1/N)\sum_u x_{iu}x_{ju}x_{ru} = [ijr], (1/N)\sum_u x_{iu}x_{ju}x_{ru}x_{su} = [ijrs]$$

$$1 \le i \le j \le r \le s \le k$$

$$M_{11.11} = \begin{bmatrix} [1111] - [11]^2 & [1122] - [11]\,[22] & \ldots & [11kk] - [11]\,[kk] \\ & [2222] - [22]^2 & \ldots & [22kk] - [22]\,[kk] \\ & & \ldots & \\ & & & [kkkk] - [kk]^2 \end{bmatrix}$$

$$M_{11.22} = \begin{bmatrix} [1122] - [12]^2 & [1123] - [12]\,[13] & \ldots & [12k-1,k] - [12]\,[k-1,k] \\ & [1133] - [13]^2 & \ldots & [13,k-1,k] - [13]\,(k-1,k] \\ & & \ldots & \\ & & & [k-1,\ k-1,\ kk] - [k-1,k]^2 \end{bmatrix}$$

$$M_{11.12} = M'_{11.21} = \begin{bmatrix} [1112] - [11]\,[12], & [1113] - [11]\,[13] & \ldots & [11,k-1,k] - [11]\,[k-1,k] \\ & [1223] - [22]\,[13] & \ldots & [22,k-1,k] - [22]\,[k-1,k] \\ & & \ldots & \\ & & & [k-1,kkk] - [kk]\,[k-1,k] \end{bmatrix}$$

$$M_{11} = \begin{bmatrix} M_{11.11} & M_{11.12} \\ M_{11.21} & M_{11.22} \end{bmatrix}$$

$$M_{22} = \begin{bmatrix} [11] - [1]^2, & [12] - [1]\,[2] & \ldots & [1k] - [1]\,[k] \\ & [22] - [2]^2 & \ldots & [2k] - [2]\,[k] \\ & & \ldots & \\ & & & [kk] - [k]^2 \end{bmatrix}$$

$$M_{12} = M'_{21} = \begin{bmatrix} [111] - [11]\,[1] & \ldots & [11k] - [11[k] \\ & \ldots & \\ [1kk] - [kk]\,[1] & \ldots & [kkk] - [kk]\,[k] \\ [112] - [12\,[1] & \ldots & [12k] - [12]\,[k] \\ & \ldots & \\ [1,k-1,k] - [k-1,k]\,[1] & \ldots & [k-1,kk] - [k-1,k]\,[k] \end{bmatrix}$$

$$M = \begin{bmatrix} M_{11} & M_{12} \\ M_{21} & M_{22} \end{bmatrix}$$

$$M_{1y} = ([11y] - [11]\,[y] \quad \ldots \quad [kky] - [kk]\,[y], [12y] - [12]\,[y], \ldots\ldots$$
$$\ldots [k-1,ky] - [k-1,k]\,[y])'$$

$$M_{2y} = ([1y] - [1]\,[y]\,, \quad \ldots, [ky] - [k]\,[y])'$$

$$M_y = (M'_{1y}, M'_{2y})'. \tag{2.15}$$

We shall assume X is such that M is p.d. Then the least squares estimates of $\underline{\beta}$ and $\underline{\tau}$ will be given by

$$\begin{bmatrix} \hat{\underline{\beta}} \\ \hat{\underline{\tau}} \end{bmatrix} = M^{-1}\underline{M}_y$$

with

$$\text{Disp} \begin{bmatrix} \hat{\underline{\beta}} \\ \hat{\underline{\tau}} \end{bmatrix} = (\sigma^2/N)\cdot M^{-1}.$$

We write $\hat{\beta}_{ij}$ $(i \leq j)$ for the elements of $\hat{\underline{\beta}}$ and $\hat{B} = (\tfrac{1}{2}(1 + \delta_{ij})\hat{\beta}_{ij})$ and remembering (2.14) take our estimate of ν as

$$\hat{\nu} = \tfrac{1}{2}\cdot\hat{B}^{-1}\hat{\underline{\tau}}$$

From (2.5) - (2.6) we see that for both the criteria we require to find $E\{(\hat{\nu}-\nu)\,(\hat{\nu}-\nu)'|\nu\cdot B\}$. Since $\underline{\nu}$ is nonlinear in B and $\underline{\tau}$, we adopt the standard δ-method and large sample approximation for this. Let

$$\Gamma_1 = \text{diag}\,(\nu_1,\ldots,\nu_k)$$

$$\Gamma_2 = \begin{bmatrix} \nu_2 & \nu_3 & \ldots & \nu_k & 0 & \ldots & 0 & \ldots & 0 \\ \nu_1 & 0 & \ldots & 0 & \nu_3 & \ldots & \nu_k & \ldots & 0 \\ 0 & \nu_1 & \ldots & 0 & \nu_2 & \ldots & 0 & \ldots & 0 \\ \cdot & \cdot & \ldots & \cdot & \cdot & \cdot\cdot & \cdot & \cdot\cdot & \cdot \\ \cdot & \cdot & \ldots & \cdot & \cdot & \cdot\cdot & \cdot & \cdot\cdot & \cdot \\ \cdot & \cdot & \ldots & \cdot & \cdot & \cdot\cdot & \cdot & \cdot\cdot & \cdot \\ 0 & 0 & \ldots & 0 & 0 & \ldots & 0 & \ldots & \nu_k \\ 0 & 0 & \ldots & \nu_1 & 0 & \ldots & \nu_2 & \ldots & \nu_{k-1} \end{bmatrix} \tag{2.16}$$

Then it can be shown that (see Chatterjee and Mandal (1981))

$$E\{(\hat{\underline{\nu}}-\underline{\nu})(\hat{\underline{\nu}}-\underline{\nu})'|\underline{\nu}, B\}$$
$$= (\sigma^2/N)B^{-1}(-\Gamma_1,-\tfrac{1}{2}\cdot\Gamma_2,\tfrac{1}{2}\ I_k)M^{-1}(-\Gamma_1,-\tfrac{1}{2}\Gamma_2,\tfrac{1}{2}\ I_k)'B^{-1}. \qquad (2.17)$$

To obtain an extended D-optimum design we have to minimize (2.5) which by (2.17) is equivalent to minimizing

$$|(\Gamma_1,\tfrac{1}{2}\Gamma_2,-\tfrac{1}{2}\ I_k)M^{-1}(\Gamma_1,\tfrac{1}{2}\Gamma_2,-\tfrac{1}{2}\ I_k)'| = D(X)\ (\text{say}) \qquad (2.18)$$

At this stage we mention that the above minimization problem does not involve B so that in the extended D-optimality approach we do not require any prior assumption about B.

In conformity with (2.15) we partition

$$M^{-1} = \begin{bmatrix} G_{11} & G_{12} \\ G_{21} & G_{22} \end{bmatrix}$$

where

$$G_{12} = G_{21}' = (g_{ijr}) \quad \text{for all } (i,j),\ i\le j=1,2,\ldots,k \ \text{ and } r=1,\ldots,k$$
$$G_{22} = (g_{ij}),\ i,j = 1,2,\ldots,k$$
$$G_{11} = \begin{bmatrix} G_{11.11} & G_{11.12} \\ G_{11.21} & G_{11.22} \end{bmatrix}$$
$$G_{11.11} = (g_{iijj}),\ i,j, = 1,\ldots,k$$
$$G_{11.22} = (g_{ijrs}) \text{ where } (i,j),(r,s) = (1,2),\ldots,(k-1,k)$$
$$G_{11.12} = G_{11.21}' = (g_{iijr}) \text{ for all } i=1,\ldots,k \text{ and } (j,r)=(1,2),\ldots,(k-1,k). \qquad (2.19)$$

Then by (2.12) and (2.16) by straightforward computation it can be shown (Mandal (1982)) that

$$D(X) = |(s_{ij})| \qquad (2.20)$$

where

$$s_{11} = \ell^{-2}\ g_{1111}+(1/4)(g_{1212}+\ldots+g_{1k1k})\} + (1/4)g_{11}$$
$$\vdots$$
$$s_{kk} = \ell^{-2}\{g_{kkkk}+(1/4)(g_{1k1k}+\ldots+g_{k-k,k,k-1,k})\} + (1/4)g_{kk}$$

$$s_{12} = \ell^{-2}\{\tfrac{1}{2}(g_{1112}+g_{2212})+(1/4)(g_{1323}+\ldots+g_{1k2k})\}$$

$$s_{k-1,k} = \ell^{-2}\{\tfrac{1}{2}(g_{k-1,k-1,k-1,k}+g_{k-1,kkk})+(1/4)(g_{1k-1,1k}+\ldots+g_{k-2,k-1,k-2,k})\} \quad (2.21)$$

For the deficiency approach we require to minimize

$$E\ \mathrm{tr}[(\Gamma_1,\tfrac{1}{2}\Gamma_2,-\tfrac{1}{2}\ I_k)M^{-1}(\Gamma_1,\tfrac{1}{2}\Gamma_2,-\tfrac{1}{2}\ I_k)'B^{-1}] \quad (2.22)$$

which by using (2.12)-(2.13) reduces to (cf. Chatterjee and Mandal (1981))

$$\ell^{-2}\ \mathrm{tr}\ G_{11.11}\Lambda + (1/4)\ell^{-2}\ \mathrm{tr}\ G_{11.22}\Lambda^* + (1/4)\mathrm{tr}\ G_{22}\Lambda = \mathrm{Def}(X)\ \text{(say)}. \quad (2.23)$$

To obtain a minimum deficiency design we have to minimize Def(X) given by (2.23) by choosing the design X.

The design X can be interpreted as a discrete distribution over (2.11) which gives mass 1/N to each of points $\underline{x}_1,\ldots,\underline{x}_N$. We now bring in the concept of continuous design and regard any distribution P whose total mass is confined to (2.11) as a design (Keifer (1958, 1959), Fedorov (1972)). Let $\xi = (\xi_1,\ldots,\xi_k)'$ be a random vector following the distribution P. Then the moments appearing in (2.15) are to be interpreted as

$$[i] = E_P\xi_i,\ [i_j] = E_P\xi_i\xi_j,\ [ijr] = E_P\xi_i\xi_j\xi_r,\ [ijrs] = E_P\xi_i\xi_j\xi_r\xi_s$$

$$1\le i\le j\le r\le s\le k$$

and we write (2.20) as D(P) and (2.23) as Def(P). Writing $\mathcal{P}$ for the class of all designs P in (2.11), our problem is to select a P in $\mathcal{P}$ which minimizes D(P) or Def(P) depending on the criterion chosen. We consider this in the next two sections. After an optimal P is obtained, in practice, we have to approximate it as closely as possible using the permitted number N of experimental points. To minimize the algebra we give the details of the solution in the case of two factors only. Extension to the case of more than two factors is indicated later.

3. EXTENDED D-OPTIMALITY CRITERION

When k=2, D(P) given by (2.20) reduces to

$$D(P) = |R_{11} + R_{22} + (1/4)\ \mathrm{diag}\ (g_{11},\ g_{22})| \quad (3.1)$$

where

$$R_{11} = l^{-2} \begin{bmatrix} g_{1111} & \frac{1}{2}\cdot g_{1112} \\ \frac{1}{2}\, g_{1112} & \frac{1}{4}\, g_{1212} \end{bmatrix}, \quad R_{22} = l^{-2} \begin{bmatrix} (1/4) g_{1212} & \frac{1}{2}\cdot g_{1222} \\ \frac{1}{2}\cdot g_{1222} & g_{2222} \end{bmatrix}$$

The following theorem represents the first step towards choosing a $P \varepsilon \mathcal{P}$ which minimizes (3.1).

<u>Theorem 3.1</u>

If P_1 and P_2 are two designs such that corresponding second and fourth order moments are same for both and if for P_1

$$[1] = [2] = [111] = [222] = [112] = [122] = 0 \tag{3.2}$$

then $D(P_1) \leq D(P_2)$.

<u>Proof</u>

Let $M^{(t)}$, $G^{(t)}$ etc. be defined as before for the design P_t, t=1,2. It is clear that

$$G_{11}^{(1)} \leq G_{11}^{(2)}, \; G_{22}^{(1)} \leq G_{22}^{(2)} \tag{3.3}$$

and equality holds iff (3.2) holds. Since

$$\begin{bmatrix} g_{1111}^{(t)} & g_{1112}^{(t)} \\ g_{1112}^{(t)} & g_{1212}^{(t)} \end{bmatrix}$$

is a principal submatrix of $G_{11}^{(t)}$ and

$$R_{11}^{(t)} = l^{-2}\, \mathrm{diag}(1,\tfrac{1}{2}) \begin{bmatrix} g_{1111}^{(t)} & g_{1112}^{(t)} \\ g_{1112}^{(t)} & g_{1212}^{(t)} \end{bmatrix} \mathrm{diag}(1,\tfrac{1}{2})$$

We see that (3.3) implies $R_{11}^{(1)} \leq R_{11}^{(2)}$. Similarly $R_{22}^{(1)} \leq R_{22}^{(2)}$. Also by (2.19), the second relation is (3.3) implies

$$\mathrm{diag}\,(g_{11}^{(1)}, g_{22}^{(1)}) \leq \mathrm{diag}\,(g_{11}^{(2)}, g_{22}^{(2)}) .$$

Using these results in (3.1) we get the conclusion of the theorem.

Q.E.D.

We denote by $\mathcal{P}_1$ the class of all designs P for which (3.2) holds. It is easy to see that given any $P \varepsilon \mathcal{P}$ we can find a $\tilde{P} \varepsilon \mathcal{P}_1$ such that P and $\tilde{P}$ have the same second and fourth order moments. Thus, in searching for an extended D-optimum design we may confine ourselves to the class $\mathcal{P}_1$. We now introduce a subclass $\mathcal{P}_2$ of $\mathcal{P}_1$ for which over and above (3.2), any mixed second and fourth order moments in which one of the variables has an odd power is zero, i.e.,

$$[1112] = [1222] = [12] = 0 \tag{3.4}$$

holds. This means for the designs in $\mathcal{P}_2$ $M_{11.12} = 0$ and M_{22} is diagonal. These designs are termed as <u>semirotatable</u>. (Recall that for a design to be rotatable we require some further conditions on the moments.) For finding an extended D-optimum design we confine ourselves heuristically to the subclass $\mathcal{P}_2$ and try to obtain a $P \varepsilon \mathcal{P}_2$ such that D(P) is a minimum. For any $P \varepsilon \mathcal{P}_2$, D(P) given by (3.1) reduces to

$$D(P) = \prod_{i=1}^{2} \{1^{-2}(g_{iiii}+1/4\cdot g_{1212}) + 1/4\cdot g_{ii}\} \tag{3.5}$$

where $g_{1212} = 1/[1122]$, $g_{ii} = 1/[ii]$, i = 1,2.

From (2.19) and (3.5) we observe that D(P) depends on P only through the distribution Q of $\underline{\eta} = (\eta_1, \eta_2)'$, $\eta_i = \xi_i^2$, i=1,2 where $\xi = (\xi_1,\xi_2)'$ follows P. Also knowing the distribution Q of $\underline{\eta}$ we can get a corresponding $P \varepsilon \mathcal{P}_2$ by taking $\xi_i = S_i\sqrt{\eta_i}$ where S_1 and S_2 are distributed independently of $\underline{\eta}$ such that prob $(S_i = \pm 1) = \frac{1}{2}$. Hence we can write D(Q) for D(P) and state our problem as that of minimization of D(Q) by choice of Q. Further, for any Q, D(Q) involves only

$$[ii] = E_Q\eta_i = \mu_i \text{ (say)}, \; i=1,2, \; [1122] = E_Q\eta_1\eta_2 = z_{12} \text{ (say)} \tag{3.6}$$

and in view of (2.11) the total probability of Q is confined to the simplex

$$Y_i \geq 0, \; i = 1,2, \; Y_1 + Y_2 \leq 1. \tag{3.7}$$

Let us denote by $\mathcal{Q}$ the class of all probability distributions Q over (3.7) for which $\mu_i > 0$, i=1,2, $z_{12} > 0$ and $M_{11.11}$ is p.d. For fixed μ_1,μ_2,z_{12}, we observe that $z_{ii} \leq \mu_i - z_{12}$, i=1,2, equality holding iff the distribution Q of $\underline{\eta}$ satisfies

$$\sum_{1}^{2} \eta_i = 0 \text{ or } 1 \text{ with probability } 1.$$

i.e., the total mass of Q is concentrated at the origin and the flat $Y_1+Y_2 = 1$ of the simplex (3.7). We call such a distribution a binary simplex or a <u>bisimplex</u> distribution. Thus, we obtain the following theorem.

Theorem 3.2

Among all distributions Q over (3.7) for which μ_1, μ_2, z_{12} are held fixed, a bi-simplex distribution (if it exists) will minimize D(Q).

Let $\underline{\eta} = (\eta_1, \eta_2)'$ follow the distribution Q. Let $\delta = (\delta_1, \delta_2)'$ be distributed such that

i) given $\underline{\eta} = \underline{0}$, we have $\delta = 0$ with probability 1 and

ii) given any fixed $\underline{\eta} \neq 0$, δ assumes the value 0, (1,0)', (0,1)', $(\sum_1^2 \eta_i)^{-1} \underline{\eta}$ with conditional probabilities $1 - \sum_1^2 \eta_i$, $\eta_1(1 - \sum_1^2 \eta_i)$, $\eta_2(1 - \sum_1^2 \eta_i)$, $(\sum_1^2 \eta_i)^2$ respectively. Let the unconditional distribution of $\underset{\sim}{\delta}$ be Q_0. It is easy to check that Q_0 is bisimplex with same μ_1, μ_2 and z_{12} as Q. Let $\mathcal{Q}_0 (\subset \mathcal{Q})$ be the sub-class of bisimplex distributions over (3.7). The upshot of the above is that we can confine our attention to $\mathcal{Q}_0$ to find an optimum design in $\mathcal{Q}$.

For a bisimplex distribution Q_0, let us write

$$M_{11.11} = A = (a_{ij}) = (a^{ij})^{-1} \tag{3.8}$$

Then writing $\mu = \mu_1 + \mu_2$, it can be seen that

$$a^{11} = 1/(\mu(1-\mu) + ((\mu_2^2/\mu^2)/((\mu_1\mu_2)/\mu)-z_{12})$$

$$a^{22} = 1/(\mu(1-\mu)) + ((\mu_1^2/\mu^2)/(\mu_1\mu_2/\mu)-z_{12})). \tag{3.9}$$

By (3.8) we can write (3.5) as

$$D(Q) = \prod_{i=1}^{2} [\ell^{-2}\{a^{ii} + 1/4z_{12}\} + 1/4\mu_i] \tag{3.10}$$

where a^{ii}, i=1,2 are given by (3.9). Now by Schwarz inequality

$$((\mu_2^2/\mu^2)/(\mu_1\mu_2/\mu-z_{12})) + 1/4z_{12} \geq (\mu_2/\mu + \tfrac{1}{2})^2/(\mu_1\mu_2/\mu),$$

$$((\mu_1^2/\mu^2)/((\mu_1\mu_2/\mu)-z_{12}) + 1/4z_{12} \geq (\mu_1/\mu+\tfrac{1}{2})^2/(\mu_1\mu_2/\mu). \tag{3.11}$$

Equality holds in (3.11) iff for some numbers f_1, f_2 we have

$$\mu_2/\mu = f_1(\mu_1\mu_2/\mu - z_{12}),\ \tfrac{1}{2} = f_1\ z_{12}$$

$$\mu_1/\mu = f_2(\mu_1\mu_2/\mu - z_{12}),\ \tfrac{1}{2} = f_2\ z_{12}\ . \tag{3.12}$$

Using (3.11) in (3.10) we get

$$D(Q) \geq \prod_{i=1}^{2} \{\ell^{-2}[1/(\mu(1-\mu)) + (\{(1-\mu_i/\mu) + 1/2\}^2)/(\mu_1\mu_2)] + 1/4\mu_i\} \quad (3.13)$$

Equality in (3.13) is attained iff (3.12) holds.

To find the minimum of (3.13) with respect to μ_1,μ_2 we write (3.13) in the form

$$(r_1/p_1 + r_2/(1-p_1) + r_3)\ (r_1/p_2 + r_2/(1-p_2) + r_3) \quad (3.14)$$

where

$$p_i = \mu_i/\mu,\ i=1,2,\ r_i=(1/4\mu)\ (1+9\ell^{-2}),\ r_2=\ell^{-2}/4\mu,\ r_3 = \ell^{-2}\ /(1-\mu). \quad (3.15)$$

We observe that (3.14) attains its maximum at $p_1 = p_2 = 1/2$, i.e., at $\mu_1 = \mu_2 = \mu/2$ and the minimum value is given by $(2r_1+2r_2+2r_3)^2$. Thus, we obtain

$$D(Q) \geq (2r_1+2r_2+r_3)^2\ . \quad (3.16)$$

The lower bound given by the right hand member of (3.16) is attained for

$$\mu_1 = \mu_2 = \mu/2,\ z_{12} = \mu/8 \quad (3.17)$$

since for these (3.12) holds with $f_1 = f_2 = 4/\mu$. We can realize the above values of μ_1,μ_2 and z_{12} by means of a simple bisimplex distribution which assigns mass $1-\mu$, $\mu/4$, $\mu/4$ and $\mu/2$ respectively to the four points 0, (1,0)', (0.1)' and $(\frac{1}{2}, \frac{1}{2})'$. We denote this bisimplex distribution by $Q^*(\frac{1}{2}; \mu/2,\mu/2)$. By direct differentiation, it is easy to see that the right hand member of (3.15) with r_1,r_2,r_3 as in (3.15) is minimized at

$$\mu = \mu^* = [1 + (\ell^2/2 + 5)^{-\frac{1}{2}}]^{-1} \quad (3.18)$$

and the corresponding minimal D(Q) is

$$D(Q) = \ell^{-4}(1-\mu^*)^{-4}\ . \quad (3.19)$$

We summarize our conclusions as follows:

<u>Theorem 3.3</u>

When k=2, within $\mathcal{Q}$, D(Q) is minimized with minimum value given by (3.19) when Q is bisimplex and μ_1,μ_2 and z_{12} are as in (3.17) with μ given by (3.18). These values of μ_1,μ_2 and z_{12} are realized, in particular, for the simple bisimplex distribution $Q^*(\frac{1}{2}, \mu^*/2,\mu^*/2)$.

In terms of P in the factor space (2.11), Theorem 3.3 can be paraphrased as follows:

Theorem 3.4

In the case k=2, among all designs P belonging to the subclass P_2, a design for which the total mass is confined to the $\underline{0}$ and the surface of the unit circle $x_1^2 + x_2^2 = 1$ and

$$\mu_1 = \mu_2 = \mu^*/2, [1111] = [2222] = [1122] = 3\mu^*/8$$

μ^* being given by (3.18), minimizes D(P). (The optimum design thus turns out to be rotatable!)

By the arguments following (3.5) and the concluding assertion of Theorem 3.3, condition of Theorem 3.4 would be realized, in particular, by a design represented by the discrete distribution which assigns mass $1-\mu^*$, $\mu^*/8$, $\mu^*/8$, $\mu^*/8$ respectively to the following points $\underline{0}$, $(\pm 1,0)'$, $(0, \pm 1)'$, $(1/\sqrt{2})(\pm 1, \pm 1)'$. Such a design is known in the literature as a central composite design.

From Theorem 3.4 we observe that the optimum design satisfying conditions of the theorem is optimum within the class P_2. Utilizing the equivalence theorem for ϕ-optimality, we now show that the design is, in fact, optimum within the entire class P.

The criterion function D(P) is not simply the classical D-optimality criterion but is equivalent to a ϕ-optimality criterion (Kiefer (1974)) where ϕ is a concave function of the complete moment matrix. Given the regression function $\eta(\underline{x}) = \underline{\theta}'\underline{f}(\underline{x})$ where $\underline{\theta} = (\theta_1,\ldots,\theta_p)'$ is the vector of unknown parameters and $\underline{f}(\underline{x}) = (f_1(\underline{x}),\ldots,f_p(\underline{x}))'$ is a mapping from a compact factor space X to R^k, for any probability distribution (design) P in X represented by the random vector δ, write $\overline{M}(P) = E_p(f(\underset{\sim}{\delta})\underline{f}'(\underset{\sim}{\delta}))$ for the complete moment matrix. The criterion $\phi(\overline{M})$ which is a real valued concave function of $\overline{M}$ is to be maximized by choice of P. The equivalence theorem for such a criterion (see e.g. Silvey (1980)) contains the following condition.

> The design P* is ϕ-optimum iff the Frechet derivative $F_\phi(\overline{M}(P^*), \underline{f}(\underline{x})\underline{f}'(\underline{x}))$ of ϕ at $\overline{M}(P^*)$ in the direction $\underline{f}(\underline{x})\underline{f}'(\underline{x})$ does not exceed 0 for all $\underline{x} \in X$.

To apply the condition to the model (2.10) we take

$$\phi(\overline{M}(P)) = -\log|E(H\overline{M}^{-1}(P)\, H')| \qquad (3.20)$$

where $\overline{M}(P)$ corresponds to $\underline{f}(\underline{x}) = (1, -x_1^2, \ldots, x_k^2, -x_1x_2, \ldots, x_{k-1}x_k, x_1, \ldots, x_k)'$ and $H = (0, \Gamma_1, \tfrac{1}{2}\cdot\Gamma_2, -\tfrac{1}{2} I_k)$.

The maximization of $\phi(\overline{M}(P))$ given by (3.20) is equivalent to the minimization of $|E(H\overline{M}^{-1}(P)H')|$ which is the same as D(P). It can be shown (cf. Mandal (1982)) that the criterion function (3.20) is concave in $\overline{M}$. Using the definition of Frechet derivative after some simplification, we get the above necessary and sufficient condition in the following form.

Theorem 3.5

A design P* is optimum iff

$$\mathrm{tr}[E(H\overline{M}^{*-1}\underline{f}(\underline{x})\underline{f}'(\underline{x})\overline{M}^{*-1}H')]\ [E(H\overline{M}^{*-1}H')]^{-1} \le k \tag{3.21}$$

for all $\underline{x}$ in (2.11) where $\overline{M}^* = \overline{M}(P^*)$.

Considering the case k-2, after some reduction it can be seen that for the P* of Theorem 3.4 $E(H\overline{M}^{*-1}H') = \ell^{-2}(1-\mu^*)^{-2}\ I_2$ and

$$\underline{f}'(\underline{x})\ \overline{M}^{*-1}[E(H'H)]\ \overline{M}^{*-1}\ \underline{f}(\underline{x})$$

$$= 2\ell^{-2}\ (1-\mu^*)^{-2} - \{4\ell^{-2}\ \mu^{*-1}(1-\mu^*)^{-2} - \mu^{*-2}\} \sum_1^2 x_i^2\ (1 - \sum_1^2 x_i^2).$$

Hence, the left hand member of (3.21) reduces to

$$2 - \ell^2(1-\mu^*)^2\{4\ell^{-2}\ \mu^{*-1}(1-\mu^*)^{-2} - \mu^{*-2}\} \sum x_i^2\ (1- \sum x_i^2)\ .$$

Since μ^* is given by (3.18) the expression within curly brackets can be seen to be positive. Hence, we readily get that (3.21) holds for all $\underline{x}$ with k=2. Thus, we have verified that the design P* is extended D-optimum over the entire class $\mathcal{P}$.

4. THE DEFICIENCY CRITERION

From (2.23), for k=2, the measure of deficiency of a design P reduces to

$$\mathrm{Def}(P) = \ell^{-2}\ \mathrm{tr}\ G_{11.11}\Lambda + (1/4)\ell^{-2}(\lambda_1+\lambda_2)g_{1212} + (1/4)\ \mathrm{tr}\ G_{22}\Lambda \tag{4.1}$$

Using the fact that $\mathrm{tr}\ G_{11.11}\Lambda \ge \mathrm{tr}\ M_{11.11}^{-1}\Lambda$, $\mathrm{tr}\ G_{22}\Lambda \ge \mathrm{tr}\ M_{22}^{-1}\Lambda > \sum_1^2 \lambda_i/[ii]$ and $g_{1212} \ge 1/(iijj]$ we get from (4.1)

$$\mathrm{Def}(P) \ge \ell^{-2}\ \mathrm{tr}\ M_{11.11}^{-1}\Lambda + (1/4)\ell^{-2}(\lambda_1+\lambda_2)/[1122] + (1/4) \sum_1^2 \lambda_i/[ii] \tag{4.2}$$

and equality holds in (4.2) iff

$$[1] = [2] = [12] = [111] = [222] = [112] = [122] = [1112] = [1222] = 0 \quad (4.3)$$

In the previous section we have denoted the subclass of designs for which (4.3) holds by $\mathcal{P}_2$. Now it is easy to see that given a design $P \in \mathcal{P}$, we can find a $\hat{\hat{P}} \in \hat{\mathcal{P}}_2$ such that P and $\hat{P}$ have same [ii], [iijj], i,j = 1,2. Thus we obtain the following theorem.

Theorem 4.1

Any design P for which Def(P) is minimized will belong to the subclass $\mathcal{P}_2$ of semi-rotatable designs.

Note that whereas in Section 3, the restriction to the subclass $\mathcal{P}_2$ was largely heuristic (so that optimality in $\mathcal{P}$ required further substantiation) in the present section the best design in $\mathcal{P}$ can justifiably be searched in the subclass $\mathcal{P}_2$. Now for any $P \in \mathcal{P}_2$

$$\mathrm{Def}(P) = \ell^{-2} \operatorname{tr} M_{11.11}^{-1}\Lambda + 1/4\cdot\ell^{-2}((\lambda_1+\lambda_2)/[1122]) + 1/4\cdot\sum_{1}^{2} \lambda_i/[ii].$$

As for the D-optimality criterion, here also we see that for $P \in \mathcal{P}_2$ Def(P) depends on P only through the distribution Q of $\underline{\eta}$ whose total mass is confined to the simplex (3.7). Arguing as in Section 3, it is easy to see that $\operatorname{tr} M_{11.11}^{-1}\Lambda$ is minimized for fixed [11], [22], [1122] iff the distribution is bisimplex. We have shown in the paragraph following Theorem 3.2 that given any distribution $Q \in \mathcal{Q}$ we can always find a bisimplex distribution Q_0 such that Q and Q_0 have the same μ_1, μ_2 and z_{12}. Hence we get the following theorem.

Theorem 4.2

For a distribution $Q \in \mathcal{Q}$, Def(Q) is minimized iff it belongs to the subclass $\mathcal{Q}_0$ of bisimplex distributions.

We note in this context that for the general problem of finding an optimal second order design over a ball, if one confines oneself to rotatable designs, it is known that the optimal design must correspond to a bisimplex distribution (see Galil and Kiefer (1977) p. 31). In our particular case, however, the above conclusion is reached without any rotatability restriction.

Now for a bisimplex distribution Q,

$$\mathrm{Def}(Q) = ((\lambda_1+\lambda_2)/(\ell^2\mu(1-\mu)) + (1/\ell^2)\{1/\mu^2\cdot(\lambda_1,\mu_2^2 + \lambda_2\mu_1^2)\,(\mu_1\mu_2/\mu - z_{12})^{-1}$$
$$+(1/4)\cdot((\lambda_1+\lambda_2)/z_{12})\} + (1/4)(\lambda_1/\mu_1 + \lambda_2/\mu_2)\ .$$

Using Schwarz inequality we see that the above expression for fixed μ_1,μ_2 attains its minimum at $z_{12} = (\mu_1\mu_2/\mu)(1+f)^{-1}$ where

$$f = 2(\lambda_1\mu_2^2 + \lambda_2\mu_1^2)^{\frac{1}{2}}\,\mu^{-1}(\lambda_1+\lambda_2)^{-\frac{1}{2}} \tag{4.4}$$

and the minimum value is

$$\ell^{-2}(\lambda_1+\lambda_2)\{\mu(1-\mu)\}^{-1} + \ell^{-2}(\mu_1\mu_2)^{-1}\{\mu^{-1}(\lambda_2\mu_2^2 + \lambda_2\mu_1^2)^{\frac{1}{2}} + (\lambda_1+\lambda_2)^{\frac{1}{2}}/2\}^2$$
$$+ (1/4)\,(\lambda_1\mu_1^{-1} + \lambda_2\mu_2^{-1}). \tag{4.5}$$

We can attain this bound by a distribution $Q(w;\mu_1,\mu_2)$ which assigns mass $1-\mu$, $w\,\mu_1$, $w\,\mu_2$, $\mu(1-w)$ respectively to the 4 points $\underline{0}$, $(1,0)'$, $(0,1)'$, $\mu^{-1}(\mu_1,\mu_2)'$, with $w = f(1+f)^{-1}$. In practice, one would need to choose μ_1,μ_2 in an optimal way. We write $\mu_1/\mu = p$ and $\mu_2/\mu = q = 1 - p$ and write (4.5) as

$$(\lambda_1+\lambda_2)/(\ell^2\,\mu(1-\mu)) + 1/\mu\cdot J$$

where

$$J = (1/pq)[(1/\ell^2)\{(\lambda_1 q^2 + \lambda_2 p^2)^{\frac{1}{2}} + \tfrac{1}{2}(\lambda_1+\lambda_2)^{\frac{1}{2}}\}^2 + (1/4)\,(\lambda_1 q + \lambda_2 p)].$$

We can find the minimum value of J with respect to p after substituting numerical values of λ_1,λ_2 and ℓ^2. Let the minimum be attained at $p = p^*$. Writing $J = J^*$ as the value of J at p^* the value of μ that minimizes

$$(\lambda_1+\lambda_2)/(\ell^2\,\mu(1-\mu)) + (1/\mu)J^*$$

is

$$\mu^* = [1+(\lambda_1+\lambda_2)^{\frac{1}{2}}\,\{(\lambda_1+\lambda_2) + \ell^2 J^*\}^{-\frac{1}{2}}]^{-1} \tag{4.6}$$

and the minimum value is given by

$$\{(\lambda_1+\lambda_2)/\ell^2\}\,\{1-\mu^*)^{-2}\} \tag{4.7}$$

We can thus conclude

<u>Theorem 4.3</u>

When k=2, the minimum value of Def(Q) is given by (4.7) and this is attained, in particular, for the bisimplex distribution $Q^*(f^*/(1+f^*),\ p^*\mu^*,\ q^*\mu^*)$ where f^* and μ^* are given by (4.4) and (4.5) respectively and p^* is as defined above.

In the factor space (2.11), in terms of P we summarize the above findings in the following form.

Theorem 4.4

In the case k=2, among all designs P, Def(P) is minimized for a design P* with minimum value given by (4.7) if for it the moments satisfy (4.5) and the total mass is confined to the point $\underline{0}$ and the boundary of the unit circle $x_1^2 + x_2^2 = 1$ with

$$[1111] = [11] - [1122],\ [2222] = [22] - [1122],\ [1122] = (([11]\ [22])/\mu^*)(1+f^*)^{-1},$$
$$[11] = \mu^* p^*, [22] = \mu^*(1-p^*) \qquad (4.8)$$

where f* and μ* are given by (4.4) and (4.6) respectively and p* is as defined above. (Thus, in general, the optimum design by this approach is not rotatable.)

The moments satisfying (4.3) and (4.8) can be realized in particular by the following discrete distribution which assigns mass to the points $\underline{0}$, $(\pm 1, 0)'$, $(0,\pm 1)'$, $(\pm p^{*\frac{1}{2}}, \pm q^{*\frac{1}{2}})$ with weights $1-\mu^*$, $w^*\mu^*_1/2$, $w^*\mu^*_2/2$, $2^{-2}(1-w^*)\mu^*$ respectively, where $w^* = f^*(1+f^*)^{-1}$.

5. A NUMERICAL EXAMPLE

We now cite a single numerical example in the case k=2 and find optimum designs under both the criteria considered in sections 3-4. The comparison would show how the optimum designs differ under differences in criterion. Let us take $\underline{c} = (50,80)'$, $\ell^2 = 25$,

$$V = \begin{bmatrix} 8 & 5 \\ 5 & 10 \end{bmatrix}, \quad W = \begin{bmatrix} 75 & 25 \\ 25 & 50 \end{bmatrix}$$

For extended D-optimality criterion, however, we do not need any assumption about W. The optimum designs under both the criteria correspond to the bisimplex distributions $Q^*(w,\mu_1,\mu_2)$. For extended D-optimality criterion we work out that $w^* = \frac{1}{2}$, $\mu_1 = \mu_2 = \mu^*/2$, $\mu^* = 7/9$ and for deficiency criterion $w = .4858$, $\mu_1 = .5561$, $\mu_2 = .3865$. Thus, we get the optimum designs under both the criteria in the unit ball (2.11). To obtain design in the original factor space we make the reverse transformation of (2.7) where L satisfies (2.8). We worked out here that

$$L = \begin{bmatrix} .426425 & -.213132 \\ -.00010 & .316289 \end{bmatrix}, \quad \lambda_1 = .4545,\ \lambda_2 = .2$$

The optimum designs in the unit ball and in the original factor space with respect to both the criterion are given below.

Extended D-Optimum Design

POINT		
Unit ball	Original factor space	Weight
(0,0)	(50.0, 80.0)	2/9
(1,0)	(61.7, 80.0)	7/72
(-1,0)	(38.3, 80.0)	7/72
(0,1)	(57.9, 95.8)	7/72
(0,-1)	(42.1, 64.2)	7/72
1/ 2(1,1)	(63.9, 91.2)	7/72
1/ 2(1,-1)	(52.7, 68.8)	7/72
1/ 2(-1,1)	(47.3, 91.2)	7/72
1/ 2(-1,-1)	(36.1, 68.8)	7/72

Minimum Deficiency Design

POINT		
Unit ball	Original factor space	Weight
(0,0)	(50.0, 80.0)	.057
(1,0)	(61.7, 80.0)	.135
(-1,0)	(38.3, 80.0)	.135
(0,1)	(57.9, 95.8)	.094
(0,-1)	(42.1, 64.2)	.094
(.7681, .6403)	(64.1, 90.1)	.121
(.7681, -.6403)	(53.9, 69.9)	.121
(-.7681, .6403)	(46.1, 90.2)	.121
(-.7681, -6403)	(35.9, 69.9)	.121

For the extended D-optimality criterion, the last column shows that if N is a multiple of 72, we can find exact optimum designs. Otherwise, in general, we have to call for some rounding off to obtain approximate frequencies.

6. GENERAL MULTIFACTOR CASE

In the general multifactor case, it is possible to extend the results of the two factor case. In this section we try to indicate the important results that have been obtained in the general case. For the extended D-optimality criterion we restrict ourselves to a subclass $\mathcal{P}_2$ of semirotatable designs in which any moment $E_p\, \xi_1^{\alpha_1} \ldots \xi_k^{\alpha_k}$, $\sum \alpha_i \leq 4$, with at least one α_i odd is zero. Then it is first shown the extended D-optimum in $\mathcal{P}_2$ must belong to a subclass of designs which correspond to a bisimplex distribution and finally obtain the following theorem.

Theorem 6.1

Among all designs P belonging to the subclass $\mathcal{P}_2$ a design for which the total mass is confined to the point $\underline{0}$ and the surface of the unit sphere $\underline{x}'\underline{x} = 1$ and

$$[ii] = \mu^*/k,\ [iiii] = 3\ [iijj] = 3\mu^*/k(k+2),\ i \neq j = 1,2,\ldots,k \tag{6.1}$$

where $\mu^* = [1 + 2/k(\ell^2/k + k + 3)^{-\frac{1}{2}}]^{-1}$ minimizes D(P), and the corresponding minimum D(P) is $\ell^{-2k}(1-\mu^*)^{-2k}$. Next using the equivalence theorem (Theorem 4.6) the optimality of P* is established. The details can be found in Mandal (1982).

For the deficiency criterion it is first shown that the minimum deficiency design must correspond to a bisimplex distribution. When λ_i's are all equal, proceeding as in the case k=2, we obtain the following theorem.

Theorem 6.2

When $\lambda_i = \lambda$, i=1,...,k, Def(P) is minimized among all designs P for P* if it belongs to P_2 and the total mass is confined to the point $\underline{0}$ and the surface of of the unit sphere satisfying (6.1) where $\mu^* = [1+\{1+ \frac{1}{4}(k-1)(k+2)^2 + (k/4)\ell^2\}^{-\frac{1}{2}}]^{-1}$ the minimum value being $(k\,\lambda/\ell^2)(1-\mu^*)^{-2}$.

When λ_i's are arbitrary, the problem of minimization of Def(P) given by (2.23) by choosing the design P appears to be algebraically intractable. To simplify matters a minimum deficiency design may be obtained within a subclass P_3 of designs. Any member of P_3 has its total mass confined only to the $1 + 2k + 2^k$ points consisting of the origin $\underline{0}$, the 2k axial points (±1,0,...,0)'... (0,0,...,±1)' and the 2^k non-axial points $\mu^{-\frac{1}{2}}(\pm\mu_1^{\frac{1}{2}},\ldots,\pm\mu_k^{\frac{1}{2}})'$ with weights $1-\mu$, $w\mu_1/2,\ldots,w\mu_k/2$ and $2^{-k}(1-w)\mu$. We call it a restricted minimum deficiency design. The question that naturally arises is whether the restricted minimum deficiency design is a minimum deficiency design in P. It is first observed that the deficiency criterion (2.23) is a linear optimality criteria (Fedorov 1971). Using the equivalence theorem (Fedorov 1971, 1972) it can be shown that the restricted minimum deficiency design P* is not optimum over P. However, using Schwarz inequality, it is possible to find a lower bound (generally unattainable) to Def(P) and on its basis the performance of the design P* can be examined. It is found that the design P* fares remarkably well in different situations. The details can be found in Mandal (1983).

7. CONCLUDING REMARKS

In the preceding sections we have considered the problem of finding the best design for hitting the bull's eye of a quadratic response surface on the basis two criteria viz. extended D-optimality criterion and the measure of deficiency. The solution is global for extended D-optimality criterion and for the deficiency criterion the solution is global in the two particular cases viz. when k=2, λ_1, λ_2 arbitrary and when $k\geq 2$, $\lambda_i = \lambda$ i=1,...,k. In all the cases the optimum design corresponds to a discrete distribution which assigns mass to $1 + 2k + 2^k$ distinct points. For the extended D-optimality criterion and for the deficiency criterion with $\lambda_i = \lambda$, i=1,...,k the optimum designs are rotatable.

We have used the above mentioned two criteria because they seem to us natural for the given problem. However, we can modify the deficiency criterion to a more general criterion by a basic modification. Let H be an k X k p.d. matrix (possibly depending on B but not on $\underline{\nu}$) and consider

$$\mathcal{E}\ E\{(\hat{\underline{\nu}}-\underline{\nu})'H(\hat{\underline{\nu}}-\underline{\nu})|\underline{\nu},\ B\} \tag{7.1}$$

$$= \text{tr}\ \mathcal{E}[E\{(\hat{\underline{\nu}}-\underline{\nu})(\hat{\underline{\nu}}-\underline{\nu})'|\underline{\nu},B\}H],$$

which, by (2.17), becomes

$$\text{tr}\ \sigma^2/N\cdot\{\mathcal{E}[(\Gamma_1,\tfrac{1}{2}\ \Gamma_2,-\tfrac{1}{2}I_k)M^{-1}(\Gamma_1,\tfrac{1}{2}\ \Gamma_2,-\tfrac{1}{2}I)'B^{-1}HB^{-1}]\} \tag{7.2}$$

Comparing (7.2) with (2.23) we see that we can obtain an optimum design with respect to the more general criterion (7.1) if we take $\mathcal{E}(B^{-1}HB^{-1})=W$ and proceed as before. The two criteria viz. the A-optimality criterion (Kiefer (1959)) and our measure of deficiency of a design are particular cases of (7.1) when H=I and when H=B respectively.

So far we have considered only the case where there is a single response and only one 'bull's eye' defined with respect to the corresponding surface. But in a multifactor experiment there may be more than one kind of response corresponding to any given combination of factors. It is possible under more restrictive assumption about the experimental region, form of the prior, etc., to extend our results to the multiresponse case by a straightforward generalization of the uniresponse case. The detailed development can be found in Mandal (1983).

Before concluding we admit that we have to impose some simplifying assumptions to reach a solution to the problem. Before using the design, in practice we must satisfy ourselves that the conditions assumed obtain up to reasonable approximation. How far the results are affected if the assumptions are violated and also whether some of these assumptions can be relaxed are questions which would require further investigation.

REFERENCES

[1] Box, G.E.P. and Hunter, J.S., Multifactor experimental designs for exploring response surfaces, Ann. Math. Statist. 28:195-241 (1957).

[2] Box, G.E.P. and Wilson, K.B., On the experimental attainment of optimum conditions, J.R. Statist. Soc. B. 13:1-45 (1951).

[3] Chatterjee, S.K. and Mandal, N.K., Response surface designs for estimating the optimal point, Cal. Statist. Assoc. Bull. 30:145-169 (1981).

[4] Draper, N.R. and Hunter, W.G., Design of experiments for parameter estimation in multiresponse situations, Biometrika 53:525-533 (1966).

[5] Draper, N.R. and Hunter, W.G., The use of prior distributions in the design of experiments for parameter estimation in nonlinear situation, Biometrika 53:525-533 (1966).

[6] Fedorov, V.V., Design of experiments for linear optimality criteria. Theory of probability and its applications 16:189-195 (1971).

[7] Fedorov, V.V., Theory of Optimal Experiments, Academic Press (1972).

[8] Galil, Z. and Kiefer, J., Comparison of rotatable designs for regression on balls (Quadratic) Jour. Stat. Plan. and Inf. 1:27-40 (1977).

[9] Kiefer, J., On the non-randomized optimality and randomized non-optimality of symmetrical designs, Ann. Math. Statist. 29:675-699 (1958).

[10] Kiefer, J., Optimum experimental designs, Jour. Roy. Stat. Soc. B 21:272-319 (1959).

[11] Kiefer, J., General equivalence theory for optimum designs (approximate theory), Ann. Stat. 2:849-879 (1974).

[12] Mandal, N.K., D-Optimal designs for estimating the optimal point in a multifactor experiment, Cal. Stat. Assoc. Bull. 31:105-130 (1982).

[13] Mandal, N.K., Response surface designs for locating the optimum. Unpublished Ph.D. Thesis, Calcutta University (1983).

[14] Silvey, S.D., Optimal Design, Chapman and Hall (1980).

BIOSTATISTICS: Statistics in Biomedical, Public Health
and Environmental Sciences, edited by P.K. Sen

ORDER STATISTICS UNDER NON-STANDARD CONDITIONS

H.A. David

Department of Statistics
Iowa State University
Ames, Iowa

A first attempt is made to review and systematize the treatment of order statistics when these arise from non-iid variates $X_1,\ldots,X_n$, where n is fixed. After an outline of the main general approaches available, special attention is given to the case when the X's are independent but non-identically distributed. This situation is motivated by (a) k-out-of-n systems of unlike independently failing components and (b) robustness against an outlier. Attention is also drawn to the recent relevant literature.

KEY WORDS: Heterogeneous distributions, k-out-of-n systems, inequalities, outliers, robustness

1. INTRODUCTION AND SUMMARY

Development of both theory and applications of order statistics received a very significant boost by the publication of Sarhan and Greenberg (1962), a multi-authored volume summarizing all aspects of order statistics as understood at the time. A little earlier one of the contributors (Gumbel, 1958) had written a book on extremes but Sarhan and Greenberg provided the first comprehensive - and still useful - account of the field of order statistics as a whole. Many tables and illustrations were included to assist the user in the immediate implementation of the techniques presented. Since then the study of order statistics has proceeded apace and the subject has again been reviewed in David (1970, 1981).

The systematic treatment of order statistics has focused on their behavior in random samples from univariate populations. We call this the standard or iid case. In recent years there has also been a good deal of emphasis on asymptotic theory under more general assumptions (e.g., Galambos, 1978; Leadbetter, Lindgren, and Rootzen, 1983). A first small attempt is made in the present article to collect some of the many approaches, scattered in the literature, for dealing with the order statistics $X_{1:n} \le \ldots \le X_{n:n}$ arising from a finite set of variously structured non-iid variates $X_1,\ldots,X_n$. Often a mere mention of relevant work will have to suffice. We proceed by increasing specialization: Section 2 reviews some general approaches, section 3 deals with independent but non-identically distributed (inid) variates, and section 4 concentrates on the special case of a single outlier. Throughout the paper attention is also drawn to recent references.

2. GENERAL APPROACHES

It is natural that most results on order statistics under non-standard conditions were obtained to meet specific needs but some general results are available. Thus Galambos (1975) treats the distribution of order statistics in samples from multivariate populations. For the bivariate case see also David (1981; Ex. 2.2.2). Except in special cases these general distributional results are necessarily rather unwieldy.

Much has been done in the important special case of the extremes, $X_{1:n}$ and $X_{n:n}$, motivated especially by tests for outliers. Since in this context interest centers usually on upper percentage points of statistics expressible as maxima (or lower percentage points of minima), the use of the principle of inclusion and exclusion provides a powerful tool. One has

$$\Pr\{X_{n:n} > x\} = S_1 - S_2 + \ldots (-1)^{n-1} S_n, \tag{1}$$

where

$$S_j = \sum \Pr\{X_{i_1} > x, \ldots, X_{i_j} > x\} \quad j=1,\ldots,n$$

with the summation extending over $1 \le i_1 < i_2 < \ldots < i_j \le n$. In (1) later terms are often negligible for sufficiently large x; indeed the first term, which always provides an upper bound to the LHS, may suffice. This last idea goes back to Pearson and Chandra Sekar (1936) who considered the distribution of the (internally) studentized extreme deviate

$$\max_{i=1,\ldots,n} \frac{Z_i - \bar{Z}}{S} = \frac{Z_{n:n} - \bar{Z}}{S},$$

where $Z_1,\ldots,Z_n$ are iid normal $N(\mu,\sigma^2)$ variates and S is the usual rms estimator of σ. Many ramifications of this approach are described or referenced in David (1981); see especially sections 5.3, 8.2, 8.4, and 8.5. Exclusive attention to tests for outliers is paid in the books by Barnett and Lewis (1978) and Hawkins (1980) and in the papers on outliers in linear models by Cook and Prescott (1981), Doornbos (1982), and Galpin and Hawkins (1981).

Clearly (1) simplifies greatly when the X_i are exchangeable variates since then

$$S_j = \binom{n}{j} \Pr\{X_1 > x, \ldots, X_j > x\}.$$

In this situation the well-known generalization of (1), viz.

$$\Pr\{X_{n-r+1:n} > x\} = \sum_{j=r}^{n} (-1)^{j-r} \binom{j-1}{r-1} S_j \quad r=1,\ldots,n$$

may become useful. See David (1981; sections 5.4 and 5.6) and Galambos (1982).

Motivated by genetic selection problems several authors have considered the distribution and moments of order statistics in normal samples of n = mk, consisting of m samples of k, with common ρ within samples and independence between samples. For a partial extension see Tong (1982).

If $K_n(x)$ denotes the number of realizations of the n events $A_i = \{X_i \le x\}$, then

$$\Pr\{X_{r:n} \le x\} = \Pr\{K_n(x) \ge r\}.$$

In the iid case $K_n(x)$ has, of course, a binomial b(p,n) distribution where $p = \Pr(A_1)$. An interesting approach outlined by Mallows (1969) for situations that are not too far removed from the iid case is to approximate $\Pr\{K_n(x) = k\}$ by

$$\binom{n}{k} p^k (1-p)^{n-k} (1 + a_1 G_1(k) + {}^1\!/_2 a_2 G_2(k) + \ldots),$$

where $G_1, G_2, \ldots$ are polynomials in k, orthogonal with respect to the binomial weight function (Krawtchouk polynomials). The parameters $p, a_1, a_2, \ldots$ may be obtained by matching moments of $K_n(x)$.

3. ORDER STATISTICS FOR INDEPENDENT NON-IDENTICALLY DISTRIBUTED (INID) VARIATES

Major motivation for the study of ordered inid variates is provided by k-out-of-n systems S which function iff at least k (k=1,...,n) of the n components function. Let $X_i (i=1,\ldots,n)$ be the lifetime of the i-th component and $p_i(x) = \Pr\{X_i > x\}$ its reliability at time x. Then the system will function at time x iff $X_{n-k+1:n} > x$. Assuming independent component failure times (X_i) and writing r=n-k+1, we may express the reliability of the system as (Pledger and Proschan, 1971)

$$\Pr\{X_{r:n} > x\} = h_k[\underset{\sim}{p}(x)],$$

where

$$\underset{\sim}{p}(x) = (p_1(x), \ldots, p_n(x))' ,$$

$$h_k(\underset{\sim}{p}) = \sum_R \prod_{i=1}^n p_i^{\Delta_i} (1-p_i)^{1-\Delta_i} ,$$

$\Delta_i = 0$ or 1, and R is the region $\sum_{i=1}^n \Delta_i \ge k$.

In general, $h_k[\underset{\sim}{p}(x)]$ is not easy to calculate but interesting inequalities can be established from the basic results of Hoeffding (1956) on independent binomial trials with unequal success probabilities p_i (i=1,...,n). From these results

Sen (1970) has obtained the following theorem relating the distribution of $X_{r:n}$ for inid variates with that for iid variates having common cdf

$$\bar{P}(x) = \frac{1}{n}\sum_{i=1}^{n} P_i(x),$$

where

$$P_i(x) = 1-p_i(x).$$

It is assumed that $\bar{\xi}_p$ is uniquely defined by $\bar{P}(\bar{\xi}_p) = p$ $(0 < p < 1)$.

<u>Theorem</u>. For $r=2,3,\ldots,n-1$ and all $x \le \bar{\xi}_{(r-1)/n} < \bar{\xi}_{r/n} \le y$

$$\Pr\{x < X_{r:n} \le y|\underset{\sim}{P}\} \ge \Pr\{x < X_{r:n} \le y|\bar{P}\}, \tag{2}$$

where equality holds only if $P_1 = \ldots = P_n = P$ at both x and y. Also, for all x

$$\Pr\{X_{1:n} \le x|\underset{\sim}{P}\} \ge \Pr\{X_{1:n} \le x|\bar{P}\} \tag{3}$$

and

$$\Pr\{X_{n:n} \le x|\underset{\sim}{P}\} \le \Pr\{X_{n:n} \le x|\underset{\sim}{\bar{P}}\} \tag{4}$$

with strict inequalities unless $P_1 = \ldots = P_n = P$ at x.

It follows from (4) that for all x

$$h_1[\underset{\sim}{p}(x)] = \Pr\{X_{n:n} > x|\underset{\sim}{P}\}$$

$$\ge \Pr\{X_{n:n} > x|\bar{P}\} = h_1[\underset{\sim}{\bar{p}}(x)], \tag{5}$$

and likewise from (2) that

$$h_n[\underset{\sim}{p}(x)] \le h_n[\underset{\sim}{\bar{p}}(x)]. \tag{6}$$

Thus a general parallel (series) system is at least (most) as reliable as the corresponding system of components each having reliability $\bar{p}(x)$. For $k=2,\ldots,n-1$ we see from (2) that, with obvious notation, $\underset{\sim}{S}$ is at least as reliable as $\bar{S}$ until time $\bar{\xi}_{(r-1)/n}$ and is at most as reliable from time $\bar{\xi}_{r/n}$ on $(r=n-k+1)$.

A lower bound for $h_k[\underset{\sim}{p}(x)]$ valid for all k (with equality for k=n) is

$$h_k[\underset{\sim}{p}(x)] \ge h_k[\underset{\sim}{p}_G(x)], \tag{7}$$

where $\underset{\sim}{p}_G = (p_G,\ldots p_G)'$, p_G being the geometric mean

$$p_G(x) = [\prod_{i=1}^{n} p_i(x)]^{\frac{1}{n}}.$$

The inequalities (7) are a special case of results obtained in Pledger and Proschan (1971) by the use of Schur functions. This line of development, particularly useful for distributions having proportional hazards, has been continued by Proschan with various co-authors, most recently in Boland and Proschan (1983a,b). The latter paper summarizes the main results.

From (5) we see that $X_{n:n}$ is stochastically larger under $\underset{\sim}{P}$ than under $\overline{P}$; by (6) the reverse holds for $X_{1:n}$. Thus, in particular

$$E(X_{n:n}|\underset{\sim}{P}) \geq E(X_{n:n}|\overline{P}), \quad E(X_{1:n}|\underset{\sim}{P}) \leq E(X_{1:n}|\overline{P}) ,$$

so that for the sample range $W_n = X_{n:n} - X_{1:n}$

$$E(W_n|\underset{\sim}{P}) \geq E(W_n|\overline{P}).$$

If the distributions of the X_i $(i=1,\ldots,n)$ are symmetric about their respective means, it can be shown (Salama and Sen, 1982) that $E(W_n|P)$ is a minimum when these means are equal but (Samuel-Cahn, 1983) there may be other minimal configurations. Samuel-Cahn establishes also that if the n distributions differ in location only, then EW_n is minimized iff the means are equal; the distributions need not be symmetric.

Before specializing the problem further we note that a formal expression for the joint pdf of $k\,(\leq n)$ order statistics stemming from n absolutely continuous populations can be given in the form of a permanent (Vaughan and Venables, 1972).

We draw attention also to a recent theoretically interesting paper by Guilbaud (1982) who shows that under rather general conditions it is possible to express functions of inid random vectors as functions of iid random vectors.

4. CASE OF A SINGLE OUTLIER

Suppose the n independent absolutely continuous variates X_j $(j=1,\ldots,n-1)$ and Y have respective cdf's $P_j(x) = F(x)$ and $G(x)$. Let $Z_{r:n}$, $r=1,\ldots,n$, denote the r-th order statistic among the n variates, with cdf $H_{r:n}(x)$. Then it is easily seen (David and Shu, 1978) that

$$H_{r:n}(x) = F_{r:n-1}(x) + \binom{n-1}{r-1} F^{r-1}(x)[1-F(x)]^{n-r}G(x), \tag{8}$$

where $F_{r:n-1}(x)$ is the cdf of $X_{r:n-1}$, the r-th order statistic among $X_1,\ldots,X_{n-1}$.

Next, suppose $G(x) = F(x-\lambda)$ and that $0 < F(x) < 1$ for all finite x.

Writing $\mu_{r:n}(\lambda) = EZ_{r:n}$ we see that $\mu_{r:n}(\lambda)$ is monotonically increasing in λ with

$$\mu_{r:n}(\infty) = EX_{r:n-1} \equiv \mu_{r:n-1} \qquad r=1,\dots,n, \tag{9}$$

where $\mu_{n:n-1} = \infty$. Likewise, with $\mu_{0:n-1} = -\infty$, we have $\mu_{r:n}(-\infty) = \mu_{r-1:n-1}$.

The result (9) may be expressed as follows: An extreme outlier on the right increases the expected value of the r-th order statistic from $\mu_{r:n}$ to $\mu_{r:n-1}$. Note that since

$$(n-r)\mu_{r:n} + r\mu_{r+1:n} = n\mu_{r:n-1}$$

we may write

$$\mu_{r:n-1} - \mu_{r:n} = (r/n)\delta_{r,n}, \tag{10}$$

where $\delta_{r,n} = \mu_{r+1:n} - \mu_{r:n}$.

Now suppose that the pdf of X is symmetric and strictly increasing up to its mean μ. Consider the estimators $M_{r,n}$ of μ:

$$M_{r,n} = \tfrac{1}{2}(Z_{n-r+1:n} + Z_{r:n}) \qquad r = [\tfrac{1}{2}n] + 1,\dots,n .$$

These two-point estimators reduce to the median $Z_{\frac{1}{2}(n+1):n}$ when n is odd and $r = \frac{1}{2}(n+1)$. We show that for $r \geq [\frac{1}{2}n] + 1$

$$EM_{r,n}(\infty) \text{ is an increasing function of } r. \tag{11}$$

Proof: From (9) and the symmetry of F we have

$$\begin{aligned} 2E[M_{r+1,n}(\infty) - M_{r,n}(\infty)] &= \mu_{r+1:n-1} + \mu_{n-r:n-1} - \mu_{r:n-1} - \mu_{n-r+1:n-1} \\ &= (\mu_{r+1:n-1} - \mu_{r:n-1}) - (\mu_{r:n-1} - \mu_{r-1:n-1}) \\ &= \delta_{r,n-1} - \delta_{r-1,n-1} > 0 , \end{aligned}$$

where the inequality follows from David and Groeneveld (1982, p. 229).

In turn, (11) means that the linear estimator

$$\sum_{i=2}^{n-1} a_i X_{i:n} \quad \text{with } \sum a_i = 1,\ a_i = a_{n+1-i}$$

is at least as biased by an extreme outlier as is

$$\sum_{1=2}^{n-1} b_i X_{i:n} \quad \text{with } \sum b_i = 1,\ b_i = b_{n+1-i}$$

if

$$\sum_{i=2}^{j} (a_i - b_i) \geq 0 \text{ for } j = 2,3,\ldots,[\tfrac{1}{2}n].$$

These results are easily generalized to the situation of u extreme outliers on the right and v on the left since in this case

$$EZ_{r:n} = \mu_{r-v:n-u-v} ,$$

where

$$\mu_{s:n} = \infty \text{ if } s > n$$

$$= -\infty \text{ if } s < 1 .$$

In particular, of the estimators in this class, the median is the most robust to extreme outliers.

Finally it may be noted that in the case of a single outlier $EZ_{r:n}$ has been tabulated when F is standard normal in David, Kennedy, and Knight (1977) for $n \leq 20$ and a range of λ-values. Variances and covariances of $Z_{r:n}$ are also tabulated in this case and for $G(x) = \Phi(x/\tau)$. For these two single-outlier models the tables permit robustness studies for all linear functions of order statistics (David and Shu, 1978; David, 1979; Hoaglin, Mosteller, and Tukey, 1983, Chapter 10).

ACKNOWLEDGEMENT: This paper was prepared with the support of the U.S. Army Research Office.

REFERENCES

[1] Barnett, V. and Lewis, T., Outliers in Statistical Data. Wiley, Chichester (1978).

[2] Boland, P.J. and Proschan, F., The reliability of k out of n systems, Ann. Probab. 11:760-64 (1983a).

[3] Boland, P.J. and Proschan, F., The impact of reliability theory on some branches of mathematics and statistics. Florida State University Statistics Report M666 (1983b).

[4] Cook, R.D. and Prescott, P., On the accuracy of Bonferroni significance levels for detecting outliers in linear models, Technometrics 23:59-64 (1981).

[5] David, H.A., Order Statistics. Wiley, New York (1970, 1981).

[6] David, H.A., Robust estimation in the presence of outliers. In: Launer, R.L. and Wilkinson, G.N. (eds). Robustness in Statistics, 61-74, Academic Press, New York (1979).

[7] David, H.A. and Groeneveld, R.A., Measures of local variation in a distribution: Expected length of spacings and variances of order statistics, Biometrika 69:227-32 (1982).

[8] David, H.A., Kennedy, W.J., and Knight, R.D., Means, variances, and covariances of normal order statistics in the presence of an outlier, Selected Tables in Mathematical Statistics 5:75-204 (1977a).

[9] David, H.A. and Shu, V.S., Robustness of location estimators in the presence of an outlier. In: David, H.A. (ed). Contributions to Survey Sampling and Applied Statistics: Papers in Honor of H.O. Hartley, 235-50, Academic Press, New York (1978).

[10] Doornbos, R., Testing for a single outlier in a linear model, Biometrics 37: 705-11 (1981).

[11] Galambos, J., Order statistics of samples from multivariate distributions, J. Amer. Statist. Assoc. 70:674-80 (1975).

[12] Galambos, J., The Asymptotic Theory of Extreme Order Statistics. Wiley, New York (1978).

[13] Galambos, J., The role of exchangeability in the theory of order statistics. In: G. Koch and F. Spizzichino (eds). Exchangeability in Probability and Statistics. North-Holland, Amsterdam (1982).

[14] Galpin, J.S. and Hawkins, D.M., Rejection of a single outlier in two- or three-way layouts, Technometrics 23:65-70 (1981).

[15] Guilbaud, O., Functions of non-iid random vectors expressed as functions of iid random vectors, Scand. J. Statist. 9:229-33 (1982).

[16] Gumbel, E.J., Statistics of Extremes. Columbia University Press, New York (1958).

[17] Hawkins, D.M., Identification of Outliers. Chapman and Hall, London (1980).

[18] *Henery, R.J., Permutation probabilities as models for horse races, J.R. Statist. Soc. B 43:86-91 (1980).

[19] Hoaglin, D.C., Mosteller, F. and Tukey, J.W. (eds). Understanding Robust and Exploratory Data Analysis. Wiley, New York (1983).

[20] Hoeffding, W., On the distribution of the number of successes in independent trials, Ann. Math. Statist. 27:713-21 (1956).

[21] Leadbetter, M.R., Lindgren, R. and Rootzen, H., Extremes and Related Properties of Random Sequences and Processes. Springer, New York (1983).

[22] Mallows, C.L., Techniques for non-standard order-statistics, Bull. Int. Statist. Inst. 43 (2):154-6 (1969).

[23] *Odeh, R.E., Tables of percentage points of the distribution of the maximum absolute value of equally correlated normal random variables, Comm. Statist. B 11:65-87 (1982).

[24] Pearson, E.S. and Chandra Sekar, C., The efficiency of statistical tools and a criterion for the rejection of outlying observations, Biometrika 28: 308-20 (1936).

[24] Pledger, G. and Proschan, F., Comparisons of order statistics from heterogeneous distributions. In: Rustagi, J.S. (ed). Optimizing Methods in Statistics, 89-113, Academic Press, New York (1971).

[25] Salama, I.A. and Sen, P.K., On expected sample range from heterogeneous symmetric distributions, J. Statist. Plann. Inf. 6:235-9 (1982).

[26] Samuel-Cahn, E., Minimizing the expected sample range, J. Statist. Plann. Inf. 8:347-54 (1983).

[27] Sarhan, A.E. and Greenberg, B.G. (eds). Contributions to Order Statistics. Wiley, New York (1962).

[28] Sen, P.K., A note on order statistics for heterogeneous distributions, Ann. Math. Statist. 41:2137-9 (1970).

[29] Tong, Y.L., Some applications of inequalities for extreme order statistics to a genetic selection problem, Biometrics 38:333-9 (1982).

[30] Vaughan, R.J. and Venables, W.N., Permanent expressions for order statistics densities, J.R. Statist. Soc. 34:308-10 (1972).

*Not cited in text.

BIOSTATISTICS: Statistics in Biomedical, Public Health
and Environmental Sciences, edited by P.K. Sen

SOME REFLECTIONS ON STRATEGIES OF MODELLING:
How, when and whether to use principal components

K. Ruben Gabriel and Charles L. Odoroff

Department of Statistics and Division of Biostatistics
University of Rochester
Rochester, NY 14642

It is argued that modelling a matrix in terms of concomitant variables for the rows and the columns should proceed by regression methods, possibly followed by simplification of the coefficient matrix by principal components. It is shown that this is superior to the common practice of first obtaining principal components of the data and then regressing scores and loadings onto the concomitants. Reanalysis of a well known data set of alcohol density is used as an example.

KEY WORDS: Principal component analysis, singular value decomposition, lower rank approximation, one degree of freedom for non-additivity, modelling strategy, resistant techniques, additive and multiplicative models, linear and bilinear models, biplot

1. INTRODUCTION

This paper is motivated by what we consider to be a widespread misuse of the technique of principal component analysis (PCA) in fitting models that relate data matrices to concomitant variables. We first describe the technique and the manner in which it is usually applied and point out that this does not lead to the best fits of models. We then discuss a case of modelling by PCA (Mandel, 1971) in detail and show an alternative strategy which leads to a closer fit. We end with some comments.

The common application of PCA to a two-way layout of data usually begins by centering the data either to the common mean or to row means and/or to column means. After that, it decomposes the resulting matrix Y, of, say, n rows and m columns, as

$$Y = \sum_{e=1}^{m} \lambda_e \underline{p}_e \underline{q}_e'. \tag{1}$$

(This is the Singular Value Decomposition which uniquely determines $\underline{p}_1,\ldots,\underline{p}_m$ and $\underline{q}_1,\ldots,\underline{q}_m$ to be orthonormal and $\lambda_k \geq \lambda_2 \geq \ldots \geq \lambda_m \geq 0$ -- m being replaced by n if n<m). The components with small λ's are usually ignored as being due to "noise" and the model becomes the rank r approximation.

$$Y = \sum_{e=1}^{r} \lambda_e \underline{p}_e \underline{q}_e', \tag{2}$$

where m-r is the number of components excluded. Note that neither the decomposition nor the reduction of rank take concomitant variables into account in any way.

The "interpretation" of such an analysis usually proceeds by inspective the "scores" $\lambda_e \underline{p}_e$ and the "loadings" $\underline{q}_e'$ individually for each included component $e = 1,\ldots,r$. These scores and loadings are used to "explain" the variation between, respectively, the rows and the columns of Y. At times, such inspection and explanation is aided by plotting and rotating scores or loadings. Thus, Gittins (1969) makes the following comments on the application of PCA to some ecological stand-by-species data.

> Although the principal components strictly speaking are purely mathematical constructions, it is frequently found that they correspond also to ecological features of the analysis. Indeed, only where the reference axes can be considered to reflect ecological factors or processes of some kind, is the ordination likely to be regarded as completely satisfying. Identification of the ecological factors involved is a matter for interpretation, and as such, is dependent in part on the judgement of the investigator. Usually it involves an attempt to evaluate the results of an ordination against information external to the analysis.

He then gives the following illustration:

> ... consider the species ...*Thymus drucei* and *Trifolium repens* ... with high loadings ... at opposite ends of the [first] axis. The preference of *T. drucei* for shallow, well-drained soils of *T. repens* for deeper soils with more balanced water relationships is well known ... using this knowledge, Component I may, as an initial hypothesis, be regarded as an expression of the possible control exerted by soil depth....

This kind of "explanation" of principal components may be appropriate when no concomitant variables are available, or when their values are known only for some rows or columns (e.g., in the above example, data were available for *T. drucei* and *T. repens*, but not for most other species.) It may also be the best that can be done when concomitant information is categorical. However, when quantitative concomitants $\underline{x}$ and $\underline{z}$ are available for the rows and columns, it is common practice to regress the component $\underline{p}$'s and $\underline{q}$'s, or rotations of them, onto the $\underline{x}$ and $\underline{z}$, respectively. Again, Gittins (1969) describes this as follows.

> Where environmental variates are included ... interpretation is simplified. ...This involves regression of the intensity of an ecological factor ... on the position along a particular [principal] axis or axes. ... An example ... a measure of light conditions is plotted against positions along the first axis... This suggests that the axis may correspond to the effect of light ...
>
> There is a second approach to the identification of components from a stand ordination. This is to plot the distribution of stand component-scores at the appropriate stand positions on a base-map of the area involved. Particular environmental variables can be treated in the same way and the resulting distribution maps contoured and compared. The maps may then suggest the existence of connections between habitat factors and components, leading to identification of the latter. A refinement of this technique... consists of fitting ... component-scores ... to polynomial functions of the map coordinates. ... The surfaces corresponding to the

> functions ... may be examined and compared for their ecological implications.

We take issue with this use of PCA because it is not an efficient way to incorporate quantitative concomitant variables into a model. It is well known that the best fit, in the least squares sense, on the concomitants is to regress each column of Y onto $\underline{x}$ and each row of Y onto $\underline{z}$. The singular value decomposition of Y is irrelevant to such fits: Regressing individual components does not lead to the best fit. Its use in PCA yields less than optimal fits.

We submit that the proper role of PCA is diagnostic, in suggesting models that may fit the data well and indicating concomitants to be introduced into the model. It can be used in that way since the model (2) which it fits to the data is quite general and includes additive row and/or column effects, as well as any number of multiplicative terms. Thus, the PCA model subsumes ANOVA additive models and Mandel's row and column regression models (Mandel, 1961; See also Yates and Cochran, 1938, and Finlay and Wilkinson, 1963, for other introductions of that model.) as well as Tukey's (1949) model for non-additivity and quadratic extensions of it (Bradu, 1984). Many of these special cases of model (2) can be diagnosed graphically by the use of various plots (Mandel, 1984) and biplots of the scores and loadings (Bradu and Gabriel, 1978). Inspection of the scores and loadings and their plots may also suggest further types of models and indicate new concomitant variables.

We hold that the role of PCA should end there and that it should play no part in the actual fitting of the suggested model that involves concomitant variables. The fitting is more appropriately done by the common least squares techniques and their resistant counterparts. We stress that it is the data that one wishes to fit the model to, and so one should not fit it to the components or anything else.

In passing, we might mention that the issue we raise applies to multiplicative models, but not to simple additive models such as those used in two-way analysis of variance. With the latter, it is readily checked that the least squares fit of a $\mu + \alpha x_i + \beta z_j$ model (with constraints $\sum_i x_i = \sum_j z_j = 0$) to the matrix elements $y_{i,j}$ is exactly the same as that obtained by first doing a two-way additive decomposition and then regressing the row "effects" onto the x_i's and the column "effects" onto the z_j's. On the other hand, if a multiplicative model $\gamma x_i z_j$ is fitted to $y_{i,j}$ by least squares, the fit is different from that obtained from first approximating Y by a multiplicative term, i.e., by the first principal component $\lambda_1 \underline{p}_1 \underline{q}_1'$, and then regressing the scores $\lambda_1 \underline{p}_1$ and loadings $\underline{q}_1'$ onto $\underline{x}$ and $\underline{z}'$, respectively. Just as we cannot see the virtue of obtaining the latter suboptimal multiplicative fit, so we also cannot see the justification for the analogous use of PCA.

It is interesting to note that the standard practice in two-way analysis of variance with quantitative factors and interaction is to fit the data directly onto the concomitants without first exploring the interaction by PCA (e.g., Snedecor and Cochran, 1965, 6th ed., Section 16.9). We see this as the correct practice and regret that practitioners of PCA often fit the components instead.

It may, at this time, be appropriate to distinguish between internal and external analyses. We consider an analysis to be internal if it uses only the data entries of Y itself, and we consider an analysis to be external if it uses concomitant information on the rows and/or columns. Thus, we regard two-way ANOVA as an internal analysis, and a regression analysis as an external one. It would seem that almost any "interpretation" or "explanation" will make an analysis external since it usually relates the findings of the analysis to other, concomitant, information. (Other terms have been used by various authors to distinguish between the two modes of analysis. Thus, Bradu and Gabriel (1978) spoke of structural - referring to the two-way structure of the data, which we now refer to as internal - versus physical - a term which was appropriate for their example. The present term has less limited connotations and is surely more appropriate to possible applications in Biostatistics.)

2. A CASE STUDY: ALCOHOL DENSITY DATA

We illustrate our point by a case study, using the data on alcohol density of Table 1, part of which was introduced into the statistical literature by Mandel (1971) in a discussion of model fitting. Later, Bradu and Gabriel (1978) showed that displaying data as a biplot gave a good idea of the appropriate type of model. More recently, Hoaglin, Wong and Emerson (1983) have suggested that resistant fitting procedures would provide more insight, since they suspected that some unusual observations may have unduly influenced the least squares fit used by Mandel. A rejoinder to their critique was given by Mandel (1984).

Mandel's 1971 analysis, and most of the subsequent reanalyses, used only a part of the original data, that is, the 6-by-7 table for the concentrations from 30% to 80% (rows 5 through 10 of Table 1). In order to make our analyses comparable, we also focused on fitting this 6-by-7 submatrix. However, we further checked our procedure against fits to the complete matrix.

In the following, we denote the data matrix by Y and the element in its i-th row and j-th column by $y_{i,j}$. We also denote the concentration for the i-th row by x_i

and the temperature for the j-th column by z_j. Thus, $y_{i,j}$ is the density observed at concentration x_i and the temperature z_j. We further use notations $x_{i,u}$ (u=1, 2, 3, 4) and $z_{j,v}$ (v=1, 2, 3) (with double subscripts) for the elements of matrices X and Z; these are defined later in a manner related to the concentrations x_i (i=1,...,6) and the temperatures z_j (j=1,...,7). Finally, $\underline{1}_m$ denotes the vector of m ones in column form, and $\underline{1}'_m$ is the same vector in row form.

Table 1. Density of Solutions of Alcohol in Water
Mean Observed Density, D_4^t

Columns are determinations at the following temperatures in Centigrade

	10	15	20	25	30	35	40	% Alcohol by Weight	
	.991108	.990468	.989534	.988312	.986853	.985167	.983266	4.907	
	.983962	.983074	.981894	.980460	.978784	.976886	.974780	9.984	
	.973490	.971765	.969812	.967642	.965290	.962750	.960034	19.122	
	.969190	.967014	.964664	.962129	.959440	.956589	.953586	22.918	
*|	.959652	.956724	.953692	.950528	.947259	.943874	.940390	30.086	|*
*|	.942415	.938851	.935219	.931502	.927727	.923876	.919946	39.988	|*
*|	.921704	.917847	.913922	.909938	.905880	.901784	.897588	49.961	|*
*|	.899323	.895289	.891202	.887049	.882842	.878570	.874233	59.976	|*
*|	.875989	.871848	.867640	.863378	.859060	.854683	.850240	70.012	|*
*|	.851882	.847642	.843363	.839030	.834646	.830202	.825694	80.036	|*
	.826442	.822174	.817866	.813515	.809120	.804672	.800171	90.037	
	.798116	.793882	.789618	.785334	.781027	.776682	.772297	99.913	

Source: Osborne, McKelvy, and Bearce, (1913).
*| Data of the six rows for concentrations 30% - 80% have also been analyzed by Mandel, (1971), Bradu and Gabriel, (1978), and Hoaglin, Wong and Emerson, (1983).

3. MANDEL'S MODELLING

Mandel's strategy of modelling the matrix Y had two stages. In the first stage he took into account only the internal structure of the data, i.e., its being arranged in rows and columns. The external factors or concomitants that were associated with these rows and columns, i.e., the levels of concentration and of temperature, were ignored at that stage. This resulted in an internal model which expressed the densities $y_{i,j}$ as a function of an overall constant and parameters which were indexed by either i or j.

In this stage of modelling, Mandel used least squares to fit a first internal model which took the form

$$y_{i,j} = \mu + \rho r_i + \zeta c_j + \theta_1 u_i v_j + \theta_2 t_i w_j + \text{error} . \qquad (3)$$

This involved additive row and column effects ρr_i and ζc_j, respectively, and two principal components of interaction. (In this notation the Greek letters indicate constants and the Latin letters parameters normalized to satisfy the usual side conditions.) Still within the stage of internal modelling, Mandel further elaborated the model by imposing quadratic relations

$$t_i = \alpha_0 + \alpha_1 u_i + \alpha_2 u_i^2 \tag{4}$$

and

$$w_j = \beta_0 + \beta_1 v_j + \beta_2 v_j^2 . \tag{5}$$

He introduced these relations into (3) and obtained his final internal model

$$\begin{aligned} y_{i,j} = {} & \mu + r_i + \zeta c_j + \theta_1 u_i v_j \\ & + \theta_2(\alpha_0+\alpha_1 u_i+\alpha_2 u_i^2)(\beta_0+\beta_1 v_j+\beta_2 v_j^2) + \text{error} . \end{aligned} \tag{6}$$

Mandel then proceeded to the second stage of modelling, in which he re-expressed the above internal model in terms of the physical measurements of concentration x_i and temperature z_j. To this end, he fitted four functions $R(\cdot)$, $C(\cdot)$, $U(\cdot)$ and $V(\cdot)$ to approximate the r_i's and u_i's by $R(x_i)$ and $U(x_i)$, respectively, and the c_j's and v_j's by $C(z_j)$ and $V(z_j)$, respectively. This yielded

$$\begin{aligned} y_{i,j} = {} & \mu + \rho R(x_i) + \zeta C(z_j) + \theta_1 U(x_i)V(z_j) \\ & + \theta_2(\alpha_0+\alpha_1 U(x_i)+\alpha_2(U(x_i))^2)\ (\beta_0+\beta_1 V(x_i)+\beta_2(V(z_j))^2) \\ & + \text{error}, \end{aligned} \tag{7}$$

which expresses the density data $y_{i,j}$ as a function of the concomitants x_i (concentration) and z_j (temperature). That, therefore, is an external, or physical model.

Mandel was not quite explicit about the reasons for his choice of internal model (3) except that he felt that an additive model was the "simplest" and a sum of multiplicative components seemed a good way to model interaction. i.e., the residuals from an additive fit. His decision to use two multiplicative components, rather than one or three or more, was purely on the grounds of goodness of fit, as was his choice of quadratic modelling (4,5) of the second component in terms of the first. Also, he did not motivate his choice of the type of functions he used for R, C, U, V, except for noting that they provided good fits.

In fitting all these models, Mandel used least squares methods. Thus, he fitted internal model (3) by stepwise least squares (which Gabriel (1978) later showed to be equivalent to overall least squares), and he fitted internal model (6) by piecewise least squares. He then grafted the separate non-linear least squares fits of R, C, U and V onto this fit of (6) and thus obtained external model (7).

The sum of squared residuals from his fit of model (3) was 217 E-12, which indicated a typical residual slightly larger than 2 E-6. We use the typographically convenient E-p for 10 to the p-th power.) For the external model (7) he obtained a sum of squared residuals of 43 E-9. Such an increase was the price paid for reducing the number of parameters from 26 for (3) to 16 for (7).

The parameter estimates of models (3), (6) and (7) appear in Mandel's paper, but his Table IV gives an erroneous constant for model (7) - it should be 0.885423.

4. OUR ALTERNATIVE MODELLING STRATEGY

Our strategy is to begin by modelling the dependence of the data $y_{i,j}$ on the external factors (or concomitants) x_i and z_j and only then attempt to impose structure. We reason that, if the data Y can be decomposed into a part which depends on the x's and z's and another part, dependent on neither x nor z, then there is no purpose in imposing structure on the latter part. Only the former part could contribute to an external model. That is why we choose to reverse the order of the stages of the modelling strategy: We begin by identifying the component of the data which depends on the external factors (concomitants) and then attempt to model the structure of that component.

In carrying out our modelling strategy on the 6-by-7 matrix of alcohol density data, we first sought functions of the concentrations x_i $(i=1,\ldots,6)$ that would carry (in the regression sense) the variability within the columns of Y, as well as functions of the temperatures z_j $(j=1,\ldots,7)$ that would carry the variability within the rows of Y. There are many ways of seeking such "carrier" functions, but we were quite successful in doing so by regressing each column of Y onto $-x_i^{-1}$, $-x_i^{-\frac{1}{2}}$, $\log(x_i)$, $x_i^{\frac{1}{2}}$, x_i, x_i^2, x_i^3 and x_i^4 by stepwise regression methods and noting which powers of x_i were chosen. With these density data, the stepwise method always led first to the choice of x_i, then to that of $-x_i^{-1}$, and often x_i^3 was chosen third; other powers were never chosen. Hence, we defined the carrier matrix as

$$X = \begin{pmatrix} 1 & (x_1-55)/20 & (x_1-55)^3/10000 & 2(1-100/x_1)/3 \\ 1 & (x_2-55)/20 & (x_2-55)^3/10000 & 2(1-100/x_2)/3 \\ \cdot & \cdot & \cdot & \cdot \\ \cdot & \cdot & \cdot & \cdot \\ \cdot & \cdot & \cdot & \cdot \\ 1 & (x_6-55)/20 & (x_6-55)^3/10000 & 2(1-100/x_6)/3 \end{pmatrix}$$

--with constants introduced to avoid excessive collinearity and greatly differing sums of squares of columns. We used this X for regressing the columns of Y onto concentrations x.

By similar stepwise regressions of the rows of Y onto powers of the temperatures z, we arrived at carrier matrix

$$Z = \begin{pmatrix} 1 & (z_1-25)/10 & (z_1-25)^2/200 \\ 1 & (z_2-25)/10 & (z_2-25)^2/200 \\ \cdot & \cdot & \cdot \\ \cdot & \cdot & \cdot \\ \cdot & \cdot & \cdot \\ 1 & (z_7-25)/10 & (z_7-25)^2/200 \end{pmatrix} .$$

We next regressed the rows and columns of the matrix Y (this consisted of the densities centered on their overall mean of 0.896765) onto carriers Z' and X, respectively, to fit the model

$$Y = XCZ' + \text{error} \tag{8}$$

by the usual least squares procedures. The resulting estimated coefficients

$$C = (X'X)^{-1}\, X'YZ\, (Z'Z)^{-1} \tag{9}$$

were

$$\hat{C} = \begin{pmatrix} .0080939 & -.0093667 & -.0000826 \\ -.0508069 & .0002825 & -.0001183 \\ -.00019 & .0000398 & .0000414 \\ .0115610 & -.0020857 & .0002912 \end{pmatrix} .$$

Thus, the constant 0.0080939 had to be added to the overall mean of 0.896765, the coefficient for (temperature-25)/10 was -.0093667, that for (concentration-55)/20 was -.0508069, that for the product of these two (an interaction term) was 0.0002825, etc.

This completed the first stage of our modelling and yielded the two-way regression model (8), whose fit was extraordinarily good; it reduced the sum of squares by 99.999977% -- see Table 2.

Table 2. Sums of Squares for fits and residuals

Source of Variation	(Model)	Sum of Squares	Independent Parameters
One Component	(11)	61,611,099 E-9	6
Residual		2,649,104 E-9	
Two Components	(4)	64,260,203 E-9	10
Residual		25 E-9	
Two-way Regression	(8)	64,260,228 E-9	12
Residual		15 E-9	
Total from Mean		64,260,243 E-9	41

In the second stage of our modelling strategy, we considered whether the coefficients $\hat{C}$ of the fit of the two-way regression model (8) could be modelled in order to obtain a simpler description of the data. That is the stage at which we introduced the idea of lower-rank approximation and principal components. However, unlike the usual way of applying these techniques directly to the (centered) data Y, we applied them to the two-way regression fit $X\hat{C}Z'$ and, more specifically, to the coefficients $\hat{C}$. Thus, we obtained lower rank approximations by using partial sums of the singular value decomposition

$$X\hat{C}Z' = \sum_{k=1}^{\ell} \lambda_k \underline{p}_k \underline{q}_k' \tag{10}$$

of the two-way fit $X\hat{C}Z'$, rather than of the data Y.

We first checked whether the rank of the coefficient matrix could be reduced. The simplest model was rank-one matrix $\hat{C} = \underline{a}\ \underline{b}'$. It could be written

$$Y = X\underline{a}\ \underline{b}'Z' \ + \text{error} \tag{11a}$$

or equivalently

$$y_{i,j} = (\sum_{u=1}^{4} x_{i,u}a_u)\ (\sum_{v=1}^{3} z_{j,v}b_v) + \text{error}\ . \tag{11b}$$

Thus, the elements of Y would be modelled as the product of one function $\sum_{u=1}^{4} x_{i,u}a_u$ of the concentrations and one function $\sum_{v=1}^{3} z_{j,v}b_v$ of the temperatures. Such a model would have been very attractive in its simplicity (and have involved only 6 independent parameters). However, it is evident from Table 2 that its fit to the alcohol density data was too poor to be useful -- its sum of squared residuals was

$$2{,}649{,}104\ E\text{-}9 + 25\ E\text{-}9 + 15\ E\text{-}9 = 2{,}469{,}144\ E\text{-}9\ .$$

Note that least squares fitting of model (11) uses the first component of (10) and solves

$$X\underline{a}\ \underline{b}'Z' = \lambda_1 \underline{p}_1 \underline{q}_1' \tag{12}$$

for $\underline{a}$ and $\underline{b}$.

We next considered a rank-two model, in which the coefficients could be expressed as

$$C = \underline{a}\ \underline{b}' + \underline{c}\ \underline{d}'. \tag{13}$$

That model could be written

$$Y = X\underline{a}\ \underline{b}'Z' + X\underline{c}\ \underline{d}'Z' + \text{error}\ , \tag{14a}$$

or equivalently,

$$y_{i,j} = (\sum_{u=1}^{4} x_{i,u}a_u)(\sum_{v=1}^{3} z_{j,v}b_v) + (\sum_{u=1}^{4} x_{i,u}c_u)(\sum_{v=1}^{3} z_{j,v}d_v) + \text{error} . \qquad (14b)$$

Here, the elements of Y were modelled as sums of two products, each of one linear function of the concentrations times one linear function of the temperatures. This was less intuitively appealing than the rank one model (11), but fit appreciably better: The sum of squared residuals was only 40 E-9 (see Table 2).

Least squares fitting of model (14) was analogous to that for model (11) except that it required two terms on both sides of an equation like (12). The resulting estimates were

$$\underline{a}\ \underline{b}' = \begin{pmatrix} .041770 \\ -.270041 \\ -.000908 \\ .061936 \end{pmatrix} (.188426, .003727, -.000313)$$

and

$$\underline{c}\ \underline{d}' = \begin{pmatrix} .111065 \\ -.015019 \\ .000422 \\ .026994 \end{pmatrix} (.003210, -.085722, -003029).$$

This model had 10 independent parameters as compared to the 16 parameters of Mandel's model (7) - and yet its fit was slightly closer. We think model (14) is simpler to understand than Mandel's model, but that may be a matter of personal taste.

We attempted further modelling of the coefficients by taking into account that the estimated $\underline{b}$ had an overwhelmingly large first element. We tried to model the other elements of $\underline{b}$ as zeroes, and obtained the model

$$y = X\underline{a}\ \underline{1}_7' + X\underline{c}\ \underline{d}'Z' + \text{error}, \qquad (15a)$$

because $\underline{b}'Z' = \underline{1}_7'$ when $\underline{b}' = (1,0,0)$. This model could be rewritten

$$y_{i,j} = \sum_{u=1}^{4} x_{i,u}a_u + (\sum_{u=1}^{4} x_{i,u}c_u)(\sum_{v=1}^{3} z_{j,v}d_v) + \text{error}. \qquad (15b)$$

The least squares fit of this model yielded coefficients

$$\underline{a} = \begin{pmatrix} .008053 \\ -.050866 \\ -.000170 \\ .011707 \end{pmatrix}$$

and

$$\underline{c}\,\underline{d}' = \begin{pmatrix} .108992 \\ -.003279 \\ .000460 \\ .024247 \end{pmatrix} (.001472, -.085925, -.002943)$$

(These were obtained by first fitting $X\underline{a}\,\underline{1}_7'$ by linear least squares, taking residuals and then obtaining $X\underline{c}\,\underline{d}'Z'$ as the first (principal) component of the singular value decomposition of the two-way regression fit of the residuals.) The coefficients of model (15) were found to be quite similar to those of the unrestricted rank two model (16). The sum of squared residuals from the fit of (15) was 104 E-9; not very much larger than that for model (14) and certainly a great deal better than that for the rank-one model (11).

Lower-rank approximation is not the only way to model the structure of the coefficients $\hat{C}$ of the two-way regression. Another approach that we tried exploits the fact that in the fit of model (8) all the large (in absolute value) coefficients were those of single factors $x_{i,u}$ or $z_{j,v}$, rather than of interaction terms $x_{i,u}z_{j,v}$. Thus, the large coefficients were $c_{1,1}$ for the constant, $c_{2,1}$ for $x_{i,2} = (x_i-55)/20$, $c_{4,1}$ for $x_{i,4} = 2(1-1/x_i)/3$, and $c_{1,2}$ for $z_{j,2} = (z_j-25)/10$. The largest interaction coefficient was $c_{4,2}$ for $(2(1-100/x_i)/3)(z_j-25)/10$. The "additive" model that makes all coefficients for interaction null is

$$Y = X\underline{a}\,\underline{1}_7' + \underline{1}_6\underline{b}'Z' + \text{error}, \tag{16a}$$

i.e.,

$$y_{ij} = \sum_{u=1}^{4} x_{i,u}a_u + \sum_{v=1}^{3} z_{j,v}b_v + \text{error}\ . \tag{16b}$$

Its coefficients' least squares estimates were

$$\underline{a}' = (.008194, -.050866, -.000170, .011707)$$

and

$$\underline{b}' = (\ \ 0\ \ , -.007936, -.000282).$$

(These were obtained by first fitting $X\underline{a}\,\underline{1}_7'$ by linear least squares and then fitting $\underline{1}_6\underline{b}'Z'$ to the residuals, but they could equally have been obtained by doing these two fits in reverse order.) This 6-parameter model (16) had sum of squared residuals 26,123 E-9, and clearly could not complete with the fits of models (14) and (15). However, it did fit very much better than the rank-one model (11). Evidently, for the alcohol density data, an additive decomposition of type (16) was much preferable to a multiplicative decomposition of type (11). It may be of interest that this finding agreed with Bradu and Gabriel's (1978) internal modelling of these data, which suggested that a rows-and-columns-additive model should fit very well.

5. SUMMARY OF MODELS AND FITS

A synopsis of the models and their fits is given in Figure 1.

Figure 1. An Array of Models for the 6-by-7 Matrix of Alcohol Density Data. (Models which are special cases of other models appear below them.)

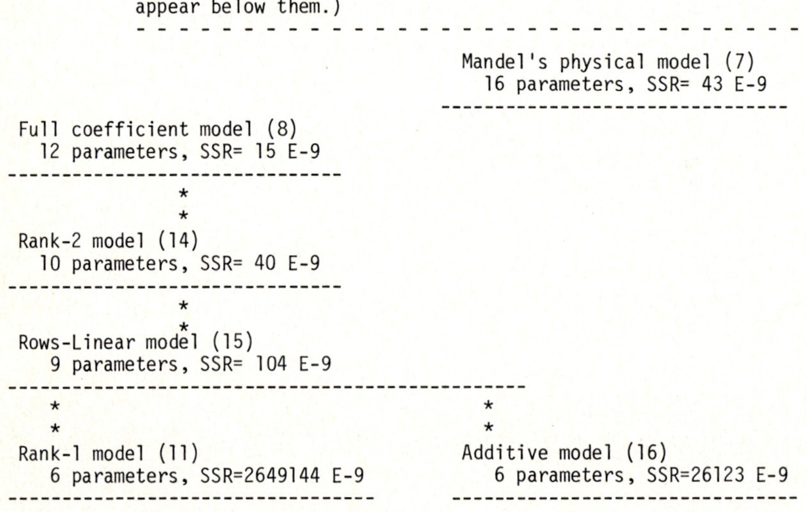

The advantage of some of the models proposed by us over Mandel's is evident, though small. One's individual preference of simplicity versus goodness of fit may lead one to opt for any one of these models ((8) or (14) or (15), or even to make do with the very simple (16)).

We believe that the reason that we found the better models is that we adopted the more logical strategy, in which external modelling precedes internal modelling, rather than the reverse strategy used earlier.

Extrapolation to the Larger 12-by-7 Data Set

When the various fitted models were extrapolated to all the levels of concentration in the original data set, Mandel's model (7) seemed to produce a much poorer fit than its newer competitors. However, the failure of model (7) occurred mostly at the very lowest concentration. When the extrapolation was limited to the remaining concentrations (in addition to the 6 used in fitting) Mandel's model acquitted itself well - fitting as well as most of the other models.

It is not quite clear what this says about the strategy of fitting: One should not really judge it by how well it does in extrapolation.

Refitting the 12-by-7 Data Set

The 6-by-7 alcohol data have served to illustrate the efficacy of our strategy of beginning with an external rather than an internal model. However, they could not illustrate the second stage of our strategy very well, because the coefficient matrix was of too low order -- 4-by-3 -- to allow effective rank reduction or other internal modelling of the coefficients. A better illustration may possibly be that of modelling the entire 12-by-7 alcohol data matrix. That may be large enough to give structural modelling of the coefficients a chance to play a real role.

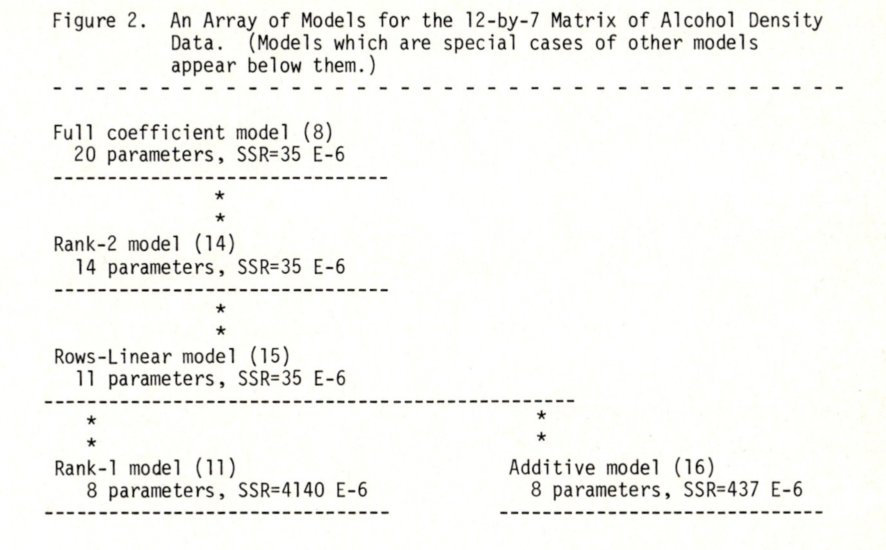

Figure 2. An Array of Models for the 12-by-7 Matrix of Alcohol Density Data. (Models which are special cases of other models appear below them.)

The goodness of fit of a number of models to the 12-by-7 data matrix is shown in Figure 2, which is analogous to Figure 1 both in definition and in methods of computation. In this wider data set the matrix of coefficients was 4-by-5 and this allowed a somewhat wider scope for modelling the coefficients. It is evident that the 20 parameter full coefficient model (8) can easily be replaced by model (14) or (15) of rank 2, or even further simplified to the additive model (16): That reduces the number of parameters by about one-half without a very large increase in the size of the residuals.

6. FINAL COMMENTS

Our case study has used least squares techniques exclusively. Some comments on the possible sensitivity of these fits to outliers have been made by Hoaglin, Wong and Emerson (1983) and resistant techniques for such fitting have been proposed by them and by Gabriel and Odoroff (1983). It may be useful to consider which parts of strategy of modelling are particularly sensitive to outliers and what resistant techniques may be needed when the presence of outliers is suspected. First, our stepwise method of selecting the carriers for the regression seems inherently resistant since it is carried out independently for each row and each column; an outlier would affect the stepwise regression only in one row and one column and thus should not influence the selection of the carriers. Secondly, however, the calculation of the two-way regression coefficients may be quite sensitive to outliers, as most least squares techniques are. In seeking a resistant alternative, we note that the minimizing criterion can be expressed in scalar as well as in matrix form, i.e.,

$$\min_{C} ||Y - XCZ'||^2 = \min_{c} ||\underline{y} - (Z \otimes X)\underline{c}||^2 , \qquad (17)$$

where y and c are obtained from Y and C by stacking the columns on top or one another (Mardia et al., 1979, p. 460). Now the right hand side of (17) is seen to be in the form of a multiple regression problem, and this suggests that resistant analogs of multiple regression could be introduced with a view to reducing sensitivity to outliers. Thirdly, the modelling of the coefficient matrix C does not require resistant techniques, since it follows the above fitting procedure which should have removed the effect of outliers (Hoaglin, 1984).

It may be of interest to consider how our strategy of two-way-regression-plus-decomposition relates to a well known method of scaling. Hirschfeld (1935) and Guttman (1941) proposed the use of the first principal component scores and loadings, or multiples of them, as "scales" for the rows and columns of Y. Such scaling may indeed be important for certain internal analyses of data; it is, however, of no interest to carry out two-way regression of Y onto these scales since that would merely recalculate the first principal component. Nor should the latter mathematical equivalence be misunderstood as a justification of PCA in terms of two-way regression, because the scales are calculated internally from the matrix Y itself, and do not use any external information on the rows and columns of that particular data matrix.

Another point is that in our example the concomitant variables had been chosen before the analysis began, but in other cases that may not be so. With alcohol density data it was clear from the beginning that rows of Y were to be modelled as functions of temperature and columns as functions of concentration, and so we used the stepwise regression procedure merely to select the powers of these given concomitants that would carry the regressions. If a wider choice of concomitants were available, the stepwise regression procedure could serve both to choose the concomitants and to select their powers (and products, where applicable) for inclusion among the carriers.

In practice, not all concomitants would be available at the outset, but inspection of the fits of a PCA model and its residuals may suggest new ones. In keeping with the basic logic of our strategy, we recommend that these be added to the pool of potential concomitants and the entire process of fitting be started afresh: First choosing carriers X and Z by stepwise regression, then fitting XCZ' to Y and finally simplifying $\hat{C}$ if possible.

The logic of this approach leads to a strategy of iterative modelling. At each iteration, some concomitants are introduced, the columns of X and of Z are determined by stepwise regression, XCZ' is fitted to Y, the coefficient matrix $\hat{C}$ is decomposed and modelled, and inspection of the fitted model and its residuals leads to some "explanation" and possibly to the suggestion of new concomitants. Iteration ends when no new concomitants are introduced.

The common usage of PCA can thus be regarded as starting with a null set of potential concomitants, so that the decomposition and modelling yield an internal analysis of Y itself. Inspection of that decomposition and its relation to external data may then lead to an indirect expression of the data -- via the components -- as functions of external concomitants for the rows and the columns. This makes inadequate use of the concomitant variables because they are not introduced into a new round of fitting and modelling to improve the fit.

Our basic contention is that the idea of modelling in terms of external or concomitant information, implies that the data can be thought of as consisting of a component that depends functionally on external concomitants, and a residual component that does not so depend. The classical strategy of modelling, as exemplified by Mandel's work and much other use of PCA, begins with an attempt to impose structure on the entire matrix, including the residual part. It then tries to relate that principal component structure to the concomitants. Surely it is more logical, when one seeks a model of the relationship to concomitants, to model only the component that depends on them. Another way of looking at this is in terms of the number of parameters used in fitting: Why introduce the additional $\underline{p}$ and $\underline{q}$ parameters at all, when one can more parsimoniously express the data directly in terms of the concomitants?

We contend that PCA is often used inappropriately. We realize that PCA can have an important role to play in "explaining" two-way data in terms of external concomitants when the latter are categorical or otherwise incomplete. In such cases it is often easier to relate the external concomitants to the principal components than to the data. However, we maintain that PCA is never appropriate as a final analysis of two-way data if numerical concomitant variables are available for the rows and the columns. For that case we recommend two-way regression of the data themselves in order to obtain a better fit.

ACKNOWLEDGEMENT: We are indebted to David Hoaglin and John Mandel for valuable comments and suggestions. The core of this paper was presented at the 1984 Annual Meeting of the American Statistical Association at Philadelphia, Pennsylvania. This work was supported in part by Office of Naval Research Contract N00014-80-C-0387.

REFERENCES

[1] Bradu, D., Description of a data matrix structure and model diagnosis by means of row and column Euclidean maps. Proceedings of the Seminar on Principal Components Analysis in the Atmospheric and Earth Sciences, Pretoria, C.S.I.R. (1983).

[2] Bradu, D. and Gabriel, K.R., The biplot as a diagnostic tool for models for two-way tables, Technometrics 20:47-68 (1978).

[3] Finlay, K.W. and Wilkinson, G.N., The analysis of adaptation in a plant breeding programme, Aust. J. Agricul. Res. 14:742-754 (1963).

[4] Gabriel, K.R., Least squares approximation of matrices by additive and multiplicative models, J. Roy. Statist. Soc. B 40:186-195 (1978).

[5] Gabriel, K.R. and Odoroff, C.L., Resistant lower rank and approximation of matrices, Computer Science and Statistics: The Interface, Gentle, J.E. (ed). Amsterdam, North-Holland, 304-308 (1983).

[6] Gittins, R., The application of ordination techniques. Ecological Aspects of the Mineral Nutrition of Plants. (British Ecol. Sco. Symp. 9, Rorison, I.H., ed)., Oxford, Blackwell Scientific Publications (1969).

[7] Guttman, L, The quantification of a class of attributes: A theory and method of scale construction. In: The Prediction of Personal Adjustment, Horst, P. (ed)., New York, Social Science Research Council, 319-348 (1941).

[8] Hirschfeld, H.O., A connection between correlation and contingency. Cambridge Philosophical Society Proceedings (Math. Proc.) 31:520-524 (1935).

[9] Hoaglin, D.C., Personal communication (1984).

[10] Hoaglin, D.C., Wong, G.Y., and Emerson, J.D., Resistant diagnosis of interaction in two-way tables, Harvard University, Statistics Memorandum AR-22 (1983).

[11] Mandel, J., Non-additivity in two-way analysis of variance, J. Amer. Statist. Assoc. 56:878-888 (1961).

[12] Mandel, J., A new analysis of variance model for non-additive data, Technometrics 11:411-429 (1971).

[13] Mandel, J., Evolution of a model for physical data appearing in two-way tables. Annual Meeting of the American Statistical Association, Section of Physical and Engineering Sciences, Philadelphia, August 15, 1984.

[14] Mardia, K.V., Kent, T., and Bibby, J.M., Multivariate Analysis, London, Academic Press (1979).

[15] Osborne, N.S., McKelvy, E.C., and Bearce, H.W., Density and thermal expansion of ethyl alcohol and of its mixtures with water, Bull. Bureau of Standards **9**:327-474 (1913).

[16] Snedecor, G.W. and Cochran, W.G., Statistical Methods (Seventh Edition), Ames, Iowa State University Press (1980).

[17] Tukey, J.W., One degree of freedom for non-additivity, Biometrics **5**:232-242 (1949).

[18] Yates, F. and Cochran, W.G., The analysis of groups of experiments, J. Agricul. Sci., Cambridge **28**:556-580 (1938).

BIOSTATISTICS: Statistics in Biomedical, Public Health
and Environmental Sciences, edited by P.K. Sen

A COMPARISON OF THE DISCRIMINATION OF DISCRIMINANT ANALYSIS AND LOGISTIC REGRESSION UNDER MULTIVARIATE NORMALITY

Frank E. Harrell, Jr. and Kerry L. Lee

Division of Biometry
Department of Community and Family Medicine
Duke University Medical Center
Durham, North Carolina

When sampling from two multivariate normal populations having equal covariance matrices, both the Fisher linear discriminant function (LDF) and logistic multiple regression model (LRM) can be used to derive valid estimates of the probability that a new observation comes from one of the two populations. In this setting, the LDF has been shown to yield asymptotically smaller relative classification error rates. When assumptions for the LDF are violated, LRM has been shown to be superior. In many situations, one is interested in using more information from a probability model than what is needed to devise a binary classification rule. In this paper we will study the relative performance of the LDF and LRM when all assumptions of the LDF are satisfied, to compare the spectrum of posterior probabilities arising from the two models. The cross-validation predictive accuracy and extent of separation (discrimination) of the posterior probabilities from the two methods will be assessed.

KEY WORDS: Discriminant analysis, logistic regression, predictive accuracy, discrimination, multivariate normality, maximum likelihood estimators, posterior probabilities

1. INTRODUCTION

Suppose that a p-dimensional random vector X can be observed from one of two p-variate normal populations with equal covariance matrices,

$$X \sim N_p\ (\mu_o, \Sigma) \text{ with probability } \Pi_o,$$

$$X \sim N_p\ (\mu_1, \Sigma) \text{ with probability } \Pi_1. \tag{1.1}$$

If a random sample under model (1.1) is available, say (X_1, Y_1), (X_2, Y_2),..., (X_n, Y_n), where Y_j is an indicator of the population being sampled so that $Y_j=1$ with probability Π_1, Bayes' theorem (Truett et al., 1967) can be used to derive a model for the probability of Y_j conditional on X_j:

$$\Pr\ (Y_j=1|X_j) = \frac{1}{1 + \exp\{-\ (a + X_j\beta)\}}\ . \tag{1.2}$$

Under the assumptions in (1.1), the maximum likelihood estimates of a and β are given by

$$\hat{\beta} = S^{-1}(\overline{X}_1,-\overline{X}_0)',$$
$$\hat{a} = \log_e n_1/n_0 - 1/2(X_0+X_1)\ \hat{\beta} \tag{1.3}$$

where

$$\overline{X}_0 = \sum_{y_j=0} X_j/n_0,\ \overline{X}_1 = \sum_{y_j=0} X_j/n_1,$$

$$S = [\sum_{y_j=0}(X_j-\overline{X}_0)(X_j-\overline{X}_0)' + \sum_{y_j=1}(X_j-\overline{X}_1)(X_j-\overline{X}_1)']/n \tag{1.4}$$

$$n_1 = \sum_{j=1}^{n} Y_j,\ n_0 = n - n_1.$$

These estimates form the basis of the LDF procedure, with the LDF being $\hat{a} + X\hat{\beta}$. The LDF has been used extensively, especially in biomedical and epidemiologic applications (see Abernathy, et al., 1966, for example). Lachenbruch (1975) has provided an excellent extensive bibliography.

By assuming only random sampling and (1.2), maximum likelihood estimates of a and β can be derived conditional on the observed X_j (Walker and Duncan, 1967). These estimates must be computed iteratively; there is no closed-form solution to the likelihood equations arising from (1.2). These conditional maximum likelihood estimates form the basis of the LRM estimation procedure.

Whichever estimation method is used, the customary binary classification rule consists of assigning an observation X to population 1 if the estimate of $\Pr(Y_j=1|X)$ from (1.2) is greater than 1/2, or to population 0 otherwise. Efron (1975) demonstrated that if (1.1) holds and n increases without bound, the LDF yields lower rates of binary classification errors relative to the LRM. He showed that the relative efficiency of the LRM also decreases when Π_1 is far from 1/2 or when the separation between populations increases. (This relative efficiency was found to range from about 1/2 to 2/3.) The separation is measured by the square root of the Mahalanobis distance, given by

$$D = [(\mu_1-\mu_0)'\ \Sigma^{-1}\ (\mu_1-\mu_0)]^{\frac{1}{2}}\ . \tag{1.5}$$

When (1.1) is not satisfied, Halperin, Blackwelder, and Verter (1971) have shown the LDF method yields biased estimates of a and β, and others (Press and Wilson, 1978, D'Agostino and Pozen, 1982) have shown that the LRM has better classification error rates. For a variety of reasons, Press and Wilson strongly advocate the use of the LRM when there is any evidence that (1.1) is violated, both in formulating a binary decision rule and in deriving absolute posterior probabilities.

Even when (1.1) holds, we believe there are several unanswered questions concerning the comparison of the LDF and LRM. 1) What is the relative behavior of the two for finite sample sizes? 2) Should one calculate relative error rates, or the difference of absolute error rates? 3) Should one even calculate error rates? Are the two models really only used to derive binary decision rules? Shouldn't a more general measure of predictive discrimination or predictive accuracy be used to judge the performance of the two models?

We believe the questions in 3) are especially important. It is apparent, especially in biological and medical applications, that these models are frequently used to derive probabilities of Y=1 rather than for making binary decisions, which can be quite arbitrary depending on the cutoff value chosen. Even when the user of the model is not accustomed to dealing with absolute probabilities, she often wishes to establish a "gray zone" for application of the model. In diagnostic modeling problems, for example, a physician may classify the disease as "present" if the estimated probability is greater than 0.90, "absent" if less than 0.10, and perform another diagnostic test if the probability is between 0.10 and 0.90. In this example, the trinary decision rule could be studied, but the element of arbitrariness of the 0.10 and 0.90 thresholds detracts from the analysis. Another major disadvantage of using simple error rates is that a predicted probability of say 0.51 carries the same penalty as a probability of 0.99 when the observation arose from population 0.

In this paper we will discuss four measures of comparing the predictive accuracy of the two models and present results of simulation studies that use these measures to compare LDF and LRM when (1.1) holds.

2. MEASURES OF PREDICTIVE ACCURACY

There are two important aspects of the predictive accuracy of a model. Reliability refers to the degree of bias of estimates. If a model estimates $Pr(Y=1|X)$ to be 0.80, 80% of observations with like values of X should have Y=1. Discrimination refers to the ability of a model to discriminate or separate values of Y. When Y is binary, discrimination is a measure of the extent to which observations with Y=1 have higher predicted probabilities of Y=1 than do the observations with Y=0. Discrimination is the more important aspect of predictive accuracy, we believe, because good discrimination is necessary for accurate predictions. Reliability can be achieved by calibration without affecting discrimination, whereas a model that does not discriminate well cannot be fixed.

For each method of assessing accuracy discussed below, it is assumed that the model is derived on a training sample and tested in a separate test sample, to obtain an unbiased estimate of the method's accuracy.

A very general measure of discrimination can be derived from the Kendall-type (Goodman and Kruskal, 1979) rank correlation between predicted probabilities and observed outcomes. The simplest way to state such correlation indices is through the concordance probability which we label c. When predicting the probability that Y=1, this index is defined by the proportion of pairs of observations, one having Y=0 and the other having Y=1, such that the one having Y=1 also had the higher predicted probability. If the two probabilities are tied, that pair receives a score of 1/2 instead of 0 or 1, i.e.

$$c = \sum_{\substack{i=1 \\ Y_i=0}}^{n} \sum_{\substack{j=1 \\ Y_j=1}}^{n} [I(P_j > P_i) + 1/2\ I(P_j = P_i)]/n_0 n_1, \tag{2.1}$$

where P_k denotes an estimate of $P(Y_k=1|X_k)$ from (1.2). The quantity 2(c-1/2) is Somer's D_{YP} rank correlation coefficient (Goodman et al., 1979). In this binary Y setting, c is proportional to the Wilcoxon-Mann-Whitney statistic for comparing P_k values from the Y=0 and Y-1 samples. The c index in this case is also the area under a "receiver operating characteristic" curve, a quantity routinely used to measure the diagnostic ability of medical tests (Hanley and McNeil, 1982). The c index takes on a value of 1 for perfect discrimination and 1/2 for random predictions. An advantage of the c index is its ease of interpretation and its generalizability to more complex problems such as ordinal response and censored survival time data (Harrell, et al., 1984).

The c index is purely a measure of discrimination. There are many measures of predictive accuracy that take discrimination into account but also penalize a predictor for being unreliable. One such measure is the logarithmic probability scoring rule (Cox, 1970), stated by Shapiro (1977) as

$$Q = \sum_{i=1}^{n} [1+\log_2 (P_i^{Y_i} (1-P_i)^{1-Y_i})]/n . \tag{2.2}$$

Q obtains a value of 1 for perfect predictions ($P_i=Y_i$ for all i), 0 for random predictions, and values less than 0 for predictions that are worse than random.

Another accuracy score similar to Q is the quadratic score of Brier (1950) considered extensively in meteorologic forecast assessment. We will state this index here as

$$B = 1 - \sum_{i=1}^{n} (P_i - Y_i)^2/n \qquad (2.3)$$

Perfect predictions receive a score of 1, and perfectly bad predictions receive a score of 0.

A fourth way to quantify the relative accuracy of two predictors, which also takes both reliability and discrimination into account, involves using both predictors as covariates in a model to predict the values of Y in the test sample. We then ask the question: do the predictions of method 1 (P_i^1) add information about predicting Y to the predictions of method 2 (P_i^2) and vice-versa? A formal statistical test can readily be made for both hypotheses, and the relative predictive ability of each method may be measured by comparing a statistic testing the strength of an individual predictor to a statistic for testing the strength of the best linear combination of the predictors. Since Y is binary, it is natural to use the LRM itself for this purpose, taking as covariates logit (P_i^1) and logit (P_i^2), where logit (q) = $\log_e[q/(1-q)]$. Then a statistic for testing whether P_i^1 adds information to that provided by P_i^2 is a likelihood ratio chi-square statistic with one degree of freedom. A measure of the "adequacy" of P_i^1 is given by

$$A(P_i^1) = \frac{L(\text{logit } P_i^1)}{L(\text{logit } P_i^1, \text{ logit } P_i^2)}, \qquad (2.4)$$

where L(logit P_i^1) is the log likelihood due to logit P_i^1 alone, and L(logit P_i^1 logit P_i^2) is the log likelihood due to the best linear combination of logit P_i^1 and logit P_i^2. It should be noted that L(logit P_i^1) would be a linear translation of Q in (2.2) had the logistic slope coefficient for logit P_i^1 been fixed at 1 and the intercept been fixed at 0. However, we are allowing these parameters to be estimated in the test sample. We define $A(P_i^2)$ in a similar way using (2.4). A disadvantage of both the Q and A measure is that if predicted probabilities of 0 or 1 exist, the measures are undefined. Since estimates of β can be infinite for the LRM, this can be a problem. However, there is no problem with the LRM predicted probabilities themselves in such cases.

3. SIMULATION STUDIES

To estimate the four measures of predictive accuracy for LDF and LRM discussed in section 2 based on independent model validation, a series of simulation experiments was conducted. The value of p (the dimension of X) was fixed at 5 for all studies, and the covariance matrix was fixed at

$$\Sigma = \begin{bmatrix} 1 & 0 & 0 & 0 & 0 \\ & 1 & 1/2 & 1/2 & 0 \\ & & 1 & 1/2 & 0 \\ & & & 1 & 0 \\ & & & & 1 \end{bmatrix}$$

and μ_0 was fixed at (0, 0, 0, 0, 0). Values of μ_1 were varied to yield different separations (1.5) as shown in Table 1.

Table 1. Values of μ_1 and Corresponding Separations

μ_1					D
.5	.5	.5	.5	.5	.94
1.0	1.0	1.0	1.0	1.0	1.87
1.4	1.4	1.4	1.4	1.4	2.62
1.75	1.75	1.75	1.75	1.75	3.27
2.0	2.0	2.0	2.0	2.0	3.74
2.5	2.5	2.5	2.5	2.5	4.68

The sample sizes consisted of n=50 and n=130. The prior probability Π_1, was set at 0.65 and 0.85 because the results of Efron (1975) suggest that LRM should have decreased efficiency as Π_1 moves farther from 0.5. For each combination of n, μ_1, and Π_1, a training sample and test sample each of size n was generated using the RANNOR and RANUNI random number generators in SAS (Ray, 1982a). The LDF coefficients were estimated using the SAS DISCRIM procedure (Ray, 1982b), and the LRM parameters were fitted using the LOGIST procedure (Harrell, 1983). Posterior probabilities were then calculated on the test sample and c, B, and Q were calculated for the test sample. These c, Q, and B values were averaged over 60 samples (replications) for each combination of parameters. For the evaluation which used LRM likelihood ratio statistics in (2.4), all test samples were combined for a particular combination of parameters and only one set of summary statistics was calculated based on [illegible] observations.

4. RESULTS

The results of the simulation study are shown in Table 2. It can readily be seen that the discrimination ability of both the LDF and LRM increases smoothly with increasing D. In the situations studied, LDF is superior to LRM, but by quite a small amount, judged by c, Q and B. The index A, where computable, quantifies the "adequacy" of LRM at 90% on the average.

Table 2. Discrimination and Predictive Accuracy Logistic Regression Model vs. Linear Discrimination Function

			c		A		Q		B		
D	Π_1	n	LDF	LRM	LDF	LRM	LDF	LRM	LDF	LRM	d
.94	.65	50	.69	.69	.99	.85	.11	.07	.79	.79	.06
		130	.72	.72	1.00	.95	.16	.15	.80	.80	.06
	.85	50	.67	.67	--	--	--	--	.87	.86	.06
		130	.71	.71	1.00	.88	.44	.42	.88	.88	.03
1.87	.65	50	.89	.88	1.00	.90	--	--	.87	.86	.06
		130	.90	.90	1.00	.95	.45	.42	.88	.88	.04
	.85	50	.88	.85	--	--	--	--	.91	.89	.06
		130	.90	.89	.99	.86	.62	.57	.92	.92	.03
2.62	.65	50	.97	.94	--	--	--	--	.93	.89	.08
		130	.96	.96	1.00	.93	.65	.59	.93	.92	.03
	.85	50	.94	.92	--	--	--	--	.94	.91	.08
		130	.96	.96	1.00	.91	.77	.71	.95	.95	.03
3.27	.65	50	.99	.97	--	--	--	--	.96	.93	.06
		130	.99	.98	--	--	--	--	.96	.95	.03
	.85	50	.98	.97	--	--	--	--	.97	.95	.05
		130	.98	.97	--	--	--	--	.97	.95	.05
3.74	.65	50	.99	.97	--	--	--	--	.97	.94	.05
4.68	.65	50	1.00	1.00	--	--	--	--	.99	.98	.02

D defined in (1.5)
Π_1 = prior probability
c defined in (2.1)
A defined in (2.4)
A defined in (2.2)
B defined in (2.3)
d = average $|P^1 - P^2|$
-- = could not be estimated due to at least one infinite estimate of β

The asymptotic results of Efron (1975) suggest that the relative performance of LRM should suffer for larger D. There is a small indication of this in Table 2. For example, c differs by more than 0.01 in many cases for larger D. However, as D exceeds 2, both LDF and LRM appear to offer excellent predictive accuracy.

Another way to study the relative quality of posterior probabilities arising from LDF and LRM is to assess how the two sets of probabilities differ on the average. Over all cases studied, the average absolute difference between P^1 and P^2 was 0.06 when the prior probability was 0.65 and 0.04 for a prior probability of 0.85. Thus the two methods yield very similar probability estimates on the average.

Examining the distributions of absolute differences according to the level of LDF posterior probability sheds further light. For small to intermediate D, LRM probabilities agreed well with LDF for all levels of LDF probabilities. Table 3 displays this distribution for a large value of D (4.68) and for n=50, $\Pi_1 = 0.65$. For this case, the overall average difference was 0.02, and 2858 of the 3000 test observations had an LDF posterior probability less than 0.05 or exceeding 0.95. Of these, the average absolute difference between LDF and LRM probabilities was 0.007. Of the remaining 142 observations, the average disagreement was 0.26. Thus, a problem with LRM may lie in an instability of estimates in the middle probability range. In samples with large separations, one or more of the maximum likelihood estimates of regression coefficients may be infinite. For such situations, some large finite regression estimates must be used, and there will not be many estimated probabilities intermediate betwen zero or one. The ones that do fall in this intermediate range are likely to be unstable. This is probably related to Efron's finding for large D regarding a dropoff in relative accuracy for LRM using a rule based on exceeding the 0.5 predicted probability cutoff. However, it is important to note that in this situation, the LRM is able to separate Y=0 from Y=1 very well with extreme probabilities. Also, the proportion of correct classifications (using a 0.5 probability cutoff) for this extreme separation case is virtually as good with LRM as with LDF (0.986 vs. 0.976). Here the ratio of <u>correct classification</u> rates is near one although the relative efficiency of LRM as measured by the ratio of classification <u>errors</u> is 0.58.

Table 3. Distribution of Absolute Differences Between LDF and LRM Probability Estimates n=50, D=4.68, Π_1=.65 All Test Samples Combined

P^1	Frequency	Average $\|P^1-P^2\|$
0 - .05	971	.01
.05 - .15	32	.17
.15 - .25	11	.32
.25 - .35	11	.36
.35 - .45	12	.47
.45 - .55	8	.34
.55 - .65	8	.47
.65 - .75	14	.31
.75 - .85	8	.21
.85 - .95	38	.16
.95 -1.00	1887	.005
Overall	3000	.02

5. DISCUSSION AND CONCLUSIONS

In assessing the predictive accuracy of a given method, it is important to study the entire spectrum of predicted values. The purpose of a probability model is generally to predict absolute probabilities. In this light, the properties of predicted probabilities from the LRM or LDF in relation to the dependent variable Y should be used to gauge the usefulness of a LRM or LDF. Relative classification rates, especially misclassification rates, are not adequate measures of relative performance for probability models.

If one accepts this premise, one must choose a measure of discrimination or predictive accuracy. All four such measures studied here indicate that even when the conditions under which LDF was optimized are satisfied, the performance of LRM is very nearly as good as that of LDF for reasonable sample sizes and values of D. Predicted probabilities from the two methods also have a small absolute difference on the average when multivariate normality holds. Since others (Halperin et al., 1971) have shown that LDF can yield arbitrarily biased probability estimates when its assumptions are violated (e.g. one of the predictor variables is dichomotous), we argue that LRM is the tool of first choice among these two competitors. With the availability of efficient computers and computer programs, the issue of the computational requirements of the LRM becomes unimportant.

ACKNOWLEDGEMENTS: This work was supported by Research Grant HL-17670 from the National Heart, Lung, and Blood Institute, Bethesda, Maryland; Research Grants HS-03834 and HS-04873 from the National Center for Health Services Research, OASH, Hyattsville, Maryland; Training Grant LM-07003 and Research Grants LM-03373 and LM-00042 from the National Library of Medicine, Bethesda, Maryland; and grants from the Prudential Insurance Company of America, Newark, New Jersey; the Kaiser Family Foundation, Palo Alto, California, and the Andrew W. Mellon Foundation, New York, New York.

REFERENCES:

[1] Abernathy, J.R., Greenberg, B.G., and Donnelly, J.F., Application of discrimination functions in perinatal death and survival, Am. J. Obstetrics and Gynecology **95**:860-7 (1966).

[2] Brier, G.W., Verification of forecasts expressed in terms of probability, Monthly Weather Rev. **78**:1-3 (1950).

[3] Cox, D.R., The Analysis of Binary Data, London, Metheun (1970).

[4] D'Agostino, R.B. and Pozen, M.W., The logistic function as an aid in the detection of acute coronary disease in emergency patients (a case study), Stat. in Med. **1**:41-8 (1982).

[5] Efron, B., The efficiency of logistic regression compared to normal discriminant analysis, J. Amer. Statist. Assoc. **70**:892-8 (1975).

[6] Goodman, L.A. and Kruskal, W.H., Measures of Association for Cross-Classifications, New York, Springer-Verlag (1979).

[7] Halperin, M., Blackwelder, W.C., and Verter, J.I., Estimation of the multivariate logistic risk function: a comparison of the discriminant function and maximum likelihood approaches, J. Chron. Dis. **24**:125-158 (1971).

[8] Hanley, J.A. and McNeil, B.J., The meaning and use of the area under a receiver operating characteristic (ROC) curve, Radiology **143**:29-36 (1982).

[9] Harrell, F.E., Lee, K.L., Califf, R.M., Pryor, D.B., and Rosati, R.A., Regression modeling strategies for improved prognostic prediction, Stat. in Med. **3**:143-152 (1984).

[10] Harrell, F.E., The LOGIST Procedure. In: SUGI Supplemental Library User's Guide, 1983 Edition, Joyner, S.P. (ed). Cary, NC, SAS Institute, 181-202 (1983).

[11] Lachenbruch, P.A., Discriminant Analysis, New York, Hafner Press, 96-126 (1975).

[12] Press, S.J. and Wilson, S., Choosing between logistic regression and discriminant analysis, JASA **73**:699-705 (1978).

[13] Ray, A.A. (ed), SAS User's Guide: Basics, Cary, NC, SAS Institute (1982a).

[14] Ray, A.A. (ed), SAS User's Guide: Statistics, Cary, NC, SAS Institute 461-73 (1982b).

[15] Shapiro, A.R., The evaluation of clinical predictions. A method and initial application, New Engl. J. Med. **296**:1509-1514 (1977)

[16] Truett, J., Cornfield, J., and Kannel, W.B., A multivariate analysis of the risk of coronary heart disease in Framingham, J. Chron. Dis. **20**:511-24 (1967).

[17] Walker, S.H. and Duncan, D.B., Estimation of the probability of an event as a function of several independent variables, Biometrika **54**:167-79 (1967).

BIOSTATISTICS: Statistics in Biomedical, Public Health and Environmental Sciences, edited by P.K. Sen

TESTING FOR INDIRECT CENSORING: A GENERAL PURPOSE TEST AND SOME MOMENT FORMULAE

Norman L. Johnson
University of North Carolina
Chapel Hill, NC

Jeremy E. Dawson
Australian National University
Canberra, A.C.T.
Australia

In an earlier paper [1], most powerful tests for certain types of indirect censoring (censoring of known numbers of extreme values of an unobserved variable, related to the observed variable(s)) was developed. In the present paper, a 'general purpose' test, which can be used when the numbers of extreme values which are censored, is constructed, using a technique analogous to that used (for direct censoring) in [2]. Formulae for moments of the test statistic, and also for correcting a conjecture made in [1], are obtained.

KEY WORDS: Censoring, multivariate problems, Farlie-Gumbel-Morgenstern distributions, likelihood ratio, combinatorics.

1. INTRODUCTION

Indirect censoring arises when data are censored in respect of one variable (or more) but observations are available only on other variables. We are concerned with the problem of using the available data to test whether censoring has, in fact, occurred.

In [1] attention was focused on the simplest case, when the available data consist of r sample values $\underset{\sim}{X}_2 = (X_{21},\ldots,X_{2r})$ of a variable X_2, which correspond to the values in a complete random sample of size n $(=r+s_0+s_r)$ from which the individuals with the s_0 least and s_r greatest values of another variable X_1, have been removed. It was assumed that the joint distribution of X_1 and X_2 is absolutely continuous, and is known. The problem is that of testing the hypothesis of no censoring, $H_{0,0}$ $(s_0=s_r=0)$ against alternatives H_{s_0,s_r}. A most powerful (likelihood ratio) test with respect to specified (s_0,s_r) was derived. The special cases of censoring from below $(s_0>0,\ s_r=0)$, or from above $(s_0=0,\ s_r>0)$, and symmetrical censoring $(s_0=s_r=s)$ when the joint distribution of X_1 and X_2 is bivariate Farlie-Gumbel-Morgenstern were considered in some detail.

In the present paper we discuss construction of a 'general purpose' test for use when the alternative values of s_o and s_r are not specified, and also provide formulae from which can be determined moments of the distribution of test criteria for the special cases discussed in [1], correcting a conjecture made in that paper.

We first recapitulate, briefly, relevant results from [1].

2. MOST POWERFUL TEST OF $H_{o,o}$ WITH RESPECT TO H_{s_o,s_r}

Using the likelihood ratio approach, the most powerful test of $H_{o,o}$ with respect to H_{s_o,s_r} was found to have a critical region of form

$$E[\{F_1(\min \underset{\sim}{X}_1)\}^{s_o}\{1-F_1(\max \underset{\sim}{X}_1)\}^{s_r}|\underset{\sim}{X}_2,H_{o,o}] > K_\alpha \tag{1}$$

where $F_1(\cdot)$ is the cumulative distribution function (CDF) of each X_1, and $X_{11},\ldots,X_{1r}$ are mutually independent with X_{ij} having the conditional distribution of X_1 given $X_2=X_{2j}$ $(j=1,\ldots,r)$.

If the (known) joint distribution of X_1 and X_2 is of Farlie-Gumbel-Morgenstern form, then without loss of generality we can take it to be given by

$$\Pr[(X_1\le x_1)\cap(X_2\le x_2)] = F_{12}(x_1,x_2) = x_1x_2\{1+\theta(1-x_1)(1-x_2)\} \tag{2}$$

$$(0\le x_j\le 1;\ |\theta|\le 1)$$

[If it is not originally of this form then the probability integral transforms $X_j = F_j(X_j')$ will make it so.]

The most powerful test then has critical region

$$L > K_\alpha \tag{3}$$

with

$$L = 1 + \sum_{h=1}^{r}(-1)^h \frac{s_o^{[h]}}{(r+s_o+1)^{[h]}}\theta^h Y_h \quad \text{(for censoring from below)} \tag{4.1}$$

$$L = 1 + \sum_{h=1}^{r} \frac{s_r^{[h]}}{(r+s_r+1)^{[h]}}\theta^h Y_h \quad \text{(for censoring from above)} \tag{4.2}$$

$$L = 1 + \sum_{1\le k\le r/2} \frac{(k+1)^{[k]}s^{[k]}}{(r+2s+1)^{[2k]}}\theta^{2k} Y_{2k} \quad \text{(for symmetrical censoring)} \tag{4.3}$$

where $a^{[b]} = a(a+1)\ldots(a+b-1)$;

$$Y_o = 1;\ Y_h = \sum_{j_1<\ldots<j_h} \Pi_{i=1}^{h}(1-2X_{2j_i})$$

3. A GENERAL PURPOSE TEST OF $H_{o,o}$

In [2], tests for direct censoring of a single variable X were considered. The most powerful test of $H_{o,o}$ with respect to H_{s_o,s_r} has critical region

$$\{F(\min X)\}^{s_o}\{1-F(\max X)\}^{s_r} > K_\alpha \tag{5}$$

where $F(\cdot)$ is the CDF of X. The similarity to (1) is clear.

A general purpose test (s_o and s_r unknown) for this situation was suggested in [2]. It has critical region

$$F(\min X) + \{1-F(\max X)\} > K_\alpha$$

Analogy with (1) and (5) suggests consideration of a criterion

$$M = E[F_1(\min \underset{\sim}{X}_1) + 1 - F_1(\max \underset{\sim}{X}_1)|\underset{\sim}{X}_2] \tag{6}$$

as a basis for a general purpose test of indirect censoring on X_1, using observed values $\underset{\sim}{X}_2$ of X_2. As in the univariate case (see [2], p. 384), this critical region is most powerful if, for given $(s_o+s_r) = s$, say, all values $(0,1,\ldots,s)$ of s_o are equally likely. We now evaluate M for the Farlie-Gumbel - Morgenstern distribution.

The conditional cumulative distribution functions, given $\underset{\sim}{X}_2$, of $C = \min \underset{\sim}{X}_1$ and $D = \max \underset{\sim}{X}_1$ are, respectively

$$1 - \prod_{j=1}^{r}\{1-F_{1|2}(c|X_{2j})\} \text{ and } \prod_{j=1}^{r} F_{1|2}(d|X_2)$$

and the corresponding density functions are:

$$\sum_{i=1}^{r} f_{1|2}(c|X_{2i}) \prod_{j\neq i}\{1-F_{1|2}(c|X_{2j})\} \text{ and } \sum_{i=1}^{r} f_{1|2}(d|X_{2i}) \prod_{j\neq i} F_{1|2}(d|X_{2j})$$

where $F_{1|2}(\cdot|X_2)$, $f_{1|2}(\cdot|X_2)$ denote the conditional cumulative distribution function and density function, respectively of X_1 given X_2. So in (6)

$$M = 1 + \sum_{i=1}^{r}\int_{-\infty}^{\infty} F_1(y) f_{1|2}(y|X_{2i}) \left[\prod_{j\neq i}\{1-F_{1|2}(y|X_{2j})\} - \prod_{j\neq i} F_{1|2}(y|X_{2j})\right]dy \tag{7}$$

If $$F_{1|2}(y|X_{2j}) = H_i(f_1(y),X_{2j}) \text{ so that}$$

$$f_{1|2}(y|X_{2j}) = f_i(y)h_1(F_1(y), X_{2j}) \tag{8}$$

with
$$h_1(u,v) = \frac{\partial H_1(u,v)}{\partial u}$$

then
$$M = 1 + \sum_{i=1}^{r} \int_0^1 zh_1(z,X_{2i}) \left[\prod_{j\neq i} \{1-H_1(z,X_{2j})\} - \prod_{j\neq i} H_1(z,X_2)\right]dz \tag{9}$$

If X_1 and X_2 have the joint distribution (2), then

$$\left.\begin{aligned} H_1(z,x_2) &= z[1+\theta(1-2x_2)(1-z)] \\ 1-H_1(z,x_2) &= (1-z)\{k-\theta(1-2x_2)z\} \\ h_1(z,x_2) &= 1+\theta(1-2x_2)(1-2z) \end{aligned}\right\} \tag{10}$$

and

$$M = 1 + \sum_{i=1}^{r} \int_0^1 z\{1+\theta(1-2X_{2i})(1-2z)\}[(1-z)^{r-1} \prod_{j\neq i} \{1-\theta(1-2X_{2j})z^{r-1}\} \prod_{j\neq i}\{1+\theta(1-2X_{2j})(1-z)\}]dz$$

$$= 2(r+1)^{-1} + 2 \sum_{1\le k\le r/2} \frac{(2k)!\ r!}{(r+2k+1)!} \theta^{2k} Y_{2k} \tag{11}$$

There are $\binom{r}{2k}$ terms in the summation for Y_{2k}. Introducing the arithmetic mean

$$\overline{Y}_{2k} = Y_{2k}/\binom{r}{2k} \tag{12},$$

(11) can be written

$$(r+1)M = 2 + 2 \sum_{1\le k\le r/2} \frac{r^{(2k)}}{(r+2)^{[2k]}} \theta^{2k}\overline{Y}_{2k} \tag{11)'$$

where $a^{(b)} = a(a-1)..(a-b+1)$.

The criterion in (4.3), for testing for symmetrical censoring, has evident points of similarity to (11)'.

Formula (11) (or (11)') requires knowledge of θ for its evaluation (though it is not necessary to know its *sign*). Since $|z|\le 1$ we will typically (though not always) have $|\overline{Y}_2| \ge |\overline{Y}_4| \ge \ldots$. As the coefficients $\theta^{2k} r^{(2k)}/(r+2)^{[2k]}$ decrease as k increases, it seems reasonable to consider using a test with critical region $Y_2 > K$ when θ is not known, as was suggested in [2] for the case $s_o = s_r$.

In [2] it was suggested that a normal distribution might be used to approximate the null hypothesis ($H_{o,o}$) distribution of Y_2, but this does not seem to be justifiable, even for large r. (For Y_1, on the other hand, it *is* a good approximation.)

The approximation (see Appendix I) $1 + 6Y_2r^{-1}$ distributed as χ^2 with one degree of freedom seems to be appropriate. The corresponding critical region would be

$$Y_2 > \frac{1}{6}r(\lambda^2_{(\frac{1}{2})\alpha} - 1)$$

where $\lambda_{\frac{1}{2}\alpha}$ is the upper $\frac{1}{2}\alpha$ point of the unit normal distribution ($\Phi(\lambda_{(\frac{1}{2})\alpha}) = 1-\frac{1}{2}\alpha$).

4. MOMENT FORMULAE UNDER H_{s_o,s_r}

Evaluation of moments of the statistics L and M in Sections 2 and 3 can be effected by finding

(i) the conditional expected value, given $\underset{\sim}{X}_1$ and then

(ii) the expected value of (i) when the joint distribution of $\underset{\sim}{X}_1$ is that of the (s_o+1)-th through (s_o+r)-th order statistics among $(r+s_o+s_r)$ independent standard uniform variables.

Since

$$E[(1-2X_{2i})^q|\underset{\sim}{X}_1] = E[(1-2X_{2i})^q|X_{1i}] = \int_0^1(1-2x)^q\{1+\theta(1-2x_{1i}(1-2x)\}dx$$

$$= \begin{cases} \theta(q+2)^{-1}(1-2X_{1i}) & \text{if q is odd} \\ (q+1)^{-1} & \text{if q is even} \end{cases} \quad (13)$$

we need only the formula

$$E[\prod_{i=1}^{p} X_{1h_i}|H_{s_o,s_r}] = \{\prod_{i=1}^{p}(s_o+h_i+i-1)\}/(r+s_o+s_r+1)^{[p]} \quad (14)$$

$(p\le r)$ in step (ii). This leads to evaluation of quantities

$$T_p(r,s_o,s_r) = \sum_{1\le h_1<h_2<\ldots<h_p\le r}^{r-p+1} \ldots \sum^{r} E[\prod_{i=1}^{p}(1-2X_{1h_i})|H_{s_o,s_r}]$$

$$= \sum_{u=0}^{p}(-2)^u\{(r+s_o+s_r+1)^{[u]}\}^{-1} \sum_{1\le h_1<\ldots<h_p\le r}^{r-p+1}\ldots\sum^{r} \sum_{1\le i_1}^{r-u+1}\ldots\sum_{i_u\le p}^{p}\prod_{\alpha=1}^{u}(s_o+h_{i_\alpha}+\alpha-1)$$

$$= \sum_{u=0}^{p} (-2)^u \{(r+s_o+s_r+1)^{[u]}\}^{-1} \binom{r-u}{p-u} \sum_{1\le h_1<\ldots<h_u\le r}^{r-u+1} \cdots \sum^{r} \prod_{\alpha=1}^{u} (s_o+h_\alpha+\alpha-1) \tag{15}$$

(since for given $(h_{i_1},\ldots,h_{i_u})$ the remaining (p-u) h's can be chosen in any was from the (r-u) remaining integers from 1 to r).

We have for example

$$\left.\begin{aligned} E[Y_q|H_{s_o},s_r] &= \left(\tfrac{1}{3}\right)^q T_q\theta^q \\ E[Y_1Y_2|H_{s_o},s_r] &= \tfrac{1}{9}(r-1)T_1\theta + \tfrac{1}{9}T_3\theta^3 \\ E[Y_2^2|H_{s_o},s_r] &= \tfrac{1}{18} r(r-1) + \tfrac{2}{27}(r-2)T_2\theta^2 + \tfrac{2}{27} T_4\theta^4 \end{aligned}\right\} \tag{16}$$

In [1] formulas for T_1, T_2, T_3 and T_4 were given, and a general formula for T_p was conjectured. Unfortunately, the formula for T_4 is erroneous and the conjecture is not valid, except when $s_0 = s_r$.

In Appendix II it is shown that, in fact

$$T_p(r,s_0,s_r) = \frac{(-1)^p r^{(p)}}{(r+s_0+s_r+1)^{[p]}} \sum_{t=0}^{p} (-1)^t \binom{s_0+p-t-1}{p-t}\binom{s_r+t-1}{t} \tag{17}$$

5. CONCLUDING REMARKS

Even for the special case considered here, quite complicated formulae are obtained. For applications to other cases - for example, multinormal joint distributions - formulae are even less elegant. However, with sufficient preparation, involving quite heavy, though straightforward, computations, application will be practicable when the necessary investment of time and effort is deemed justifiable.

ACKNOWLEDGEMENTS: The authors wish to thank Dr. David Culpin for some very helpful suggestions.

REFERENCES

[1] Johnson, N.L., Extreme sample censoring problems with multivariate data: Indirect censoring and the Farlie-Gumbel-Morgenstern distribution, J. Multiv. Anal. 10:351-362 (1980).

[2] Johnson, N.L., A general purpose test of censoring of extreme sample values, S.N. Roy Memorial Volume, University of North Carolina Press. pp. 377-384 (1970).

APPENDIX I

We first show that if $Y_2 = \sum_{i<j}^{r-1}\sum^{r} Z_iZ_j$ and the Z's are independent and identically distributed, with zero expected value and all moments ($\mu_q'=\mu_q$) finite, then

(I) the limit of the moment ratio $\{\mu_q(Y_2)\}/\{\mu_2(Y_2)\}^{\frac{1}{2}q}$ as $r \to \infty$ is finite and does not depend on the common distribution, and

(II) if the Z's are normal then the limiting distribution of

$Y_2/\{\frac{1}{2}r(r-1)\mu_2{}^2\}^{\frac{1}{2}}$ is that of $\frac{1}{\sqrt{2}}\{(\chi^2$ with 1 d.f.$)-1\}$.

Proof of (I) $\quad E[Y_2) = 0;$

$$\mu_q(Y_2) = E[(\sum_{i<j}^{r-1}\sum^{r} Z_iZ_j)^q] \text{ and so, if } r > q,$$

$$\mu_q(Y_2) = \binom{r}{q}g_q\mu_2{}^q + \text{terms in } r^{q-1}, r^{q-2}, \ldots r, 1$$

where g_q is the coefficient of $Z_1^2Z_2^2\ldots Z_q^2$ in $(\sum_{i<j}^{r-1}\sum^{r} Z_iZ_j)^q$ - the same for any $r > q$. Hence

$$\lim_{r\to\infty} \frac{\mu_q(Y_2)}{\{\mu_2(Y_2)\}^{\frac{1}{2}q}} = \frac{2^q}{q!}\;\frac{g_q}{g_2^{\frac{1}{2}q}} \tag{A1}$$

Proof of (II) If the Z's are normal N(0,1), then

$$Y_2 = \tfrac{1}{2}\{(\sum_{i=1}^{r} Z_i)^2 - \sum_{i=1}^{r} Z_i^2 = \tfrac{1}{2}\{(r-1)(\bar{Z}\sqrt{r})^2 - \sum_{i=1}^{r}(Z_i - \bar{Z})^2\} \text{ with } \bar{Z} = r^{-1}\sum_{i=1}^{r} Z_i$$

$$= \tfrac{1}{2}\{(r-1)\chi_1^2 - \chi_{r-1}^2\}$$

where χ_1^2, χ_{r-1}^2 are mutually independent, so

$$\frac{Y_2}{\sqrt{\{\frac{1}{2}r(r-1)\}}} = \frac{1}{\sqrt{2}}\left[\chi_1^2\sqrt{\frac{r-1}{r}} - \frac{\chi_{r-1}^2}{\sqrt{\{r(r-1)\}}}\right] \tag{A2}$$

and the limiting distribution as $r \to \infty$ is that $\frac{1}{\sqrt{2}}(\chi_1^2-1)$. Result (I) suggests that (II) applies whether or not Z is N(0,1), and this is so. Dr. Hoeffding has pointed out to us that Y_2 is proportional to a *degenerate* U-statistic, and that the general result can be established by a similar analysis, as follows:

$$\frac{Y_2}{r\ \mathrm{var}(Z)} = \frac{1}{2}\left[\frac{(r^{-\frac{1}{2}}\sum_{i=1}^{n} Z_i)^2}{\mathrm{var}(Z)} - \frac{r^{-1}\sum_{i=1}^{n} Z_i^{\ 2}}{\mathrm{var}(Z)}\right] ;$$

as $r \to \infty$ the distribution of $(r^{-\frac{1}{2}}\sum_{i=1}^{n} Z_i^{\ 2})^2/\mathrm{var}(Z)$ tends to χ_1^2, and $r^{-1}\sum_{i=1}^{n} Z_i^2/\mathrm{var}(Z) \to 1$ in probability. The distribution of $Y_2/\{r\ \mathrm{var}(Z)\}$ therefore tends to that of $\frac{1}{2}(\chi_1^2 - 1)$. In our case, when $Z = 1-2F_1(X_1)$ the common distribution of the Z's (under $H_{0,0}$) is uniform (-1,1) and

$$E[Z] = 0,\ \mathrm{var}(Z) = \frac{1}{3} \tag{A3}$$

This suggests using the approximation

$$\frac{3Y_2}{r} \text{ approximately distributed as } \tfrac{1}{2}(\chi_1^2-1)$$

or equivalently

$$1 + \frac{6Y_2}{r} \text{ approximately distributed as } \chi_1^2 . \tag{A4}$$

The first four moments of Y_2 are, in general

$$E[Y_2] = 0 \tag{A5.1}$$

$$\mathrm{var}(Y_2) = \mu_2(Y_2 = \binom{r}{2}\mu_2^2 = \tfrac{1}{2}r^{(2)}\mu_2^2 \tag{A5.2}$$

$$\mu_3(Y_2) = \binom{r}{3}\binom{3}{1,1,1}\mu_2^3 + \binom{r}{2}\mu_3^2 = r^{(3)}\mu_2^3 + \tfrac{1}{2}r^{(2)}\mu_3^2 \tag{A5.3}$$

$$\mu_4(Y_2) = \binom{r}{4}\{\binom{3}{2}\binom{4}{1,1,1,1} + 3\binom{4}{2,2}\}\mu_2^4 + \binom{r}{3}3\binom{4}{2,1,1}\mu_3^2\mu_2$$

$$+ \binom{r}{3}3\binom{4}{2,2}\mu_4\mu_2^2 + \binom{r}{2}\mu_4^2$$

$$= \frac{15}{4}\,r^{(4)}\mu_2^4 + 6r^{(3)}\mu_3^2\mu_2 + 3r^{(3)}\mu_4\mu_2^2 + \tfrac{1}{2}r^{(2)}\mu_4^2 \tag{A5.4}$$

Inserting the special values (A3), we obtain

$$E[Y_2|H_{o,o}] = 0 \tag{A6.1}$$

$$\mathrm{var}(Y_2|H_{o,o}) = \frac{1}{18}r^{(2)} \tag{A6.1}$$

$$\mu_3(Y_2|H_{o,o}) = \frac{1}{27}r^{(3)} \tag{A6.2}$$

$$\mu_4(Y_2|H_{o,o}) = \frac{1}{50}r^{(2)} + \frac{1}{15}r^{(3)} + \frac{5}{108}r^{(4)} \tag{A6.3}$$

$$= \frac{1}{2700}r^{(2)}(125r^2-445r+444) \tag{A6.4}$$

APPENDIX II

Lemma A $\sum_{1\le h_1<\ldots<h_u\le r} \Pi_{\alpha=1}^{u}(s+h_\alpha+\alpha-1) = \binom{r}{u}\sum_{t=0}^{u} 2^{-t}(r+1)^{[t]}\binom{u}{t}s^{[u-t]}$ (A7)

Proof (By induction on u)

Taking u=1, left-hand side of (A7) is $\sum_{h=1}^{r}(s+h) = rs + \frac{1}{2}r(r+1)$

right-hand side is $r\{s+2^{-1}(r+1)\} = rs + \frac{1}{2}r(r+1)$

so the equality holds for u=1. Assuming it holds for (u-1), the left-hand side is

$$\sum_{h_u=u}^{r}\{\sum_{1\le h_1<\ldots<h_{u-1}\le h_u-1} \Pi_{\alpha=1}^{u-1}(s+h_\alpha+\alpha-1)\}(s+h_u+u-1)\}$$

$$= \sum_{h_u=u}^{r}\binom{h_u-1}{u-1}\sum_{t=0}^{u-1}2^{-t}h_u^{[t]}\binom{u-1}{t}s^{[u-1-t]}\{(s+u-1-t) + (h_u+t)\}$$

$$= \sum_{h_u=u}^{r}\binom{h_u-1}{u-1}\sum_{v=0}^{u}2^{-v}h_u^{[v]}s^{[u-v]}\{\binom{u-1}{v} + 2\binom{u-1}{v-1}\}$$

(interpreting $\binom{u-1}{u}$ and $\binom{u-1}{-1}$ as zero)

$$= \sum_{v=0}^{u}\sum_{h_u=u}^{r}2^{-v}\frac{(h_u-1)!h_u^{[v]}}{(u-1)!(h_u-u)!}\cdot\frac{(u-1)!(u+v)}{v!(u-v)!}s^{[u-v]}$$

$$= \sum_{v=0}^{u}2^{-v}(u+1)^{[v]}\{\sum_{h_u=u}^{r}\binom{h_u+v-1}{u+v-1}\}\binom{u}{v}s^{[u-v]}$$

$$= \sum_{v=0}^{u} 2^{-v}\{(u+1)^{[v]}\binom{r+v}{u+v}\}\binom{u}{v}s^{[u-v]}$$

$$= \binom{r}{u}\sum_{v=0}^{u} 2^{-v}(r+1)^{[v]}\binom{u}{v}s^{[u-v]} \quad , \quad \text{as in (A7).}$$

This completes the proof.

<u>Lemma B</u> If $U_p = T_p(r, s_o, s_r)\cdot(r+s_0+s_r+1)^{[p]}/r^{(p)}$ then

$$\sum_{p=0}^{\infty} U_p x^p = (1+x)^{-s_o}(1-x)^{-s_r}$$

<u>Proof</u> Writing $(r+s_0+s_r+1)^{[u]}$ as $(r+s_0+s_r+1)^{[p]}/(r+s_0+s_r+p)^{(p-u)}$ in (15) we have

$$T_p = \frac{1}{(r+s_0+s_r+1)^{[p]}}\sum_{u=0}^{p}(r+s_0+s_r+p)^{(p-u)}(-2)^u\binom{r-u}{p-u}\sum_{1\le h_1<\ldots<h_u\le r}^{r-u+1}\cdots\sum^{r}\Pi_{\alpha=1}^{u}(s_0+h_\alpha+\alpha-1)$$

$$= \frac{1}{(r+s_0+s_r+1)^{[p]}}\sum_{u=0}^{p}(r+s_0+s_r+p)^{(p-u)}(-2)^u\binom{r-u}{p-u}\binom{r}{u}\sum_{t=0}^{u}2^{-t}(r+1)^{[t]}\binom{u}{t}s_0^{[u-t]}$$

(by Lemma A)

Hence

$$U_p = \sum_{u=0}^{p}(-1)^u\binom{r+s_0+s_r+p}{p-u}\sum_{t=0}^{u}2^{u-t}\binom{r+t}{t}\binom{s_0+u-t-1}{u-t}$$

Putting $q = p-u$ and $v = u-t$,

$$\sum_{p=0}^{\infty}U_p x^p = \sum_{t=0}^{\infty}\sum_{v=0}^{\infty}\binom{r+t}{t}\binom{s_0+v-1}{v}2^v(-x)^{t+v}\sum_{q=0}^{\infty}\binom{r+s_0+s_r+v+t+q}{q}x^q$$

$$= \sum_{t=0}^{\infty}\sum_{v=0}^{\infty}\binom{r+t}{t}\binom{s_0+v-1}{v}2^v(-x)^{t+v}(1-x)^{-r-s_0-s_r-v-t-1}$$

$$= \{\sum_{t=0}^{\infty}\binom{r+t}{t}\left(\frac{-x}{1-x}\right)^t\sum_{v=0}^{\infty}\binom{s_0+v-1}{v}\left(\frac{-2x}{1-x}\right)^v\}(1-x)^{-r-s_0-s_r-1}$$

$$= \left(1+\frac{x}{1-x}\right)^{-r-1}\left(1+\frac{2x}{1-x}\right)^{-s_0}(1-x)^{-r-s_0-s_r-1}$$

$$= (1+x)^{-s_0}(1-x)^{-s_r}$$

This completes the proof.

Combining the results of Lemma A and Lemma B we find

$$T_p(r, s_0, s_r) = \frac{r^{(p)}}{(r+s_0+s_r+1)^{[p]}} \times [\text{coefficient of } x^p \text{ in } (1+x)^{-s_0}(1-x)^{-s_r}]$$

$$= \frac{(-1)^p r^{(p)}}{(r+s_0+s_r+1)^{[p]}} \sum_{t=0}^{p} (-1)^t \binom{s_0+p-t-1}{p-t}\binom{s_r+t-1}{t}$$

In particular

$$T_1 = \frac{r}{r+s_0+s_r+1}(s_r-s_0)$$

$$T_2 = \frac{r^{(2)}}{2(r+s_0+s_r+1)^{[2]}}\{(s_r-s_0)^2+s_0+s_r\}$$

$$T_3 = \frac{r^{(3)}}{6(r+s_0+s_r+1)^{[3]}}(s_r-s_0)\{(s_r-s_0)^2+3(s_0+s_r)+2\}$$

$$T_4 = \frac{r^{(4)}}{24(r+s_0+s_r+1)^{[4]}}\{(s_r-s_0)^4+6(s_r-s_0)^2(s_0+s_r)+8(s_r-s_0)^2+3(s_0+s_r)^2+6(s_0+s_r)\}$$

The formulae for T_1, T_2 and T_3 agree with those in [1]; that for T_4 differs by $3[(s_0+s_r)^2-(s_r-s_0)^2] = 12s_0s_r$ in the terms in braces.

BIOSTATISTICS: Statistics in Biomedical, Public Health
and Environmental Sciences, edited by P.K. Sen

A TWO-STAGE PROCEDURE FOR THE ANALYSIS OF ORDINAL CATEGORICAL DATA

Gary G. Koch and Ingrid A. Amara

Department of Biostatistics
University of North Carolina
Chapel Hill, North Carolina

Julio M. Singer

Departamento de Estatistica
Universidade de Sao Paulo, Brazil

A method of fitting linear models to logits of cumulative proportions is considered for the analysis of ordinally scaled categorical data. It consists of a two-stage procedure which allows for partial proportional odds models in the sense that such structures may be imposed on some explanatory variables but not on others. Asymptotic properties of the resulting estimators are discussed and a computational strategy is proposed. Aspects of application are illustrated through several examples.

KEY WORDS: Functional asymptotic regression methodology (FARM), logistic model, ordinal data, proportional odds model, residual score statistic, seemingly unrelated regression, weighted least squares analysis of categorical data.

1. INTRODUCTION

Research data from many fields can frequently be summarized by contingency tables where rows represent the subpopulations corresponding to the classification of the study units by one or more dimensions of the study design (or explanatory variables) and columns represent the ordinal categories of a response variable. Examples of such situations are:

(i) clinical trials in which patients with different diagnoses of a certain disease are treated with one of several drugs by one of several investigators and are classified according to the amount of pain relief as either poor, fair, moderate, good or excellent;

(ii) observational studies in which subjects belonging to different risk groups are classified according to the increasing severity of some health condition.

The objectives of analyses of this type of data usually focus at among subpopulation comparisons of the ordinal distribution of the response variable and at the study of the relationship of such comparisons to the research design (explanatory variables) structure.

Among the several approaches that may be considered for such purposes, the one requiring relatively minimal assumptions is based on randomization methods for testing hypotheses of no association between the response and one or more dimensions of the subpopulation classification. They are discussed in Landis et al. (1978) and Koch et al. (1982), for example. Although such methods have a broad flexibility with respect to adjustment for covariables or stratification, they are limited by the fact that the respective conclusions apply only to the subjects under study. Furthermore, they are primarily directed at hypothesis testing and for these reasons they usually must be supplemented by other analyses for descriptive purposes and generalizability considerations; see Koch et al. (1982), Koch and Gillings (1983), and Koch and Sollecito (1984) for further discussion along these lines.

Another approach for dealing with ordinal response variables involves summary statistics; e.g., mean scores, as discussed in Grizzle et al. (1969) and Koch et al. (1978), and some rank function measures, as considered in Forthofer and Koch (1973), Forthofer and Lehnen (1981), Agresti (1980) and Semenya et al. (1983). These methods have the advantage of no assumptions about an underlying scale for the ordinal categories, nor a formal probability structure for the probabilities π_{ij} associated with the cells of the contingency table. However, substantial computational effort can be required if the number of subpopulations is not small (i.e. ≥ 9).

A third approach for the analysis of contingency tables with an ordinal response variable involves methods which assume parametric models for some functions of the probabilities π_{ij}. These methods have the advantage of providing a comprehensive framework for the analysis; however, they must be employed with caution, especially when small sample sizes may weaken the extent to which goodness of fit can be assessed effectively. Among these methods, the equal adjacent odds ratios model considered in Haberman (1974), Andrich (1979), Clogg (1982) and Koch et al. (1982), for example, is of some interest because of its log-linear structure. More specifically, it is directed at log ratios of frequencies of adjacent categories and this leads to computational advantages which include the existence of statistical software for both Maximum Likelihood (ML) and Weighted Least Squares (WLS) procedures. However, their specification is tied to the response categories under consideration, and this property may not be appealing for situations where the ordinal response categories can be viewed as falling along an underlying continuum. For this reason, another model of interest is based on the proportional odds structure which is focused at logits of cumulative proportions. As noted in Walker and Duncan (1967) and McCullagh (1980), the major assumption for this type of model is that the difference between the cumulative logits corresponding to different subpopulations is independent of the response category involved. The reader is referred to Agresti (1983) for a survey on these and other related methods as well as for other references.

In this paper, we present a more general class of models for such cumulative proportions which allows for the possibility of a partial proportional odds structure (in the sense that such structure may hold for some explanatory variables, but not for others) and includes the model described in McCullagh (1980) as a special case. Such models have been previously discussed by Williams and Grizzle (1972) for contingency tables where nearly all counts are at least moderately large (e.g., ≥ 5), Here, computational strategies are described for more general data arrays. We introduce some notation and present some theoretical results related to the asymptotic properties of a two-stage estimation procedure in Section 2. Aspects of the application of such procedures are illustrated in Section 3 through a series of examples. In Section 4 we present a general discussion concerning the application of the proposed methods. The more applications-oriented reader may scan Section 2 for a general impression and concentrate principal attention on Section 3 since the most practical aspects of the methodology are discussed in the context of the examples.

2. METHODOLOGY

We assume that there are s subpopulations of elements from which independent random samples of fixed sizes $n_1,\ldots,n_s$ respectively are selected. The responses of the n_i elements from the ith subpopulation are classified into $(r + 1)$ ordinally scaled categories for which n_{ij}, $j = 1,\ldots,r+1$ denotes the number of elements classified into the jth response category. The vector $\underset{\sim}{n}_i = (n_{i1},\ldots,n_{i,r+1})'$ will be assumed to follow a multinomial distribution with parameters n_i and $\underset{\sim}{\pi}_i = (\pi_{i1},\ldots,\pi_{i,r+1})'$ where π_{ij} denotes the probability of response j in subpopulation i. Thus, the relevant product multinomial model is

$$\Phi = \prod_{i=1}^{s} \left\{ \frac{n_i!}{\prod_{j=1}^{r+1} n_{ij}!} \prod_{j=1}^{r+1} \pi_{ij}^{n_{ij}} \right\} \tag{1}$$

where $\sum_{j=1}^{r+1} \pi_{ij} = 1$ for $i = 1,\ldots,s$.

A suitable way of taking the ordinality of the response variable into account is by modeling a set of cumulative probabilities $\theta_{i\ell} = \sum_{j=1}^{u_\ell} \pi_{ij}$, $1 \leq u_1 < u_2 < \ldots < u_q \leq r$, $\ell = 1,\ldots,q$ by logistic models of the form:

$$\theta_{i\ell} = \exp(\underset{\sim}{x}'_{i\ell}\underset{\sim}{\beta}_\ell)/\{1 + \exp(\underset{\sim}{x}'_{i\ell}\underset{\sim}{\beta}_\ell)\} \tag{2}$$

or equivalently $\log\{\theta_{i\ell}/(1 - \theta_{i\ell})\} = \underset{\sim}{x}'_{i\ell}\underset{\sim}{\beta}_{\ell}$, where $\underset{\sim}{x}_{i\ell}$ is a $(t_{\ell} \times 1)$ vector of known constants and $\underset{\sim}{\beta}_{\ell}$ is a $(t_{\ell} \times 1)$ vector of unknown parameters, $i = 1,\ldots,s$, $\ell = 1,\ldots,q$. In most situations, we take $u_{\ell} = \ell$ and $q = r$. In such a case, if $\underset{\sim}{x}'_{i\ell} = (1,\underset{\sim}{x}'_{i})$ and $\underset{\sim}{\beta}_{\ell} = (\xi_{\ell},\underset{\sim}{\beta}')'$, $\ell = 1,\ldots,q$, then model (2) has the "proportional odds" structure discussed in McCullagh (1980). More generally, the model (2) is analogous to the seemingly unrelated regression (SUR) model which has been given attention in the literature for econometrics and multivariate analysis; see Zellner (1962), Revankar (1976), Srivastava and Dwivedi (1979), and Conniffe (1982). Methodology for SUR involves the simultaneous fitting of possible different regression models to a set of correlated response variables. Another related framework is the generalized multivariate model considered by Kleinbaum (1973). In a similar spirit, the "proportional odds" structure can be viewed as expressing no (response x subpopulation) interaction in a sense analogous to that addressed through "multivariate profile analysis;" see Morrison (1967, Ch. 4).

WLS procedures may be applied to fit such models through the use of the general methodology described in Grizzle, Starmer and Koch (1969); such analysis has been described by Williams and Grizzle (1972). Statistical software for the computation of the WLS estimate of $\underset{\sim}{\beta} = (\underset{\sim}{\beta}'_1,\ldots,\underset{\sim}{\beta}'_q)'$ and of test statistics for hypotheses

$$H_0: \underset{\sim}{C}\underset{\sim}{\beta} = \underset{\sim}{0}, \tag{3}$$

where $\underset{\sim}{C}$ is a full rank $(c \times \sum_{\ell=1}^{q} t_{\ell})$ matrix of known constants, is available in several packages; e.g. GENCAT (see Landis et al. (1976)). Application of such methods, however, is only recommended if the cumulative cell counts $\{\sum_{j=1}^{u_{\ell}} n_{ij}\}$ are large enough to support the assumption that they follow a multivariate normal distribution. Alternatively, a ML approach might be considered. However, for the general structure described by (2), it may require specialized iterative algorithms formulated on an individual model basis; also, the design of such algorithms may be further complicated by the need to avoid zero or negative probabilities at each iteration. These considerations make ML methods computationally less attractive than their WLS counterparts even though they might be expected to have more stable asymptotic properties as in the log-linear model situation (for which Imrey et al. (1981) as well as other references note that the asymptotic normality is induced through linear combinations of the cell counts).

Here we describe an alternative estimation strategy based on functional asymptotic regression methodology (FARM). It requires less stringent assumptions concerning the sizes of the cell counts $\{n_{ij}\}$ than the WLS procedures and also has computational advantages over their ML counterparts. As described in Imrey et al. (1981), it is a two-stage procedure which consists of obtaining separate ML estimates of the parameter vectors β_{ℓ}, $\ell = 1,\ldots,q$ described in (2) together with a consistent estimate of their covariance matrix and then proceeding to fit reduced models (e.g., proportional odds models) to the compound vector $\beta = (\beta_1',\ldots,\beta_q')'$ via WLS methods at the second stage.

Since the models (2) are log-linear in each $\theta_{i\ell}$, $\ell = 1,\ldots,q$ considered individually, solutions to the corresponding likelihood equations can be obtained with existing computer software. In this regard we note that the likelihood equations obtained by substituting (2) into the marginal distribution of $\sum_{i=1}^{u_\ell} n_{ij}$ from (1), differentiating the result with respect to β_ℓ and subsequently setting it equal to zero yields:

$$\sum_{i=1}^{s}\sum_{j=1}^{u_\ell} n_{ij}x_{i\ell} = \sum_{i=1}^{s} n_{i}x_{i\ell}\left\{\frac{\exp(x_{i\ell}'\hat{\beta}_\ell)}{1+\exp(x_{i\ell}'\hat{\beta}_\ell)}\right\} . \tag{4}$$

These equations may be re-expressed in matrix notation as

$$X_\ell' D_{n_{*+}} f_{*\ell} = X_\ell' D_{n_{*+}} \theta_{*\ell}(\hat{\beta}_\ell)$$

Where $X_\ell' = (x_{1\ell},\ldots,x_{s\ell})$ is of full rank $t_\ell \le s$, $n_{*+} = \sum_{j=1}^{r+1} n_{*j}, n_{*j} = (n_{1j},\ldots, n_{sj})'$, $D_{n_{*+}}$ is a diagonal matrix with the elements of n_{*+} down the main diagonal, $f_{*\ell} = (f_{1\ell},\ldots,f_{s\ell})' = \sum_{j=1}^{u_\ell} p_{*j}$, $p_{*j} = (p_{1j},\ldots,p_{sj})' = (n_{*j}/n_i)$, $j = 1,\ldots,r+1$ and $\theta_{*\ell}(\hat{\beta}_\ell) = \hat{\theta}_{*\ell} = (\hat{\theta}_{1\ell},\ldots,\hat{\theta}_{s\ell})'$ is the estimate of the vector of cumulative probabilities $\theta_{*\ell} = (\theta_{1\ell},\ldots,\theta_{s\ell})'$ under model (2). Iterative models such as the Newton-Raphson method or other numerical approximation algorithms may be used to obtain $\hat{\beta}_\ell$. It is of interest to note that the asymptotic results for the first stage model require the less stringent multinormality of the statistics $\{X_\ell' D_{n_{*+}} f_{*\ell}\}$ as opposed to the multinormality of the set of cumulative counts $\{\sum_{j=1}^{u_\ell} n_{ij}\}$ required for the application of WLS techniques.

The asymptotic distribution of the first stage estimators may be obtained in the usual way via linear Taylor series methods. For this purpose, we first compute $\partial\hat{\beta}_\ell/\partial f_{i\ell}$, $i = 1,\ldots,s$ through implicit differentiation of (4), obtaining:

$$n_i x_{i\ell} = \sum_{k=1}^{s} [n_k \hat{\theta}_{k\ell}(1-\hat{\theta}_{k\ell}) x_{k\ell} x_{k\ell}' \{\partial\hat{\beta}_\ell/\partial f_{i\ell}\}] \tag{5}$$

where $\hat{\theta}_{k\ell} = \exp(x_{k\ell}'\hat{\beta}_\ell)/\{1+\exp(x_{k\ell}'\hat{\beta}_\ell)\}$. In matrix notation, (5) may be re-expressed as:

$$X_\ell' D_{n_{*+}} = \{X_\ell' \hat{D}_{\ell\ell} X_\ell\}\{\partial\hat{\beta}_\ell / \partial f_{*\ell}\}$$

where $\hat{D}_{\ell\ell} = \text{diag}\{n_1\hat{\theta}_{1\ell}(1 - \hat{\theta}_{1\ell}),\ldots,n_s\hat{\theta}_{s\ell}(1 - \hat{\theta}_{s\ell})\}$. Therefore:

$$\partial\hat{\beta}_\ell / \partial f_{*\ell} = (X_\ell' \hat{D}_{\ell\ell} X_\ell)^{-1} X_\ell' D_{n_{*+}}$$

and the related first order Taylor series is given by:

$$\hat{\beta}_\ell \cong \beta_\ell + (X_\ell' D_{\ell\ell} X_\ell)^{-1} X_\ell' D_{n_{*+}} (f_{*\ell} - \theta_{*\ell}) \tag{6}$$

where $D_{\ell\ell}$ corresponds to $\hat{D}_{\ell\ell}$ with the $\{\hat{\theta}_{i\ell}\}$ replaced by the $\{\theta_{i\ell}\}$. An expression for the asymptotic covariance matrix of $\hat{\beta}_\ell$ is obtained through the δ-method; it is given by:

$$\text{Var}_A(\hat{\beta}_\ell) = V_\ell = (X_\ell' D_{\ell\ell} X_\ell)^{-1}$$

Also the asymptotic covariance structure between $\hat{\beta}_\ell$ and $\hat{\beta}_{\ell'}$, $\ell' > \ell$ may be obtained from:

$$\text{Cov}_A(\hat{\beta}_\ell,\hat{\beta}_{\ell'}) = (X_\ell' D_{\ell\ell} X_\ell)^{-1} X_\ell' D_{n_{*+}} \text{Cov}_A(f_{*\ell},f_{*\ell'}) D_{n_{*+}} X_{\ell'} (X_{\ell'}' D_{\ell'\ell'} X_{\ell'})^{-1}$$

$$= V_\ell X_\ell' D_{\ell\ell'} X_{\ell'} V_{\ell'}.$$

where $D_{\ell\ell'} = \text{diag}\{n_1\theta_{1\ell}(1 - \theta_{1\ell'}),\ldots,n_s\theta_{s\ell}(1 - \theta_{s\ell'})\}$. We note from (6) that the estimator $\hat{\beta} = (\hat{\beta}_1',\ldots,\hat{\beta}_q')'$ can essentially be represented as a linear function of the statistics $\{X_\ell' D_{n_{*+}} f_{*\ell}\}$. Therefore, for sufficiently large samples it follows a multivariate normal distribution with mean vector β and covariance matrix

$$\text{Var}_A(\hat{\beta}) = VX'DXV = V_{\hat{\beta}}$$

where V and X are block diagonal matrices with block diagonal components V_ℓ and X_ℓ, $\ell = 1,\ldots,q$ respectively,

$$D = \begin{bmatrix} D_{11} & D_{12} & \cdots & D_{1q} \\ & D_{22} & \cdots & D_{2q} \\ & & & \vdots \\ \text{symmetric} & & & \vdots \\ & & & D_{qq} \end{bmatrix} = D_{n_*} \begin{bmatrix} D_{\theta_{*1}}(I_s - D_{\theta_{*1}}) & D_{\theta_{*1}}(I_s - D_{\theta_{*2}}) & \cdots & D_{\theta_{*1}}(I_s - D_{\theta_{*q}}) \\ & D_{\theta_{*2}}(I_s - D_{\theta_{*2}}) & \cdots & D_{\theta_{*2}}(I_s - D_{\theta_{*q}}) \\ \text{symmetric} & & & \vdots \\ & & & D_{\theta_{*q}}(I_s - D_{\theta_{*q}}) \end{bmatrix}$$

$D_{\theta_{*\ell}} = \text{diag}(\theta_{1\ell},\ldots,\theta_{s\ell})$, $\ell = 1,\ldots,q$ and $D_{n_*} = D_{n_{*+}} \otimes I_q$.

Hypotheses of the form (3) may be tested via Wald statistics given by:

$$Q_C = \hat{\beta}'C'(C\hat{V}_{\hat{\beta}}C')^{-1}C\hat{\beta} \tag{7}$$

which under the null nypothesis follows a chi-squared distribution with c degrees of freedom (d.f.).

Goodness of fit for the first stage models can be assessed by verifying that various expanded models (X,W) can be reduced to the model X through the use of the following statistic directed at linear functions of the residuals:

$$Q_S = (n_c - D_{n_*}\hat{\theta}_*)'W\{W'\text{Var}_A(n_c - D_{n*}\hat{\theta}_*)W\}^{-1}W(n_c - D_{n*}\hat{\theta}_*) \tag{8}$$

Here $n_c = \left(\sum_{j=1}^{u_1} n'_{*j}, \sum_{j=1}^{u_2} n'_{*j}, \ldots, \sum_{j=1}^{u_q} n'_{*j}\right)'$, $\hat{\theta}_* = (\hat{\theta}'_{*1},\ldots,\hat{\theta}'_{*q})'$, and W is a block diagonal matrix with block diagonal components W_ℓ $(s \times w_\ell)$ such that rank $(X_\ell, W_\ell) = t_\ell + w_\ell$, $\ell = 1,\ldots,q$. The asymptotic covariance matrix of the residuals $r_* = n_c - D_{n*}\hat{\theta}_*$ may be obtained via the δ-method through the following linear Taylor series expansion around $n_c = D_{n*}\theta_*$ where $\theta_* = (\theta'_{*1},\ldots,\theta'_{*q})'$:

$$r_* \cong \left\{ I_{qs} - \left[\frac{\partial D_{n*}\hat{\theta}_*}{\partial \hat{\beta}}\bigg|_{\hat{\beta}=\beta}\right]\left[\frac{\partial \hat{\beta}}{\partial n_c}\bigg|_{n_c=D_{n*}\theta_*}\right]\right\}(n_c - D_{n*}\theta_*)$$

Using implicit differentiation methods as done previously we obtain:

$$\left[\frac{\partial \hat{\beta}}{\partial n_c}\bigg|_{n_c=D_{n*}\theta_*}\right] = VX' \quad \text{and} \quad \left[\frac{\partial D_{n*}\hat{\theta}_*}{\partial \hat{\beta}}\bigg|_{\hat{\beta}=\beta}\right] = D^*X$$

where D^* is a block diagonal matrix with $D_{\ell\ell}$, $\ell = 1,\ldots,q$ as the block diagonal components. Therefore

$$\begin{aligned}\text{Var}_A(n_c - D_{n_*}\hat{\theta}_*) &= (I_{qs} - D^*XVX')\text{Var}_A(n_c - D_{n_*}\theta_*)(I_{qs} - D^*XVX')' \\ &= (I_{qs} - D^*XVX')D(I_{qs} - D^*XVX')'\end{aligned}$$

Given that θ_* is compatible with the model X, Q_S has an approximate chi-squared distribution with $w = \sum_{\ell=1}^{q} w_\ell$ d.f.; also, since Q_S is directed at linear functions of residuals, it is called a residual score statistic in the subsequent discussion.

The second stage corresponds to fitting models of the form $E(\hat{\beta}) = Z\gamma$ to the vector of first stage estimated parameters $\hat{\beta}$ via WLS methods, here Z is a $(\sum_{\ell=1}^{q} t_{\ell} \times t)$ known matrix of full rank t and γ is a $(t \times 1)$ vector of unknown parameters. The corresponding WLS estimates are given by:

$$\hat{\gamma} = (Z'\hat{V}_{\hat{\beta}}^{-1}Z)^{-1}Z'\hat{V}_{\hat{\beta}}^{-1}\hat{\beta}$$

where $\hat{V}_{\hat{\beta}}$ is a consistent estimate of $V_{\hat{\beta}}$. If the sample sizes are large enough to support the approximate multinormality of the first stage estimates $\hat{\beta}$, it follows that $\hat{\gamma}$ is approximately normally distributed with mean vector γ and covariance matrix

$$\text{Var}_A(\hat{\gamma}) = (Z'V_{\hat{\beta}}^{-1}Z)^{-1}$$

As a related issue, it can be noted that this FARM procedure may not be as efficient as its WLS or ML counterparts since no use is made of the covariance structure of the cumulative cell counts $\left\{\sum_{j=1}^{u_{\ell}} n_{ij}\right\}$ at the first stage. Nevertheless, such tendencies were not evident in the illustrative examples examined in Section 3.

3. EXAMPLES

In this section we present a series of examples to illustrate data analysis using the FARM procedure outlined in Section 2. We intend to emphasize the flexibility of the method with respect to the evaluation of the proportional odds model discussed in McCullagh (1980), the comparison of the results with those obtained through WLS or ML methods where possible, and the identification of alternative models for situations where the proportional odds model is not applicable.

Usage of the suggested methodology requires a two-stage modeling strategy which can be outlined as follows. Initially we attempt to fit a preliminary model to each of the different cumulative logits of interest through separate ML procedures; usually, these preliminary models correspond to main effects models, but they could be extended to modular models in which each module corresponds to levels of one or more explanatory variables within which main effects models for other explanatory variables are used. Simultaneous goodness of fit for these preliminary models is assessed through residual score statistics of the form (8), corresponding to expanded models that include other components not considered initially, such as higher order interactions. If the proposed set of preliminary models fits, we proceed to the second stage where we identify those explanatory variables (or interactions) for which proportional odds seems appropriate. For this purpose we can either use Wald statistics of the form (7) with conveniently chosen $\underset{\sim}{C}$ matrices or use WLS methods to fit suitable models of the form $E(\hat{\underset{\sim}{\beta}}) = \underset{\sim}{Z}\underset{\sim}{\gamma}$ to the vector of parameter estimates $\hat{\underset{\sim}{\beta}} = (\hat{\underset{\sim}{\beta}}_1', \ldots, \hat{\underset{\sim}{\beta}}_r')'$; here $\underset{\sim}{Z}$ is a known $(\sum_{\ell=1}^{g} t_\ell \times 1)$ orthocomplement to $\underset{\sim}{C}$ and $\underset{\sim}{\gamma}$ is a $(t \times 1)$ vector of unknown parameters. Therefore, the resulting (FARM) final model specification matrix is $\underset{\sim}{X}\underset{\sim}{Z}$.

In the following examples, computations for the first stage were undertaken through a modified version of SAS macro CATMAX (see Stokes and Koch (1983)); computations for the second stage as well as those needed to obtain WLS estimates were performed through GENCAT (see Landis et al. (1976)); the ML estimates for the proportional odds models parameters were obtained with LOGIST (see Harrell (1983)).

Example 1

Table 1 displays results from a surgical clinical trial pertaining to treatment of duodenal ulcer. The response variable is dumping syndrome severity, an undesirable sequela of surgery. The subpopulations correspond to four operations, Vagotomy and Drainage (V + D), Vagotomy and Antrectomy (V + A), Vagotomy and Hemigastrectomy (V + H) and Gastric Resection (GR); these involve removal of 0, 25, 50 and 75% of gastric tissue respectively. The ordinally scaled response variable corresponds to the relatively subjective categorization of severity into none, slight and moderate.

Table 1. Observed and Model Predicted Proportions and Estimated Standard Errors for the Dumping Syndrome Data

		Observed Proportions with Dumping Syndrome Severity			FARM Model Predicted Proportions with Dumping Syndrome Severity		
Operation	Number of Patients	None	Slight	Moderate	None	Slight	Moderate
V + D	96	0.635 (0.049)	0.292 (0.046)	0.073 (0.027)	0.659 (0.039)	0.259 (0.029)	0.082 (0.016)
V + A	104	0.654 (0.047)	0.221 (0.041)	0.125 (0.032)	0.607 (0.027)	0.292 (0.023)	0.101 (0.015)
V + H	110	0.527 (0.048)	0.364 (0.046)	0.109 (0.030)	0.552 (0.026	0.325 (0.024)	0.123 (0.017)
GR	107	0.495 (0.048)	0.355 (0.046)	0.150 (0.034)	0.495 (0.040)	0.355 (0.029)	0.150 (0.024)

The objectives of analysis for this example are

i. the evaluation of among subpopulation comparisons of the ordinal distributions of the response variable

ii. the formulation of a framework for the relationship of such comparisons to the quantitative structure of the subpopulations.

Several authors have addressed such objectives in previous analyses of these data; they include Imrey et al. (1982) who considered equal adjacent odds ratios log-linear models and Semenya et al. (1983) who considered linear models for some rank functions. Here, attention is given to an alternative approach based on the methods described in Section 2. For this purpose, the patients with each type of surgical procedure are regarded as conceptually representative of corresponding infinite subpopulations in a sense equivalent to stratified simple random sampling.

The preliminary models considered for this example have specification matrices

$$\underset{\sim}{X}_1 = \underset{\sim}{X}_2 = \begin{bmatrix} 1 & 0 \\ 1 & 1 \\ 1 & 2 \\ 1 & 3 \end{bmatrix}$$

This structure represents separate linear trends for the cumulative logits corresponding to (none vs. at least slight) and to (at most slight vs. moderate). Goodness of fit for the (simultaneous) preliminary models was assessed through the residual score statistic Q_S (8) corresponding to extended (saturated) models with specification matrices $(\underset{\sim}{X}_\ell, \underset{\sim}{W}_\ell)$, $\ell = 1,2$, where:

$$\underset{\sim}{W}_1 = \underset{\sim}{W}_2 = \begin{bmatrix} 0 & 0 \\ 0 & 0 \\ 1 & 0 \\ 0 & 1 \end{bmatrix}$$

Since under the null hypothesis, Q_S approximately follows a chi-squared distribution with 4 d.f., the observed value $Q_S = 4.158$ supports the preliminary model by its non-significance ($p = 0.385$). The corresponding (logitwise) ML parameter estimates and their estimated standard errors are presented in Table 2. The hypothesis that a proportional odds model fits the data was tested through Wald's statistic (7) with $\underset{\sim}{C} = [0 \;\; 1 \;\; 0 \;\; -1]$. The observed value was $Q_{\underset{\sim}{C}} = 0.020$ which is non-significant ($p = 0.889$) relative to a chi-squared distribution with 1 d.f. The reduced model which incorporates the proprotional odds structure has the (second stage) specification matrix

$$\underset{\sim}{Z} = \begin{bmatrix} 1 & 0 & 0 \\ 0 & 0 & 1 \\ 0 & 1 & 0 \\ 0 & 0 & 1 \end{bmatrix} .$$

It was fitted to the vector of preliminary parameter estimates via WLS methods. The resulting (proportional odds) FARM parameter estimates and their estimated standard errors are displayed in Table 2, together with their WLS and ML counterparts for comparison purposes. The corresponding (proportional odds) FARM predicted proportions and their estimated standard errors are presented in Table 1. These predicted proportions are of interest because they reflect the model structure of proportional odds and linear increase of log-odds with increasing amounts of gastric tissue removal. Also, their estimated standard errors are substantially smaller than those for the observed proportions. This gain in efficiency for the predicted values is a consequence of their formulation with respect to the $t = 3$ estimated model parameters.

We note that the FARM parameter estimates are comparable to those obtained by the other two methods. In particular, there appears to be no indication of any loss of precision of the proposed (logitwise) procedure with respect to either the ML or to the WLS methods, both of which are asymptotically efficient via the linkage of their theoretical properties to the general framework in Neyman (1949).

Table 2. Preliminary and Final Model Estimated Parameters and Estimated Standard Errors for the Dumping Syndrome Data

Preliminary Model (Logitwise)		Final Model (Proportional Odds)			
Parameters	Logitwise ML Estimates (Standard Errors)	Parameters	Estimates (Standard Errors) FARM	WLS	ML
Intercept for none vs. at least slight (β_{11})	0.664 (0.176)	Intercept for none vs. at least slight(γ_1)	0.659 (0.173)	0.657 (0.172)	0.657 (0.173)
Slope for none vs. at least slight (β_{12})	-0.229 (0.091)	Intercept for at most slight vs. moderate(γ_2)	2.413 (0.215)	2.368 (0.214)	2.411 (0.215)
Intercept for at most slight vs. moderate (β_{21})	2.386 (0.291)	Common slope(γ_3)	-0.226 (0.088)	-0.222 (0.088)	-0.225 (0.088)
Slope for at most slight vs. moderate (β_{22})	-0.211 (0.142)				

Q_{WLS} = 4.565 with 5 d.f. is the WLS residual (or Wald) goodness of fit chi-squared statistic for the proportional odds model with respect to the WLS estimates ($p = 0.471$).

Example 2.

The data for this example are from a study concerned with the effects of air pollution, job exposure and smoking status on chronic respiratory status of individuals. They are from a study described in Lan and Shy (1979); and aspects of their analysis via ordinal data methods have been previously discussed by Semenya and Koch (1980). The twelve subpopulations correspond to the cross classification of the factors of air pollution, job exposure and smoking status as indicated in Table 3. The ordinally scaled response variable corresponds to the following categories of a measure of chronic respiratory disease:

Level I: No symptoms;
Level II: Cough or phlegm for less than three months per year;
Level III: Cough or phlegm for more than three months per year;
Level IV: Cough and phlegm accompanied by shortness of breath for more than three months per year.

We illustrate here how the FARM methods described in Section 2 compare with both ML and WLS procedures for fitting a proportional odds model to the data. For this purpose, the individuals in the respective (air pollution x job exposure x smoking status) groups are considered to be conceptually representative of corresponding large subpopulations in a sense equivalent to stratified simple random sampling.

The preliminary models considered for this example have specification matrices

$$\underset{\sim}{X}_1 = \underset{\sim}{X}_2 = \underset{\sim}{X}_3 = \begin{bmatrix} 1 & 1 & 1 & 1 & 1 & 1 & 1 & 1 & 1 & 1 & 1 & 1 \\ 0 & 0 & 0 & 0 & 0 & 0 & 1 & 1 & 1 & 1 & 1 & 1 \\ 0 & 0 & 0 & 1 & 1 & 1 & 0 & 0 & 0 & 1 & 1 & 1 \\ 0 & 1 & 0 & 0 & 1 & 0 & 0 & 1 & 0 & 0 & 1 & 0 \\ 0 & 0 & 1 & 0 & 0 & 1 & 0 & 0 & 1 & 0 & 0 & 1 \end{bmatrix}' = \begin{bmatrix} \underset{\sim}{x}_1' \\ \underset{\sim}{x}_2' \\ \underset{\sim}{x}_3' \\ \underset{\sim}{x}_4' \\ \underset{\sim}{x}_5' \end{bmatrix}' \tag{9}$$

which represent (reference cell) main effects models for the cumulative logits corresponding to level I vs. levels II, III or IV, levels I or II vs. levels III or IV and levels I, II or III vs. level IV. Goodness of fit for the proposed preliminary model was assessed by verifying that five different expanded models could be reduced to it via residual score statistics Q_S of the form (8). The related W matrices, the observed values, the d.f. and the respective P-values are indicated in Table 4; the (logitwise) ML estimates of the preliminary model parameters and their estimated standard errors are presented in Table 5.

The hypothesis that a proportional odds model fits the data was tested via Wald's statistic (7) with

Table 3. Observed and Model Predicted Proportions and Estimated Standard Errors for the Chronic Respiratory Disease Data

Air Pollution	Job Exposure	Smoking Status	Sample Size	Observed Proportions with Response				FARM Model Predicted Proportions with Response			
				I	II	III	IV	I	II	III	IV
LOW	NO	NON	172	0.919 (0.021)	0.052 (0.017)	0.029 (0.013)	0.000 -	0.890 (0.016)	0.061 (0.009)	0.029 (0.005)	0.021 (0.004)
		EX	194	0.861 (0.025)	0.098 (0.021)	0.026 (0.011)	0.015 (0.009)	0.844 (0.013)	0.084 (0.009)	0.041 (0.006)	0.031 (0.005)
		CUR	560	0.548 (0.021)	0.182 (0.016)	0.148 (0.015)	0.121 (0.014)	0.561 (0.018)	0.192 (0.010)	0.129 (0.009)	0.118 (0.009)
	YES	NON	37	0.703 (0.075)	0.135 (0.056)	0.135 (0.056)	0.027 (0.027)	0.775 (0.031)	0.117 (0.015)	0.061 (0.010)	0.047 (0.008)
		EX	58	0.655 (0.062)	0.207 (0.053)	0.069 (0.033)	0.069 (0.033)	0.697 (0.031)	0.149 (0.014)	0.085 (0.011)	0.069 (0.010)
		CUR	248	0.379 (0.031)	0.194 (0.025)	0.185 (0.025)	0.242 (0.027)	0.352 (0.022)	0.213 (0.011)	0.197 (0.013)	0.238 (0.018)
HIGH	NO	NON	107	0.879 (0.032)	0.065 (0.024)	0.047 (0.020)	0.009 (0.009)	0.893 (0.016)	0.060 (0.009)	0.028 (0.005)	0.020 (0.004)
		EX	82	0.817 (0.043)	0.098 (0.033)	0.049 (0.024)	0.037 (0.021)	0.847 (0.019)	0.083 (0.010)	0.040 (0.006)	0.030 (0.005)
		CUR	318	0.578 (0.028)	0.203 (0.023)	0.105 (0.017)	0.114 (0.018)	0.567 (0.022)	0.191 (0.011)	0.127 (0.010)	0.115 (0.011)
	YES	NON	42	0.762 (0.066)	0.071 (0.040)	0.143 (0.054)	0.024 (0.024)	0.779 (0.030)	0.115 (0.014)	0.060 (0.009)	0.046 (0.008)
		EX	56	0.696 (0.061)	0.196 (0.053)	0.071 (0.034)	0.036 (0.025)	0.702 (0.031)	0.147 (0.014)	0.083 (0.011)	0.067 (0.010)
		CUR	215	0.358 (0.033)	0.223 (0.028)	0.181 (0.026)	0.237 (0.029)	0.358 (0.022)	0.213 (0.011)	0.195 (0.013)	0.234 (0.019)

$$\underset{\sim}{C} = \begin{bmatrix} \underset{\sim}{0}_4 & \underset{\sim}{I}_4 & \underset{\sim}{0}_4 & -\underset{\sim}{I}_4 & \underset{\sim}{0}_4 & \underset{\sim}{0}_{44} \\ \underset{\sim}{0}_4 & \underset{\sim}{I}_4 & \underset{\sim}{0}_4 & \underset{\sim}{0}_{44} & \underset{\sim}{0}_4 & -\underset{\sim}{I}_4 \end{bmatrix} \tag{10}$$

where 0_m and 0_{mm} respectively denote a (m x 1) vector and a (m x m) matrix with all elements equal to zero. The observed value was $Q_{\underset{\sim}{C}} = 10.594$ which is non-significant (p = 0.226) with respect to a chi-squared distribution with 8 d.f., and so this result indicates that a proportional odds model adequately fits the data. The specification matrix for the corresponding second stage model is

$$\underset{\sim}{Z} = \begin{bmatrix} 1 & 0 & 0 & 0 & 0 & 0 & 0 & 0 & 0 & 0 & 0 & 0 & 0 & 0 & 0 \\ 0 & 0 & 0 & 0 & 0 & 1 & 0 & 0 & 0 & 0 & 0 & 0 & 0 & 0 & 0 \\ 0 & 0 & 0 & 0 & 0 & 0 & 0 & 0 & 0 & 0 & 1 & 0 & 0 & 0 & 0 \\ 0 & 1 & 0 & 0 & 0 & 0 & 1 & 0 & 0 & 0 & 0 & 1 & 0 & 0 & 0 \\ 0 & 0 & 1 & 0 & 0 & 0 & 0 & 1 & 0 & 0 & 0 & 0 & 1 & 0 & 0 \\ 0 & 0 & 0 & 1 & 0 & 0 & 0 & 0 & 1 & 0 & 0 & 0 & 0 & 1 & 0 \\ 0 & 0 & 0 & 0 & 1 & 0 & 0 & 0 & 0 & 1 & 0 & 0 & 0 & 0 & 1 \end{bmatrix}' \tag{11}$$

FARM, WLS and ML parameter estimates correponding to the final (proportional odds) model are presented in Table 5 together with their estimated standard errors; FARM final model predicted proportions and their estimated standard errors are displayed in Table 3.

Table 4. Goodness of Fit for the Preliminary Model for the Chronic Respiratory Disease Data

Extra Terms Included in the Extended Model	Columns of the $\underset{\sim}{W}_1 = \underset{\sim}{W}_2 = \underset{\sim}{W}_3$ Matrices†	Observed Statistic Q_S	d.f.	P-value
All interactions	$[\underset{\sim}{w}_{23}\underset{\sim}{w}_{24}\underset{\sim}{w}_{25}\underset{\sim}{w}_{34}\underset{\sim}{w}_{35}\underset{\sim}{w}_{234}\underset{\sim}{w}_{235}]$	12.069	21	0.938
All pairwise interactions	$[\underset{\sim}{w}_{23}\underset{\sim}{w}_{24}\underset{\sim}{w}_{25}\underset{\sim}{w}_{34}\underset{\sim}{w}_{35}]$	7.166	15	0.953
Air Pollution x Job Exposure	$[\underset{\sim}{w}_{23}]$	0.652	3	0.884
Air Pollution x Smoking Status	$[\underset{\sim}{w}_{24}\underset{\sim}{w}_{25}]$	3.429	6	0.753
Job Exposure x Smoking Status	$[\underset{\sim}{w}_{34}\underset{\sim}{w}_{35}]$	3.771	6	0.708

† $\underset{\sim}{w}_{ij} = \underset{\sim}{x}_i \# \underset{\sim}{x}_j$, (i = 2,3; j = 3,4,5, i ≠ j); $\underset{\sim}{w}_{23k} = \underset{\sim}{w}_{23} \# \underset{\sim}{x}_k$, (k = 4,5) where $\underset{\sim}{f} \# \underset{\sim}{g}$ denotes the elementwise multiplication of the vectors $\underset{\sim}{f}$ and $\underset{\sim}{g}$; also the $\underset{\sim}{x}_i$ are the respective columns of $\underset{\sim}{X}_1 = \underset{\sim}{X}_2 = \underset{\sim}{X}_3$ in (9).

Finally, it can be noted from Table 5 that there is a reasonable agreement among the estimates obtained through the three methods.

Table 5. Preliminary and Final Model Estimated Parameters and Estimated Standard Errors for the Chronic Respiratory Disease Data

Preliminary Model (Logitwise)

Parameters	Logitwise ML Estimates (Standard Errors)
Intercept for I vs II,III,IV (β_{11})	2.098 (0.165)
Slope due to high air pollution for I vs II,III,IV (β_{12})	0.004 (0.100)
Slope due to job exposure for I vs II,III,IV (β_{13})	-0.848 (0.103)
Slope due to previous smoking for I vs II,III,IV (β_{14})	-0.437 (0.202)
Slope due to current smoking for I vs II,III,IV (β_{15})	-1.838 (0.166)
Intercept for I,II vs III,IV (β_{21})	2.840 (0.221)
Slope due to high air pollution for I,II vs III,IV (β_{22})	0.097 (0.114)
Slope due to job exposure for I,II, vs III,IV (β_{23})	-0.856 (0.114)
Slope due to previous smoking for I,II vs III,IV (β_{24})	-0.037 (0.288)
Slope due to current smoking for I,II vs III,IV (β_{25})	-1.758 (0.221)
Intercept for I,II,III vs IV (β_{31})	5.029 (0.586)
Slope due to high air pollution for I,II,III vs IV (β_{32})	0.025 (0.149)
Slope due to job exposure for I,II,III vs IV (β_{33})	-0.873 (0.146)
Slope due to previous smoking for I,II,III vs IV (β_{34})	-1.253 (0.651)
Slope due to current smoking for I,II,III vs IV (β_{35})	-3.025 (0.585)

Final Model (Proportional Odds)

Parameters	Estimates (Standard Errors) FARM	WLS	ML
Intercept for I vs II, III,IV (γ_1)	2.091 (0.164)	2.044 (0.166)	2.089 (0.163)
Intercept for I,II vs III,IV (γ_2)	2.962 (0.169)	2.902 (0.172)	2.969 (0.169)
Intercept for I,II,III vs IV (γ_3)	3.862 (0.178)	3.796 (0.180)	3.893 (0.177)
Common slope due to high air pollution (γ_4)	0.026 (0.094)	0.039 (0.094)	0.036 (0.094)
Common slope due to job exposure (γ_5)	-0.855 (0.096)	-0.851 (0.095)	-0.862 (0.096)
Common slope due to previous smoking (γ_6)	-0.405 (0.201)	-0.371 (0.202)	-0.400 (0.201)
Common slope due to current smoking (γ_7)	-1.846 (0.165)	-1.801 (0.166)	-1.854 (0.165)

Q_{WLS} = 22.090 with 29 d.f. is the WLS residual (or Wald) goodness of fit chi-squared statistic for the proportional odds model with respect to the WLS estimates ($p = 0.816$).

Example 3

The data in Table 6 are from a multi-center clinical trial pertaining to the comparison of an active treatment with placebo for a curable gastrointestinal condition. The twelve subpopulations correspond to the different (Investigator x Treatment) groups indicated in Table 6 and the ordinally scaled response variable related to the categorization of the cure-status as cured at 2 weeks, cured at 4 weeks but not at 2 weeks, or not cured by 4 weeks.

For the purpose of the analysis, the patients in each (Investigator x Treatment) group are regarded as conceptually representative of corresponding infinite subpopulations in a sense equivalent to stratified simple random sampling. Main effects models for the cumulative logits corresponding to cured at 2 weeks vs. not cured at 2 weeks and cured by 4 weeks vs. not cured by 4 weeks were considered at the first stage. The specification matrices are

$$\underset{\sim}{X}_1 = \underset{\sim}{X}_2 = \begin{bmatrix} 1 & 1 & 1 & 1 & 1 & 1 & 1 & 1 & 1 & 1 & 1 & 1 \\ 0 & 0 & 1 & 1 & 0 & 0 & 0 & 0 & 0 & 0 & 0 & 0 \\ 0 & 0 & 0 & 0 & 1 & 1 & 0 & 0 & 0 & 0 & 0 & 0 \\ 0 & 0 & 0 & 0 & 0 & 0 & 1 & 1 & 0 & 0 & 0 & 0 \\ 0 & 0 & 0 & 0 & 0 & 0 & 0 & 0 & 1 & 1 & 0 & 0 \\ 0 & 0 & 0 & 0 & 0 & 0 & 0 & 0 & 0 & 0 & 1 & 1 \\ 1 & 0 & 1 & 0 & 1 & 0 & 1 & 0 & 1 & 0 & 1 & 0 \end{bmatrix}' = \begin{bmatrix} \underset{\sim}{x}_1' \\ \underset{\sim}{x}_2' \\ \underset{\sim}{x}_3' \\ \underset{\sim}{x}_4' \\ \underset{\sim}{x}_5' \\ \underset{\sim}{x}_6' \\ \underset{\sim}{x}_7' \end{bmatrix}' \qquad (12)$$

Goodness of fit for this preliminary model was assessed by verifying that six different expanded models could be reduced to it via residual score statistics Q_S of the form (8). The related $\underset{\sim}{W}$ matrices, the observed values, the d.f. and the respective P-values are indicated in Table 7.

The hypothesis that a proportional odds model fits the data was tested via Wald's statistic (7) with $\underset{\sim}{C} = [\underset{\sim}{0}_6 \;\; \underset{\sim}{I}_6 \;\; \underset{\sim}{0}_6 \;\; -\underset{\sim}{I}_6]$. Since $Q_C = 11.332$ approached significance ($p = 0.079$) relative to its approximate chi-squared distribution with 6 d.f. under the hypothesis, usage of the proportional odds model was considered questionable. Additional Wald tests with $C = [\underset{\sim}{0}_6' \;\; 1 \;\; \underset{\sim}{0}_6' \;\; -1]$, ($Q_C = 0.138$, d.f. = 1, $p = 0.710$) and $\underset{\sim}{C} = [\underset{\sim}{0}_5 \;\; \underset{\sim}{I}_5 \;\; \underset{\sim}{0}_{52} \;\; -\underset{\sim}{I}_5 \;\; \underset{\sim}{0}_5]$ ($Q_C = 11.061$, d.f. = 5, $p = 0.050$) indicated that proportional odds was applicable for Treatment but not for Investigator. Also, examination of the estimated parameters for the preliminary model suggested that the departures from proportional odds were similar for the effects of investigators 2-6 relative to investigator 1. Thus, we proceeded to fit an alternative model corresponding to the following second stage specification matrix:

Table 6. Observed and Model Predicted Proportions and Estimated Standard Errors for the Gastrointestinal Condition Clinical Trial Data

			Observed Proportions with Response			FARM Model Predicted Proportions with Response		
Investigator	Treatment	Number of Patients	Cured at 2 Weeks	Cured at 4 but not at 2 Weeks	Not Cured	Cured at 2 Weeks	Cured at 4 but not at 2 Weeks	Not Cured
1	Active	7	0.714 (0.171)	0.286 (0.171)	0.000 -	0.820 (0.088)	0.094 (0.065)	0.087 (0.061)
1	Placebo	10	0.800 (0.126)	0.000 -	0.200 (0.126)	0.726 (0.116)	0.134 (0.088)	0.140 (0.092)
2	Active	27	0.370 (0.093)	0.556 (0.096)	0.074 (0.050)	0.423 (0.072)	0.415 (0.043)	0.162 (0.044)
2	Placebo	29	0.241 (0.079)	0.586 (0.091)	0.172 (0.070)	0.299 (0.062)	0.451 (0.037)	0.250 (0.059)
3	Active	28	0.357 (0.091	0.393 (0.092)	0.250 (0.082)	0.327 (0.064)	0.447 (0.038)	0.226 (0.053)
3	Placebo	25	0.280 (0.090)	0.320 (0.093)	0.400 (0.098)	0.221 (0.053)	0.445 (0.039)	0.334 (0.066)
4	Active	17	0.000 -	0.412 (0.119)	0.588 (0.119)	0.125 (0.043)	0.377 (0.057)	0.498 (0.092)
4	Placebo	16	0.125 (0.083)	0.375 (0.121)	0.500 (0.125)	0.077 (0.029)	0.293 (0.062)	0.630 (0.087)
5	Active	23	0.304 (0.096)	0.435 (0.103)	0.261 (0.092)	0.240 (0.059)	0.449 (0.038)	0.311 (0.068)
5	Placebo	22	0.045 (0.044)	0.500 (0.107)	0.455 (0.106)	0.155 (0.045)	0.408 (0.046)	0.437 (0.077)
6	Active	16	0.438 (0.124)	0.375 (0.121)	0.188 (0.098)	0.408 (0.087)	0.421 (0.048)	0.171 (0.053)
6	Placebo	15	0.333 (0.122)	0.333 (0.122)	0.333 (0.122)	0.286 (0.075)	0.452 (0.037)	0.262 (0.071)

$$\tilde{Z} = \begin{bmatrix} 1 & 0 & 0 & 0 & 0 & 0 & 0 & 0 & 0 & 0 & 0 & 0 & 0 & 0 \\ 0 & 0 & 0 & 0 & 0 & 0 & 0 & 1 & 0 & 0 & 0 & 0 & 0 & 0 \\ 0 & 1 & 0 & 0 & 0 & 0 & 0 & 0 & 1 & 0 & 0 & 0 & 0 & 0 \\ 0 & 0 & 1 & 0 & 0 & 0 & 0 & 0 & 0 & 1 & 0 & 0 & 0 & 0 \\ 0 & 0 & 0 & 1 & 0 & 0 & 0 & 0 & 0 & 0 & 1 & 0 & 0 & 0 \\ 0 & 0 & 0 & 0 & 1 & 0 & 0 & 0 & 0 & 0 & 0 & 1 & 0 & 0 \\ 0 & 0 & 0 & 0 & 0 & 1 & 0 & 0 & 0 & 0 & 0 & 0 & 1 & 0 \\ 0 & 0 & 0 & 0 & 0 & 0 & 1 & 0 & 0 & 0 & 0 & 0 & 0 & 1 \\ 0 & 0 & 0 & 0 & 0 & 0 & 0 & 0 & 1 & 1 & 1 & 1 & 1 & 0 \end{bmatrix}'$$

The last parameter for this model expresses the common amount by which the log odds of cured by 4 weeks vs. not cured by 4 weeks is greater than the log odds of cured at 2 weeks vs. not cured at 2 weeks for investigators 2-6. The Wald goodness of fit chi-squared statistic with 5 d.f. for reduction to this model was $Q = 7.975$ indicating an adequate fit ($p = 0.158$). FARM parameter estimates corresponding to this final model are presented in Table 8 together with their estimated standard errors; FARM final model predicted proportions and their estimated standard errors are displayed in Table 6. For purposes of comparison,

Table 7. Goodness of Fit for the Preliminary Model for the Gastrointestinal Condition Clinical Trial Data

Extra Terms Included in the Extended Model	Columns of the $W_1 = W_2$ matrices†	Observed Statistic Q_S	d.f.	P-Value
All interactions	$[\tilde{w}_{27}\ \tilde{w}_{37}\ \tilde{w}_{47}\ \tilde{w}_{57}\ \tilde{w}_{67}]$	11.928	10	0.290
Investigator 2 x Treatment	$[\tilde{w}_{27}]$	0.181	2	0.913
Investigator 3 x Treatment	$[\tilde{w}_{37}]$	0.151	2	0.927
Investigator 4 x Treatment	$[\tilde{w}_{47}]$	4.908	2	0.086
Investigator 5 x Treatment	$[\tilde{w}_{57}]$	3.385	2	0.184
Investigator 6 x Treatment	$[\tilde{w}_{67}]$	0.068	2	0.966

†$\tilde{w}_{i7} = \tilde{x}_i \# \tilde{x}_7$, $(i = 2,\ldots6)$ where $\tilde{f} \# \tilde{g}$ denotes the elementwise multiplication of the vectors $\tilde{f}$ and $\tilde{g}$. Also, the $\tilde{x}_i$ are the respective columns of $\tilde{X}_1 = \tilde{X}_2$ in (12).

results from WLS analysis of the final model are also given in Table 8. However, these quantities should be viewed rather cautiously because the sample sizes for many of the (investigator x treatment) groups are not really large enough to support their asymptotic rationale; moreover, 0-counts corresponding to some of the outcomes in Table 6 had to be changed to 0.5's in order for the computation of the WLS estimates to be feasbile. Aside from these considerations, it is worthwhile to note that the FARM estimates and their WLS counterparts are relatively similar.

Another issue that merits attention here is that ML results have not been given for the final model. The principal reason for this is the lack of convenient procedures for their straightforward computation. This comment does not mean that ML results are difficult to obtain, but rather that specialized algorithms need to be formulated for them on an individual model basis either directly in their own right or in conjunction with a general optimization procedure such as MAXLIK (see Kaplan and Elston (1972)). Also, as noted in Section 2, the iterative nature of such procedures can be further complicated by the need to avoid 0 or negative predicted proportions at the respective steps, and this consideration can be an awkward source of difficulty when models without proportional odds structure are being used. In contrast, the computational advantage of FARM is that it allows results for general linear specifications of reduced models to be obtained directly without iterations so that their nature can be conveniently evaluated.

Table 8. Final Model Estimated Parameters and Estimated Standard Errors for the Gastrointestinal Condition Clinical Trial Data

Parameters	Estimates (Standard Errors) FARM	WLS†
Intercept for cured at 2 weeks vs not cured at 2 weeks (γ_1)	0.974 (0.581)	0.775 (0.541)
Intercept for cured at 2 or 4 weeks vs not cured (γ_2)	1.815 (0.761)	1.195 (0.626)
Increment for cured at 2 weeks vs not cured at 2 weeks due to investigator 2 (γ_3)	-1.826 (0.635)	-1.568 (0.592)
Increment for cured at 2 weeks vs not cured at 2 weeks due to investigator 3 (γ_4)	-2.236 (0.637)	-1.958 (0.593)
Increment for cured at 2 weeks vs not cured at 2 weeks due to investigator 4 (γ_5)	-3.457 (0.694)	-3.056 (0.647)
Increment for cured at 2 weeks vs not cured at 2 weeks due to investigator 5 (γ_6)	-2.669 (0.654)	-2.302 (0.612)
Increment for cured at 2 weeks vs not cured at 2 weeks due to investigator 6 (γ_7)	-1.887 (0.670)	-1.619 (0.628)
Common slope due to active treatment (γ_8)	0.541 (0.256)	0.464 (0.253)
Common increment for cured at 2 or 4 weeks vs not cured due to investigator 2-6 (γ_9)	1.110 (0.606)	1.444 (0.420)

†Q_{WLS} = 14.134 with 15 d.f. is the WLS residual (or Wald) goodness of fit chi-squared statistic for the final model with respect to the WLS estimates; its non-significance ($p = 0.515$) supports this model. However, the WLS results for this example should be viewed cautiously because the sample sizes are not sufficiently large to support their asymptotic rationale.

Example 4

Table 9 contains data pertaining to a randomized clinical trial designed to compare an active treatment with placebo for an aspect of the condition of patients with chronic pain. The sixteen subpopulations correspond to the cross classification of four types of diagnostic status, two investigators and the two treatments; the ordinally scaled response variable corresponds to the categorization of pain condition into poor, fair, moderate, good or excellent. Also, the patients in the respective (diagnostic status x investigator x treatment) groups are regarded as conceptually representative of corresponding large subpopulations in a sense equivalent to stratified simple random sampling.

The FARM procedures described in Section 2 have a special appeal for this example; here the small counts involved do not support the assumptions required for a WLS analysis; on the other hand, the use of ML methods if a proportional odds model does not hold is not computationally straightforward for reasons given in the discussion of Example 3. The preliminary models considered here have specification matrices

$$\underset{\sim}{X}_1 = \underset{\sim}{X}_2 = \underset{\sim}{X}_3 = \underset{\sim}{X}_4 = \begin{bmatrix} 1&1&1&1&1&1&1&1&1&1&1&1&1&1&1&1 \\ 0&0&0&0&1&1&1&1&0&0&0&0&0&0&0&0 \\ 0&0&0&0&0&0&0&0&1&1&1&1&0&0&0&0 \\ 0&0&0&0&0&0&0&0&0&0&0&0&1&1&1&1 \\ 0&0&1&1&0&0&1&1&0&0&1&1&0&0&1&1 \\ 1&0&1&0&1&0&1&0&1&0&1&0&1&0&1&0 \end{bmatrix}' = \begin{bmatrix} \underset{\sim}{x}_1' \\ \underset{\sim}{x}_2' \\ \underset{\sim}{x}_3' \\ \underset{\sim}{x}_4' \\ \underset{\sim}{x}_5' \\ \underset{\sim}{x}_6' \end{bmatrix}' \tag{13}$$

which represent (reference cell) main effects models for the cumulative logits corresponding to poor vs. at least fair, at most fair vs. at least moderate, at most moderate vs. at least good and at most good vs. excellent. Goodness of fit for such a model was assessed through the residual score statistic Q_S (8) for 6 different extended models. The related $\underset{\sim}{W}$ matrices, the observed values of the statistic, the corresponding d.f. and P-values are indicated in Table 10; the (logitwise) ML estimates of the preliminary model parameters and their estimated standard errors are displayed in Table 11.

A proportional odds structure does not seem to fit the data since the corresponding Wald test statistic $Q_{\underset{\sim}{C}} = 35.019$ is significant ($p = 0.002$) relative to chi-squared distribution with 15 d.f. Additional tests similar to those described for Example 3 were conducted to identify the explanatory variables for which a proportional odds structure holds; the results indicated that such a structure

Table 9. Observed and Model Predicted Proportions and Estimated Standard Errors for the Chronic Pain Clinical Trial

Diagnostic Status	Inves-tigator	Treat-ment	Sample Size	Observed Proportions with Response					FARM Model Predicted Proportions with Response				
				Poor	Fair	Moderate	Good	Excellent	Poor	Fair	Moderate	Good	Excellent
I	A	Active	8	0.375	0.250	0.250	0.125	0.000	0.426	0.042	0.122	0.345	0.065
				(0.171)	(0.153)	(0.153)	(0.117)	-	(0.058)	(0.022)	(0.024)	(0.046)	(0.020)
I	A	Placebo	10	0.700	0.000	0.100	0.100	0.100	0.534	0.042	0.114	0.267	0.043
				(0.145)	-	(0.095)	(0.095)	(0.095)	(0.060)	(0.023)	(0.023)	(0.042)	(0.014)
I	B	Active	16	0.063	0.375	0.063	0.313	0.188	0.203	0.266	0.122	0.242	0.168
				(0.061)	(0.121)	(0.061)	(0.116)	(0.098)	(0.047)	(0.041)	(0.024)	(0.040)	(0.040)
I	B	Placebo	17	0.294	0.235	0.118	0.176	0.176	0.282	0.294	0.114	0.194	0.116
				(0.111)	(0.103)	(0.078)	(0.092)	(0.092)	(0.056)	(0.045)	(0.023)	(0.036)	(0.031)
II	A	Active	6	0.167	0.000	0.167	0.333	0.333	0.240	0.033	0.108	0.480	0.140
				(0.152)	-	(0.152)	(0.192)	(0.192)	(0.048)	(0.017)	(0.023)	(0.046)	(0.038)
II	A	Placebo	4	0.250	0.250	0.000	0.250	0.250	0.329	0.039	0.120	0.418	0.095
				(0.217)	(0.217)	-	(0.217)	(0.217)	(0.059)	(0.021)	(0.024)	(0.049)	(0.029)
II	B	Active	9	0.000	0.111	0.111	0.111	0.667	0.098	0.175	0.108	0.298	0.321
				-	(0.105)	(0.105)	(0.105)	(0.157)	(0.028)	(0.034)	(0.023)	(0.047)	(0.060)
II	B	Placebo	10	0.300	0.100	0.100	0.500	0.000	0.144	0.224	0.120	0.278	0.234
				(0.145)	(0.095)	(0.095)	(0.158)	-	(0.038)	(0.040)	(0.024)	(0.045)	(0.051)

Table 9 (Continued)

Diagnostic Status	Inves-tigator	Treat-ment	Sample Size	Observed Proportions with Response					FARM Model Predicted Proportions with Response				
				Poor	Fair	Moderate	Good	Excellent	Poor	Fair	Moderate	Good	Excellent
III	A	Active	10	0.200 (0.126)	0.000 -	0.300 (0.145)	0.300 (0.145)	0.200 (0.126	0.240 (0.048)	0.033 (0.017)	0.108 (0.023)	0.480 (0.046)	0.140 (0.038)
III	A	Placebo	14	0.357 (0.128)	0.000 -	0.000 -	0.571 (0.132)	0.071 (0.069)	0.329 (0.059)	0.039 (0.021)	0.120 (0.024)	0.418 (0.049)	0.095 (0.029)
III	B	Active	20	0.100 (0.067)	0.200 (0.089)	0.050 (0.049)	0.500 (0.112)	0.150 (0.080)	0.098 (0.028)	0.175 (0.034)	0.108 (0.023)	0.298 (0.047)	0.321 (0.060)
III	B	Placebo	14	0.143 (0.094)	0.357 (0.128)	0.071 (0.069)	0.286 (0.121)	0.143 (0.094)	0.144 (0.038)	0.224 (0.040)	0.120 (0.024)	0.278 (0.045)	0.234 (0.051)
IV	A	Active	16	0.500 (0.125)	0.063 (0.061)	0.188 (0.098)	0.250 (0.108)	0.000 -	0.426 (0.058)	0.042 (0.022)	0.122 (0.024)	0.345 (0.046)	0.065 (0.020)
IV	A	Placebo	11	0.455 (0.150)	0.000 -	0.273 (0.134)	0.273 (0.134)	0.000 -	0.534 (0.060)	0.042 (0.023)	0.114 (0.023)	0.267 (0.042)	0.043 (0.014)
IV	B	Active	12	0.083 (0.080)	0.417 (0.142)	0.167 (0.108)	0.250 (0.125)	0.083 (0.080)	0.203 (0.047)	0.266 (0.041)	0.122 (0.024)	0.242 (0.040)	0.168 (0.040)
IV	B	Placebo	16	0.188 (0.098)	0.250 (0.108)	0.188 (0.098)	0.250 (0.108)	0.125 (0.083)	0.282 (0.056)	0.294 (0.045)	0.114 (0.023)	0.194 (0.036)	0.116 (0.031)

Table 10. Goodness of Fit for the Preliminary Model for the Chronic Pain Condition Clinical Trial Data

Extra Terms Included in the Extended Model	Columns of the $\underset{\sim}{W}_1 = \underset{\sim}{W}_2 = \underset{\sim}{W}_3 = \underset{\sim}{W}_4$ Matrices†	Observed Statistic Q_S	d.f.	P-value
All interactions	$[\underset{\sim}{w}_{25}\underset{\sim}{w}_{35}\underset{\sim}{w}_{45}\underset{\sim}{w}_{26}\underset{\sim}{w}_{36}\underset{\sim}{w}_{46}\underset{\sim}{w}_{56}\underset{\sim}{w}_{256}\underset{\sim}{w}_{356}\underset{\sim}{w}_{456}]$	17.391	40	0.999
All pairwise interactions	$[\underset{\sim}{w}_{25}\underset{\sim}{w}_{35}\underset{\sim}{w}_{45}\underset{\sim}{w}_{26}\underset{\sim}{w}_{36}\underset{\sim}{w}_{46}\underset{\sim}{w}_{56}]$	25.756	28	0.586
Diagnostic Status x Investigator	$[\underset{\sim}{w}_{25}\underset{\sim}{w}_{35}\underset{\sim}{w}_{45}]$	9.580	12	0.653
Diagnostic Status x Treatment	$[\underset{\sim}{w}_{26}\underset{\sim}{w}_{36}\underset{\sim}{w}_{46}]$	13.332	12	0.345
Investigator x Treatment	$[\underset{\sim}{w}_{56}]$	3.929	4	0.416

†$\underset{\sim}{w}_{ij} = \underset{\sim}{x}_i \# \underset{\sim}{x}_j$, $(i = 2,3,4,5; j = 5,6, i \neq j)$; $\underset{\sim}{w}_{k56} = \underset{\sim}{x}_k \# \underset{\sim}{w}_{56}$, $(k = 2,3,4)$ where $\underset{\sim}{f} \# \underset{\sim}{g}$ denotes the elementwise multiplication of the vectors $\underset{\sim}{f}$ and $\underset{\sim}{g}$. Also, the $\underset{\sim}{x}_i$ are the respective columns of $\underset{\sim}{X}_1 = \underset{\sim}{X}_2 = \underset{\sim}{X}_3 = \underset{\sim}{X}_4$ in (13).

fits for diagnostic status ($Q_{\underset{\sim}{C}} = 13.332$, d.f. = 12, p = 0.345) and for treatment ($Q_{\underset{\sim}{C}} = 3.505$, d.f. = 3, p = 0.320) but not for investigator ($Q_{\underset{\sim}{C}} = 22.982$, d.f. = 3, $p < 0.001$). We then proceeded to fit an intermediate model corresponding to the following second stage specification matrix

$$\underset{\sim}{Z}_1 = \begin{bmatrix} 1&0&0&0&0&0&0&0&0&0&0&0&0&0&0&0&0&0&0&0&0&0&0&0 \\ 0&0&0&0&0&0&1&0&0&0&0&0&0&0&0&0&0&0&0&0&0&0&0&0 \\ 0&0&0&0&0&0&0&0&0&0&0&0&1&0&0&0&0&0&0&0&0&0&0&0 \\ 0&0&0&0&0&0&0&0&0&0&0&0&0&0&0&0&0&0&1&0&0&0&0&0 \\ 0&1&0&0&0&0&0&1&0&0&0&0&0&1&0&0&0&0&0&1&0&0&0&0 \\ 0&0&1&0&0&0&0&0&1&0&0&0&0&0&1&0&0&0&0&0&1&0&0&0 \\ 0&0&0&1&0&0&0&0&0&1&0&0&0&0&0&1&0&0&0&0&0&1&0&0 \\ 0&0&0&0&1&0&0&0&0&0&0&0&0&0&0&0&0&0&0&0&0&0&1&0 \\ 0&0&0&0&0&1&0&0&0&0&0&1&0&0&0&0&0&1&0&0&0&0&0&1 \end{bmatrix}'$$

The Wald goodness of fit statistic Q = 15.735 with d.f. = 15 was non-significant (p = 0.400) and thus supported this model. The corresponding FARM predicted proportions and estimated standard errors are diaplayed in Table 11. A further refinement was achieved by fitting a reduced model with specification matrix

$$\underset{\sim}{Z}_2 = \begin{bmatrix} 1&0&0&0&0&0&0&0&0 \\ 0&1&0&0&0&0&0&0&0 \\ 0&0&1&0&0&0&0&0&0 \\ 0&0&0&1&0&0&0&0&0 \\ 0&0&0&0&1&1&0&0&0 \\ 0&0&0&0&0&0&0&1&0 \\ 0&0&0&0&0&0&0&0&1 \end{bmatrix}$$

to the parameter vector $\tilde{\gamma}$ corresponding to the intermediate model mentioned previously. The (third stage) model is supported by the non-significance (p = 0.457) of the corresponding Wald goodness of fit statistic, Q = 16.963 with 17 d.f. FARM final model (with specification martix $\tilde{X}\tilde{Z}_1\tilde{Z}_2$ and parameter vector $\tilde{\tau}$) predicted proportions and estimated standard errors are presented in Table 11.

It can be verified from either the observed or the final model predicted proportions that Investigator A had a tendency to classify as poor the responses corresponding to patients which Investigator B could classify as either poor or fair.

Example 5

Here we illustrate an application of the methods discussed in Section 2 to the case-record data displayed in Table 12. These data pertain to a randomized clinical trial on patients with rheumatoid arthritis where the ordinally scaled outcome is patient response status, categorized as excellent, good, moderate or poor and the explanatory variables of interest are treatment (active or placebo), sex and age. The major difference between this setting and the ones in the previous examples relates to the fact that the sample sizes for the respective subpopulations are very small (in most cases, equal to 1).

The preliminary models considered here have specification matrices of the form $\tilde{X}_1 = \tilde{X}_2 = \tilde{X}_3 = (\tilde{x}_1\tilde{x}_2\tilde{x}_3\tilde{x}_4)$, where $\tilde{x}_1$ is a (84 x 1) vector of 1's, $\tilde{x}_2$ is a (84 x 1) vector with the ith element equal to 0 if the corresponding patient received the active treatment and equal to 1 otherwise, $\tilde{x}_3$ is a (84 x 1) vector with the ith element equal to 0 if the corresponding patient is female and equal to 1 otherwise and $\tilde{x}_4$ is a (84 x 1) vector with the ith element corresponding to the age of the ith patient. Goodness of fit for such models was assessed via the residual score statistic given in (8) corresponding to extended models $(\tilde{X}_\ell, \tilde{W}_\ell)$ where $\tilde{W}_\ell = \tilde{x}_5$, ℓ = 1,2,3 and $\tilde{x}_5$ is a (84 x 1) vector with the ith element equal to 0 if the ith patient's age is less than 55 and equal to (age-55) otherwise. This extended model includes different slopes for patients whose age is greater than 55. The observed value Q_S = 6.845 approaches significance (p = 0.077) relative to its approximate chi-squared distribution with d.f. = 3 and thus does not fully support the preliminary model. The extended model described here was then considered as an alternative preliminary model, and its goodness of fit was verified by the further expansions indicated in Table 13. Since the related goodness of fit residual score statistics indicate an adequate fit, we proceeded to the second stage by testing the hypothesis that a proportional odds model fits the data, via Wald's statistic (7) with the $\tilde{C}$ matrix given in (10). Since the observed

Table 11. Preliminary, Intermediate and Final Model Parameter Estimates and Estimated Standard Errors

Parameter		Poor vs. at Least Fair		At Most Fair vs. at Least Moderate		At Most Moderate vs. at Least Good		At Most Good vs. Excellent
Intercept	β_{11}	0.432 (0.416)	β_{21}	0.496 (0.376)	β_{31}	1.092 (0.397)	β_{41}	2.668 (0.612)
Slope due to Diag. Status II	12	-0.808 (0.608)	β_{22}	-1.154 (0.504)	β_{32}	-1.185 (0.488)	β_{42}	-1.047 (0.581)
Slope due to Diag. Status III	13	-0.842 (0.486)	β_{23}	-0.841 (0.397)	β_{33}	-0.997 (0.402)	β_{43}	-0.006 (0.565)
Slope due to Diag. Status IV	14	-0.186 (0.452)	β_{24}	-0.244 (0.394)	β_{34}	0.066 (0.421)	β_{44}	0.946 (0.727)
Slope due to Investigator B	15	-1.442 (0.365)	β_{25}	-0.173 (0.305)	β_{35}	-0.404 (0.312)	β_{45}	-0.702 (0.481)
Slope due to Active Treatment	16	-0.833 (0.364)	β_{26}	-0.392 (0.299)	β_{36}	-0.271 (0.303)	β_{46}	-0.634 (0.443)

Intermediate Model (FARM) Parameter	Estimate (Standard Errors)	Final Model (FARM)	Estimate (Standard Errors)
Intercept for Poor vs. at Least Fair (γ_1)	0.129 (0.303)	Intercept for Poor vs. at Least Fair (τ_1)	0.137 (0.240)
Intercept for at most Fair vs. at Least Moderate (γ_2)	0.302 (0.294)	Intercept for at most Fair vs. at Least Moderate (τ_2)	0.309 (0.229)
Intercept for at most Moderate vs. at Least Good (γ_3)	0.795 (0.300)	Intercept for at most Moderate vs. at Least Good (τ_3)	0.801 (0.236)
Intercept for at most Good vs. Excellent (γ_4)	3.120 (0.398)	Intercept for at most Good vs. Excellent (τ_4)	3.106 (0.353)
Common slope Due to Diag. Status II (γ_5)	-1.154 (0.429)	Common Slope Due to Diag. Status II and Diag. Status III (τ_5)	-0.851 (0.270)
Common Slope Due to Diag. Status III (γ_6)	-0.693 (0.355)	Common slope for Poor vs. at Least Fair and at Most Good vs. Excellent Due to Investigator (τ_6)	-1.071 (0.223)
Common Slope due to Diag. Status IV (γ_7)	0.011 (0.356)	Common Slope Due to Treatment B (τ_7)	-0.436 (0.266)
Common Slope Due to Investigator B (γ_8)	-1.070 (0.224)		
Common Slope Due to Active Treatment (γ_9)	-0.440 (0.266)		

Table 12. Data Listing for the Rheumatoid Arthritis Clinical Trial

T	Sex	Age	PRS	T	Sex	Age	PRS	T	Sex	Age	PRS	T	Sex	Age	PRS
1	0	23	4	1	0	67	2	2	0	30	4	2	0	59	1
1	0	32	4	1	0	68	3	2	0	30	4	2	0	61	4
1	0	37	3	1	0	68	1	2	0	31	3	2	0	63	3
1	0	41	2	1	0	69	3	2	0	32	4	2	0	64	4
1	0	41	4	1	0	69	4	2	0	33	2	2	0	65	1
1	0	48	4	1	0	70	3	2	0	37	4	2	0	66	4
1	0	48	2	1	1	27	3	2	0	44	4	2	0	66	4
1	0	55	2	1	1	29	4	2	0	45	4	2	0	66	3
1	0	55	1	1	1	30	4	2	0	46	4	2	0	68	3
1	0	56	2	1	1	32	1	2	0	48	4	2	0	74	2
1	0	57	2	1	1	46	2	2	0	49	4	2	1	37	4
1	0	57	2	1	1	58	2	2	0	51	4	2	1	44	4
1	0	57	2	1	1	59	2	2	0	53	4	2	1	50	4
1	0	58	4	1	1	59	4	2	0	54	4	2	1	51	4
1	0	59	2	1	1	63	4	2	0	54	2	2	1	52	4
1	0	59	1	1	1	63	4	2	0	54	4	2	1	53	4
1	0	60	1	1	1	64	3	2	0	55	2	2	1	59	4
1	0	61	2	1	1	64	4	2	0	57	3	2	1	59	4
1	0	62	3	1	1	69	4	2	0	57	4	2	1	62	4
1	0	62	2	1	1	70	2	2	0	58	3	2	1	62	4
1	0	66	1	2	0	23	4	2	0	59	3	2	1	63	2

PRS = Patient Response Status (1 = Excellent, 2 = Good, 3 = Moderate, 4 = Poor)
T = Treatment (1 = Active, 2 = Placebo)
Sex (0 = Female, 1 = Male)

value Q_C = 9.300 with d.f. = 8 supports the hypothesis (p = 0.318), we fitted the second stage model using the specification matrix (11). The final (proportional odds) model parameter estimates and the respective estimated standard errors obtained by the FARM procedure as well as by ML methods are presented in Table 14. Also, displayed in Table 15 are the FARM final model predicted proportions for ages 35, 45, 55 and 65 together with the corresponding estimated standard errors.

Table 13. Goodness of Fit for the Preliminary Model for the Rheumatoid Arthritis Clinical Trial Data

Extra Terms Included in the Extended Model	Columns of the $W_1 = W_2 = W_3$ Matrices†	Observed Statistic Q_S	d.f.	P-Value
Sex x Treatment	$[w_{23}]$	0.774	3	0.856
Treatment x Age	$[w_{24} w_{25}]$	10.006	6	0.124
Sex x Age	$[w_{34} w_{35}]$	10.363	6	0.110

†$w_{ij} = x_i \# x_j$, (i = 2,3; j = 3,4,5, i ≠ j) where $f \# g$ denotes the elementwise multiplication of the vectors f and g; also the x_i are the respective columns of the model specification matrices.

Table 14. Final Model Estimated Parameters and Estimated Standard Errors for the Rheumatoid Arthritis Clinical Trial Data

Parameters	Final Model (Proportional Odds) Estimates (Standard Errors) FARM	ML
Intercept for Excellent vs. at most Good (γ_1)	-3.764 (1.535)	-4.494 (1.560)
Intercept for at Least Good vs. at most Moderate (γ_2)	-2.123 (1.490)	-2.651 (1.498)
Intercept for at Least Moderate vs. Poor (γ_3)	-1.554 (1.478)	-1.736 (1.476)
Common Slope Due to Active Treatment (γ_4)	-1.459 (0.480)	-1.746 (0.471)
Common Slope due to Males (γ_5)	-1.119 (0.545)	-1.263 (0.530)
Common Slope Due to Age (γ_6)	0.049 (0.031)	0.061 (0.031)
Common Slope Increment Due to Age > 55(γ_7)	-0.038 (0.072)	-0.060 (0.070)

Table 15. Final Model Estimated Proportions and Estimated Standard Errors for the Rheumatoid Arthritis Clinical Trial Data

		Active Treatment (1) Patient Response Status						Placebo Treatment (2) Patient Response Status			
Sex	Age	Excellent	Good	Moderate	Poor	Sex	Age	Excellent	Good	Moderate	Poor
0	35	0.114 (0.061)	0.285 (0.088)	0.141 (0.045)	0.459 (0.135)	0	35	0.029 (0.019)	0.105 (0.051)	0.081 (0.036)	0.785 (0.091)
0	45	0.174 (0.070)	0.347 (0.077)	0.137 (0.044)	0.342 (0.100)	0	45	0.047 (0.024)	0.155 (0.052)	0.107 (0.038)	0.691 (0.084)
0	55	0.256 (0.105)	0.384 (0.076)	0.119 (0.043)	0.242 (0.099)	0	55	0.074 (0.039)	0.218 (0.072)	0.130 (0.044)	0.578 (0.113)
0	65	0.278 (0.094)	0.387 (0.075)	0.113 (0.040)	0.221 (0.076)	0	65	0.083 (0.042)	0.234 (0.071)	0.133 (0.044)	0.550 (0.109)
1	35	0.040 (0.028)	0.138 (0.071)	0.099 (0.045)	0.722 (0.126)	1	35	0.009 (0.008)	0.038 (0.027)	0.034 (0.022)	0.918 (0.053)
1	45	0.064 (0.038)	0.198 (0.078)	0.124 (0.045)	0.614 (0.130)	1	45	0.016 (0.012)	0.061 (0.035)	0.051 (0.028)	0.873 (0.068)
1	55	0.101 (0.063)	0.266 (0.098)	0.139 (0.046)	0.494 (0.158)	1	55	0.025 (0.019)	0.093 (0.055)	0.074 (0.039)	0.808 (0.102)
1	65	0.122 (0.060)	0.282 (0.084)	0.141 (0.045)	0.466 (0.130)	1	65	0.028 (0.021)	0.103 (0.056)	0.079 (0.039)	0.789 (0.103)

4. DISCUSSION

In this paper we consider a two-stage FARM procedure for analyzing ordinal categorical data using linear models. In the general case defined in (2), the method

(i) has computational advantages over its ML counterpart in the sense that it may be undertaken with a combination of existing statistical software and appropriate matrix operations as opposed to requiring formulation of some model based optimization algorithms;

(ii) has somewhat less stringent requirements than the corresponding WLS procedure with respect to sample sizes.

Although theoretically the proposed FARM technique may be less efficient than either the ML or the WLS method, such tendencies were not evident in the examples; in fact, the results were comparable in the cases where the methods were contrasted. It is of interest to note that for comparison purposes, the modeling strategy adopted in Section 3 was directed at fitting proportional odds models or variations thereof (in which such structure was imposed on some explanatory variables, but not on others) to the data. However, the proposed method allows extensions in other directions; first, different specification matrices may be assigned to the different cumulative logits and second, the more general cumulative proportions described at the beginning of Section 2 may be considered.

Finally, as noted in the Introduction, parametric model methods for ordinal data should be applied cautiously to small sample situations because they may not allow an effective assessment of goodness of fit. Such considerations apply to the extensions of proportional odds methods discussed in this paper, and a careful effort was directed at dealing with them reasonably.

ACKNOWLEDGEMENTS: This research was partially supported by the U.S. Bureau of the Census through Joint Statistical Agreement JSA 84-1. Work of Julio M. Singer was done while he was on leave at the Department of Biostatistics, University of North Carolina at Chapel Hill. The authors would like to thank Ann Thomas for editorial assistance; and Alan Agresti and Peter Imrey for helpful comments.

REFERENCES

[1] Agresti, A., Generalized odds ratios for ordinal data, Biometrics 36:59-67 (1980).

[2] Agresti, A., A survey of strategies for modeling cross-classifications having ordinal variables, J. Amer. Statist. Assoc. 78:184-198 (1983).

[3] Andrich, D., A model for contingency tables having an ordered response classification, Biometrics 35:403-415 (1979).

[4] Clogg, C.C., Some models for the analysis of association in multiway cross-classifications having ordered categories, J. Amer. Statist. Assoc. 77:803-815 (1982).

[5] Conniffe, D., Covariance analysis and seemingly unrelated regressions, Amer. Statist. 36:169-171 (1982).

[6] Forthofer, R.N. and Koch, G.G., An analysis for compounded functions of categorical data, Biometrics 29:143-157 (1973).

[7] Forthofer, R.N. and Lehnen, R.G., Public Program Analysis: A New Categorical Data Approach, Wadsworth, Inc., Belmont, CA (1981).

[8] Grizzle, J.E., Starmer, C.F., and Koch, G.G., Analysis of categorical data by linear models, Biometrics 25:489-504 (1969).

[9] Haberman, S.J., Log-linear models for frequency tables with ordered classifications, Biometrics 30:589-600 (1974).

[10] Harrell, F., SAS Supplemental Library User's Guide, 1983 Ed., SAS Institute Inc., Cary, NC (1983).

[11] Imrey, P.B., Koch, G.G., and Stokes, M.E., Categorical data analysis: Some reflections on the log linear model and logistic regression, Internat. Statist. Rev. 49:265-283 (Part I); 50:35-63 Part II (1981, 1982).

[12] Kaplan, E.B. and Elston, R.C., A subroutine package for maximum likelihood estimation (MAXLIK), Inst. Statist. Mimeo Ser. No. 823, Chapel Hill, University of North Carolina (1972).

[13] Kleinbaum, D.G., Testing linear hypotheses in generalized multivariate linear models, Commun. Statist. 1:433-457 (1973).

[14] Koch, G.G., Amara, I.A., Davis, G.W., and Gillings, D.B., A review of some statistical methods for covariance analysis of categorical data, Biometrics 38:563-595 (1982).

[15] Koch, G.G. and Gillings, D.B., Inference, design based versus model based, Encyclopedia of Statistical Sciences 4, N.L. Johnson and S. Kotz (eds.), 84-88, Wiley, New York (1983).

[16] Koch, G.G., Grizzle, J.E., Semenya, K.A., and Sen, P.K., Statistical methods for evaluation of mastitis treatment data, J. Dairy Sci. 61:830-847 (1978).

[17] Koch, G.G. and Sollecito, W.A., Statistical considerations in the design, analysis, and interpretation of comparative clinical studies, Drug Infor. J. 18:131-151 (1984).

[18] Lan, S.P. and Shy, C.M., The effect of suspended particular pollution on chronic respiratory disease: A natural experiment in the Southeast Region, 1971-1972. Unpublished manuscript. (1979).

[19] Landis, J.R., Heyman E.R., and Koch, G.G., Average partial association in three-way contingency tables; A review and discussion of alternative tests, Internat. Statist. Rev. 46:237-254 (1978).

[20] Landis, J.R., Stanish, W.M., Freeman, J.L., and Koch, G.G., A computer program for the generalized chi-square analysis of categorical data using weighted least squares (GENCAT), Computer Programs in Biomedicine 6:196-231 (1976).

[21] McCullagh, P., Regression models for ordinal data, J.R. Statist. Soc. B 42:109-142 (1980.

[22] Morrison, D.F., Multivariate Statistical Methods, McGraw Hill Book Company, New York (1967).

[23] Neyman, J., Contribution to the theory of the χ^2-test. In Proceedings of the Berkeley Symposium on Mathematical Statistics and Probability, J. Neyman (ed.), 239-273, Berkeley, University of California Press (1949).

[24] Revankar, N.S., Use of restricted residuals in SUR Systems: some finite sample results, J. Amer. Statist. Assoc. 71:183-188 (1976).

[25] Semenya, K.A. and Koch, G.G., Compound function and linear model methods for the multivariate analysis of ordinal categorical data, Inst. Statist. Mimeo Ser. No. 1323, Chapel Hill, University of North Carolina (1980).

[26] Semenya, K.A., Koch. G.G., Stokes, M.E., and Forthofer, R.N., Linear models methods for some rank function analysis of ordinal categorical data, Commun. Statist. A 12:1277-1298 (1983).

[27] Srivastava, V.K. and Dwivedi, T.D., Estimation of seemingly unrelated regression equations, J. Econometrics 10:15-32 (1979).

[28] Stokes, M.E. and Koch, G.G., A macro for maximum likelihood fitting of log-linear models to Poisson and multinomial counts with contrast matrix capability for hypothesis testing. In Proceedings of the Eighth Annual SAS Users Group International Conference, 795-800 (1983).

[29] Walker, S.H. and Duncan, D.B., Estimation of the probability of an event as a function of several independent variables, Biometrika 54:167-179 (1967).

[30] Williams, O.D. and Grizzle, J.E., Analysis of contingency tables having ordered response categories, J. Amer. Statist. Assoc. 67:55-63 (1972).

[31] Zellner, A., An efficient method of estimating seemingly unrelated regressions and tests for aggregation bias, J. Amer. Statist. Assoc. 57:348-368 (1962).

BIOSTATISTICS: Statistics in Biomedical, Public Health
and Environmental Sciences, edited by P.K. Sen

THE GENERALIZED SIGNED RANK TEST, THE GENERALIZED SIGN TEST AND THE STRATIFIED LOG RANK TEST

Peter A. Lachenbruch and Robert F. Woolson

University of Iowa
Iowa City, Iowa

The Generalized Signed Rank (GSR), Generalized Sign (GS), and Stratified Log Rank (SLR) tests are compared for analyzing paired data. Two models are considered: one in which randomization is done by pairs, and one in which randomization is done within strata for the Stratified Log Rank test but within pairs for the Generalized Signed Rank test. It was shown that the Stratified Log Rank test is equivalent to the Generalized Sign test if the strata coincide with the pairs. If strata include more than one pair of observations, the SLR performs better than the GSR if balanced randomization is done, but the GSR is substantially better if randomization is done within strata.

KEY WORDS: Paired Survival Tests, Generalized Signed Rank, Log Rank, Power

INTRODUCTION AND SUMMARY

In 1980, we introduced the generalized signed rank (GSR) and generalized sign (GS) tests (Woolson and Lachenbruch, 1980) for paired survival data. Do these linear rank tests offer any significant advantage over the commonly used log-rank (LR) test, or a stratified version of the log-rank (SLR) test? The stratified log-rank test is a straightforward extension of the log-rank test using the usual Mantel-Haenszel arguments for combining 2x2 tables. These are described in the third section. This study examines the power of these tests when matching (i.e., pairing) is important.

Information on pairing may be explicit, as in the case of determining pairs on the basis of a set of readily observed variables such as age, sex, or weight; or it may be implicit, as we might find in the case of pairing on the basis of group membership. Examples might be littermates, members of a group which lives together such as religious orders, and so forth. Strata may be formed, and a randomization within strata done with the goal of performing a stratified log-rank test. Alternatively, a matched analysis may be done. Still another alternative might include ignoring any stratification variables and performing a log-rank test. The natural question is which is to be preferred? This paper addresses these questions.

The balanced randomization study refers to the case in which patients are randomized to treatments within strata might be created from covariate values, or information used in creating pairs. The unbalanced randomization study refers to the case in which this information is not used to allocate patients to treatments. We compare the behavior of the GS, GSR, LR, and SLR tests in these two contexts. We find that if balanced randomization is used, the GS, GSR and SLR have approximately the correct size (the LR is conservative) and that the power of the SLR is greater than that of the SGR or LR for heavy censoring, but for light or no censoring there is little to choose among them. For the unbalanced randomization case, the GSR is more powerful than any of the others.

One of the requirements of the GS and GSR tests is that there is a paired observation for each randomized patient. This is not always possible, especially if the matching criteria are stringent. If this leads to a large number of unmatched observations, a separate unmatched analysis may be performed (assuming, of course, that lack of matching is not related to treatment group). This will not be discussed further in this paper. The remainder of the paper describes the details of the two studies.

DATA

The usual data from a paired survival study consists of a set of paired observations. It is assumed that the maximum possible time of observation is the same for both members of the pair, although this time may vary among pairs. If not, the minimum of the two individual censoring times is commonly used as the common censoring time. Let the survival time from observations on the first treatment be denoted as $t_{11} \ldots t_{1n}$ and those for the second treatment as $t_{21} \ldots t_{2n}$ (we assume all observations are matched) and their differences are $d_1 = t_{11}-t_{21} \ldots d_n = t_{1n}-t_{2n}$. Because any observation may be a failure or censored (i.e., the individual has not failed), the pairs fall into four types:

I. Both t_{1i} and t_{2i} are failure times, and thus d_i is known. There are n_I such pairs.

II. t_{1i} is censored, t_{2i} is a failure time. Thus d_i is known to exceed the given value. It is right censored. There are n_{II} Type II pairs.

III. t_{1i} is a failure time, t_{2i} is censored. The difference is left censored and is known to be less than the negative value. There are n_{III} of these pairs.

IV. Both observations are censored. There are n_{IV} Type IV pairs.

We always know the sign of the difference unless we have a Type IV pair since the censoring time is the same for each member of the pair.

TESTS

We wish to test $H_0:F_1=F_2$, where F_i is the distribution in group i. We consider the following three tests for matched data:

a. The generalized sign test (GS) is identical to the usual sign test after discarding all pairs in which both members were censored observations. This is intuitively obvious, and the behavior of this test is the same as the usual sign test based on the number of pairs which are not Type IV. Let $n=N-n_{IV}$ and p be the number of positive differences among the n pairs. Then

$$T_{GS} = (p-n/2)/(n/4)^{\frac{1}{2}}$$

is asymptotically N(0,1) under the null hypothesis.

b. The generalized signed rank (GSR) test is the generalization of the signed rank test to censored data. Again, Type IV pairs are eliminated in computing this statistic. Let $s_{(i)}$ be the sign of the ith ordered difference (=± if positive, -1 if negative), $g_{(i)}=1$ if both observations are failures (i.e., a Type I pair), and 0 otherwise. Let

$$r_i = \sum_{j=1}^{i} (n-j+1)/(n-j+1+g_{(j)}),$$

$$T = \sum (s_{(i)}(1-r_i)g_{(i)} + s_{(i)}(1-.5*r_i)(1-g_{(i)}),$$

and $$V = \sum (1-r_i)^2 g_{(i)} + (1-.5*r_i)^2(1-g_{(i)}),$$

then $T_{GSR} = T/V^{\frac{1}{2}}$ is asymptotically N(0,1).

c. The generalized signed rank test may also be performed on the logarithms of the observations. The test is identical to the GSR in form, but the ordering is done on the differences of logs. In fact, if the differences of logs are logistically distributed the GSR test is the score test based on the likelihood for the generalized rank vector.

Alternatively, one might ignore whatever matching has been introduced in the design of the study. If the pairing effects are not large (or are nonexistent), this may lead to a more powerful test than one which accounts for matching. Thus, a natural competitor to the matched methods is the log-rank test (LR). A middle ground is to use the matching information to stratify the data. If one stratifies so that each pair is a stratum, one has obtained a paired test based on the log-rank statistic. This reduces to the generalized sign test. Most stratification schemes will use fewer strata than there are pairs. The stratification scheme we used was to break the groups into two strata with 12 pairs in one group and 13 in the other. The stratified log-rank test is then given as

$$T = \sum_{i,s} (D_{1is} - ED_{1is})$$

$$V = \sum_{i,s} \mathrm{Var}(D_{1is})$$

where s indexes the stratum,

$$ED_{1is} = n_{1is}D_{is}/N_{is},\ \mathrm{Var}(D_{1is}) = n_{1is}n_{2is}D_{is}(N_{is}-D_{is})/N_{is}^{2}(N_{is}-1).$$

N_{is} is the number at risk at the ith death time in stratum s, D_{is} is the number of deaths at the ith death time in the sth stratum, n_{jis} is the number at risk in the sth stratum, i^{th} death time, and j^{th} group, and D_{1is} is the observed number of deaths in group 1, at i^{th} death time and s^{th} stratum. The stratified log-rank (SLR) test statistic is

$$T_{SLR} = T/V^{\frac{1}{2}}.$$

For S = 1, this is the usual log rank test.

THE STUDY

We assume the statistician has knowledge of whatever pairing of stratifying information is present. If this information is ignored, the log-rank test (or some other two-sample test such as Generalized Wilcoxon, G rho, gamma, etc.) may be used. If it is used, the matched analysis methods (Woolson and Lachenbruch) may test the hypothesis of common survival distributions, or a stratified version of the log-rank test may be employed. Still another option arises if the randomizations were to occur ignoring the stratifying information, but the investigator wished to incorporate that data into the analysis. In this case, one would find the possibility of an unbalanced randomization in which heterogeneity of the subjects occurred in the treatment groups.

Various comparisons may be made. First, when the stratifying information is available and used, how do the five candidate tests compare? The randomization occurs as follows. Strata are formed on the basis of the "covariate." They consist of a number of pairs based on the underlying parameters of the survival distributions. Thus, in this study, stratum 1 consists of all observations whose exponential parameter is less than .7, and stratum 2 is the remainder. For the matched pairs analyses, subjects are randomly allocated within pairs, while for the log-rank and stratified log-rank tests, they are allocated within strata. This is the balanced randomization study. There is at most limited heterogeneity in the subjects in the log-rank or stratified log-rank tests (i.e., the distribution of the underlying parameters of the survival distributions is the same in both groups). Second, if post-randomization stratification occurs for the stratified log-rank test, and randomization does not depend on stratum for the log-rank test, how do these compare to the matched analyses? In this case randomization is again performed within pairs for the matched analyses. This is the unbalanced randomization study. Pairing information is used for the matched analyses. If the only information we have on matching is the identity of the pair members, the stratified log-rank and generalized sign test are identical. In this case, the comparison of interest is only between the GS and GSR tests. Since the underlying distributions are assumed to be exponential, the log-rank and stratified log-rank are optimal if there is no difference among the pairs (i.e., pairing makes no difference).

We report on two studies. The first, denoted as the balanced randomization study, assumes the statistician has knowledge of the pairing information and allocates subjects to treatments so that one member of each pair is allocated to each treatment. This, of course, is not usually the case and minimizes whatever effects heterogeneity in pairs has on the tests. The second study ignores this information for the log-rank and stratified log-rank tests. Thus, it is possible for an imbalance to occur which may affect the performance of the tests. For the balanced randomization study, we generated pairs of observations having an exponential distribution with parameters a function of the pair number and the "treatment." Thus, the observations in the ith pair have parameters

$$A_iD \quad \text{in group 1}$$

$$A_i/D \quad \text{in group 2}$$

where A_i is the expected value of the order statistics from a sample of n exponential observations with parameter equal to one, i.e.,

$$A_i = \sum_{1}^{i} 1/(n+1-j), \text{ for } i = 1,\ldots,n.$$

A set of the values of A_i is given in Table 1. Thus A_i is a block effect and D is the treatment effect.

An observation could be matched with probability p_u (=.1 or .2). A random number, $U(0<U<1)$, was generated and if $0<U<p_u/2$, the observation from group 1 was considered missing (only group 2 was observed); if $p_u/2<U<p_u$ the observation from group 2 was "missing." We did not consider fractions of missingness different from .5 as we assumed that missingness was random between the treatments. This would correspond to not being able to find a suitable member of a matched pair,

rather than loss after treatment assignment. Clearly, asymmetric missing observations change the performance of any test by changing significance levels and altering the power function.

Table 1. Expected Values of Unit Exponential (A_i)

i	A_i	i	A_i
1	.040	14	.796
2	.082	15	.887
3	.125	16	.987
4	.171	17	1.098
5	.218	18	1.223
6	.268	19	1.366
7	.321	20	1.533
8	.376	21	1.733
9	.435	22	1.983
10	.498	23	2.316
11	.564	24	2.816
12	.636	25	3.816
13	.713		

For each observation, a censoring time was chosen. This observation was uniformly distributed between 0 and T_0, where T_0 is chosen to give various probabilities of censoring an individual observation. For paired observations, the minimum of the two censoring times was chosen. The values of T_0 and the corresponding probabilities of censoring are:

T_0	P(One Observation Censored)	P(Both Observations Censored)
1.61	.50	.30
2.30	.39	.22
3.00	.32	.17
5.00	.20	.10
10.00	.10	.05
20.00	.05	.025
∞	.000	.000

The last case corresponds to the no censoring situation.

For the second, unbalanced randomization study, randomization was made using the pairing information for the GS and GSR tests. For the LR test, no concomitant information was used and for the SLR, only the stratum was used to make pairs. Thus, considerable within-stratum heterogeneity was possible.

RESULTS

Results for the balanced randomization study are given in Tables 2 to 4 for the size and power of the tests. We note that the log-rank and stratified log-rank tests do not have the appropriate significance levels. The log rank rejects too rarely under the null hypothesis which should lead to too conservative a test, and thus have too low power. The rate of rejection seems to decrease as the censoring rate decreases. The stratified log-rank test rejects too often at high censoring levels and too rarely at low censoring levels. The GS, GSR, and GSR based on the logarithms of the observations all reject at the correct level based on 1,000 replications of samples of size 25 in each group.

Table 2. Balanced Randomization. Size of Tests for Nominal α = .05

Number of rejections of H_0 out of 1000 replications
p_u = .1, D = 1 (null hypothesis value)

T_o	GS	GSR	GSR-log	LR	SLR
1.61	49	41	41	22	68
2.30	43	54	46	24	65
3.00	56	46	50	22	62
5.00	58	48	52	22	68
10.00	50	49	55	11	41
20.00	42	42	39	11	35
∞	56	38	39	2	9

In Table 3 we have p_u=.1, D=1.5. For the set of T_o, we see the power of the LR is lower than the GSR and about that of the GS. The SLR test is the highest for finite sample sizes. The GSR on logs and GSR tests behave similarly; however, it should be stressed that the results arise under the balanced randomization model. Any effects that pairing might have is taken care of by assigning different treatments to the members of each pair.

Table 3. Balanced Randomization. Power of Tests at α = .05

Number of rejections of H_o out of 1000 replications
p_u = .1, D = 1.5

T_o	GS	GSR	GSR-log	LR	SLR
1.61	185	261	268	269	423
2.30	223	320	317	261	482
3.00	272	367	365	267	528
5.00	261	379	375	299	575
10.00	302	444	454	315	619
20.00	368	525	502	328	624
∞	428	542	550	277	503

Note: The LR and SLR lose power as T tends to infinity. The SLR may be gaining the high power because it is rejecting too many at the null value; it is anticonservative. See text.

Table 4. Balanced Randomization. Power of Tests at α = .05

Number of rejections of H_0 out of 1000 replications
p_u = .1, D = 2

T_0	GS	GSR	GSR-log	LR	SLR
1.61	483	604	611	639	797
2.30	553	703	711	693	858
3.00	611	756	768	736	905
5.00	653	820	813	785	922
10.00	715	869	872	840	961
20.00	754	903	895	835	955
∞	855	949	945	828	950

An interesting finding is that both the LR and SLR lose power as the censoring point goes to infinity. This occurs because the ratio of hazards is no longer a constant, but tends to 1 as t tends to infinity. Thus, for large censoring times, we have the opportunity to observe "similar" observations from both groups which reduces the power. Specifically, for group 1 the conditional distribution given A_i is $f(t\ A_i{*}D) = A_i{*}D{*}\exp(-A_i{*}D{*}t)$ and the marginal distribution (integrated over the "exponential" A_i is

$$f(t) = D/(D(t+1))^2$$

$$S(t) = 1/(D(t+1))$$

$$h(t) = D/(D(t+1)).$$

For group 2, a similar derivation gives a hazard function of

$$h(t) = (1/D)/(t/D+1)$$

so the ratio of the hazards is

$$h_1(t)/h_2(t) = (D^2)(t/D+1)/D(t+1)).$$

This has value D^2 at t=0 but tends to 1 as t tends to infinity. Thus we no longer have a proportional hazards model.

The results for the study in which pairing information is ignored in randomizing for LR and SLR tests are given in Tables 5 and 6 (unbalanced randomization study). We have not tabulated results for the GSR-logs or for intermediate censoring points. In this situation, the matched procedures use the pairing information while the LR ignores it. The SLR partially ignores this information. The additional information used by the paired procedures gives them a substantial advantage. The apparent superiority of the SLR test no longer holds. The GSR is the more powerful for D=1.5 and D=2. For heavy censoring (50% of pairs have one

Table 5. Unbalanced Randomization

Number of rejections of H_0 out of 1000 replications
alpha = .05, p_u = .2

D	T_0	GS	GSR	LR	SLR
1	1.61	58	42	57	62
	5.00	61	57	67	60
	∞	47	45	82	76
1.5	1.61	176	260	282	279
	5.00	271	365	317	305
	∞	328	486	348	349
2	1.61	418	562	596	581
	5.00	622	773	684	655
	∞	743	891	708	699

An average of 45 observations was available for the log rank procedures. There were 9.5, 14.1, and 20.0 pairs available on the average for the paired procedures.

Table 6. Unbalanced Randomization

Number of rejections of H_0 out of 1000 replications
alpha = .05, p_u = .1

D	T_0	GS	GSR	LR	SLR
1	1.61	49	41	79	72
	5.00	58	48	77	66
	∞	56	38	85	74
1.5	1.61	185	261	269	257
	5.00	261	379	335	325
	∞	428	542	353	336
2	1.61	483	604	625	601
	5.00	653	820	675	672
	∞	855	949	746	726

An average of 47.5 observations was available for the log rank procedures. For the paired procedures, there were 10.8, 15.8, and 22.5 pairs available for each case.

censored observation), it is not significantly different from the LR and SLR. The advantage of the GSR is clear in uncensored trials and quite substantial in all but the very heavy censoring. The GS test outperforms the LR and SLR in the uncensored cases. The LR and SLR are anticonservative at low levels of censoring. We adjusted the critical value so that exactly 50 rejections would occur and found that the power was much less than the GSR and GS. McNemar's test was used to compare results within the parameter combination. These showed the GSR superior for all cases. These results have been found for underlying exponential distributions which lead to optimal properties for the GSR test and the LR test (in the unpaired case). For other distributions, the results may be different. However, other work (Lachenbruch and Woolson, 1984) suggests the GSR and GS tests are quite robust to failure of the distributional assumptions.

DISCUSSION

If pairing information is based on covariates which are observable explicitly, the stratified log-rank test has greater power than any of the other tests in the situation we created here. It becomes conservative (as does the log-rank) test and loses power relative to the matched analyses when there is little or no censoring. If post - stratification is used (i.e., strata are formed at the time of analysis), the generalized signed rank test is preferable. In particular, if no information about strata is available other than the pair membership, the generalized signed rank test is the method of choice, since in this case the stratified log rank test is the same as the generalized sign test.

REFERENCES

[1] Lachenbruch P.A. and Woolson, R.F., On small sample properties of the Generalized Signed Rank and Generalized Sign Tests, accepted subject to revision by Communication in Statistics (1984).

[2] Woolson, R.F. and Lachenbruch, P.A., Rank tests for Censored Matched Pairs, Biometrika **67**:597-606 (1980).

BIOSTATISTICS: Statistics in Biomedical, Public Health
and Environmental Sciences, edited by P.K. Sen

KOLMOGOROV-SMIRNOV STATISTICS -- WEIGHTINGS, MODIFICATIONS, AND VARIATIONS

Nathan Mantel

Department of Mathematics, Statistics and Computer Science
The American University
Bethesda, Maryland

Critical values of the Anderson-Darling weighted single-sample Kolmogorov-Smirnov (K-S) statistic are extravagantly large because of the possibility that at least one observation will arise when the expectation is still very small. The weighted K-S statistic cannot be interpreted as a maximal normal deviate due to the non-asymptotic situations in which the very large values occur. If the actual null distribution of the weighted K-S is employed, a fresh difficulty arises. It is that in the low probability event that no observations have yet occurred when the expectation is moderate or large, the weighted K-S statistic attains only moderate values. A procedure for basing the K-S test on the null tail probabilities for the actual number of events observed is outlined. Relations to the statistics of Rényi, of Kuiper, and of Greenwood are discussed.

KEY WORDS: Kolmogorov-Smirnov tests; Rényi test; Logrank procedures; Pseudo-sequential tests; Goodness-of-fit tests; Maximal chi squares; Mantel-Haenszel tests; Non-parametric tests; Kuiper statistic; Greenwood statistic.

1. INTRODUCTION

The Kolmogorov-Smirnov (K-S) statistics (Kolmogorov, 1933; Smirnov, 1939) have been the stimulus of much interesting statistical research and theory. Durbin (1973) gives an excellent review of the theory of K-S tests, and also of the Cramér-von Mises tests, together with an extensive bibliography. Limiting null distributions of the K-S statistics, D^+ or D^-, for example, are shown or indicated to follow, when squared, the exponential form, in both the single-sample and two-sample cases. More recently, Niederhausen (1981) has shown how Sheffer polynomials could be employed to get exact distributions of K-S-related statistics, citing various key references not available in Durbin (1973). Miller and Siegmund (1980) consider the problem of evaluating a maximally selected chi square when all possible cut points are considered in comparing two survival distributions. This maximal chi square is the two-sample version of the Anderson-Darling (1952) test statistic for the one sample goodness-of-fit problem. Anderson and Darling (1952), whom Niederhausen does not specifically cite, had described a weighted variant of the one-sample K-S statistics.

If $F(x)$ is the hypothetical cumulative distribution of a continuous random variable X, while $F_N(x)$ is its empirical cumulative distribution based on a sample of size N, the usual K-S single-sample statistic $\sup|F_N(x) - F(X)|$ becomes, with the Anderson-Darling weighting, $\sup[N^{\frac{1}{2}}|F_N(x) - F(x)|/F(x)(1-F(X))]$, the supremum being taken over the region appropriate for x. Essentially, the difference between $F_N(x)$ and $F(x)$ is re-expressed in standard deviation units.

Theoretical appeal of the K-S statistics stems from the easily understandable formulation to which the K-S approach gives rise. In the single-sample case, the hypothetical distribution maps onto the unit uniform distribution, so that the ordered observations become order statistics from the uniform distribution. For the two-sample case, a permutational approach to the data serves the same function as does reference to the uniform distribution in the single-sample case. (Conceptually, the one-sample problem is a special case of the two-sample problem, but with the second sample of indefinitely large size.) In either situation, the K-S statistics can be employed in pseudosequential manner when applied to response time data, as I have previously suggested (Mantel, 1966), an added desirable feature. If one knows what is the magnitude of a critically significant departure, then significance might be declared early on in observing a time-to-response distribution (or distributions).

Actually, in Mantel (1966), I had indicated certain weaknesses in the K-S procedures and had noted the Anderson-Darling (1952) approach for modifying and strengthening the K-S procedures. Subsequently, in Mantel (1968) I had occasion to fault a remedial approach by Rényi (1953) to the K-S one-sample problem. Again, I suggested that the Anderson-Darling approach might be remedial. But I have become aware that, for the one-sample problem, the Anderson-Darling method has basically the same weakness as the Rényi method. What that weakness is, and what K-S variant should instead be used is my subject here.

2. THE RÉNYI MODIFICATION

A basic flaw I had noted with the original K-S one-sample approach was that it could fail to reject the hypothesized distribution in the face of data incompatible with that distribution. A single, or even two, observed quantities outside the range, say, 10-20, would be incompatible with the possibility that the distribution is some one with density or probability only inside that range, say the uniform, 10-20. Yet those few incompatible observations would not be sufficient to make the K-S statistic significantly large. While no reasonable statistician would be misled in such an extreme situation, the possibility of being misled in less extreme situations remains. In any case, Rényi's suggestion was to consider the relative difference between the observed cumulative distribution and the hypothetical relative distribution, determining where that relative difference was maximal. An infinite relative difference would arise if any actual observations were below the permitted range, were the hypothetical distribution a bounded one.

The difficulty with this approach is that the critical value of the Rényi statistic must be extraordinarily high. For one-sided testing at test size α, the critical relative difference must be at least $1/\alpha-1$, or 19 for $\alpha=0.05$, 99 for $\alpha=0.01$, and irrespective of sample size. A consequence I bring out (Mantel 1968) is that statistical power may not rise, and could even fall, with sample size. Also, if statistical significance has not yet occurred by the time point at which the hypothetical cumulative probability is α, it cannot occur however much out of line subsequent observations might be with the hypothetical distribution.

That the critical value for the Rényi statistic is on of the order of $1/\alpha-1$ is readily established. Consider the time point at which the hypothetical cumulative probability is α/N, so that the expected number of responses among a group of size N is α. Then for small α, with approximate probability α a response would already have occurred. At the time of occurrence, the cumulative observed probability would have $1/N$, the theoretical less than α/N, for a relative excess not less than $1/\alpha-1$.

Niederhausen (1981) does indicate a modification of the Rényi test so as to consider calculation of the statistic only over some central range of the hypothetical cumulative distribution. While that would minimize some of the difficulties alluded to, it would re-introduce the possibility that clearly incompatible data would not lead to rejection of the hypothetical distribution.

3. THE ANDERSON-DARLING MODIFICATION

The appeal of the Anderson-Darling modification lies in its expressing the difference between the empirical and the hypothetical cumulated distribution relative to the standard error of the observed cumulated distribution.

Concern with the results of the Anderson-Darling weighted one-sample K-S procedure was expressed by Canner (1975), who properly identified the difficulty as due to marked sensitivity to the extreme order statistics. Following the work of Noé (1972; see also Noé and Vandewiele, 1968), Canner gave some few critical values of the Anderson-Darling weighted K-S statistics. Then, appalled by how extreme these were, he suggested modifications limited to the central range of the distribution, giving results for simulations limited alternatively to the central 98% and the central 95% of the distribution. The critical values were then much better behaved, but the possibility for rejecting clearly incompatible data would have disappeared.

Actually, Anderson and Darling (1962) had suggested the possibility for restricting their weighted K-S statistic to the central range. And Niederhausen (1981) illustrates the results of his use of Sheffer polynomials with both restricted and unrestricted ranges on the distribution functions, covering both one-sample and two-sample situations. Where Niederhausen (1981) overlaps with Canner (1975), there is rather good agreement between the exact results of the first and the simulated results of the second. (The single gross disagreement is a misprint. Niederhausen shows a calculated critical point of 3.9829 for N=10, α=0.05, 98% central range, incorrectly showing Canner's value as 3.33 rather than 4.00.)

Actual critical values cited by Canner (1975) for the weighted K-S statistic, and which he considered extreme, ranged from 6.43-6.45 at 2α=0.05 (N=10,20, or 50) and were all at 14.19 at 2α=0.01 (N=10,20, or 50). Noé and Vandewiele (1968) show a much larger range of critical weighted K-S statistics. For sample sizes ranging from 1-100, their critical values ranged: for 2α=0.20, from 3.00 to 3.56; for 2α=0.10, from 4.36 to 4.72; for 2α=0.05, from 6.24 to 6.49; for 2α=0.02, from 9.95 to 10.10; and for 2α=0.01 from 14.11 to 14.21.

Consider now the value of the Anderson-Darling weighted K-S statistic which would arise if, as considered above in the Rényi modification case, by the time point corresponding to cumulative probability α/N, a single response had occurred, that event having approximate probability α. The weighted K-S statistic would then take on the value

$$(1 - \alpha)/N \times \frac{\alpha}{N}(1 - \frac{\alpha}{N})^{\frac{1}{2}} = (1 - \alpha)/[\alpha(1 - \frac{\alpha}{N})]^{\frac{1}{2}}$$

which, for small α, would be approximated by $1/\alpha^{\frac{1}{2}}$, effectively the square root of the critical Rényi measure when that critical measure has unity added back to it.

It follows that the critical weighted K-S measure must be on the order of $1/\alpha^{\frac{1}{2}}$. For 2α=0.05, this would be 6.3246, according well with the range above shown of 6.24 to 6.49, while for 2α=0.01, the $1/\alpha^{\frac{1}{2}}$ value of 14.1421 would accord with the range 14.11 to 14.21.

In Table I, I show these limiting values for the weighted K-S statistic for α ranging downward from 0.10 to 0.001. It is clear that these values are far more extreme than the normal deviates that would correspond to those probabilities. Thus, the normal deviate for a right tail area of 0.01 would be 2.326, while the right tail area for a normal deviate of 10.00 would be 7.6×10^{-24}. A rather inadequate partial correction of this difficulty would occur if we considered use of a 0.5 continuity-correction factor. The limiting critical values of Table I would then be approximately halved, since the deviation from expectation would be reduced from $(1-\alpha)$ to $(0.5-\alpha)$. But the normal right tail area for a normal deviate of 5.00 would be 2.9×10^{-7}, still not close to 0.01. Inconsistency with a normal deviate interpretation of the critical weighted K-S values is yet more extreme for still smaller values of α.

Table 1. Limiting Values of the Weighted Single-Sample K-S Statistic When There Is an Excess at Probability Level α in the Tail of a Distribution and its Contrast with the Limiting Value of the K-S Statistic When There Is an Equally Improbable Deficit

α	$1/\alpha^{1/2}$ Limiting value of the weighted K-S statistic, if at least one observation occurs when the expectation does not yet exceed α	$-\ln\alpha$ The Poisson expectation for which the probability of a zero observation is α	$(-\ln\alpha)^{1/2}$ Limiting value of the weighted K-S statistic when the observation is zero, the expectation is $-\ln\alpha$
0.10	3.1627	2.3026	1.5174
0.05	4.4721	2.9957	1.7308
0.025	6.3246	3.6889	1.9206
0.0125	8.9443	4.3820	2.0933
0.01	10.0000	4.6052	2.1460
0.005	14.1421	5.2983	2.3018
0.0025	20.0000	5.9915	2.4477
0.002	22.3607	6.2146	2.4929
0.001	31.6228	6.9078	2.6283

Table I shows additionally, in its right half, limiting values for the weighted K-S statistic in the situation where there is a deficiency rather than an excess in the left tail of the observed distribution. For this purpose I consider the sample size to be so large and the true probability for the left tail to be so small that the number of observations falling within the left tail would follow a Poisson distribution. For any given expectation, λ, in the left tail, the greatest dificiency would occur were the observation zero, which would have associated probability $e^{-\lambda}$, and associated weighted K-S statistic $\lambda^{1/2}$. Accordingly, for each value, Table I provides the $\lambda=-\ln\alpha$ for which, then, the zero observation would have probability α, and shows also the corresponding weighted K-S statistic.

It is clear that the weighted K-S statistic when there is a low-probability deficiency in the tail is nowhere near so great in absolute value as arises when there is an equally probable excess. For α=0.001, the event of no observations with 6.91 expected yields a weighted K-S statistic of only 2.63 compared with that of 31.62 were there an excess with associated probability of 0.001, or even of 3.16

were the associated probability 0.10. As yet more extreme, consider that the extreme deficiency had associated probability $\alpha=10^{-12}$ ($\lambda=27.63$) -- the weighted K-S statistic would still take on only the moderate value of 5.26, not even attaining the critical value of 6.32 for $\alpha=0.025$ corresponding to excess observations.

However, in all this, it is evident that the single-sample weighted K-S statistic does not show the inconsistent behavior of the Rényi modification. If the true cumulative distribution differs in any way from the hypothesized distribution then, with certainty, as sample size increases indefinitely any finite critical value for the weighted K-S statistic will be surpassed. The issue though is that those critical values are extraordinarily large. As examples, the limiting value (as shown in Table I) for $\alpha=0.005$ of 4.4721 has a normal tail probability of 3.9×10^{-6}, that of 10.0000 for $\alpha=0.01$ has a normal tail probability of 1.12×10^{-23} and finally, that of 31.6228 for $\alpha=0.001$ has a normal tail probability of 1.05×10^{-219}.

All of these suggest that we would be paying just too high a price for the privilege of taking repeated looks at the data, as we are essentially doing anytime we use a K-S statistic. The number of looks we are taking should not exceed the sample size N, but the equivalent number of independent looks would be even less. Elsewhere it has been suggested that repeated look-sees are roughly equivalent to six independent looks while a rule for using $N^{\frac{1}{2}}$ as the equivalent number of independent looks has also been proposed.

4. A QUESTION OF TAILS

If we are to remedy weaknesses in K-S procedures, we must understand the nature of the difficulty just brought out relative to deficiencies in the tails of the observed distribution.

Two-tailed employment of the one-sample K-S procedure can be thought of in two ways. At any point in the cumulative distribution, there can be an excess or a deficiency -- one-tailed we can test either the properly-scaled extreme excess or the extreme deficiency, while two-tailed we would test the extreme deviation. Or we can look for deviations only in the left tail of our distribution, or only in the right tail, or in both tails. In a way, there are four ways of violating the null hypothesis that the hypothesized distribution is the true distribution -- excess or deficiency in the left tail and excess or deficiency in the right tail -- but the weighted K-S procedure allows us to detect only excesses in one or the other tail, not deficiencies.

Of course, if there is a deficiency in either tail, say the left tail, there would have to be a compensating excess in the balance of the distribution, which, in a sense, is also a tail, i.e. the right tail. But I will distinguish between these -- the balance from a left tail is a right balance, from a right tail, a left balance, as distinguished from more extreme forms of tails. If there is an excess in either balance, any statistical test on the excess would only be a test of the deficiency in the opposite tail -- the excess in the balance of the distribution would not result in statistical significance unless the excess managed to get reflected in a suitably large excess in the tail proper.

It would follow that even if the weighted K-S test had power to detect slippage alternatives, and would likely retain power for detecting increases in dispersion, it would likely be ineffective at detecting decreases in dispersion. Issues like these of power should be resolved by remedial K-S approaches.

5. A REMEDIAL APPROACH

In proposing their weighting scheme, Anderson and Darling (1952) had in mind that the unusualness of a departure from expectation could be gauged by expressing that departure in units of its standard deviation. True, they would be considering all possible deparutres and then selecting the largest standardized departure or the largest standardized absolute departure, but that would be coped with by restricting attention to the distribution of just such maximal departures.

The difficulty, however, was that a fair proportion of maximal standardized departures arose in situations where asymptotic normality was grossly violated. If the expected number of events is α, with α usually 0.05 or less, and the actual number of events is either zero or some positive integer, we haven't even a remote resemblance to the required normal distribution if the Anderson-Darling weighting is to be reasonable.

Once proposed, however, the Anderson-Darling K-S statistic took on a life of its own. The emphasis focused on what was the distribution of that statistic and what were its critical points, rather than on what the problem was which Anderson and Darling were addressing. These were rather interesting mathematical exercises, whether or not they were good practical statistics.

Even in the approach taken by Canner (1975) and Niederhausen (1981) of limiting the weighted K-S test to the central range, the emphasis was on what then happened to the critical points of the K-S statistic. In using 98% and 90% central ranges. Canner was requiring, with N=50, that the expectation should be at least 0.5 in the one case, 2.5 in the other. When in one situation Niederhausen restricted himself to the 50% central range with N=100, he was assuming that the tail expectation was at least 25. The weighted K-S procedure should then work nicely, but at the price of overlooking leads occurring in the earlier part of the distribution.

With the advent of computational facilities not available at the time of their work, we can go back to the motivation underlying the Anderson-Darling proposal. Their motivation was to identify the most unusual departure from expectation, and to provide some evaluable index of its unusualness.

It is immediate that if we map our hypothesized distribution onto the unit uniform distribution (0-1), then our actual observations will map out as order statistics from that uniform distribution. The distribution of any single order statistic is readily manageable, with unusual values of individual order statistics identifiable. Our problem, however, is to consider the joint distribution of all the order statistics in a way as to identify which is the most unusual in any single set of order statistics. A Monte-Carlo approach to this is feasible, whether or not the problem is mathematically tractable.

For a particular value of N, we draw samples of size N from the unit uniform. For any point X in the interval 0-1, we know that the number of observations among the N which are $\leq$ X (actually, the probability of equality should be zero) would be given by the binomial with sample size N, probability X. We can thus determine the probability that our sample contains as few or fewer values below X, or alternatively as many or more values above X, than it does contain.

This operation need not be performed for every possible X, but only for the X values arising in our sample. Suppose our fifth order statistic takes on the value X_5. We consider an X infinitesimally below X_5, and ascertain the probability of 4 or fewer successes in N trials with $P=X_5$; call it P_L. Then, by considering an X infinitesimally above X_5, ascertain the probability of 5 or more successes in N trials with $P=X_5$; call it P_U. Repeat this kind of operation for all the observations in the sample and then ascertain three quantities, P_L', the smallest of the P_L, P_U', the smallest of the P_U, and, finally P', the lesser of (P_L', P_U').

P_L', P_U', and P' now represent, respectively, the most unusual negative departure, positive departure, and absolute departure in our sample of N. If we repeat the sampling process many times we will obtain empirical distributions of P_L', P_U' and P' from which we can get empirical determinations of the values corresponding to alternative specifications for α. For rather small α, we should have a care that our simulations are numerous. But since, from symmetry, P_L' and P_U' are identically distributed, we can merge their empirical distributions in getting critical values corresponding to our α value, i.e. $P_L'(\alpha)=P_U'(\alpha)$. Additionally, we would have $P'(\alpha)$.

In application, we would not have to analyze the data from any situation, but would need only to plot our sample order statistics, expressed in the uniform scaling, on a chart with certain reference curves. One curve would show the theoretical lower percentage point at probability level $P_L'(\alpha)=P_U'(\alpha)$ for each of the order statistics, while another curve would show the corresponding upper percentage points at probability level $1-P_L'(\alpha)$. For one-tailed testing, any order statistic falling below the lower curve would indicate a significant positive departure, in terms of number of events, at probability level α. Order statistics above the upper curve would reflect a significant negative departure. Conceivably, significance in both directions could occur.

For two-tailed testing we would have two other reference curves, but possibly on the same chart -- these would be the lower and upper percentage points in the order statistics at respective probability levels $P'(\alpha/2)$ and $1-P'(\alpha/2)$. Significance at level α would obtain if any of the sample order statistics fell outside the range defined by the two reference curves.

While use of the above approach would be particularly appropriate for slippage alternatives, i.e. left or right shifts of the distribution, there would be a reasonable degree of power for alternatives involving changes in dispersion or of scale parameters. If power is particularly sought for a dispersion or scale alternative, that could be achieved by considering the folded uniform distribution. The unit uniform would be folded at 0.5, then stretched back out to the unit uniform. It would be this folded distribution onto which the hypothesized distribution and actual observation would be mapped. The procedure would be sensitive to increases or decreases in dispersion, and could be even more appropriate for changes in kurtosis. Of course, the option of using the procedure in a pseudosequential manner would be lost if the folded-uniform approach were applied. What would remain feasible would be to employ the unfolded approach in a pseudosequential manner as the data were being generated, then re-examining the data by the folded-uniform approach once the data were all in. Since the folded-uniform approach does retain power for slippage alternatives, a finding of significance employing it should not be interpreted as necessarily a change in scale or dispersion. The actual data should be examined to determine just what is the nature of the departure from the hypothesized distribution.

(Use of the folded-uniform approach could require some careful thought. Consider the distinct situations: an exponential distribution alternative to the LaPlace distribution, which would not be detectable on folding and a LaPlace distribution alternative to the exponential, which would likely be detected.)

6. TWO-SAMPLE TESTS AND OTHER CONSIDERATIONS

Reasonable results are yielded when the Anderson-Darling weighting scheme is used to modify the two-sample K-S procedure. There would, thus, be little need to implement an approach parallel to that described in the preceding section in which the apparent most extreme significance level at any point (in time) is taken only as a basis for establishing the significance level.

Special case situations apparently do not dominate the behavior of the weighted two-sample K-S critical values. Thus if the two sample sizes are rather large and equal, there would be about a $1/2^{10}=1/1024$ chance that the first 10 observations all came from a particular one of the two samples. That outcome would yield a chi value of 3.1623, or, continuity-corrected, 2.8460. The associated normal tail probabilities for these chi values are 0.00078 and 0.00201, respectively. One instance that Canner shows of a two-sided critical value obtained by simulation at the 0.005 level using sample sizes of 500+500 and 1000 simulations is 4.20. But the 10-0 split can occur, effectively, in four ways. It could be 10-0 or 0-10 in the left tail and, almost independently, 10-0 or 0-10 in the right tail. Reasonably large numbers of events must occur before significant values for the two-sample weighted K-S statistic can arise, so that normal approximation approaches will be suitable. Neiderhausen (1981) in presenting his results for the two-sample case notes that they are the same when restriction is to the central 98% of the data as when no such restriction is made, reflecting the need for some minimal number of actual events before significance could be attained.

For pseudosequential handling of the two-sample situation, a competitor to K-S statistics was broached by myself (Mantel, 1966), a suggestion which was followed through on by Muenz, Green, and Byar (1977). In applying the procedure pseudosequentially, various options available through use of the logrank approach could be lost, e.g. handling of right-censored data and taking into account possible stratification of the data. The logrank approach would take into account not only the sample cumulative probabilities at a point in time, but also how they got there. That logrank approaches can be applied in the single-sample case is suggested in the work of Oleinick and Mantel (1970), who compared mortality experience of relatives of lupus erythematosus patients with the anticipated experience for the general population, but taking into account each person's own birth cohort. The relationship of Mantel's 1966 work to that of Mantel and Haenszel (1959) was indicated.

Among the procedures to which Niederhausen (1981) applies his approach is that of Kuiper (1960). Kuiper's statistic is given by $D^{+}+D^{-}$ and is the sum of the absolute values of the largest positive difference and the largest negative difference between two cumulative distributions; it is applicable both to single-sample and two-sample cases.

While directed originally to examining distributions on the circle, the Kuiper statistic is applicable, as noted by Durbin (1973), to distributions on the line -- one has only to think of the line as turning full circle. The essential feature of the Kuiper statistic is that its value is not influenced by the point on the circle from which one begins cumulating.

When applied to distributions on the line, the Kuiper approach would have behavior somewhat similar to the approach indicated above for folding the uniform distribution -- there would be power for slippage alternatives, altered scale or dispersion alternatives, and for altered kurtosis alternatives. But the problem would remain with the Kuiper approach in the single-sample case (with some parallels in the two-sample case) that outcomes completely incompatible with the hypothesized distribution would not lead to its outright rejection. The remedial approach which I proposed above, whether in original form or in the folded-uniform form would lead to rejection if the incompatible data arose at either tail of the distribution. Where even that approach could fail would be where the hypothesized distribution had interior intervals with zero density, yet in which intervals actual observations occurred.

Ordinarily, the Kuiper statistic should be applied in a single-tail approach, the single tail of the Kuiper statistic encompassing both tails of the K-S statistics. Yet, should the Kuiper statistic take on an extraordinarily low value, suggesting too good an agreement between observed and expected, one might wish to consider that possibly the data had been fudged. In the same vein, if the data obtained are grossly incompatible (by whatever statistical approach used), as when some observations are in a zero-density region, the possibility that such observations represent blunders arises -- the offending observations are checked out, then corrected or removed, the resulting or remaining data re-analyzed. In large-scale investigations, like national censuses, some blunders are almost inevitable, while in small-scale laboratory investigations, mix-ups are not at all uncommon.

For departures of any kind from a hypothesized distribution, the Greenwood (1946) approach is applicable. Recent publications (Currie, 1981 and Stephens, 1981) extending the range of sample sizes for which critical points of the Greenwood statistic are available have revived interest in this novel procedure.

The Greenwood statistic is rather simple. If N points are distributed on the unit interval, they divide the unit interval into N+1 subintervals. The sum of the squared widths of the subintervals is then the sample statistic of interest, and it is referred back to the null distribution of just such a sum if the N points represented independent selections from the unit uniform distribution. (The

Greenwood statistic for N points on the line would follow the same distribution as that for N+1 points on the circle.)

Some qualitative description of the behavior of the Greenwood statistic is in order. Suppose that in some region of hypothetically high density, the actual data are rather sparse. This would translate into large gaps relative to the uniform distribution, and although there would be some compensating effects elsewhere, the process of squaring the gaps would serve to make the Greenwood statistic large.

Reverse the situation. Let the actual data be abundant in a region of hypothetically low density. This translates into small gaps relative to the unit uniform. But the Greenwood statistic would tend towards increase because of the compensating generally increased gaps in the remainder of the unit interval. But whether that tendency for an increased value for the Greenwood statistic would be sufficient to meet some critical requirement would be uncertain.

What seems to be the case then is that the Greenwood approach has power for detecting alternatives in which there are regions of sharply reduced density, but it may have only limited power for detecting alternatives with regions of increased density. The situation of any actual observations in a region of zero density would give rise to only some few zero gaps, and the resulting increase in yet other gaps would not likely be sufficient to give rise to a critically large value for the Greenwood statistic.

An interesting possibility would be to consider the reciprocals of the widths of the gaps. A zero gap would give rise to an infinite reciprocal and since, for continuous distributions, ties have zero probability of occurring, we would be duly alerted. A first difficulty, however, is that in real life our distributions are discrete because of the need to express value to a limited number of decimal places -- ties are then a real possibility. The next difficulty is that the reciprocals of the gaps do not have bounded expectations to begin with. The reciprocal of a uniform 0-1 observation has no second moment and, intimately related to this, neither has the reciprocal of a two degree of freedom chi square, nor the reciprocal of an exponential waiting time, nor the actual value of an observation from an F distribution with two denominator degrees of freedom. Adding a small positive quantity to each of the gaps before taking reciprocals would make for bounded moments, and that approach could bear investigating.

ACKNOWLEDGEMENT: This work was supported by Grant CA-34096 from the National Cancer Institute.

REFERENCES:

[1] Anderson, T.W. and Darling, D.A., Asymptotic theory of certain "goodness of fit" criteria based on stochastic processes, Ann. Math. Statist. 23:193-212 (1952).

[2] Canner, P.L., A simulation study of one- and two-sample Kolmogorov-Smirnov statistics with a particular weight function, J. Amer. Statist. Assoc. 70: 209-211 (1975).

[3] Currie, I.D., Further percentage points of Greenwood's statistic, J. Roy. Statist. Soc. A 144:360-363 (1981).

[4] Durbin, J, Distribution theory for tests based on the sample distribution function. Regional conference series in applied mathematics, SIAM, 9 (1983).

[5] Greenwood, M., The statistical study of infectious diseases, J. Roy. Statist. Soc. A 109:85-109 (1946).

[6] Kolmogorov, A., Sulla determinazione di una legge di distribuzione, G. Ist. Ital. Attuari. 4:83-91 (1933).

[7] Kuiper, N.H., Tests concerning random points on a circle, Nederl. Akad. Wetensch. Proc. Ser. A. 63:38-47 (1960).

[8] Mantel, N., Evaluation of survival data and two new rank order statistics arising in its consideration, Cancer Chemotherapy Reports 50:163-170 (1966).

[9] Mantel, N., Kolmogorov-Smirnov tests and Rényi's modification, Biometrics 24:1018-1023 (1968).

[10] Mantel, N. and Haenszel, W., Statistical aspects of the analysis of data from retrospective studies of disease, J. Natl. Cancer Instit. 22:719-748 (1959).

[11] Miller, R. and Siegmund, D., Maximally selected chi-squares. Technical Report No. 64, Division of Biostatistics, Stanford University (1980).

[12] Muenz, L.R., Green, S.B., and Byar, D.P., Application of the Mantel-Haenszel statistic to the comparison of survival distributions, Biometrics 33:617-626 (1977).

[13] Niederhausen, H., Sheffer polynomials for computing exact Kolmogorov-Smirnov and Rényi type distributions, Ann. Statist. 9:923-944 (1981).

[14] Noé, M., The calculation of distributions of two-sided Kolmogorov-Smirnov type statistics, Ann. Math. Statist. 43:58-64 (1972).

[15] Noé, M. and Vandewiele, G., The calculation of distributions of Kolmogorov-Smirnov type statistics including a table of significance points for a particular case, Ann. Math. Statist. 39:233-241 (1968).

[16] Oleinick, A. and Mantel, N., Family studies in systemic lupus erythematosus - II. Mortality among siblings and offspring of index cases with a statistical appendix concerning life table analysis, J. Chron. Dis. 22:617-625 (1970).

[17] Rényi, A., On the theory of order statistics, Acta. Math. Acad. Sci. Hungar 4:191-231 (1953).

[18] Smirnov, N.V., On the estimation of the discrepancy between empirical curves of distribution for two independent samples, Bull. Math. Univ. Moscou, Série Int. 2:3-14 (1939).

[19] Stephens, M.A., Further percentage points for Greenwood's statistic, J. Roy. Statist. Soc. A 144:364-366 (1981).

BIOSTATISTICS: Statistics in Biomedical, Public Health
and Environmental Sciences, edited by P.K. Sen

ON SOME PATTERN MATRICES

A.E. Sarhan and Ibrahim A. Salama

Cairo University, Egypt and The University of
North Carolina at Chapel Hill

In this paper we discuss matrices whose inverse is diagonal type 2 matrix. We compute the eigenvalues of some 0-1 matrices related to the Jacobi matrix. A class of matrices related to the Frobenius matrices are introduced and their eigenvalues and inverses are discussed.

KEY WORDS: Pattern matrices, graphs, Frobenius matrices

1. INTRODUCTION

Pattern matrices appear in a variety of mathematical problems. The questions of interest for such matrices are to find determinants, inverses (given its existence), eigenvalues, and eigenvectors. In statistics, for example, it is of interest to find the inverse of some variance-covariance matrix which possesses some specific structure. Other areas, for example symbolic dynamics, it is of interest to find the largest eigenvalue and the corresponding right and left eigenvectors for some pattern matrices. A nice review of some of the results in this area may be found in Graybill (1969 - Chapter 8) and the references given there.

In this paper we consider nonsymmetric diagonal type 2 matrices, some 0-1 matrices related to the Jacobi matrix J_n, and some matrices related to the class of Forbenius matrices.

2. ON DIAGONAL TYPE 2 MATRICES

An $n \times n$ matrix $A = (A(ij))$, is diagonal of type 2 if $A(ij) \neq 0$ for $|i - j| \leq 1$ and $A(ij) = 0$ otherwise. Ukita (1955) showed if V is symmetric, then V^{-1} is diagonal of type 2 iff $V(ij)/V(1j) = \theta_i$, $2 \leq i \leq j \leq n$. In this case V^{-1} is given in Greenberg and Sarhan (1960). We consider the nonsymmetric case as follows.

<u>Assertion 1</u>:

For a matrix V, V^{-1} is diagonal of type 2 iff

$$V(ik)/V(1k) = \theta_i \quad , \quad 2 \leq i \leq k \leq n$$

and

$$V(kj)/V(k1) = \hat{\theta}_j \quad , \quad 2 \leq j \leq k \leq n \tag{2.1}$$

where the θ_i and $\hat{\theta}_j$ are suitable constants.

Proof: The if part is shown by evaluating the elements of adj V using condition (2.1). For the only if part, assume V^{-1} is diagonal of type 2. Let $V = [V_1 \ldots V_n]$, where V_j is the jth column of V. Similarly, let $V^{-1} = [V_1^{-1} \ldots V_n^{-1}]$. Consider the product VV_1^{-1}, then we have $V^{-1}(11)V_1 + V^{-1}(21)V_2 = e_1$, and

$$V(i2) = -(V^{-1}(11)/V^{-1}(21))V(i1) \equiv \hat{\theta}_2 \, V(i1), \; i = 2,\ldots, n \tag{2.2}$$

Assuming that we have $\hat{\theta}_2,\ldots, \hat{\theta}_J$ such that

$$V(ij) = \hat{\theta}_j \, V(i1) \quad , \quad 2 \le j \le i \le n \text{ and } 2 \le j \le J \tag{2.3}$$

Now, consider VV_J^{-1}, then we have

$$V(i, J + 1) = - \frac{(V^{-1}(J - 1, J)\hat{\theta}_{J-1} + V^{-1}(J, J)\hat{\theta}_J)}{V^{-1}(J+1,J)} V(i1) \equiv \hat{\theta}_{J+1}V(i1),$$

$$i = J + 1,\ldots, n \tag{2.4}$$

Similar relations for the rows of V are obtained by considering $V^t(V^{-1})^t$.

Assertion 2: If V satisfies condition (2.1), then V^{-1} is given by

$$V^{-1}(i, i + 1) = 1/(V(i + 1, i) - \theta_{i+1}V(1, i)), \; 1 \le i \le n - 1,$$

$$V^{-1}(i - 1, i) = 1/(V(i, i + 1) - \hat{\theta}_{i+1}V(i, 1)), \; 1 \le i \le n - 1,$$

$$V^{-1}(1, 1) = (1 - V^{-1}(1, 2) \, V(2,1))/V(1, 1),$$

$$V^{-1}(i, i) = -(V^{-1}(i, i - 1) \, V(i - 1, 1) + V^{-1}(i - 1, i)V(i + 1, 1))/$$
$$V(i, 1), \; 2 \le i \le n - 1,$$

and

$$V^{-1}(n, n) = -(V^{-1}(n, n - 1) \, V(n - 1, 1))/V(n, 1).$$

3. ON SOME 0-1 MATRICES

Consider the matrices J_n and K_n, where

$$J_n(ij) = \begin{cases} 1 & \text{if } |i - j| = 1 \\ 0 & \text{otherwise} \end{cases} \tag{3.1}$$

and

$$K_n(ij) = \begin{cases} 1 & \text{if } |1 - j| = 1 \quad \text{or} \quad i = j = 1 \\ 0 & \text{otherwise} \end{cases} \tag{3.2}$$

These are well known matrices and the eigenvalues of $J_n(K_n)$ are given by $2 \cos \frac{\pi r}{n+1}$, $r = 1, \ldots, n$ $(2 \cos \frac{2\pi r}{2n+1}$, $r = 1, \ldots, n)$. In what follows we consider some matrices which are closely related to J_n and K_n.

Assertion 3: Le A_n be given by

$$A_n(ij) = \begin{cases} 1 & \text{if } i + j = n \quad \text{or} \quad i + j = n + 2 \\ 0 & \text{otherwise} \end{cases} \tag{3.3}$$

Then, (i) for n odd, the eigenvalues of A_n are those of J_n.
(ii) for n even, the eigenvalues of A_n are those of $K_{n/2}$, each with multiplicity two.

Proof: Let σ_1 and σ_2 be the permutations over $1, \ldots, n$ given by

$$\sigma_1(i) = \begin{cases} i & \text{if } i \text{ is odd} \\ n - i + 1 & \text{if } i \text{ is even} \end{cases} \tag{3.4}$$

and

$$\sigma_2(i) = \begin{cases} i & \text{if } i \text{ is odd, } i \leq n/2 \text{ or } i \text{ is even, } i > n/2 \\ n - i + 1 & \text{otherwise} \end{cases} \tag{3.5}$$

Let P_1 and P_2 be the permutation matrices generated by σ_1 and σ_2 respectively. If n is odd, then A_n is similar to J_n under P_1. If n is event, the A_n is similar to $K^*_{n/2} \oplus K_{n/2}$ under P_2, where

$$K^*_{n/2}(ij) = \begin{cases} 1 & \text{if} |i - j| = 1 \quad \text{or} \quad i = j = n \\ 0 & \text{otherwise.} \end{cases} \tag{3.6}$$

Although K_n is not strictly diagonal type 2 matrix, K_n^{-1} satisfies conditions (2.1).

Assertion 4: For $1 \leq i \leq n$, let

$$\theta_i = \begin{cases} 1 & \text{if } i = 4\ell \quad \text{or} \quad 4 \;\; + 1 \quad , \;\; \ell \geq 0 \\ -1 & \text{if } i = 4\ell + 2 \quad \text{or} \quad 4\ell + 3 \; , \; \ell \geq 0 \; . \end{cases} \tag{3.7}$$

(i) if n is odd,

$$K_n^{-1}(ij) = \begin{cases} (-1)^{\frac{j-1}{2}} & \text{if } i = 1,\ j \text{ odd}, \\ 0 & \text{if } i = 1,\ j \text{ even} \\ K_n^{-1}(1j)\theta_j & \text{if } 1 \le j \le i \le n. \end{cases} \tag{3.8}$$

(ii) if n is even

$$K_n^{-1}(ij) = \begin{cases} (-1)^{\frac{j-2}{2}} & \text{if } i = 1,\ j \text{ even} \\ 0 & \text{if } i = 1,\ j \text{ odd} \\ K_n^{-1}(1j)\theta_j & \text{if } 1 \le j \le i \le n. \end{cases} \tag{3.9}$$

This example along with the proof of assertion 1 suggests to define: A_n is diagonal type 2 matrices if $A_n(ij) = 0$ for $|i - j| > 2$, $A(ij) \neq 0$ if $|i - j| = 1$, and $A_n(ii)$ is free.

Assertion 5: For $r \ge 2$, let $J_n^{(r)}$ be given by

$$J_n^{(r)}(ij) = \begin{cases} 1 & \text{if } |i - j| = r \\ 0 & \text{otherwise} \end{cases} \tag{3.10}$$

then, the eigenvalues of $J_n^{(r)}$ are given by

$$\lambda_{ij} = 2\cos\frac{j\Pi}{c_i + 1},\ j = 1,\dots, c_i,\ i = 1,\dots, r \text{ where } c_i = \left\lceil\frac{n - i + 1}{r}\right\rceil$$

Proof: Consider the directed graph G associated with $J_n^{(r)}$ with the states indexed by $\{1,\dots, n\}$. Let

$$C(i) = \{j \mid \text{there is a path in G from } i \text{ to } j\},\ i = 1,\dots, n \tag{3.11}$$

clearly we have r commuting classes $C(1),\dots, C(r)$. For $1 \le i \le r$, let k_i be the largest integer $\le n$ and satisfying $k_i = i \bmod r$. Then $c_i \equiv \# C(i) = 1 + \frac{k_i - i}{r} = \left\lceil\frac{n - i + 1}{r}\right\rceil$ (where $\lceil x\rceil$ denotes the smallest integer $\ge x$). Let $C(i) = \{j_{i_1},\dots j_{ic_i}\}$, and assume $j_{i_1} < \dots < j_{ic_i}$. Let σ be the permutation over $\{1,\dots, n\}$ given by

$$\sigma(j_{ik}) = \sum_{\ell=1}^{i-1} c_\ell + k,\ 1 \le k \le c_i,\ i = 1,\dots, r \tag{3.12}$$

If P_σ is the permutation matrix generated by σ, then $J_n^{(r)}$ is similar to $J_{c_1} \oplus \ldots \oplus J_{c_r}$ under P_σ. Hence the result.

In studying pattern matrices, we can generate new classes from old ones. For example, if A and B are two pattern matrices, then in studying AB we use the known information about A and B.

Another natural way to generate new classes is by considering matrices which are elements of the vector space generated by $\{A^i,\ i \geq 0\}$.

<u>Assertion 6</u>: Let A_n be given by

$$A_n(ij) = \begin{cases} 1 \text{ if } |i - j| = 1 \text{ and } i \neq 1, 2, n - 1 \text{ or } n. \\ \quad \text{or } |i - j| = 3 \\ 0 \text{ otherwise} \end{cases} \tag{3.13}$$

Then, the eigenvalues of A_n are

$$\lambda_i = 4 \cos\left(\frac{\pi i}{n + 1}\right)\left[2 \cos^2\left(\frac{\pi i}{n + 1}\right) - 1\right],\ i = 1,\ldots, n. \tag{3.14}$$

<u>Proof</u>: By noting $A_n = J_n^3 - 2J_n$.

4. FROBENIUS AND RELATED MATRICES

Let X, Y be vectors, $I_n' \equiv \left[\frac{I_{n-1}}{\quad}\Big|_0\right]$ and $c \neq 0$. Consider the following matrices:

$$L(X)(ij) = \begin{cases} x_i & \text{if } i = 1 \\ c & \text{if } i - j = 1 \\ 0 & \text{otherwise}, \end{cases} \tag{4.1}$$

$$L_1(X)(ij) = \begin{cases} x_i & \text{if } i = n \\ c & \text{if } j - i = 1 \\ 0 & \text{otherwise}, \end{cases} \tag{4.2}$$

$$T(X, Y)(ij) = \begin{cases} X^tY & \text{if } i = j = 1 \\ x_{i-1} & \text{if } i = 1,\ j = 2,\ldots, n \\ y_{i-1} & \text{if } j = 1,\ i = 2,\ldots, n \\ c & \text{if } i = j,\ i = 2,\ldots, n \\ 0 & \text{otherwise} \end{cases} \tag{4.3}$$

and

$$T_1(X,Y) = XY^t + c\ I_n'\ . \tag{4.4}$$

In what follows and without loss of generality, we assume c = 1. Also, the following notation will be used. If A is a matrix then $\overset{\circ}{A}$ denotes the matrix obtained by reflecting A through the opposite diagonal. (e.g., if $A = \begin{pmatrix} 1 & 1 \\ 1 & 0 \end{pmatrix}$, then $\overset{\circ}{A} = \begin{pmatrix} 0 & 1 \\ 1 & 1 \end{pmatrix}$). If $X = (x_1, \ldots, x_n)^t$, then $\overset{\circ}{X} = (x_n, \ldots, x_1)^t$ and $X_1 = x_n^{-1}(1, -x_1, \ldots, -x_{n-1})^t$.

Assertion 7: (i) $\det(T(X, Y)) \equiv w = X^tY - \sum_{i=1}^{n-1} x_i y_i$,

$$\text{(ii)} \quad f_T(\lambda) = (1 - \lambda)^{n-2}[\lambda^2 - (1 + X^tY)\lambda + w] , \tag{4.5}$$

and

$$\text{(iii)} \quad T^{-1}(X, Y) = \overset{\circ}{T}_1(\overset{\circ}{X}_1, \overset{\circ}{Y}_1).$$

Proof: (i) Let $w = X^tY - \sum_{i=1}^{n-1} x_i y_i$ and choose x_n, y_n satisfying $x_n y_n = w$. Then $T(X, Y) = L(X)L^t(Y)$. Hence $\det(T(X,Y)) = x_n y_n = w$.

(ii) Clearly, 1 is an eigenvalue of T with algebraic multiplicity n - 2. If λ_1, λ_2 are the other two eigenvalues, then $\lambda_1\lambda_2 = w$ and $\lambda_1 + \lambda_2 = 1 + X^tY$. Hence they are the roots of the equation $\lambda^2 - (1 + X^tY)\lambda + w = 0$.

(iii) Noting that $L^{-1}(X) = L_1(X_1)$, we have

$$\begin{aligned} T^{-1}(X, Y) &= L^{t^{-1}}(Y)L^{-1}(X) = [Y_1X_1^t + \overset{\circ}{I}{}'_n] \\ &= [(\overset{\circ}{X}_1\overset{\circ}{Y}{}_1^t)^\circ + \overset{\circ}{I}{}'_n] = [(\overset{\circ}{X}_1\overset{\circ}{Y}{}_1^t) + I'_n]^\circ \\ &= \overset{\circ}{T}_1(\overset{\circ}{X}_1, \overset{\circ}{Y}_1). \end{aligned} \tag{4.6}$$

Similarly, we have

Assertion 8: (i) $\det(T_1(X, Y) = x_n y_n$,

$$\text{(ii)} \quad f_{T_1}(\lambda) = (1 = \lambda)^{n-2}[\lambda^2 - (1 + X^tY) + x_n y_n] , \tag{4.7}$$

and

$$\text{(iii)} \quad T_1^{-1}(X,Y) = \overset{\circ}{T}(\overset{\circ}{X}_1, \overset{\circ}{Y}_1).$$

NOTE: If X = Y, (i) and (ii) may be obtained directly from Roy, Greenberg, and Sarhan (1960).

In what follows we consider a class of matrices which, in some sense, contains the Frobenius matrices. Let A_n be a matrix, and G be the corresponding directed graph with the states indexed by {1,..., n} and the edges labelled by the elements of A_n. A path in G of length m from i_0 to i_m, $P_{i_0i_m}(m)$, is a sequence

$<i_0,\ldots, i_m>$ in $\{1,\ldots, n\}$ such that $\prod_{j=1}^{m} A_n(i_{j-1}i_j) \neq 0$. If $i_o = i_m$, then $P_{i_0 i_0}(m)$ is a loop of length m based at i_0, and is denoted by $\ell_{i_0}(m)$. Let

$$L = \{\ell_{i_0}(m) | i_0 \in \{1,\ldots, n\}, m \geq 1 \text{ and } i_j \neq i_0, j = 1,\ldots, m - 1\} \qquad (4.8)$$

A set $B \subset \{1,\ldots, n\}$ is a <u>Base</u> (or a <u>Rome</u>) for G iff $\ell_i(m) \in L$ implies $i_j \in B$ for some $j = 1,\ldots, m$.

<u>Definition</u>: A matrix A_n is a generalized Frobenius matrix, if the corresponding graph G has a base B with # B = 1.

Finding the characteristic polynomial, $f_{A_n}(\lambda)$, of such matrices is simple as follows. Let $B = \{1\}$, and define:

$$L_1(m) = \{\ell_1(m) | \ i_j \neq 1, j = 1,\ldots, m - 1\}, m = 1, 2,\ldots \qquad (4.9)$$

Note that there exists an integer $M \leq n$ such that $L_1(M) \neq \phi$ and $L_1(m') = \phi$ for $m' > M$. For $m \leq M$, and $\ell \in L_1(m)$, define $w(\ell) = \prod_{j=1}^{m} A_n(i_{j-1}, i_j)$ and let $f_m(\lambda) = \sum_{\ell\in\ell_1(m)} w(\ell)\lambda^{-m}$.

Then

$$f_{A_n}(\lambda) = (-1)^{n-1}\lambda^n[\sum_{m=1}^{M} f_m(\lambda) - 1]. \qquad (4.10)$$

For a proof of this result see Block et al. (1980). We give an example to illustrate this procedure for the following 10 x 10 matrix A.

A =

	1	2	3	4	5	6	7	8	9	10
1	5	1	0	0	0	0	-4	0	0	0
2	0	0	2	0	0	0	0	0	0	0
3	0	0	0	-3	2	0	0	0	0	0
4	4	0	0	0	0	0	0	0	0	0
5	0	0	0	0	0	1	0	0	0	0
6	1	0	0	0	0	0	0	0	0	9
7	0	0	0	0	0	0	0	2	1	0
8	0	0	0	0	0	0	0	0	0	2
9	2	0	0	0	0	0	0	0	0	0
10	3	0	0	0	0	0	0	0	0	0

the corresponding graph G is given by

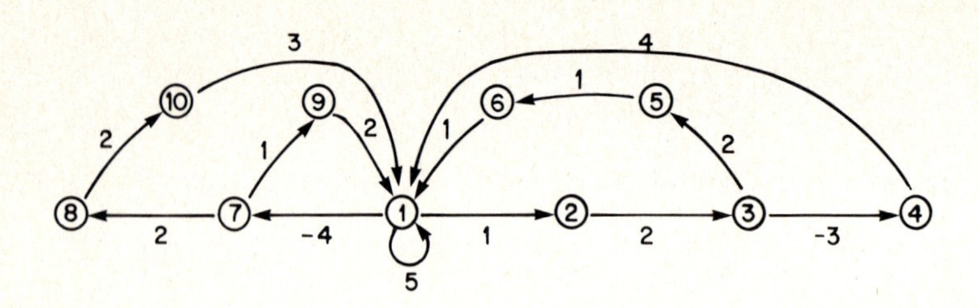

now, $L_1(1) = \{<1, 1>\}$, $L_1(2) = \phi$, $L_1(3) = \{<1, 7, 9, 1>\}$, $L_1(4) = \{<1, 2, 3, 4, 1>, <1, 7, 9, 10, 1>\}$, $L_1(5) = \{<1, 2, 3, 5, 6, 1>\}$ and $L_1(m) = \phi$ for $m > 5$. Thus, $f_1(\lambda) = 5\lambda^{-1}$, $f_2(\lambda) = 0$, $f_3(\lambda) = -8\lambda^{-3}$, $f_4(\lambda) = [-24 - 48]\lambda^{-4}$ and $f_5(\lambda) = 4\lambda^{-5}$. Hence

$$\begin{aligned} f_A(\lambda) &= (-1)^9\lambda^{10}[5\lambda^{-1} - 8\lambda^{-3} - 72\lambda^{-4} + 4\lambda^{-5} - 1] \\ &= \lambda^5(\lambda^5 - 5\lambda^4 + 8\lambda^2 + 72\lambda - 4). \end{aligned} \tag{4.11}$$

Note that A is nonsingular iff M = n.

ACKNOWLEDGEMENT: This research was partially sponsored by the grant 5-732-ES07018-01 form the National Institute of Environmental Health Sciences.

REFERENCES

[1] Block, L., Gluckenheimer, J, Misiurewicz, M., and Young, L.S., Periodic points and topological entropy of one dimensional maps. Globol theory of dynamical systems. Spring-Verlag, Lecture notes in mathematics **819**:18-34 (1980).

[2] Graybill, F., Introduction to Matrices with Applications in Statistics, Wadsworth Publishing Company, Inc. (1969).

[3] Greenberg, B. and Sarhan, A.E., Generalization of some results for inversion and partitioned matrices. Contributions to Probability and Statistics, Essays in Honor of Harald Hotelling, Olkin, I., et al. (eds). Stanford University Press, 216-223 (1960).

[4] Ukita, Y., Characterization of 2-type diagonal matrices with an application to order statistics, *Journal Hokkaido College of Art and Literature* **6**:66-75 (1955).

[5] Roy, S.N., Greenberg, B., and Sarhan, A.E., Evaluation of determinants, characteristics equation and their roots for a class of patterned matrices, *JRSS*, *Series B* **22**:348-359 (1960).

BIOSTATISTICS: Statistics in Biomedical, Public Health
and Environmental Sciences, edited by P.K. Sen

MULTIVARIATE LINEAR MODELS FOR LONGITUDINAL DATA: A BOOTSTRAP STUDY OF THE GLS ESTIMATOR

James H. Ware and Victor De Gruttola

Department of Biostatistics
Harvard University School of Public Health
Boston, Massachusetts

Comparative studies of alternative approaches to the multivariate linear regression analysis of longitudinal data have shown that generalized least squares (GLS) estimation with estimated covariance matrices is an attractive procedure in many applications (Ware, 1983). Recently, however, Freedman and Peters (1984) showed that the usual asymptotic formula for the variance of this estimator can be seriously biased in small samples. This paper reports a bootstrap simulation study of the performance of the GLS estimator in data sets taken from two longitudinal studies of environmental risk factors. In these examples, the asymptotic variance estimator has surprisingly small bias in samples of size 15 to 50 with 3 or 4 repeated measurements. Although these results are encouraging regarding the use of GLS estimation in the analysis of longitudinal data, further work is needed to identify the determinants of the magnitude of the bias of the asymptotic variance estimator.

KEY WORDS: Bootstrap, generalized least squares, longitudinal studies, E-M algorithm

1. INTRODUCTION

In most longitudinal studies, the primary objective is the collection and analysis of repeated measurements of participant characteristics. The multivariate linear model is an important tool for the analysis of such data. When the study is characterized by irregular examinations, missing visits, or time-varying covariables, however, linear models motivated by the hypotheses of primary interest may not be of standard form. Faced with this situation and uncertain about the proper methodology for fitting of nonstandard models, many investigators either restrict their analyses to models supported by readily available software or utilize ad hoc estimation procedures. Recent work, however, has shown that satisfactory estimation procedures can be defined for multivariate linear models with arbitrary design matrices and that the implementation of these procedures is straightforward (Ware, 1984).

One recommended procedure is generalized least squares (GLS) estimation with an estimated covariance matrix. The variances of the GLS estimators are estimated from the conventional asymptotic formula, once again with an estimate substituted for the population covariance matrix. For simplicity, we call this variance estimator the asymptotic estimator. Recently, however, Freedman and Peters (1984) raised the possibility that this approach can perform poorly in small data sets. They reported that, in a related problem from an econometric setting, this approach resulted in substantial underestimation of the variance of the GLS estimator. The discrepancy has two sources: 1) the use of an estimated covariance matrix in the GLS estimator increases its variance, and 2) the use of the estimated covariance matrix in the formula for the asymptotic variance produces downward bias in the estimation of that variance.

To assess the implications of the work of Freedman and Peters for the analysis of longitudinal studies in biomedical settings, we investigated the performance of the GLS estimator in data sets taken from two longitudinal studies of the health effects of environmental exposures. Two versions of the GLS estimator were considered, maximum likelihood (ML) and a two-stage estimator, called GLS,1 by Freedman and Peters.

The estimators were evaluated by simulation studies using the "bootstrap" (Efron, 1979). For each data set studied, we first specified a multivariate linear model and then estimated the regression parameters of that model by ML. The estimated regression coefficients were used as the population values in the simulations. The covariate values from the original data set were also used for the simulation. The error vectors for the simulated data sets were chosen by sampling with replacement from the set of multivariate residuals defined by the original analysis. The covariates, parameter values, and resampled errors defined sets of "pseudo-data," to each of which the two estimators were applied. Estimates calculated from the successive samples are independent observations from the bootstrap distribution of the estimator and can be used to estimate the variance of the estimator in the usual way. This empirical variability was compared to the average value of the estimated aysmptotic variance.

The first of the two longitudinal studies is an ongoing study of the effects of blood lead on cognitive development. The second is a continuing study of pulmonary function development in a population-based sample of preadolescent school children. Three and four repeated measurements, respectively, were available for study. Sample sizes of 15, 25, and 50 were considered. We found that the discrepancy between the actual and estimated variance of the GLS estimator was much smaller than that reported by Freedman and Peters, even for very small sample sizes. We also found that the performance of GLS,1 compares favorably to that of ML.

This article is organized as follows: Section 2 defines the family of multivariate linear models to be investigated, the estimation procedures, and the methods used in the simulation studies. Section 3 describes the two data sets and gives the results of the simulations. The implications of this investigation are discussed in Section 4.

2. METHODS

A General Multivariate Model for Repeated Measurements

Let $\underset{\sim}{y}_i$, i=1,...,n be a vector of p_i responses for the ith subject. Assume that p_i occasions of measurement are a subset of p occasions on which observations can be obtained. We consider the following general multivariate model:

$$\underset{\sim}{y}_i = \underset{\sim}{X}_i \underset{\sim}{\beta} + \underset{\sim}{\varepsilon}_i \tag{1}$$

In this model, $\underset{\sim}{X}_i$ is a $p_i$xk design matrix of independent variables and $\underset{\sim}{\beta}$ is a kx1 vector of regression parameters. The error vector, $\underset{\sim}{\varepsilon}_i$, is assumed to have a multivariate normal distribution with mean $\underset{\sim}{0}$ and covariance matrix $\underset{\sim}{\Sigma}_i$, where $\underset{\sim}{\Sigma}_i$ is a submatrix from the pxp covariance matrix $\underset{\sim}{\Sigma}$ which characterizes the covariance structure for the total of p occasions available for study. An important feature of this model is the allowance for arbitrary design matrices, $\underset{\sim}{X}_i$.

Estimation

If $\underset{\sim}{\Sigma}$ were known, maximum likelihood (ML) estimates of the regression coefficients could be obtained from the Aitken or generalized least squares (GLS) estimator:

$$\hat{\beta} = \left(\sum_{i=1}^{n} \underset{\sim}{X}_i' \underset{\sim}{\Sigma}_i^{-1} \underset{\sim}{X}_i\right)^{-1} \left(\sum_{i=1}^{n} \underset{\sim}{X}_i' \underset{\sim}{\Sigma}_i^{-1} \underset{\sim}{y}_i\right) \tag{2}$$

When $\underset{\sim}{\Sigma}$ is not known, an iterative approach is required. One algorithm suitable for small data sets combines features of the Gauss-Seidel (Dahlquist and Bjork, 1974) and E-M (Dempster et al., 1977) algorithms. The likelihood function for $\underset{\sim}{\beta}$ and $\underset{\sim}{\Sigma}$ yields two sets of gradient equations. Given an estimate, $\hat{\underset{\sim}{\Sigma}}^{(k)}$, of $\underset{\sim}{\Sigma}$, the solution to the gradient equations for $\underset{\sim}{\beta}$ is given by (2) with $\hat{\underset{\sim}{\Sigma}}^{(k)}$ substituted for Σ. If each subject has the full number (p) of observations, the solution to the gradient equations for $\underset{\sim}{\Sigma}$ given the current estimate, $\hat{\underset{\sim}{\beta}}^{(k)}$, of $\underset{\sim}{\beta}$ is given by $\hat{\underset{\sim}{\Sigma}}^{(k)} = \sum_{i=1}^{n} (\underset{\sim}{y}_i - \underset{\sim}{X}_i \hat{\underset{\sim}{\beta}}^{(k)})'(\underset{\sim}{y}_i - \underset{\sim}{X}_i \underset{\sim}{\beta}^{(k)})/n$. Otherwise, the solution to the gradient equations for $\underset{\sim}{\Sigma}$, given $\hat{\underset{\sim}{\beta}}^{(k)}$, can be obtained by assuming that the residuals, $\underset{\sim}{y}_i - \underset{\sim}{X}_i \hat{\underset{\sim}{\beta}}^{(k)}$, represent an incomplete sample from the multivariate normal distribution with mean $\underset{\sim}{0}$ and covariance matrix $\underset{\sim}{\Sigma}$ and applying the method of Beale and Little (1975) based on the E-M algorithm. Combining the two maximization steps into a Gauss-Seidel algorithm, one alternately solves the first set of gradient equations in $\underset{\sim}{\beta}$ by applying (2), and the second set of equations in $\underset{\sim}{\Sigma}$ by applying the E-M algorithm to the incomplete sample of residuals.

To begin the iterative procedure, we used an initial estimate of $\underset{\sim}{\Sigma}$, $\hat{\underset{\sim}{\Sigma}}^{(0)}$, based on the residuals, $\underset{\sim}{e}_0$, from the ordinary least squares (OLS) fit. In the simulations using incomplete data sets, one iteration of the E-M algorithm was used to calculate $\hat{\underset{\sim}{\Sigma}}^{(0)}$. The estimate obtained from substituting $\hat{\underset{\sim}{\Sigma}}^{(0)}$ into equation (2) is the first GLS estimate of $\underset{\sim}{\beta}$, denoted by $\hat{\underset{\sim}{\beta}}^{(1)}$. Freedman and Peters call this the GLS,1 estimator. To obtain the ML estimate of $\underset{\sim}{\beta}$, iteration proceeds until the changes in $\hat{\underset{\sim}{\beta}}^{(k)}$ and $\hat{\underset{\sim}{\beta}}^{(k)}$ are small.

The covariance matrix of $\hat{\underset{\sim}{\beta}}$ can be estimated by

$$\hat{COV} = \left(\sum_{i=1}^{n} \underset{\sim}{X}_i \hat{\underset{\sim}{\Sigma}}_i^{-1} \underset{\sim}{X}_i\right)^{-1} \qquad (3)$$

where $\hat{\underset{\sim}{\Sigma}}_i$ is the appropriate submatrix from the estimate of $\underset{\sim}{\Sigma}$. This is the usual asymptotic formula with an estimate substituted for $\underset{\sim}{\Sigma}$. Although this estimator is consistent, the work of Freedman and Peters (1984) suggests that it may be severaly biased in small samples.

The Simulation Study

To assess the performance of these estimators and their estimated variances, we performed simulation studies using data sets taken from two longitudinal studies. From each study, we selected random samples ranging in size from 15 to 50 from the set of subjects with complete data for the period of interest. We then fitted the model of interest to each data set by ML. The models had been developed through analyses of the complete data set from each study. The residual vectors from the fitted model defined an empirical distribution, from which samples equal in size to the initial data set were chosen by sampling with replacement. This procedure creates sets of "pesudodata" defined by the design matrices, $\underset{\sim}{X}_i$, the initial estimates of the regression coefficients (represented simply as $\hat{\underset{\sim}{\beta}}$), and the error vectors, $\underset{\sim}{e}^{(i)}$, as follows:

$$\underset{\sim}{y}^{(i)} = \underset{\sim}{X}_i \hat{\underset{\sim}{\beta}} + \underset{\sim}{e}^{(i)} \qquad (4)$$

Here, $\underset{\sim}{X}_i$ is the design matrix for the ith subject and $\underset{\sim}{e}^{(i)}$ is the ith randomly selected residual vector. The ML and GLS,1 estimators were then applied to the sets of pseudodata.

This procedure corresponds to the bootstrap (Efron, 1979) except in two details. First, the model fitted by ML was used in the simulation studies of both the GLS,1 and ML estimator and the two estimators were applied to the same sets of pseudodata. This permits direct comparison of the two estimators. Second, the focus of the study is performance of the estimators in the bootstrap universe defined by the fitted model and the empirical distribution of residuals. In this universe, the performance of the estimators can be observed directly. Evaluation of the performance of the estimators in the analysis of the initial data set is of secondary interest.

One hundred replications were performed for each study, sample size, and estimation procedure. Each replication produced estimates $\hat{\underset{\sim}{\beta}}^*_{ML}$ and $\hat{\underset{\sim}{\beta}}^*_{GLS,1}$. The sample means and standard deviations of these values were computed over the set of replicates. For each replication, we also obtained estimated standard errors from expression (3). The root mean square of these standard errors (RMSSE), that is, the square root of the mean of the estimated variances, was calculated to estimate the expected value of the asymptotic estimator of the variance of $\hat{\underset{\sim}{\beta}}^*$.

As noted previously, the initial ML estimates of the linear model were used in the simulation studies of both estimators. This ensured that the 'population' parameters were the same for both sets of simulations, justifying comparisons between the means and variances of the two estimators.

Construction of 'Incomplete' Data Sets

Rubin (1976) considered three possible assumptions about the relationship between the occurrence of missing values and the data values themselves. Data are "missing completely at random" when failure to observe a data point is independent of all data values. Data are "missing at random" or "observed at random" when the failure to observe a data point depends on the observed or unobserved values, respectively. Presumably, the proper definition or definitions of the bootstrap in the setting of incomplete multivariate observations will depend on the assumed mechansim for the occurrence of missing values. In this investigation, we chose not to investigate this complex issue, but instead to emphasize the implementation of ML and GLS,1 in incomplete data sets and the sensitivity of these estimation procedures to loss of information.

To accomplish this, we considered the simplest kind of missing data, single visits missing completely at random. To simulate this situation, we began with the fitted model and set of residuals based on a complete data set, generated the pseudo-data, and then randomly deleted a fixed number of observations at the last occasion. This procedure was applied only in one of the data sets, and there only with the smaller sample sizes. Thus, the results reported here represent only a first step in the study of the bootstrap for incomplete data.

3. RESULTS

Example 1

The first simulation study was based on data obtained in a prospective study of the relationship between blood lead concentrations during pregnancy and infancy and cognitive development in the first years of life (Bellenger et al., 1983). Newborn infants were screened for blood lead level by examination of blood from the umbilical cord. Those whose cord blood lead concentration fell into one of three intervals, described here as HIGH, LOW, and MIDdle, were enrolled for longitudinal follow-up. At six month intervals, cognitive development of the infants was measured using the Mental Developmental Index (MDI) of the Bayley Scales of Infant Development. On the same occasions, blood specimens were collected from the children, and questionnaires designed to elicit information about other environmental factors which influence mental development were administered to parents. The outcome variables were the Bayley scores from tests administered to study infants at ages 6, 12, and 18 months. These scores had been adjusted for gestational age and for the HOME score--a measure of the quality of the environment in the home. The independent variables considered included cord and infant blood lead concentration as well as the age of the infant at examination.

Analysis of the entire study sample of 210 children suggested that the serial measurements could be described by the model

$$y_{ij} = \beta_0 + \beta_1 LOW_i + \beta_2 MID_i + \beta_3 * AGE_{ij} + \beta_4 * AGE_{ij} * LEAD_i + \varepsilon_{ij}$$

where y_{ij} represents the adjusted Bayley score for individual i at occasion j, AGE_{ij} is the age at that examination, and $LEAD_i$ is the logarithm of 1 plus the blood concentration of the infant at age 6 months. The variables LOW_i and MID_i are indicator variables referring to the low and middle category of cord blood lead concentration. The 3x1 error vectors, ε_i, are assumed to have a multivariate distribution with mean 0 and unknown covariance matrix, $\underset{\sim}{\Sigma}$.

For the simulation study, samples of size 15, 25, and 50 were chosen at random from a pool of all subjects who had a complete set of observations at ages 6, 12, and 18 months. The ML estimates of the five-parameter model were obtained for each data set, producing a set of 15, 25, or 50 trivariate residuals. These residuals defined a multivariate empirical distribution, from which observations were selected at random with replacement to form sets of pseudodata. These samples of pseudodata were then analyzed by fitting the same five-parameter model, first by GLS,1 and then by ML. The results for the two estimation procedures over 100 bootstrap replicates are given in Tables 1 and 2.

Columns 1 and 2 give the results from the original analysis of each sample. Since all simulations were based on the model obtained by ML estimation, the two columns are identical for Tables 1 and 2. Columns 3 through 5 give the results of 100 replications of GLS,1 (Table 1) and ML estimation (Table 2). Column 3 gives the mean of the 100 estimates, column 4 the sample standard deviation, and column 5 the RMSSE.

The mean values of $\hat{\underset{\sim}{\beta}}^*$ for the GLS,1 estimator (Table 1, Col. 3) agree well with the parameters used in the bootstrap simulations (Col. 1). None of these means is more than 2 standard errors away from the parameter value, suggesting that the GLS,1 estimator has negligible bias even with the very non-normal error distributions used in these simulations. Comparison of the observed and estimated variability of the estimator (Cols. 4 and 5) shows remarkably little bias in the estimator based on the asymptotic variance. For samples of size 15, some downward bias is seen, but the average estimated bias over the 5 parameters is only 14.0%. For samples of size 25, the average estimated bias is 6.0% and one estimator has a positive estimated bias. Finally, for samples of size 50, the average estimated bias is 8.6%.

Similar results were found with the ML estimator (Table 2). To limit computing costs, only samples of size 15 and 25 were studied. For both of these sample sizes, there was no evidence of bias in the estimators. In fact, the means for the ML estimator were very close to those obtained with the GLS,1 estimator in every instance. The RMSSE agreed well with the empirical variability of the ML estimator. For samples of size 15, the average percentage downward bias was 20.2%, while for samples of size 25, the bias was 9.1%. These values are slightly larger than the corresponding values for the GLS,1 estimator because of a modest increase in observed variability of the ML estimator relative to the GLS,1 estimator. As we shall see more clearly in the incomplete data sets, this reflects a tendency of the ML estimator toward instability in very small sample sizes.

To further assess the agreement between the GLS,1 and ML estimators, we computed the correlations between the estimates obtained by the two methods over the 100 bootstraps replicates. These correlations (Table 1, Col. 6) uniformly exceed .90 for samples of size 15 and .98 for samples of size 25.

To study the performance of the two estimators in incomplete data sets, we began with the data sets of size 15 and 25, generated complete sets of pseudovalues, and then randomly deleted a subset of the observations on the third occasion. For samples of size 15, 7 observations were deleted and, for samples of size 25, 12 observations were deleted. Thus, approximately 50% of the observations at

Table 1. Analysis of Complete Bayley Score Data
Results for GLS,1 Estimator

	Fitted Model		Bootstrap Results		
	(1) Estimate	(2) SE	(3) Mean	(4) SD	(5) RMSSE
			Sample Size 15		
INT	-5.63	4.19	-5.56	4.14	3.79
LOW	-2.21	4.73	-2.20	5.51	3.97
MID	3.76	4.21	4.38	4.16	3.53
TIME	.66	5.51	-.37	5.16	4.56
TIME * LEAD	2.26	5.17	3.11	4.76	4.23
			Sample Size 25		
INT	-3.56	5.29	-4.01	4.52	4.82
LOW	.26	5.13	.99	4.95	4.54
MID	-2.05	5.06	-1.96	4.53	4.49
TIME	3.05	2.93	3.44	2.95	2.66
TIME * LEAD	-.11	2.60	-.41	2.80	2.31
			Sample Size 50		
INT	-2.57	3.05	-3.03	2.97	2.95
LOW	4.53	2.75	4.60	2.65	2.57
MID	-.53	2.82	-.54	2.92	2.63
TIME	3.18	2.00	3.45	2.07	1.89
TIME * LEAD	-1.77	1.70	-1.75	1.98	1.57

Table 2. Analysis of Complete Bayley Score Data
Results for ML Estimator

	Fitted Model		Bootstrap Results			
	(1) Estimate	(2) SE	(3) Mean	(4) SD	(5) RMSSE	(6) CORR
			Sample Size 15			
INT	-5.63	4.19	-5.59	4.32	3.65	.98
LOW	-2.21	4.73	-2.09	5.43	3.82	.92
MID	3.76	4.21	4.46	4.33	3.40	.96
TIME	.66	5.51	-.35	5.41	4.52	.97
TIME * LEAD	2.26	5.17	3.15	5.11	4.19	.96
			Sample Size 25			
INT	-3.56	5.29	-3.83	4.75	4.80	.98
LOW	.26	5.13	.87	5.13	4.52	.99
MID	-2.05	5.06	-2.11	4.65	4.46	.99
TIME	3.05	2.93	3.42	2.98	2.65	.99
TIME * LEAD	-.11	2.60	-.35	2.86	2.30	.99

eighteen months were missing completely at random.

The results for the incomplete data sets are given in Tables 3 and 4. Columns 1 and 2 of these tables are identical to the same columns in Tables 1 and 2, reflecting the use of the same initial data sets and parameter estimates in the simulations using complete and incomplete data.

Once again, there is little evidence for bias of the GLS,1 estimator. In fact, the simulations based on samples of size 25 show remarkably good agreement between the parameters used in the bootstrap simulations and the average estimates of these parameters over the bootstrap replicates (Col. 3). As would be expected, the parameter estimates are more variable in the incomplete data sets, reflecting the loss of information. The average percentage increase in standard deviation is 11.2% for samples of 15 and 23.8% for samples of 25. The bias in the estimated asymptotic variance remains small. The average percentage bias is 23.7% for samples of size 15 and 16.5% for samples of size 25.

The ML estimator proved to be somewhat less stable in these small incomplete data sets (Table 4). Although this instability was not apparent in the means of the parameter estimates, it was noticeable in their observed variability. The bootstrap standard deviation of the ML estimator was, on average, 17.1% greater for samples of size 15 but only 3.5% greater for samples of size 25 than the corresponding standard deviation for the GLS,1 estimator. Further, this instability was observed in individual bootstrap samples. In 8 of the incomplete samples of size 15, the algorithm for the ML estimator required more than 35 iterations. In 2 samples, more than 75 iterations were required. The parameter estimates obtained in these 2 cases were strikingly different than those obtained in the 92 samples for which convergence occurred in fewer than 35 iterations. With incomplete samples of size 25, only 3 data sets required more than 20 iterations and none required more than 25. The results for these three data sets were consistent with the results in the other samples.

This instability in the ML estimator produced greater discrepancies between the observed and estimated variance than seen with the GLS,1 estimator. The average percentage bias of the estimated asymptotic variance was 39.1% for samples of size 15 and 20.9% for samples of size 25. The discrepancy was greatest for the 'longitudinal' parameters, AGE and AGE*LEAD, presumably because these estimators are functions of within-individual differences and are subject to a greater loss of information from the missing data pattern used in these simulations. Finally, we note that the correlations between the ML and GLS,1 estimators are relatively poor for samples of size 15 (Table 3), while all correlations are .99 (after rounding) for samples of size 25.

Example 2

The Harvard Study of Air Pollution and Health (Ferris et al., 1979), is a longitudinal study of the respiratory health effects of air pollutants generated by combustion of fossil fuels. In each community, a sample of first and second grade school children was enrolled for long-term follow-up. Each year, participating children were given a pulmonary function examination and their parents were asked to complete a questionnaire about respiratory health and family characteristics. To analyze the data obtained from the serial spirometric examinations, the investigators developed a mathematical model for normal growth (Dockery et al., 1983).

Table 3. Analysis of Incomplete Bayley Score Data
Results for GSL,1 Estimator

	Fitted Model		Bootstrap Results		
	(1) Estimate	(2) SE	(3) Mean	(4) SD	(5) RMSSE
Sample Size 15 (7 subjects missing final observation)					
INT	-5.63	4.19	-6.17	4.75	3.95
LOW	-2.21	4.73	-2.05	5.01	3.78
MID	3.76	4.21	3.12	4.58	3.36
TIME	.66	5.51	1.63	6.04	4.59
TIME * LEAD	2.26	5.17	1.80	5.56	4.19
Sample Size 25 (12 subjects missing final observation)					
INT	-3.56	5.29	-3.87	5.71	5.03
LOW	.26	5.13	.63	5.72	4.97
MID	-2.05	5.06	-2.06	5.60	4.70
TIME	3.05	2.93	2.97	3.76	3.10
TIME * LEAD	-.11	2.60	-.02	3.54	2.70

Analysis of data from 44,664 examinations of 13,299 children showed that the normative model for Forced Expiratory Volume in One Second (FEV_1) was best expressed as a linear relation between the logarithms of FEV_1 and height (HT), weight (WT), and AGE. Although the expected response also depends on the sex of the respondent, the data set for the simulation study was limited to examinations of girls living in one of the study communities, Portage, WI.

For preadolescent girls, the prediction equation developed by Dockery and colleagues can be expressed as

$$\log(FEV_{1,ij}) = \beta_0 + \beta_1 \log(HT_{ij}) + \beta_2 \log(WT_{ij}) + \beta_3 \log(AGE_{ij}) + \varepsilon_{ij}$$

where FEV_1 is measured in liters, HT in meters, WT in kilograms, and age in years.

Table 4. Analysis of Incomplete Bayley Score Data Results for ML Estimator

	Fitted Model		Bootstrap Results			
	(1) Estimate	(2) SE	(3) Mean	(4) SD	(5) RMSSE	(6) CORR
Sample Size 15 (7 subjects missing final observation)						
INT	-5.63	4.19	-6.58	5.16	3.79	.94
LOW	-2.21	4.73	-2.02	5.74	3.42	.93
MID	3.76	4.21	3.18	5.21	3.04	.90
TIME	.66	5.51	1.98	7.30	4.29	.88
TIME * LEAD	2.26	5.17	1.78	7.10	3.87	.87
Sample Size 25 (12 subjects missing final observation)						
INT	-3.56	5.29	-3.74	5.97	4.98	.99
LOW	.26	5.13	.65	5.52	4.70	.99
MID	-2.05	5.06	2.16	5.88	4.64	.99
TIME	3.05	2.93	2.88	3.94	3.06	.99
TIME * LEAD	-.11	2.60	.05	3.78	2.66	.99

Samples of 25 and 50 participants were selected from the pool of all Portage girls who had a complete set of four observations between the ages of 6 and 10. As before, the linear model was fitted by ML and 100 samples of pseudodata were generated from each data set. The GLS,1 estimate was obtained for samples of size 25 and 50, but the ML estimates were obtained only for samples of size 25 to limit computing costs.

The pattern of relationships between bootstrap SD and RMSSE and between the GLS,1 and MLE estimators in these data sets was very similar to that found in the first investigation (Tables 5 and 6). As before, Columns 1 and 2 of Tables 5 and 6 are identical. For samples of size 25, both the GLS,1 and ML estimators gave results which agreed well with the parameters used to generate the pseudosamples. The correlations between the GLS,1 and ML estimates were .94 or greater for each parameter (Table 5). Further, little bias was seen in the estimated asymptotic variance. The average percentage bias was 17.5% for GLS,1 and 22.6% for ML. In samples of size 50, the RMSSE was actually .9% greater on average than the observed standard deviations.

Table 5. Analysis of Complete Pulmonary Function Data
Results for GLS,1 Estimator

	Fitted Model		Bootstrap Results		
	(1) Estimate	(2) SE	(3) Mean	(4) SD	(5) RMSSE
			Sample Size 25		
INT	.354	.305	.359	.340	.276
WT	2.256	.318	2.461	.357	.282
HT	-.131	.081	-.153	.090	.073
AGE	.058	.087	.045	.090	.080
			Sample Size 50		
INT	-.386	.248	-.359	.231	.215
WT	1.912	.234	1.923	.209	.214
HT	.016	.062	.013	.060	.057
AGE	.131	.072	.123	.063	.067

Table 6. Analysis of Complete Pulmonary Function Data
Results for ML Estimator

	Fitted Model		Bootstrap Results			
	(1) Estimate	(2) SE	(3) Mean	(4) SD	(5) RMSSE	(6) CORR
			Sample Size 25			
INT	.354	.305	.379	.336	.257	.94
WT	2.256	.318	2.468	.374	.267	.94
HT	-.131	.081	-.150	.087	.068	.94
AGE	.058	.087	.031	.091	.076	.95

4. DISCUSSION

In their paper on the performance of GLS estimators with estimated variances, Freedman and Peters raised the possibility that the estimated variance of the GLS estimator calculated in the usual way by substituting the estimated variance of the response vectors into the formula for the asymptotic variance could be seriously biased relative to the true variance of the estimator. They provided an example of this phenomenon in a simulation study taken from an econometric setting. In their example, where 15 samples of size 10 were available for analysis, the nominal standard errors underestimated the true variability by a factor as great as 3.

Our simulation results, based on data sets taken from longitudinal studies of environmental risk factors, show that the magnitude of this bias varies greatly from problem to problem. In the data set taken from a study of the health effects of lead exposure, the average percentage bias of the asymptotic variance estimator was less than 20% for complete samples of size 15, and less than 10% for complete samples of size 25 and 50. For incomplete samples from the same data set, the bias was consistently larger for both the GLS,1 and ML estimator, as would be expected because of the reduction in effective sample size. The average bias still did not exceed 25% for the GLS,1 estimator, but the ML estimator proved to be unstable in incomplete samples of size 15, resulting in an average bias of nearly 40%. In the data sets taken from a study of pulmonary function in preadolescent children, the estimated bias was again small. For samples of size 25, the average percentage bias was approximately 20% for both estimators, while an average upward bias of 1% was seen for the GLS,1 estimator in samples of size 50.

These results show that the bias of estimated variances of generalized least squares estimators derived from the asymptotic formula can be small even for very small sample sizes. The length of the observation vector is likely to be a major determinant of the magnitude of this bias. The two examples were characterized by relatively small numbers of repeated measurements, 3 and 4, so that the numbers of variance parameters to be estimated were only 6 and 10. More importantly, however, longitudinal studies in medical research usually have substantially more than 25 to 50 participants. The study of environmental lead considered here had over 200 participants and the study of outdoor air pollution had over 10,000 participants. Thus, although caution is indicated in the analysis of very small data sets and studies with many repeated measurements, these results are very encouraging regarding the performance of GLS estimators with estimated covariance matrices in the analysis of longitudinal data. As noted by Freedman and Peters, the determinants of the adequacy of the estimated asymptotic variance are not well understood, though the number of estimated parameters is certain to be one factor. Characterization of the circumstances in which the estimated asymptotic variance can safely be used is an important but technically difficult research question.

ACKNOWLEDGEMENT: Supported by grant GM-29745 from the National Institute of General Medical Sciences and by grant HL-07427 from the National Heart, Lung, and Blood Institute.

REFERENCES:

[1] Beale, E.M.L. and Little, R.J.A., Missing values in multivariate analysis, J. Roy. Statist. Soc., Series B **37**:129-145 (1975).

[2] Bellenger, D., Leviton, A., Needleman, H., and Waternaux, C., Early sensory-motor development and prenatal exposure to lead. To appear in J. Neurobehavioral Toxicology and Teratology (1984).

[3] Dahlquist, G. and Bjork, A., Numerical Methods, New York, Prentice-Hall (1974).

[4] Dempster, A.P., Laird, N.J., and Rubin, D.B., Maximum likelihood from incomplete data via the EM algorithm, J. Roy. Statist. Soc., Series B **39**:1-38 (1977).

[5] Dockery, D.W., Berkey, C.S., Ware, J.H., Speizer, F.E., and Ferris, B.G. Jr., Distribution of forced vital capacity and forced expiratory volume in one second in children 6 to 11 years of age, Amer. Rev. Respir. Dis. **128**:405-412 (1983).

[6] Efron, B., Bootstrap methods: Another look at the Jackknife, Ann. Statist. **7**:1-26 (1979).

[7] Ferris, B.G. Jr., Speizer, F.E., Spengler, J.D., Dockery, D.W., Bishop, Y.M.M., Wolfson, M., and Humble, C., Effects of sulfur oxides and respirable particles on human health: Methodology and demography of populations in study, Amer. Rev. Respir. Dis. **120**:767-779 (1979).

[8] Freedman, D.A. and Peters, S.C., Bootstrapping a regression equation: Some empirical results, J. Amer. Statist. Assoc. **79**:97-106 (1984).

[9] Rubin, D.B., Inference and missing data (with discussion), Biometrika **63**: 581-592 (1976).

[10] Ware, J.H., Linear models for the analysis of longitudinal studies. To appear in The American Statistician (1984).

HUMAN GENETICS

BIOSTATISTICS: Statistics in Biomedical, Public Health
and Environmental Sciences, edited by P.K. Sen

QUANTITATIVE TRAITS IN COMPLEX PEDIGREES: LIKELIHOOD CALCULATIONS

George E. Bonney and Robert C. Elston

Department of Biometry
Louisiana State University Medical Center
New Orleans, Louisiana 70112

A simple definition and classification of pedigrees are given, and the results of Bonney and Elston (1984b) used to derive an algorithm for calculating pedigree likelihoods under polygenic and combined oligogenic and polygenic (mixed) models of genetic inheritance. By analytically evaluating the integrals, simple recurrence formulas are obtained for a p-variate phenotype measured on individuals in a pedigree of arbitrary structure. In the case of the purely polygenic model, the calculation requires no more time for a complex pedigree than for a simple pedigree of the same size.

KEY WORDS: Pedigree, likelihood, oligogenic inheritance, polygenic inheritance, multivariate phenotype

INTRODUCTION

The analysis of pedigree data affords a means of unraveling the hereditary mechanisms underlying human traits. This analysis is often based on maximum likelihood methods, whether for parameter estimation or hypothesis testing (see e.g. Elston and Rao, 1978, and Elston, 1981, for reviews), and so efficient methods of calculating pedigree likelihoods are of interest. Lange and Elston (1975) and Cannings, Thompson and Skolnick (1978) have presented methods for calculating the likelihood of univariate data over pedigrees of arbitrary structure. Both methods are extensions of the original Elston and Stewart (1971) method for calculating the likelihood of a simple pedigree. However, the underlying genetic models assumed were restricted to oligogenic inheritance, i.e., the case in which segregation at only a finite set of loci controls the trait, or <u>phenotype</u> being analyzed. For a polygenic model, in which the phenotype is controlled by an indefinitely large number of separately indistinguishable loci, the classical method is the one based on normality originated by Fisher (1918) and described specifically for human pedigrees by Lange et al. (1976). To allow for a mixed model of inheritance, incorporating both oligogenic and polygenic components, an attempt at integrating the two approaches in the same computational scheme has been made by Ott (1979), but his results are not practical.

In this paper we shall generalize the method of Lange and Elston (1975) to polygenic and mixed models of inheritance, using the results of Bonney and Elston (1984a) to derive simple recurrence formulas. Our results are presented for a p-variate phenotype, assuming this to be multinormally distributed conditional on the major (oligogenic) genotype. To facilitate the development we shall first define a pedigree and present a simple scheme for classifying pedigrees.

DEFINITION AND CLASSIFICATION OF PEDIGREES

A graph theoretic definition and classification of pedigrees were presented by Lange and Elston (1975). A different classification system, also formulated in graph theoretic terms, has been provided by Cannings, Thompson and Skolnick (1978). We shall present simple definitions which are virtually equivalent to those of Lange and Elston (1975).

Definition 1

A pedigree is a pictorial representation of a set of related individuals in which parental lines of descent are indicated; either both parents or no parent must be indicated for each individual in the pedigree.

Definition 2

A spouse-pair is a set of two individuals who have had a biological union resulting in offspring in the pedigree.

Definition 3

Two spouse-pairs are connected if they have one member in common, i.e., there are only three distinct individuals involved. If a spouse-pair is not connected to any other spouse-pair it is isolated.

Definition 4

A spouse-connected set is a set of connected spouse-pairs that cannot be subdivided into two sets such that no individual in either set forms a spouse-pair with an individual in the other set. We shall assume there are no loops in a spouse-connected set, i.e., every subset of n distinct spouse-pairs in it contains at lease n+1 individuals.

Let us note that however complex the pedigree structure, there are only two sets of individuals in a pedigree: those whose parental lines are indicated and those whose parental lines are not indicated. We shall use this fact to classify pedigrees.

Definition 5

A pedigree is simple if the parental lines are indicated for only one or no member of every isolated spouse-pair and every spouse-connected set. The simplest pedigree is a nuclear family, consisting of two parents with at least one child (Fig. 1a).

If for at least one spouse-pair or one spouse-connected set parental lines are

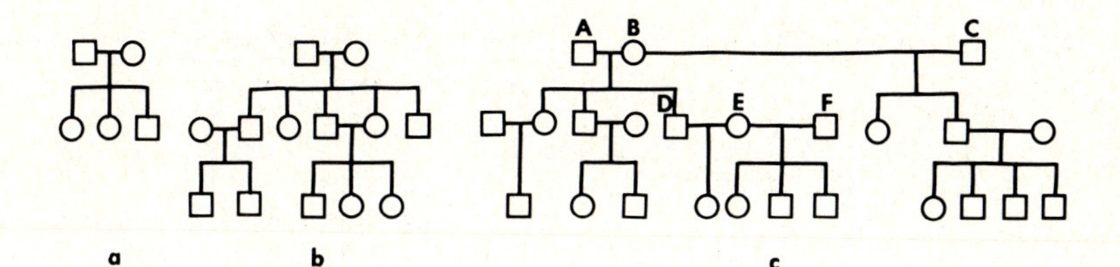

FIGURE 1. EXAMPLES OF SIMPLE PEDIGREES

Square indicates males, Circle females

a Nuclear family

b Simple extended family with no half-sibs

c Extended families including half-sibs.

Note that (A,B) and (B,C) are connected spouse-pairs so that {A,B,C} is a spouse-connected set. Similarly, {D,E,F} is a spouse-connected set.

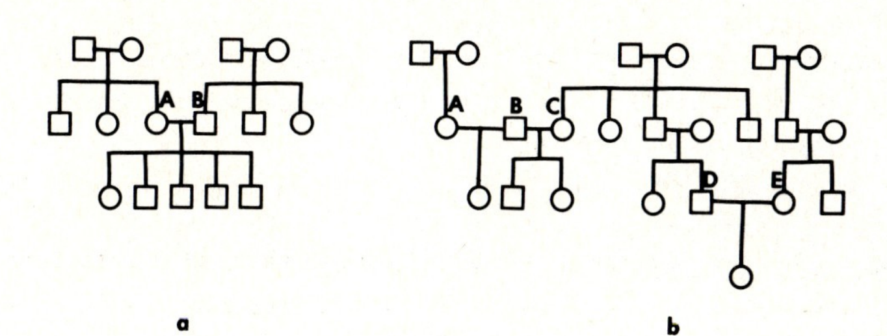

FIGURE 2. EXAMPLES OF COMPLEX PEDIGREES WITHOUT LOOPS

a Ancestors of both numbers of spouse-pair (A,B) are indicated

b Ancestors of more than one member of spouse-connected set {A,B,C} and spouse-pair (D,E) are indicated.

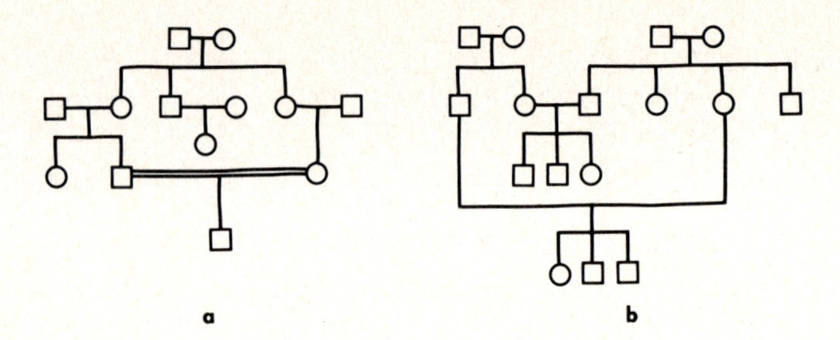

FIGURE 3. EXAMPLES OF COMPLEX PEDIGREES WITH LOOPS

a Consanguineous loop; the consanguineous mating is indicated by a double line.

b Non-consanguineous loop.

indicated for more than one member, the pedigree is complex. We can distinguish between two types of complex pedigrees: those in which the lines of descent form loops as in Fig. 3 and those in which they do not (Fig. 2). A loop may be due to a consanguineous mating (i.e., a biological union between related individuals - Fig. 3a), or not (Fig. 3b).

For simple pedigrees, we let I be the set of individuals related to someone in the previous generation, i.e., whose parental lines are indicated, M the set of "marry ins" and original parents, i.e., individuals whose parental lines are not indicated, and T the set of all individuals; thus T = I + M.

SIMPLE PEDIGREES

We summarize here some results for simple pedigrees, which are then extended to complex pedigrees. The likelihood of a pedigree having observed values of a p-variate quantitative phenotype $\underset{\sim}{z}$ depends on the underlying genetic mechanism. Consider oligogenic inheritance, the case with only a few major loci. Let k be the number of major genotypes underlying the variation in $\underset{\sim}{z}$ (arranged in some specified order so that it is meaningful to talk of the u-th genotype, u = 1,2, ...,k), ψ_u the probability that a randomly chosen individual in the population has the u-th genotype, and $g_u(\underset{\sim}{z})$ be the conditional probability density function, given the u-th genotype, of observing $\underset{\sim}{z}$. Also let $p_{u_F u_M u_i}$ be the conditional probability that individual i has genotype u_i, given that the genotypes of the parents of i are u_F and u_M (u_F, u_M = 1,2,...,k).

Under additive polygenic inheritance the genotype u_i of individual i is replaced by the p-vector $\underset{\sim}{s}_i$, the vector polygenic value, and the associated probability distributions are specified by probability density functions. Let $\phi(\underset{\sim}{z}-\underset{\sim}{u},V)$ denote the ordinate at $\underset{\sim}{z}$ of the p-variate multinormal density with mean $\underset{\sim}{\mu}$ and covariance matrix V, i.e.

$$\phi(\underset{\sim}{z}-\underset{\sim}{u},V) = (2\pi)^{-p/2}|V|^{-\frac{1}{2}} e^{-\frac{1}{2}(\underset{\sim}{z}-\underset{\sim}{\mu})'V^{-1}(\underset{\sim}{z}-\underset{\sim}{\mu})} .$$

Then ψ_{u_i}, the probability that a random individual i has genotype u_i, is replaced by $\phi(\underset{\sim}{s}_i,2\Phi_{ii}V_g)$, where V_g is the additive genetic covariance matrix and Φ_{ii} is the kinship coefficient between individual i and himself (for a non-inbred individual, $\Phi_{ii} = \frac{1}{2}$). The conditional probability density function $g_{u_i}(\underset{\sim}{z})$ is replaced by $g_{\underset{\sim}{s}_i}(z)$. Lastly, the conditional probability $p_{u_F u_M u_i}$ is replaced by $\phi(\underset{\sim}{s}_i-[\alpha_{Fi}\underset{\sim}{s}_F+\alpha_{Mi}s_M],\alpha_i V_g)$, where $\underset{\sim}{s}_F$ and $\underset{\sim}{s}_M$ are the polygenic values of the father and mother of i respectively, and α_{Fi}, α_{Mi} and α_i are the associated relatedness coefficients (Bonney and Elston, 1984b); under random mating and in the absence of consanguineous matings $\alpha_{Fi} = \alpha_{Mi} = \alpha_i = \frac{1}{2}$.

Using the above notation the likelihood for an entire simple pedigree of n individuals under the three major modes of inheritance can be written as follows.

Oligogenic model:

$$L_o = \sum_{u_i}\sum_{u_2}\cdots\sum_{u_n}\prod_{i=1}^{n} p_{u_i}g_{u_i}(\underset{\sim}{z}_i) \tag{1}$$

Polygenic model:

$$L_p = \int_{\underset{\sim}{s}_1}\int_{\underset{\sim}{s}_2}\cdots\int_{\underset{\sim}{s}_n}\prod_{i=1}^{n} f_i(\underset{\sim}{s}_i)g_{\underset{\sim}{s}_i}(\underset{\sim}{z}_i) \tag{2}$$

Mixed model:

$$L_m = \sum_{u_i}\sum_{u_2}\cdots\sum_{u_n}\int_{\underset{\sim}{s}_1}\int_{\underset{\sim}{s}_2}\cdots\int_{\underset{\sim}{s}_n}\prod_{i=1}^{n} p_{u_i}f(\underset{\sim}{s}_i)g_{u_i\underset{\sim}{s}_i}(\underset{\sim}{z}_i), \tag{3}$$

where

$$p_{u_i} = \begin{cases} \psi_{u_i} & , \text{ if } i \in M \\ \rho_{u_F u_M u_i} & , \text{ if } i \in I \end{cases}$$

$$f(s_i) = \begin{cases} \phi(s_i,2\Phi_{ii}V_g) & \text{if } i \in M \\ \phi(\underset{\sim}{s}_i-\alpha_{1i}\underset{\sim}{s}_{Fi}-\alpha_{2i}\underset{\sim}{s}_{Mi},\alpha_{3i}V_g) & \text{if } i \in I, \end{cases}$$

and $g_{u_i \underset{\sim}{s}_i}(\underset{\sim}{z}_i)$ is the density of $\underset{\sim}{z}_i$ conditional on u_i and $\underset{\sim}{s}_i$. We shall further assume multinormal density functions $g_{u_i}(\underset{\sim}{z}_i) = \phi(\underset{\sim}{z}_i - \underset{\sim}{\mu}_{u_i}, V_e)$, $g_{\underset{\sim}{s}_i}(\underset{\sim}{z}_i) = \phi(\underset{\sim}{z}_i - \underset{\sim}{s}_i, V_e)$ and $g_{u_i \underset{\sim}{s}_i}(\underset{\sim}{z}_i) = \phi(\underset{\sim}{z}_i - \underset{\sim}{\mu}_{u_i} - \underset{\sim}{s}_i, V_e)$.

Evaluating the integrals, Bonney and Elston (1984b) obtained the following result for simple pedigrees.

Let

$$a_i = \begin{cases} 0 & \text{if } i \text{ is not observed} \\ 1 & \text{if } i \text{ is observed} \end{cases}$$

$$b_i = \begin{cases} \tfrac{1}{2}\Phi_{ii}^{-1} & \text{if } i \in M \\ \alpha_i^{-1} & \text{if } i \in I \end{cases}$$

$$c_i = \begin{cases} \sum_\ell \alpha_{F\ell}^2 \alpha_\ell^{-1} & \text{if } i \text{ is male}, \\ \sum_\ell \alpha_{M\ell}^2 \alpha_\ell^{-1} & \text{if } i \text{ is female, where the summations are over all children } \ell \text{ of } i \\ 0 & \text{if } i \text{ has no children} \end{cases}$$

$$c_{ij} = \begin{cases} \sum_\ell \alpha_{F\ell}\alpha_{M\ell}\alpha_\ell^{-1} & \text{, where the summation is over all children } \ell \text{ of } i \text{ and } j \\ 0 & \text{if } i \text{ and } j \text{ have no children in common} \end{cases}$$

$$d_{ij} = \begin{cases} \alpha_{Fi}\alpha_i^{-1} & \text{if } j \text{ is the father of } i \\ \alpha_{Mi}\alpha_i^{-1} & \text{if } j \text{ is the mother of } i \\ \alpha_{Fj}\alpha_j^{-1} & \text{if } i \text{ is the father of } j \\ \alpha_{Mj}\alpha_j^{-1} & \text{if } i \text{ is the mother of } j \\ 0 & \text{otherwise.} \end{cases}$$

Then

$$\left.\begin{aligned} L_p &= |V_g|^{-n/2} \prod_{i=1}^{n} Y_i e^{-\frac{1}{2}w_i} \\ \text{and} \quad & \\ L_m &= |V_g|^{-n/2} \sum_{u_1} \sum_{u_2} \cdots \sum_{u_n} \prod_{i=1}^{n} P_{u_{i-}} Y_i e^{-\frac{1}{2}w_i(u_i)} \end{aligned}\right\} \quad (4)$$

where

$$Y_i = (2\pi)^{-a_i p/2} b_i^{\frac{1}{2}} |V_e|^{-a_i/2} |D_{i,i}^{(i)}|^{-\frac{1}{2}}, \quad (5)$$

$$w_i(u_i) = (\underset{\sim}{z}_i - \underset{\sim}{\mu}_{u_i})' V_e^{-1} (\underset{\sim}{z}_i - \underset{\sim}{\mu}_{u_i}) - D_{i,0}^{(i)'} D_{i,i}^{(i)-1} D_{i,0}^{(i)}, \quad (6)$$

and the D's satisfy the recurrence relation

$$D_{i,j}^{(m+1)} = D_{i,j}^{(m)} - D_{m,i}^{(m)'} D_{m,m}^{(m)-1} D_{m,j}^{(m)} \quad (7)$$

$$(m = 1,2,\ldots,n-1, \; i,j = 0,1,2,\ldots,n),$$

with initial conditions defined by

$$\left.\begin{aligned} D_{i,i}^{(1)} &= a_i V_e^{-1} + (b_i + c_i) V_g^{-1} \\ D_{i,j}^{(1)} &= (c_{ij} - d_{ij}) V_g^{-1}, \; i \neq j \\ D_{i,0}^{(1)} &= -V_e^{-1} (\underset{\sim}{z}_i - \underset{\sim}{\mu}_{u_i}), \end{aligned}\right. \quad (8)$$

where for the polygenic model $\underset{\sim}{\mu}_{u_i}$ in (6) and (8) is replaced by $\underset{\sim}{\mu}$.

COMPLEX PEDIGREES WITHOUT LOOPS

The method of Lange and Elston (1975) for calculating the likelihood of a complex pedigree without loops involves breaking up the complex pedigree into simple ones, by splitting one member of every spouse-pair or spouse-connected set for which two or more parental lines are indicated; i.e., in place of one member of every such spouse-pair or spouse-connected set, we create two separate but phenotypically and genotypically identical individuals. Designate one of the resulting simple pedigrees as the "root" pedigree. Then if n is the number of individuals in the root pedigree, the likelihood of the root pedigree and the adjacent "bits" of pedigree, under oligogenic inheritance, is given by Lange and Elston (1975), using our notation, by

$$L_o = \sum_{u_1} \sum_{u_2} \cdots \sum_{u_n} \prod_{i=1}^{n} \{p_{u_i} g_{u_i}(z_i) Q_i(u_i)\} \ , \tag{9}$$

where

$$Q_i(u_i) = \begin{cases} \prod_j \dfrac{L_j(u_i)}{\psi_{u_i} g_{u_i}(z_i)} \ , & \text{if } i \text{ is split} \\ 1 \ , & \text{if } i \text{ is not split,} \end{cases}$$

and $L_j(u_i)$ is the likelihood of the j-th broken bit of pedigree given that the individual identical to i has genotype u_i. If we compare (9) to the likelihood of a simple pedigree (1), we see immediately that the process is one of calculating $\prod_j \frac{L_j(u_i)}{\psi_{u_i} g_{u_i}(z)}$ and attaching the result as a factor to the term for individual i in the root pedigree.

For concreteness, let us consider the pedigree in Fig. 4. The pedigree may be broken at A or B. Let us break it at B, i.e., in place of B create two separate

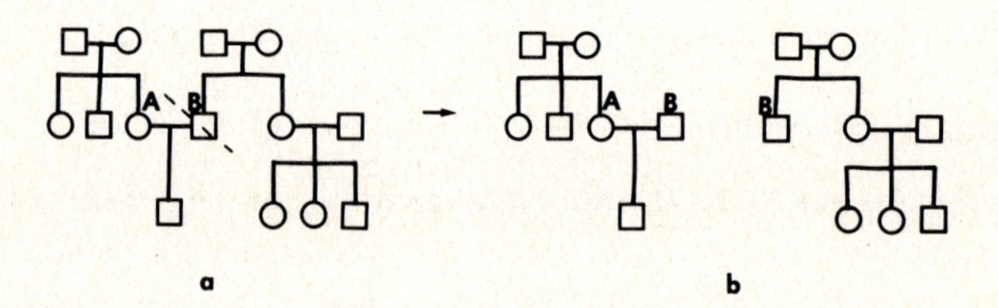

FIGURE 4 . BREAKING UP A COMPLEX PEDIGREE.
The complex pedigree shown in a is broken up at B into the two simple pedigrees shown in b .

but genotypically and phenotypically identical individuals. We can take pedigree 1 as the root, and calculate the likelihood of pedigree 2 given B has a particular genotype u. We thus obtain $L_2(u)$, which we attach to the term for B in the root pedigree. Note here that for this case (9) is equivalent to

$$L_0 = \sum_u \frac{L_1(u)L_2(u)}{\psi_u g_u(x)} \qquad (10)$$

where x is the phenotype of B.

<u>The Polygenic Model</u>

The analogue of (10) for a multivariate polygenic model is obtained by writing

$$L_p = \int_{\tilde{s}} \frac{L_1(\tilde{s})\cdot L_2(\tilde{s})}{\phi(\tilde{s},V_g)g_{\tilde{s}}(\tilde{x})} \qquad (11)$$

where $\tilde{s}$ is the vector polygenic value of B and $\tilde{x}$ is the corresponding vector phenotype. We shall now evaluate (11). We first calculate $L_1(\tilde{s})$ and $L_2(\tilde{s})$ by the recurrence formula for polygenic inheritance over simple pedigrees ((4) and following; see Bonney and Elston (1984a, 1984b) for details). This is easily done if we first do the integrations for all individuals except B, separately for each of the two simple pedigrees. Then, for the simple case with no missing data, and marry-ins not inbred, letting

$$\beta_1^{(n_1)}(\tilde{s}) = \tilde{s}'D_{1,n_1,n_1}^{(n_1)}\tilde{s} + 2D_{1,n_1,0}^{(n_1)'}\tilde{s} + (\tilde{x}-\tilde{\mu})'V_e^{-1}(\tilde{x}-\tilde{\mu})$$

and

$$\beta_2^{(n_2)}(\tilde{s}) = \tilde{s}'D_{2,n_2,n_2}^{(n_2)}\tilde{s} + 2D_{2,n_2,0}^{(n_2)'}\tilde{s} + (\tilde{x}-\tilde{\mu})'V_e^{-1}(\tilde{x}-\tilde{\mu}),$$

where the first subscript (1 or 2) on the D's denotes simple pedigree 1 or 2, we obtain

$$L_1(\tilde{s}) = K_1|V_g|^{-\frac{n_1}{2}}|V_e|^{-\frac{n_1}{2}}\prod_{i=1}^{n_1-1}\{|D_{1,i,i}^{(i)}|^{-\frac{1}{2}}e^{-\frac{1}{2}w_{1i}}\}e^{-\frac{1}{2}\beta_1^{(n_1)}(\tilde{s})}$$

and

$$L_2(\tilde{s}) = K_2|V_g|^{-\frac{n_2}{2}}|V_e|^{-\frac{n_2}{2}}\prod_{i=1}^{n_2-1}\{|D_{2,i,i}^{(i)}|^{-\frac{1}{2}}e^{-\frac{1}{2}w_{2i}}\}e^{-\frac{1}{2}\beta_2^{(n_2)}(\tilde{s})},$$

where K_1 and K_2 are constants independent of the parameters, and w_{ri} is given by (6) for pedigree r(= 1,2), with $\tilde{\mu}_{\tilde{u}_i} = \tilde{\mu}$. Substituting in (11) we obtain

$$L_P = K_1K_2(|V_e||V_g|)^{-\frac{1}{2}(n_1+n_2)} \cdot |V_g|^{\frac{1}{2}}|V_e|^{\frac{1}{2}} \prod_{i=1}^{n_1-1} |D^{(i)}_{1,i,i}|^{-\frac{1}{2}} e^{-\frac{1}{2}w_{1i}}$$

$$\times \prod_{j=1}^{n_2-1} |D^{(j)}_{2,j,j}|^{-\frac{1}{2}} e^{-\frac{1}{2}w_{2j}} \int_{\underset{\sim}{s}} e^{-\frac{1}{2}\{\beta_1^{(n_1)}(\underset{\sim}{s})+\beta_2^{(n_2)}(\underset{\sim}{s})-\underset{\sim}{s}'V_g^{-1}\underset{\sim}{s}-(\underset{\sim}{s}+\underset{\sim}{\mu}-\underset{\sim}{x})'V_e^{-1}(\underset{\sim}{s}+\underset{\sim}{\mu}-\underset{\sim}{x})\}} .$$

Minus twice the exponent in the integrand equals

$$\{\ \} = \underset{\sim}{s}'[D^{(n_1)}_{1,n_1,n_1} + D^{(n_2)}_{2,n_2,n_2} - V_g^{-1} - V_e^{-1}]\underset{\sim}{s} + 2[D^{(n_1)}_{1,n_1,0} + D^{(n_2)}_{2,n_2,0}$$

$$+ V_e^{-1}(\underset{\sim}{x}-\underset{\sim}{\mu})]'s + (\underset{\sim}{x}-\underset{\sim}{\mu})'V_e^{-1}(\underset{\sim}{x}-\underset{\sim}{\mu})$$

$$= \underset{\sim}{s}'D^{(n)}_{n,n}\underset{\sim}{s} + 2D^{(n)'}_{n,0}\underset{\sim}{s} + (\underset{\sim}{x}-\underset{\sim}{\mu})'V_e^{-1}(\underset{\sim}{x}-\underset{\sim}{\mu}), \text{ say.}$$

Thus

$$L_P = K_1K_2(|V_g||V_e|)^{-\frac{n}{2}}|D^{(n)}_{n,n}|^{-\frac{1}{2}} e^{-\frac{1}{2}w_n} \prod_{i=1}^{n_1-1} |D^{(i)}_{1,i,i}|^{-\frac{1}{2}} e^{-\frac{1}{2}w_{1i}}$$

$$\times \prod_{j=1}^{n_2-1} |D^{(j)}_{2,j,j}|^{-\frac{1}{2}} e^{-\frac{1}{2}w_{2j}}, \qquad (12)$$

where $N = n_1 + n_2 - 1$ is the total size of the complex pedigree,

$$D^{(n)}_{n,n} = D^{(n_1)}_{1,n_1,n_1} + D^{(n_2)}_{2,n_2,n_2} - V_g^{-1} - V_e^{-1}$$

$$D^{(n)}_{n,0} = D^{(n_1)}_{1,n_1,0} + D^{(n_2)}_{2,n_2,0} + V_e^{-1}(\underset{\sim}{x}-\underset{\sim}{\mu})$$

and

$$w_n = (\underset{\sim}{x}-\underset{\sim}{\mu})'V_e^{-1}(\underset{\sim}{x}-\underset{\sim}{\mu}) - D^{(n)'}_{n,0}D^{(n)-1}_{n,n}D^{(n)}_{n,0} .$$

As a first step toward generalization, suppose B had m_B spouses whose parental lines were indicated, then breaking the pedigree at B results in m_B simple pedigrees each containing B as a marry-in and one simple pedigree in which B is an offspring. Each pedigree containing B as a marry-in contributes to the likelihood a factor similar to that of pedigree 1 above. In each pedigree integrate first over all individuals except the one identical to B. Then we find for the last integration, analogous to the case above,

$$D^{(n)}_{n,n} = \sum_r D^{(n_r)}_{r,n_r,n_r} - m_B(V_g^{-1}+V_e^{-1})$$

$$D^{(n)}_{n,0} = \sum_r D^{(n_r)}_{r,n_r,0} + m_B V_e^{-1}(\underset{\sim}{x}-\underset{\sim}{\mu}) \ .$$

Although we have assumed in the above that the integration over B is done last, this is not in fact necessary. Suppose, without loss of generality, we designate the first simple pedigree as the root pedigree, and denote by B_n the set of remaining simple pedigrees containing B, then

$$D^{(n)}_{n,n} = D^{(n_1)}_{1,n_1,n_1} + \sum_{r\varepsilon B_n} D^{(n_r)}_{r,n_r,n_r} - m_B(V_g^{-1} + V_e^{-1}).$$

Now by repeated use of the recurrence relation (7), we find

$$D^{(n_1)}_{1,n_1,n_1} = D^{(1)}_{1,n_1,n_1} - \sum_{\ell=1}^{n_1-1} D^{(\ell)'}_{1,n_1,\ell} D^{(\ell)-1}_{1,\ell,\ell} D^{(\ell)}_{1,\ell,n_1} .$$

From the original definition (8), for simple pedigrees,

$$D^{(1)}_{1,n,n_1} = a_{n_1} V_e^{-1} + (b_{n_1}+c_{n_1})V_g^{-1} \ .$$

But if $D^{(1)}_{1,n_1,n_1}$ is redefined as

$$D^{(1)}_{1,n_1,n_1} = a_{n_1} V_e^{-1} + (b_{n_1}+c_{n_1})V_g^{-1} + \sum_{r\varepsilon B_n} D^{(n_r)}_{r,n_r,n_r} - m_B(V_g^{-1}+V_e^{-1}) \qquad (13)$$

then $D^{(n_1)}_{1,n_1,n_1}$ derived from the repeated use of (7) is equal to $D^{(n)}_{n,n}$, which can therefore be dispensed with. Similarly, $D^{(1)}_{1,n_1,0}$ needs to be redefined as

$$D^{(1)}_{1,n_1,0} = -V_e^{-1}(\underset{\sim}{x}-\underset{\sim}{\mu}) + \sum_{r\varepsilon B_n} D^{(n_r)}_{r,n_r,0} + m_B V_e^{-1}(\underset{\sim}{x}-\underset{\sim}{\mu}). \qquad (14)$$

All other $D^{(1)}$'s in (8) remain the same, and the integrations may now be done in any order.

Generalization of the above procedure to more complex pedigrees without loops is not difficult. Break up the pedigree into simple pedigrees as above, and select one of the resulting simple pedigrees containing n individuals as root. For an individual in the root pedigree at which the pedigree was broken, compute the likelihood of each adjacent pedigree given the polygenotype of this individual as above, and attach all their likelihoods as factors, the procedure being repeated for each such individual until only the root pedigree remains. The analogue of (9) for the polygenic model is then

$$L_p = \int_{\underset{\sim}{s}_1}\int_{\underset{\sim}{s}_2}\cdots\int_{\underset{\sim}{s}_n} \prod_{i=1}^{n} \{f_i(\underset{\sim}{s}_i) g_{\underset{\sim}{s}_u}(\underset{\sim}{z}_i) Q_i(\underset{\sim}{s}_i)\}, \qquad (15)$$

where

$$Q_i(\underset{\sim}{s}_i) = \begin{cases} \prod_{j=1}^{m} \dfrac{L_j(\underset{\sim}{s}_i)}{\phi(\underset{\sim}{s}_i, b_i^{-1}V_g) g_{\underset{\sim}{s}_i}(\underset{\sim}{z}_i)} & \text{if } i \text{ is split} \\ 1 & \text{if } i \text{ is not split .} \end{cases}$$

The root pedigree is simple, so the likelihood of the whole complex pedigree can be computed from the recurrence formulas for simple pedigrees provided the D's in (8) are redefined as follows

$$\begin{aligned} D_{i,i}^{(1)} &= a_i V_e^{-1} + (b_i + c_i)V_g^{-1} + \sum_{r\varepsilon B_i} D_{r,1,1}^{(n_r)} - m_i(b_i V_g^{-1} + V_e^{-1}) \\ D_{i,j}^{(1)} &= (c_{ij} - d_{ij})V_g^{-1}, \qquad i \neq j \qquad\qquad (16) \\ D_{i,0}^{(1)} &= -V_e^{-1}(\underset{\sim}{z}_i - \underset{\sim}{\mu}_i) + \sum_{r\varepsilon B_i} D_{r,i,0}^{(n_r)} + m_i V_e^{-1}(\underset{\sim}{z}_i - \underset{\sim}{\mu}_i) , \end{aligned}$$

where B_i denotes simple pedigrees, other than the root, that result from breaking the complex pedigree at individual i in the root pedigree, and m_i denotes the number of such pedigrees. Note that the last two terms of $D_{i,i}^{(1)}$ and $D_{i,0}^{(1)}$ are zero for individuals that are not broken. The modifications to the D's are obvious extensions of (13) and (14). $D_{i,j}^{(1)}(i \neq j)$ remains the same as in (8) since it is the matrix of the bilinear form in $\underset{\sim}{s}_i$ and $\underset{\sim}{s}_j$, and i and j refer only to individuals in the root pedigree. With these modifications, we find that the likelihood becomes

$$L_p = |V_g|^{-N/2} \prod_{i=1} Y_i e^{-\frac{1}{2}W_i} \prod_{r\varepsilon B_i} \prod_{\ell} Y_{r\ell} e^{-\frac{1}{2}W_{r\ell}} \qquad (17)$$

where N is the total number of individuals in the complex pedigree, and $Y_{r\ell}$ and

$w_{r\ell}$ are as defined in (5) and (6) but for the ℓ-th individual in the r-th simple pedigree connected to individual i in the root pedigree.

Formulas (15) and (17) are strictly speaking meant for the case where a complex pedigree without loops can be broken as indicated into a simple "root" pedigree and simple "adjacent" pedigrees. However, they can be used for the case where some of the adjacent pedigrees themselves are complex pedigrees without loops; we need only calculate the $L_j(u_i)$ and $L_j(\underset{\sim}{s}_i)$ for such adjacent pedigrees as for complex pedigrees.

It is computationally advantageous to choose the order of integration such that as many of the D's as possible are zero. The order suggested for simple pedigrees-unmarried individuals first followed by spouse-pairs (see Bonney and Elston, 1984b) - seems to be the best for complex pedigrees as well.

The Mixed Model

By combining (9) and (15), the likelihood under the mixed model of a complex pedigree with no loops is seen to be

$$L_m = \sum_{u_1} \sum_{u_2} \cdots \sum_{u_n} \int_{\underset{\sim}{s}_1} \int_{\underset{\sim}{s}_2} \cdots \int_{\underset{\sim}{s}_n} \prod_{i=1} \{p_{u_i} f_i(\underset{\sim}{s}_{i-}) g_{u_i \underset{\sim}{s}_i}(\underset{\sim}{z}_i) Q_i(u_i, \underset{\sim}{s}_i)\} \tag{18}$$

where

$$Q_i(u_i \underset{\sim}{s}_i) = \begin{cases} \displaystyle\prod_{j=1}^{m_i} \frac{L_j(u_i, \underset{\sim}{s}_i)}{\psi_{u_i} \phi(\underset{\sim}{s}_i, b_i^{-1} V_g) g_{u_i \underset{\sim}{s}_i}(\underset{\sim}{z}_i)} & \text{if } i \text{ is split} \\ 1 \quad , & \text{if } i \text{ is not split,} \end{cases}$$

where $L_j(u_i, \underset{\sim}{s}_i)$ is the likelihood of the j-th broken bit of pedigree given all individuals identical to i have major genotype u_i and polygenic value $\underset{\sim}{s}_i$. As expected, the integrals have the same form as for the polygenic model (15), and can therefore be evaluated by (17), with $\underset{\sim}{\mu}_{u_i}$ replacing $\underset{\sim}{\mu}$. After evaluating the integrals, the likelihood function reduces to virtually the form for oligogenic inheritance (9), with the recurrence relations already developed. Thus (18) becomes

$$L_m = |V_g|^{-N/2} \sum_{u_1} \sum_{u_2} \cdots \sum_{u_n} \prod_{i=1} P_{u_{i-}} Y_i e^{-\frac{1}{2} w_i(u_{i-})} \prod_{j \in B_i} (\prod_{\ell} Y_{j\ell}) \frac{L_j(u_i)}{\psi_{u_i}} \tag{19}$$

where $w(u_i) = (\underset{\sim}{z}_i - \underset{\sim}{\mu}_{u_i})' V_e^{-1} (\underset{\sim}{z}_i - \underset{\sim}{\mu}_{u_i}) - D_{i,0}^{(i)'} D_{i,i}^{(i)^{-1}} D_{i,0}^{(i)}$, and, in the calculation of

$L_j(u_i)$, the g-functions are replaced by the corresponding $e^{-\frac{1}{2}w(u_i)}$.

COMPLEX PEDIGREES WITH LOOPS

When there are loops in the pedigree, Lange and Elston [2] suggest breaking them as above, i.e., replacing one member of every spouse pair with both parental lines observed by two separate but phenotypically and genotypically identical individuals. Breaking all loops this way results in either a simple pedigree or a complex pedigree without loops; in the latter case further individuals are replicated in the same fashion. Suppose there are a total of b breaks and at the j-th break n_j identical individuals are created, and let $L(v_1,v_2,\ldots,v_b)$ be the likelihood of the pedigree without loops when all n_j identical individuals resulting from the j-th break have their genotypes fixed at $v_j (j = 1,2,\ldots,b)$. Then Lange and Elston calculate the likelihood under oligogenic inheritance by

$$L_o = \sum_{v_1} \sum_{v_2} \cdots \sum_{v_b} \frac{L(v_1,v_2,\ldots,v_b)}{\prod_j (\psi_{v_j} g_{v_j}(z))^{n_j-1}} , \qquad (20)$$

in our notation. We can extend this to polygenic inheritance and for multivariate traits with multinormal phenotypes simply by writing

$$L_p = \int_{\underset{\sim}{s}_1^0} \int_{\underset{\sim}{s}_2^0} \cdots \int_{\underset{\sim}{s}_b^0} \frac{L(\underset{\sim}{s}_1^0,\underset{\sim}{s}_2^0,\ldots,\underset{\sim}{s}_b^0)}{\prod_j [\phi(\underset{\sim}{s}_j, b_j^{-1} V_g)\ \phi(\underset{\sim}{s}_j + \underset{\sim}{\mu}_j - \underset{\sim}{z}_j, V_e)]^{n_j-1}} \qquad (21)$$

where $L(s_1^0,s_2^0,\ldots,s_b^0)$ is the likelihood of the broken pedigree given that the genotypic values of the b individuals broken are fixed at $s_1^0,s_2^0,\ldots,s_b^0$. The likelihood function (21) looks complicated, but it is equal to the likelihood function without loops (17). We can most easily see this if we consider the case where breaking a looped pedigree of size n results in a simple pedigree of size $n'(= n + \sum_j n_j - b)$, all individuals are observed, and marry-ins are not inbred so that $b_i = 1$. For this case (21) becomes

$$L_p = \kappa \prod_{j=1}^{b} [(2\pi)^{p/2} |V_g|^{\frac{1}{2}} |V_e|^{\frac{1}{2}}]^{n_j-1} \int_{\underset{\sim}{s}_1^0} \int_{\underset{\sim}{s}_2^0} \cdots \int_{\underset{\sim}{s}_b^0} \int_{\underset{\sim}{s}_1} \int_{\underset{\sim}{s}_2} \cdots \int_{\underset{\sim}{s}_{n'}}$$

$$e^{-\frac{1}{2}\{\beta - \sum_{j=1}^{b} [\underset{\sim}{s}_j^{0\prime} V_g^{-1} \underset{\sim}{s}_j^0]^{n_j-1} - \sum_{j=1}^{b} [(\underset{\sim}{z}_j - \underset{\sim}{\mu}_j - \underset{\sim}{s}_j^0)' V_e^{-1} (\underset{\sim}{z}_j - \underset{\sim}{\mu}_j - \underset{\sim}{s}_j^0)]^{n_j-1}\}}, \qquad (22)$$

where n'' is the number of individuals not fixed (n''+b=n),

$$\{\underset{\sim}{s}_1,\underset{\sim}{s}_2,\ldots,\underset{\sim}{s}_{n''}\} \text{ excludes } \underset{\sim}{s}_1^0,\underset{\sim}{s}_2^0,\ldots,\underset{\sim}{s}_b^0\},$$

$$\beta = \sum_{i\varepsilon I}\alpha_i^{-1}(\underset{\sim}{s}_i-[\alpha_{Fi}\underset{\sim}{s}_{Fi}+\alpha_{Mi}\underset{\sim}{s}_{Mi}])'V_g^{-1}(\underset{\sim}{s}_i-[\alpha_{Fi}\underset{\sim}{s}_{Fi}+\alpha_{Mi}\underset{\sim}{s}_{Mi}])$$

$$+\sum_{i\varepsilon M}\underset{\sim}{s}_i'V_g^{-1}\underset{\sim}{s}_i+\sum_{i\varepsilon T}(\underset{\sim}{x}_i-\underset{\sim}{\mu}_i-\underset{\sim}{s}_i)'V_e^{-1}(\underset{\sim}{x}_i-\underset{\sim}{\mu}_i-\underset{\sim}{s}_i), \tag{23}$$

the sets I, M and T are as defined above, but for the simple pedigree of size n', and $\kappa = \text{constant} \times (|V_g||V_e|)^{-\frac{n'}{2}}$. Now note that when we break the complex pedigree at the j-th position, resulting in n_j identical individuals, n_j-1 of these are artificially created marry-ins. These contribute n_j-1 terms to the second summation in (23) and n_j-1 terms to the third summation. But the terms added by the artificially created marry-ins are subtracted from β in the exponent term of (22). We also note that

$$\kappa\prod_{j=1}^{b}[(2\pi)^{p/2}|V_g|^{\frac{1}{2}}|V_e|^{\frac{1}{2}}]^{n_j-1} = \text{constant} \times (|V_g||V_e|)^{-n/2} .$$

There is no trouble changing the order of integrations in (22) to correspond to $\underset{\sim}{s}_1,\underset{\sim}{s}_2,\ldots,\underset{\sim}{s}_n$. We find, therefore, that the likelihood of the complex pedigree is the same as the likelihood of the simple pedigree resulting from breaking the loops, but without the artificially created marry-ins. Thus we need not break up the loops at all, conceptually. We can start from any computationally convenient individual and "eliminate" until we have covered all individuals in the pedigree, using the recursive formulas already given. The observation we have just made is true for the more general case where breaking all loops results in a complex pedigree without loops. Thus formula (17) can be used for all complex pedigrees.

Let us now briefly consider combined oligogenic and polygenic inheritance in complex pedigrees with loops. By combining (20) and (21), we can write down the desired likelihood as

$$L_m = \sum_{v_1}\sum_{v_2}\cdots\sum_{v_b}\int_{\underset{\sim}{s}_1^0}\int_{\underset{\sim}{s}_2^0}\cdots\int_{\underset{\sim}{s}_b^0}\frac{L(\underset{\sim}{s}_1^0v_1,\underset{\sim}{s}_1^0v_2,\ldots,\underset{\sim}{s}_b^0v_b)}{\Pi[\psi_{v_j}\phi(\underset{\sim}{s}_j^0,b_i^{-1}V_g)\phi(\underset{\sim}{s}_j^0+\underset{\sim}{\mu}_{v_j}-\underset{\sim}{z}_j,V_e)]^{n_j-1}} . \tag{24}$$

Using (17) to evaluate the integrals, we have

$$L_m = |V_g|^{-N/2} \sum_{v_1} \sum_{v_2} \dots \sum_{v_b} \frac{L(v_1, v_2, \dots, v_b)}{\prod_j (\Psi_{v_j})^{n_j - 1}} \qquad (25)$$

where in the calculation of $L(v_1, v_2, \dots, v_b)$ the g-function for individual i is replaced by $Y_i \, e^{-\frac{1}{2}w(u_i)}$ and modified as in (19), if the pedigree is further broken at i into simple pedigrees.

CONCLUSION

We have presented a simple definition of a pedigree, introduced an easy scheme for classifying pedigrees, and derived formulas for calculating the likelihood under the polygenic and mixed models of genetic inheritance. Although the method of breaking up the pedigree is used, the form of the results shows that, conceptually at least, we need not break up the pedigree at all. The formulas can therefore be easily adapted for use with computer algorithms such as those of Cannings, Thompson and Skolnick (1978) that do not break up the pedigrees. In the case of the purely polygenic model, the calculation of the likelihood for an arbitrarily complex pedigree is seen to be no more time-consuming than that for a simple pedigree of the same size. For the mixed model, however, as for the oligogenic model, it is necessary to perform a multiple summation that can in some instances be impractical with current computer technology. Nevertheless, in view of the speed with which computer technology is advancing, what appears to be formidable now may well become quite manageable in the very near future.

ACKNOWLEDGEMENT: This work was supported in part by U.S. Public Health Service research grants GM 28356 from the National Institute of General Medical Sciences and CA 28198 from the National Cancer Institute.

REFERENCES

[1] Bonney, G.E. and Elston, R.C., Integrals of products of multinormal mixtures, J. Appl. Math. Comp., in press (1984a).

[2] Bonney, G.E. and Elston, R.C., Likelihood models for multivariate traits in human genetics, Biometrical J., in press (1984b).

[3] Cannings, C. Thompson, E.A. and Skolnick, M.H., Probability functions on complex pedigrees, Adv. Appl. Prob. 10:26-61 (1978).

[4] Elston, R.C., Segregation analysis, Adv. Hum. Genet. 11:63-120 (1981).

[5] Elston, R.C. and Rao, D.C., Statistical modeling and analysis in human genetics, Ann. Rev. Biophys. Bioeng. 7:253-286 (1978).

[6] Elston, R.C. and Stewart, J.,A general model for the genetic analysis of pedigree data, Hum. Hered. 21:523-542 (1971).

[7] Fisher, R.A., The correlation between relatives on the supposition of Mendelian inheritance, Trans. Roy. Soc. Edin. 52:399-433 (1918).

[8] Lange, K. and Elston, R.C., Extensions to pedigree analysis. I. Likelihood calculations for simple and complex pedigrees, Hum. Hered. 25:95-105 (1975).

[9] Lange, K., Westlake, J. and Spence, M.A., Extensions to pedigree analysis. III. Variance components by the scoring method, Ann. Hum. Genet. 39:485-491 (1976).

[10] Ott, J., Maximum likelihood estimation by counting methods under polygenic and mixed models in human genetics, Amer. J. Hum. Genet. 31:161-175 (1979).

BIOSTATISTICS: Statistics in Biomedical, Public Health
and Environmental Sciences, edited by P.K. Sen

EVALUATION OF GENETIC MODELS TO DELINEATE THE MODE OF INHERITANCE OF COMPLEX HUMAN TRAITS

Kadambari K. Namboodiri and Ellen B. Kaplan

Department of Biostatistics
University of North Carolina
Chapel Hill, N.C.

Genetic models to delineate the mode of transmission and some of their recent applications to complex human traits are reviewed to assess the successes and limitations of these approaches to real data. These methods are also applied to a well documented simulated data set from the Genetic Analysis Workshop II to illustrate some of the problems as well as to highlight the usefulness of these applications to improved genotypic resolution and hence risk assessment. It seems reasonable to conclude that until clear biochemical clues on gene action are forthcoming, segregation and linkage analyses will prevail as important tools to gain insights and resolve the mode of inheritance of complex human traits.

KEY WORDS: Genetic models; Segregation, Linkage; Association; Complex traits.

INTRODUCTION

In the field of Public Health, both the geneticist and the epidemiologist strive to improve predictions regarding the health/disease status of the populations and related individuals. In recent decades, with most infectious diseases brought largely under control, there has been greater confluence of interest between these two fields, and an increased awareness of the roles played by genetic factors in many human diseases. This resulted in the emergence of a hybrid field of science - genetic epidemiology - integrating principles from the two major disciplines. The practitioners of genetic epidemiology use appropriate population and family data for genetic screening, delineating modes of transmission, genetic counseling and resolving the determinants of variations. The ultimate aim is to improve prediction and/or early detection of carrier status to facilitate appropriate health care/medical intervention strategies as indicated.

Considering the complexity of the genetic material - the human genome, the search for major genetic determinants in diseases is an arduous task. The estimates of the number of genes in man have been largely by indirect approaches and range from 50,000 to over 200,000 structural genes. McKusick (1983) has catalogued a total of 3368 confirmed or presumptive autosomal and 243 X-linked Mendelizing loci, perhaps only less than 2% of the total number of estimated loci. Among the nearly 1500 loci confirmed to exist on autosomes, chromosome localization and mapping information are available on only about 20%. Thus the human gene map is still a vast uncharted territory and the delineation of genetic factors governing human traits and their mapping, although gravely challenging, are extremely important in accumulating and utilizing basic knowledge for the long-range benefit of our species.

In order to understand the etiology of diseases, one has to view the full spectrum of underlying causation due to genetic and environmental factors, with infectious diseases at the environmental and inborn errors of metabolism at the genetic end, and common diseases such as cardiovascular and neoplastic diseases spanning the middle range. The challenging problem of genetic epidemiology is to devise study designs and analytic methods for the latter type of disorders and traits (i.e. those with multiple etiologies).

The aim of the present paper is to give an overview of the state of the art methodologies in delineating the mode of inheritance of complex traits, summarize some important applications of these methods to real data to assess their advantages and limitations, and examine a set of simulated data critically to illustrate some of the insights gained as well as potential problems with the use of these methods.

OVERVIEW OF METHODOLOGY TO DELINEATE THE MODE OF TRANSMISSION OF COMPLEX TRAITS

All inherited variations - both normal and abnormal - show familial aggregation as well as segregation. The former is a feature of many diseases and could be due to shared genes and/or environment, but segregation is considered to be a powerful tool to delineate the mode of transmission. Classic segregation methods, although extended to accommodate ascertainment problems, new mutations and phenocopies (Elandt-Johnson 1971, Morton 1969) are largely restricted to sibships and are found to be inadequate in dealing with complex traits. Expectations based on precise Mendelian segregation are not generally realized in these complex traits due to (a) lack of one to one correspondence between genotype and phenotype, (b) heterogeneity of the trait with an interplay of more than one factor, genetic and/or environmental, (c) late onset or incomplete penetrance, and (d) poor definition or clinical diagnosis of the disease, or a host of similar problems.

Statistical-genetic models to handle these traits are basically probabilistic representations of complex processes and are designed as analytical tools in interpreting familial resemblance. Two apparently divergent earlier approaches, qualitative traits by Mendelian segregation and quantitative by biometric (correlation-regression type) methods, were unified by Fisher (1918) when he demonstrated that genetic contribution of the latter is compatible with the transmission of many small, independent, additive effects, each controlled by a Mendelizing factor. With the increased availability of powerful computer hardware and software, segregation analyses of both qualitative and quantitative traits of complex etiology have progressed in the last decade to accommodate complicated family structures (i.e. from sibships to multigenerational kindreds) and a variety of refined statistical models. These models reviewed in Elston (1981) incorporate generalized transmission, multiple thresholds, variation in age of onset, segregation distortion and most importantly, major locus transmission with a multifactorial background variation. Thus these are aimed at answering the important question of whether the trait variation is due to a single locus with a large effect, or a number of factors with small effects, or both.

The three widely used approaches in this context are: (a) the generalized major locus (GML) model (Elston and Stewart 1971, Cannings et al. 1978), (b) the multifactorial (MF) model (Falconer 1965), and (c) the mixed (MM) model (Morton et al. 1974, Lalouel and Morton 1981). All three share the basic probability components defining the likelihood of observations, namely

i) $f(a)$ = Pr[random person has genotype a],

ii) $t(c|a,b)$ = Pr[offspring is c|genotypes of parents are a and b], and

iii) $g_a(z)$ = Pr[observing z|genotype is a],

where a, b, and c index the genotypes (1,2,...,k) and z is the phenotype. The first represents the genotypic proportions, the second the transmission rules between generations, and the third a penetrance or risk function. Using these, the likelihood of a random observation, i.e.

Pr[random person has phenotype z] is proportional to $\sum_{a=1}^{k} f(a)g_a(z)$.

Also, assuming that conditional on the genotype, the phenotypes of all family members are independent, the likelihood (L) for a nuclear family with phenotypes z_f, z_m, and z_i, i=1,2,...,s, for the father, mother and the ith child respectively is proportional to:

$$L = \sum_{a=1}^{k} f(a)g_a(z_f) \sum_{b=1}^{k} f(b)g_b(z_m) \prod_{i=1}^{s} \sum_{c=1}^{k} t(c|a,b)g_c(z_i) \qquad (1)$$

where the summations are over all major locus genotypes (replaced by integrations for polygenotypes) and the product is over all children. Since analytical solutions are difficult, parameter estimation of the model is done by maximizing the log likelihood using numerical iterative techniques, and hypothesis tests are performed using the likelihood ratio criterion.

The operational differences in the three approaches - GML, MF and MM - stem mainly from the complexities introduced in the postulation of the underlying genetic mechanism, the usage of different pedigree structures, the finer details of computer implementation of algorithms and maximization procedures. Each approach has some flexibility in introducing additional complexities but also has its limitations. The salient features of each, the parameters involved, and the usual tests of hypotheses are summarized below and also in Table I.

a. Generalized Major Locus (GML)

Under this model, a large part of the variation in the trait is determined by a major locus with two alleles (say, A,A') and three genotypes (designated here as 1, 2, and 3 for types AA, AA', and A'A' respectively). Then, the parameterization for the three component probabilities in the likelihood formulation are as follows: (i) the genotypic probabilities are expressed in terms of the parameter ψ_i, (i=1-3) with the constraint $\sum\psi_i=1$. Under Hardy-Weinberg equilibrium (HWE), the ψ_i become functions of just one parameter, q, the allele frequency of A, taking values q^2, $2q(1-q)$, and $(1-q)^2$ respectively for the three types with the constraint $\psi_2 = 2\sqrt{\psi_1\psi_3}$; (ii) the transmission rules are parameterized in terms of three transmission probabilities or τ_i, (i=1-3). Thus τ_i is the conditional probability that an offspring receives an A allele, given the parent has the ith genotype. The model allows the transmission probabilities to be arbitrary, but under Mendelian transmission rules the τ's take on values 1, .5 and 0 respectively. No vertical transmission is postulated when all the τ's are forced to be equal; (iii) the third component or the conditional probability/density of the phenotype z, given the ith genotype $g_i(z)$, (i=1-3) is specified by a penetrance/susceptibility function for qualitative traits or a normal density for quantitative traits. Penetrances are usually defined in terms of affection status. Thus for a dichotomous phenotype z (0=normal, 1=affected), a fully penetrant dominant allele A causing the disease, these probabilities $g_i(z)$ are $g_1(1)=1$, $g_2(1)=1$, and $g_3(1)=0$ with complementary probabilities $(1-g_i(1))$ for the normal phenotype. Incomplete penetrance results when $g_i(z)$ take on values different from 0 and 1. When the generally 'normal' genotype manifests affection, such cases are termed sporadics or phenocopies. The prevalence of trait/disease (P), if known, can be incorporated into the model as a function of genotypic frequencies and penetrances (i.e. $P=\sum\psi_i g_i(1)$). For a quantitative phenotype z, the conditional densities or the g-functions are specified in terms of normal probability densities, i.e. $g_i(z) = \phi(\mu_i,\sigma^2)$, where μ_i, (i=1-3) are the genotype specific means and σ^2 the common variance.

The simplest GML model thus has nine parameters: ψ_i, τ_i, and g_i, (i=1-3) (see also Table I). More parameters are allowed to specify sex-specific penetrances or mean values and other complications in the model. For instance, when the families are not randomly sampled, but selected via probands, the likelihood of the sample is adjusted in ways such as taking into account the sampling frame for probands and possible probands (Elston and Sobel 1979a). The ascertainment probability π can be jointly estimated in the model. The GML also has the advantage of extending the formulation to complex pedigree structures and multigenerational families.

The fit of the model to the data is assessed by maximizing the log likelihood LL (equation (1)) of the pedigree with respect to the unknown parameters. Tests of hypotheses are performed by computing likelihoods for the general (LL_g) and also a number of restricted hypotheses (LL_r). Twice the difference in the likelihoods, $2(LL_g-LL_r)$, is asymptotically distributed as a χ^2, with the degrees of freedom corresponding to the number of independent constraints imposed by the hypotheses. Some of the common hypotheses tested are: $\tau_1=1$, $\tau_2=.5$, $\tau_3=0$ or Mendelian transmission; $\tau_1=\tau_2=\tau_3=\tau$ or no vertical transmission; $\psi_2=2\sqrt{\psi_1\psi_3}$ or H-W Equilibrium; and $g_1=g_2$ or $g_3=g_2$ for dominant or recessive mode of inheritance.

b. Multifactorial Model

This model assumes the trait (for e.g. a quantitative measure x) to be determined by the combined action of numerous genetic (g) and nongenetic (e) factors (i.e. $x=g+e$), the genetic effect being due to a large number of unlinked loci, each with equal and additive effects. No dominance or epistatic effects are assumed. For a dichotomous trait (for e.g. x=1 for affected, 0 for unaffected), a normally distributed underlying liability due to genetic and nongenetic factors is assumed, with the affection manifested when the liability exceeds a certain threshold. The affection status is thus modelled using a risk function approach. The exact form of the risk function is arbitrary (Elston 1981) and it defines the probability that an individual of a given genotype is affected.

Under such a model, the three component probabilities for the likelihood formulation are normal densities as shown in Table I. The first component, the population genotypic distribution, is regarded as effectively continuous due to the large number of loci (i.e. polygenotypes) involved and normally distributed with mean equal to the polygenotype value and variance σ_g^2. The second component, the distribution of the offspring conditional on parental polygenotypes, is also a normal density with mean equal to the mid-parental value and variance $\sigma_g^2/2$. The third component, or the penetrance function, $g_A(z)$, is also a normal density with mean μ and variance σ_e^2 due to nongenetic factors. The heritability of the trait (h^2) is then defined as the variance ratio: $\sigma_g^2/(\sigma_g^2 + \sigma_e^2)$. Analogous to expression (1), the likelihood under MF can be written for a nuclear family with phenotypes z_f, z_m and z_i, i=1,2,...,s for father, mother and ith child as:

$$L = \int_{-\infty}^{\infty} \phi(A,\sigma_g^2)g_A(z_f) \int_{-\infty}^{\infty} \phi(B,\sigma_g^2)g_B(z_m) \prod_{i=1}^{s} \int_{-\infty}^{\infty} \phi(C-(A+B)/2,\sigma_g^2/2)\ g_C(z_i)dCdBdA \quad (2)$$

where A, B and C index the polygenotypes. The likelihood is maximized to estimate the heritability (h^2) and to test its significance. Although a measure of the genetic involvement in the determination of that trait can be obtained using this approach, it will not lead to insights on any specific gene actions.

c. <u>Mixed Model</u>

This model assumes the the phenotype z results from the additive contributions of a major locus effect m, a multifactorial transmissible effect g, distributed $\sim N(0,\sigma_g^2)$ and a random nontransmissible effect e, distributed $\sim N(0,\sigma_e^2)$, such that z=m+g+e. Assuming independence of effects, the overall variance $\sigma^2=\sigma_m^2+\sigma_g^2+\sigma_e^2$, where σ_m^2 is due to the variation of the major locus. The three basic component probabilities in the likelihood formulation are analogous to expression (1) and are essentially derived from the corresponding components in GML and MF and are given in Table I. Thus likelihood (L) for a nuclear family with phenotypes z_f, z_m and z_i, i=1,...,s, major genotypes indexed a, b and c, and polygenotypes indexed as A, B and C for the father, mother and child respectively is proportional to:

$$L = \sum\psi_a \int_{-\infty}^{\infty} \phi(A,\sigma_g^2)\ g_{aA}(z_f) \sum\psi_b \int_{-\infty}^{\infty} \phi(B,{}_g^2)\ g_{bB}(z_m) \prod_{i=1}^{s} \sum t(c|a,b) \int_{-\infty}^{\infty} \phi(C-(A+B)/2,\ \sigma_g^2/2)$$

$$g_{cC}(z_i)\,dCdBdA \qquad (3)$$

where summation is over major genotypes, integration over polygenotypes and product over all s children. The g-functions use probability densities for quantitative trait and risk functions as in MF for qualitative traits.

Lalouel et al. (1981) used reparameterization of the model in their expositions and computer implementations of MM. These are briefly, the gene frequency q of the high valued or deleterious gene (r) at the major locus (frequency (1-q) for its allele r'), distance in standard deviation units between the means of the two major locus homozygotes ($\mu_{rr}-\mu_{r'r'}$), termed displacent parameter t, degree of dominance d, defining penetrance determined as $(\mu_{rr'}-\mu_{r'r'})/t$, with d=0 for recessive and d=1 for dominance of the high valued homozygote and incomplete dominance for other values of d. The heritability (h^2) is parameterized as σ_g^2/σ^2. Appropriate changes are made in the formulations for qualitative traits. HWE is generally assumed and if the families are selected via probands, the probability of ascertainment is assumed to be known. Recently, MM has been unified with GML allowing for tests of departures from Mendelian transmission (Lalouel et al. 1983a).

The method is applied mostly to nuclear families with information about ascertainment incorporated through probands or pointers. The basic parameters of the model besides the overall mean and variance of the trait are q, t, d and h^2 with τ_i recently added to the list. The hypothesis tests follow the likelihood ratio principles, but here major gene is detected by rejecting the H_0:q=t=d=0 and multifactorial transmissibility by rejecting H_0:h^2=0. In addition, dominance relationships and departures from Mendelian transmission can also be tested. The likelihood for nuclear families can be specified either as overall joint likelihood or likelihood conditional on the parental phenotypes.

The exact likelihood calculation under the mixed model is formidable and the computational load increases exponentially with increase in sample size such that analyses of large date sets are still impractical. To circumvent these problems, several approximations have been used (Morton et al. 1974, Hasstedt 1982) in the currently computer implemented versions of the method.

d. Linkage Analysis

Segregation analysis can only suggest the presence of a major locus. There is no test, statistical or otherwise, that can provide definitive evidence other than locating the appropriate DNA segment. However, indication of linkage (i.e. greater co-segregation of genes at two loci than expected from independent assortment) is an extremely pursuasive argument for the confirmation of a major locus for a trait. Linkage analysis leads to the construction of a genetic map, where the distance between loci depend both on the extent of physical separation and the rate of occurrence of crossing over of chromosome segments and their recombination (i.e. odd numbers of points of exchange between loci). Linkage may also serve to clarify the mode of inheritance as in the case of hemochromatosis (Kravitz et al. 1979) and ankylosing spondylitis (Kidd et al. 1977).

Either using the parameter estimates from segregation analysis or jointly estimating them, linkage between a trait and a marker can be tested using likelihood ratios. The likelihood (L) is computed generally for a number of recombination values (θ) and the logarithm (to base 10) of the odds (LOD) score for linkage, Z, is defined as $Z(\theta)=\mathrm{Log}(L(\theta=\hat{\theta})/L(\theta=.5))$, contrasting the likelihood at a given θ to $\theta=.5$ or no linkage. Since a priori, the chance of two loci being on the same chromosome is low because of the large number (22 pairs of autosomes), the significance levels are held by convention to be conservative; $Z \geq 3$, or odds 1000:1 favoring linkage is assumed to be conclusive evidence for linkage while $Z < -2$ (odds 100:1 against linkage) is held as proof against tight linkage between loci. The efficiency of linkage analysis depends on the mode of inheritance and tightness of linkage between loci and also the family structures used. Operationally, linkage analyses have been extended from sibships to multigenerational pedigrees and loci with multiple alleles by Ott (1974) using the Elston and Stewart (1971) algorithm.

At present there are about 30-40 polymorphic blood markers available for linkage analysis. Elston and Lange (1975a) estimate under certain simplifying assumptions that the prior probability of finding a linkage between a trait and any one of 30 markers is approximately .5. With over 200 DNA restriction fragment length polymorphisms brought to light in recent years, linkage may be a powerful tool in delineating and confirming the mode of inheritance and also aiding prediction of genotypes of complex traits and locating them on the genome.

Association of a disease/complex trait to an established genetic marker also suggests genetic involvement in the etiopathogenesis of the trait. The association may be due to linkage disequilibrium as in many Human Leukocyte Antigen (HLA)-associated diseases or epistasis. However, unless the marker itself is clearly of etiologic significance, the interpretation of the association in genetic terms will be unclear as in many HLA associated diseases. The exception is the case of ankylosing spondylitis where Kidd et al. (1977) have utilized the information on association in gaining a clearer understanding of the genetics of the disease.

APPLICATION OF THE GENETIC MODELS TO REAL DATA

In recent years, many complex traits have been investigated by the methods reviewed above and it is appropriate at this time to evaluate the performance of these methods in gaining greater insights to the mode of transmission of these traits. Consistency of the findings for the same trait in different populations using diverse methods adds strength to the arguments and detection of linkage with a putative major gene virtually confirms its presence.

A summary of recent applications of GML, mostly on multigenerational families, is given in Table II. Among the quantitative traits studied, the two successes were for hypercholesterolemia (Elston et al. 1975b) and hemochromatosis (Kravitz et al. 1979), both delineated as major genes and corroborated by linkage with genetic markers. Among the others, some were suggestive of specific modes of inheritance, but had no success in finding linkage and for others the results were inconclusive, with competing hypotheses giving similar likelihoods.

For the discrete traits/diseases, the findings were mostly equivocal. Incorporation of the age of onset information, sex differences in susceptibility and complex modes of ascertainment correction did not improve the situation. The exceptions are retinitis pigmentosa favoring x-linked inheritance and ataxia determined by a dominant gene linked to HLA. Panic disorder was also suggestive of a dominant mode of inheritance, but additional data analyzed using the same techniques (KKN, personal communication) failed to corroborate the earlier claims. Thus, in general, genetic analysis of qualitative traits had not met with much success.

Some of the applications of mixed models to quantitative traits are given in Table III. In most instances, the data analyzed were nuclear families. As model discrimination depends on gene frequencies and displacement of means in standard deviation units, some parameter estimates are also given to illustrate their spread. Fairly consistent results are obtained from different samples for cholesterol, corroborating the genetic mechanism revealed by linkage and also by biochemical approaches (Goldstein and Brown 1973). Both major locus and multifactorial background were significant in many traits. Traits which showed inconsistent results were triglycerides and IgE. A summary of the application of MM to discrete traits is given in Table IV. Genetic heterogeneity is well established for diabetes and even with evidence of association of insulin dependent diabetes with HLA, its mode of inheritance was unclear. For multiple sclerosis also, a range of estimates gave equally good (or poor) fit to the data. With mental disorders the experience was no different and in most instances, when competing models gave equal fit, the parsimonious multifactorial model was suggested as the likely mechanism.

Thus, while an impressive list of complex traits have been subjected to segregations analysis and some important breakthrough has emerged for some quantitative traits, the approach in general has not been very fruitful with discrete traits. In view of the associated complications of trait definition, difficulties in diagnosis, variable ages of onset, possible genetic heterogeneity, reduced penetrance and occurrence of sporadic cases, this is not very surprising.

APPLICATIONS OF SEGREGATION AND LINKAGE ANALYSES TO SIMULATED DATA

One of the concerns that surfaced with the analysis of real data was the contradictory findings on the same trait such as IgE or congenital glaucoma by different investigators using different computer-implemented packages of some of the basic models. This concern prompted the initiation of a series of annual Genetic Analysis Workshops sponsored by the National Institutes of Health and held at the annual American Society of Human Genetics meetings, aimed at comparison and quality control of the applications of these heavily computer-oriented methodologies. As part of the workshop, family data were simulated under specific conditions and distributed to participants and the results of the different groups are compared and discussed at the annual meeting of the Society. These data have limited scope in assessing the power and robustness of these methods, since that would require extensive simulation experiments with hundreds of replicates. However, these well publicized data are an important resource, liable to be used widely for illustrative purposes and validation of computer packages. Hence it is important to document some of the associated problems and highlight some of the resolutions achieved with the genetic analysis of these data.

A. Materials and Methods

The details of the simulation procedures and checks on the accuracy of simulations are given in MacCluer et al. (1984). Here we focus on the second data set of Workshop II, consisting of 60 pedigrees and 606 individuals ascertained through affected persons aged 30-50 and extended using a sequential sampling scheme (Cannings and Thompson 1977). The dichotomous phenotype was determined on the basis of whether or not the underlying liability fell above or below a threshold, the liability modelled as the sum of the effects of a major locus, 32 polygenic loci and random environment. The major locus with near recessive mode of inheritance was closely linked to two markers and with significant association with one of the linked markers. Because of the sampling scheme, a mixture of pedigree structures (i.e. 20, 64 and 16 percents with two, three and more than three generations respectively) were involved. The population prevalence, the probability of ascertainment and the sampling frame were provided, but the participants were blinded as to the true mode of inheritance and linkage relationships when the data were distributed.

The segregation analyses were performed using GML as implemented in the package GENPED (Kaplan and Elston, unpublished program documentation) and MM using the program POINTER (Lalouel and Morton 1981). The linkage screening was done using LIPED (Ott 1974) and for the linked locus, linkage disequilibrium estimation was performed using GENPED. Since the ascertainment correction implemented for intact pedigrees in GENPED is inappropriate for sequential sampling, the data were analyzed in a variety of ways: (1) Intact pedigress with every one designated as a possible proband in the sampling frame; (2) Intact pedigrees with only those aged 30-50 designated as possible probands in the sampling frame; (3) Pedigrees trimmed in such a way that it is not extended beyond the first degree relatives of the proband; (4) Pedigrees split into nuclear families; (5) Pedigrees split into nuclear families ensuring each family having either a proband or a 'pointer'. For #1 and 2, there were no changes in the number of families or individuals, but for #3 the data comprised only 503 individuals, for #4, 145 nuclear families with 728 individuals of whom 36 were pointers to families without probands. Data sets 1-4 were used with GML, #5 with MM, and #1 with linkage analyses.

The ultimate aim of genetic analysis is the improved classification of genotypes and since the correct classification is available for this data, this offered an opportunity not only for the usual parameter estimation and hypothesis tests for segregation and linkage, but also to compute the genotypic probabilities of individuals given varying levels of information and assess their relative resolution. The latter was done as follows: Under a specified model, the likelihood of all other persons in the pedigree can be described as the sum of three terms corresponding to the designated person being one of the three possible genotypes, i.e. the conditional likelihood $L(g_i)$ of the data, given that person has specific genotype g_i, is obtained by replacing that individual's summation in the likelihood (for e.g. expression (1)) by a single term corresponding to a specific g_i. Then the probability that the designated person has genotype g_i, conditional on what is known about himself and/or his relatives, is $L(g_i)/\sum L(g_i)$, where the summation is over the three genotypes of the individual. Since these conditional likelihoods are proportional to the genotypic probabilities of the individual, this procedure is used to compute the genotypic distribution of individuals, given the model and various levels of information such as (a) only phenotype information, (b) only pedigree information, (c) phenotype and pedigree information, (d) pedigree, phenotype and marker linkage information, and (e) all the information in (d) as well as information about linkage disequilibrium. These are then graphically represented using triangular coordinate systems.

B. Results

1. Segregation and linkage analyses. The results of segregation analyses for different pedigree structures and hypotheses are summarized in Table V. The specification of sampling frame including all ages or only ages 30-50 gave similar results and among Mendelian hypotheses, codominant and recessive modes gave better fit than dominant hypotheses. Hence only the results of the parsimonious recessive model are presented in the table. Panels A-C give the results with different pedigree structures (intact, trimmed and nuclear families respectively) using GENPED. For intact pedigrees, Mendelian hypothesis was rejected and no transmission accepted, suggesting no major gene involvement. The analyses also suggested significant departure from the assumption of HWE. For trimmed pedigrees, even though the ascertainment correction was more appropriate, both Mendelian and no transmission hypotheses were rejected. In contrast, for nuclear families, there was suggestion of a major locus with the rejection of the no transmission model and a better fit for the Mendelian hypothesis. This illustrates how different inferences are possible from the same basic data using different data structures when the structures themselves are determined by the sampling frame. The wide range of estimates of "sporadics" in these analyses (about 40% for intact to 7% in nuclear families), also provides a glimpse of how familial resemblance is accommodated with different pedigree structures and model restrictions.

Panel D of the table gives the results using POINTER. There was no significant multifactorial contribution to the trait and the results favored a major locus with recessive mode of inheritance. The estimates of genotypic proportions and penetrances were very similar in panels C and D, attesting the comparability of the two program packages, even with the use of different likelihoods (joint with the former and conditional with the latter) and maximization procedures. The "true" estimates are also listed (panel E) for comparison.

Linkage analysis using the segregation parameter estimates from nuclear families showed the trait to be linked with three loci - F, G, and B. The disease allele X1 was found to be in linkage disequilibrium with the G locus with the parameter estimates for Δ (deviation from expected haplotype frequency) being X1G1=.01, X1G2=.05, and X1G3=-.06.

2. Comparison of genotypic distributions. A graphic representation of the estimated genotypic probabilities of individuals of the sample, grouped on the basis of their "true" genotypes X1X1 (affected), X1X2, or X2X2 (normal) and using different levels of genetic information, is given in Figure 1. In the figure, in an equilateral triangle with an altitude of unity, each point can be regarded as depicting an individual, with the perpendicular distances from the point to the three sides of the triangle summing up to unity and representing the genotypic probabilities of that individual. Using such a coordinate system, let the lower left corner of each triangle represent the affected genotype X1X1 (i.e. the perpendicular distance to the opposite side is equal to the altitude, unity); the lower right corner represent the heterozygote, X1X2; and the apex, the normal homozygote X2X2. Under this scheme, points near the center of the triangle represent individuals equiprobable of being any of the three gneotypes, while those closer to their respective corners indicate sharper genotypic resolution. Thus with no misclassification, all X1X1 would be crowded to the lower left, X1X2 to the lower right, and X2X2 to the apex of the triangle.

Subjects were grouped into their "true" genotypes X1X1, X1X2 and X2X2 in columns one to three respectively in Figure 1. Note that overprinting is allowed and therefore while some points represent single individuals, others may represent several subjects. Each row depicts a different level of genetic information used in the genotypic classification. Thus row (a) uses familial information of each subject (but not his phenotype information). This is analogous to the risk estimates given for a consultand of unknown phenotype (like an unborn child) based only on the family information. Without the individuals' phenotype information, few individuals were classified as X1X1 and misclassification appears to be rampant. Row (b) gives the genotypic distribution using both pedigree and phenotype information. Better resolution than (a) is evident for all genotypes, especially for X2X2. Row (c) depicts the genotypic distribution when, in addition to segregation at the major locus, linkage information with the marker locus G is also incorporated into the computations. The resolution appears to have improved for both X1X1 and X1X2 with a tendency for aggregation to their respective corners. Additional information on marker locus and major locus association and hence linkage disequilibrium is shown in row (d). While this improved the classification of some individuals, others were worse than in the previous case for both X1X1 and X1X2. Thus, these diagrams illustrate the use of genetic-epidemiologic techniques in providing sharper genotypic resolution and at the same time the imperfections in these approaches and the need for more precise tools for unequivocal classifications of individuals.

D. Discussion

Several investigators have examined the power, the robustness and other properties of GML and MM (Go et al. 1978, MacLean et al. 1975, Goldin et al. 1981, Reich et al. 1981) and found desirable features in both, as well as limitations. The former as presently implemented cannot distinguish between major gene and polygenic transmission. Under GML, Mendelian transmission is accepted by default if not rejected when tested against a much broader alternative. Since the acceptance can also result from insufficient information, it is recommended that the no transmission model be rejected as well. The transmission model is flexible, permitting resolution of environmentally induced transmission not confounded with polygenic mechanism and allowing genotypic distributions to be different among generations, and thus detecting segregation distortion and possible heterogeneity. The unified model (Lalouel et al. 1983a) combines the desirable properties of GML and MM, mainly for nuclear families. More experience is needed to assess its performance. However, the present analyses illustrate that different conclusions can be reached with slightly different treatments of the same basic data and bring to focus the pressing need to further investigate the sensitivity and behavior of transmission parameters, especially with regard to different pedigree structures, and misspecification of sampling.

The multifactorial background was not a significant factor in these data and therefore the very similar results using different models and program packages were gratifying. The contrast between GML and MM application would have been pronounced in the presence of significant multifactorial transmission since such a component in the former would be interpreted as major gene segregation or it may affect the transmission probabilities to such an extent as to mask the evidence of the major gene. However, any polygenic mechanism which gives rise to a non-normal distribution will tend to be interpreted as major gene by either model.

The relative advantages of using joint likelihood or likelihoods conditional on parents' phenotypes have been discussed (Rao et al. 1980, Go et al. 1978). Conditional likelihoods are considered to be less sensitive to selection, etiologic heterogeneity and temporal or other trends which tend to induce systematic generational differences or violations in distributional assumptions. In the present analysis for GML we used both joint and conditional likelihoods and the results were very similar.

The current analyses also highlighted the importance of having marker information to improve our knowledge of the genetic mechanism and genotype classifications. Kravitz et al. (1979) and Moll et al. (1984) have used genotypic probability distributions for better prediction and also to reveal heterogeneity among families. In the present data, the additional knowledge of linkage and the linkage disequilibrium enhanced the classification of genotypes, illustrating improved prediction and hence more precise risk assessment with increased knowledge of the biological features of the trait.

In conclusion, segregation analysis is just one tool to understand the mode of inheritance of complex traits. It can seldom give unequivocal results regarding the presence of major genes. The detection of linkage with markers can sharpen the identification of informative families for biochemical analysis and lead to better resolution of the genotype classifications. Thus, although high optimism regarding genetic analyses in providing simple answers to complex questions has to be moderated, until clear biochemical answers are forthcoming, segregation and linkage analyses, in spite of their limitations will prevail as the most valuable tools in gaining insights to the elusive mode of transmission of complex traits.

ACKNOWLEDGEMENT: This work was supported in part by NHLBI contract number N01-HV1-2243-L. The authors wish to thank Dr. Jean MacCluer for providing us with additional information on simulated data sets, Ernestine Bland for typing the manuscript, and Evelyn Smith for technical assistance.

REFERENCES:

[1] Borecki, I.B. and Ashton, G.C., Evidence for a major gene influencing performance on a vocabulary test, Beh. Genet. 14:63-79 (1984).

[2] Blumenthal, M.N., Namboodiri, K., Mendell, N., Gleich, G., Elston, R.C., et al., Genetic transmission of serum IgE levels, Am. J. Med. Genet. 10: 219-228 (1981).

[3] Bucher, K.D., The genetics of manic depressive illness: A pedigree and linkage study. Inst. Stat. Mimeo Series 1141, University of North Carolina, Chapel Hill (1977).

[4] Cannings, C. and Thompson, E.A., Ascertainment in the sequential sampling of pedigrees, Clin. Genet. 12:208-212 (1977).

[5] Cannings, C., Thompson, E.A., and Skolnick, M.H., Probability function on complex pedigrees, Adv. Appl. Prob. 10:26-61 (1978).

[6] Carter, C.L. and Chung, C.S., Segregation analysis of schizophrenia under a mixed genetic model, Hum. Hered. 30:350-356 (1980).

[7] Chung, C.S., Rao, D.C., and Ching, G.H.S., Population and family studies of cleft lip and cleft palate. In: Etiology of Cleft Lip and Cleft Palate, Alan R. Liss, New York, 325-352 (1980).

[8] Crowe, R.R., Namboodiri, K., Ashby, H.B., and Elston, R.C., Segregation and linkage analysis of a large kindred of unipolar depression, Neuropsychobiology 7:20-25 (1981).

[9] Demenais, F., Further analysis of familial transmission of congenital glaucoma, Am. J. Hum. Genet. 35:1156-1160 (1983).

[10] Elandt-Johnson, R.C., Probability Models and Statistical Methods in Genetics, Wiley, New York (1971).

[11] Elston, R.C. and Stewart, J., A general model for the genetic analysis of pedigree data, Hum. Hered. 21:523-542 (1971).

[12] Elston, R.C. and Lange, K., The prior probability of autosomal linkage, Ann. Hum. Genet. 38:341-350 (1975a).

[13] Elston, R.C., Namboodiri, K., Glueck, C.J., Fallat, R., Tsang, R., et al., Study of the genetic transmission of hypercholesterolemia and hypertriglyceridemia in a 195-member kindred, Ann. Hum. Genet. 39:67-87 (1975b).

[14] Elston, R.C., Namboodiri, K., Spence, M.A., and Rainer, J.D., A genetic study of schizophrenia peidgrees. II. One locus hypothesis. Neuropsychobiology 4:193-206 (1978).

[15] Elston, R.C. and Sobel, E., Sampling considerations in the gathering and analysis of pedigree data, Am. J. Hum. Genet. 31:62-69 (1979a).

[16] Elston, R.C., Namboodiri, K., and Hames, C.G., Segregation and linkage analyses of dopamine-β-hydroxylase activity, Hum. Hered. 29:284-292 (1979b).

[17] Elston, R.C., Segregation analysis, Adv. Hum. Genet. 11:63-120 (1981).

[18] Falconer, D.C., The inheritance of liability to certain diseases estimated from the incidence among relatives, Ann. Hum. Genet. 29:51-76 (1965).

[19] Fisher, R.A., The correlation between relatives on the supposition of Mendelian inheritance, Trans. Roy. Soc. (Edinburgh) 52:399-433 (1918).

[20] Go, R.C.P., Elston, R.C., and Kaplan, E.B., Efficiency and robustness of pedigree segregation analysis, Am. J. Hum. Genet. 30:28-37 (1978).

[21] Go, R.C.P., King, M.C., Baily-Wilson, J., Elston, R.C., and Lynch, H.T., Genetic epidemiology of breast cancer and associated cancers in high risk families. I. Segregation analysis, J. Natl. Cancer Inst. 71:455-461 (1983).

[22] Goldin L.R., Elston, R.C., Graham, J.B., and Miller C.H., Genetic analysis of von Willebrand's disease in two large pedigrees: A multivariate approach, Am. J. Med. Genet. 6:279-293 (1980).

[23] Goldin, L.R., Kidd, K.K., Matthysse, S., and Gershon, E.S., The power of pedigree segregation analysis for traits with incomplete penetrance. In: Genetic Research Strategies for Psychobiology and Psychiatry, Gershon, E.S., Matthysse, S., Breakfield, X.O., and Ciaranello, R.D. (eds). Boxwood Press, California, 305-317 (1981).

[24] Goldin, L.R., Gershon, E.S., Lake, C.R., Murphy, D.L., McGinniss, M., et al. Segregation and linkage studies of plasma dopamine β-hydroxylase (DBH), erythrocyte catechol-O-methyl transferase (COMT) and platelet monoamine oxidase (HAO): Possible linkage between ABO locus and a gene controlling DBH activity, Am. J. Hum. Genet. 34:250-262 (1982).

[25] Goldin, L.R., Gershon, E.S., Targum, S.D., Sparkes, R.S., and McGinniss, M., Segregation and linkage analysis of families of patients with bipolar, unipolar and schizo affective mood disorders, Am. J. Hum. Genet. 35:274-287 (1983)

[26] Goldstein, J.L. and Brown, M.S., Familial hypercholesterolemia: Identification of a defect in the regulation of 3-hydroxy-3-methyl-glutaryl-coenzyme, A reductase associated with over-production of cholesterol, Proc. Natl. Acad. Sci. 70:2804-2808 (1973).

[27] Greenberg, L.J., Bradley, P.W., Chopyk, R., Lalouel, J.M., Immunogenetics of response to purified antigen from group A streptococci, Immunogenetics 11: 145-160 (1980).

[28] Haile, R.W., Iselius, L., Hodge, S.E., Morton, N.E., and Detels, R., Segregation and linkage analysis of 40 multiplex multiple sclerosis families, Hum. Hered. 31:252-258 (1981).

[29] Hasstedt, S.J., A mixed model likelihood approximation for large pedigrees, Comp. Biomed. Res. 15:295-307 (1982).

[30] Hasstedt, S.J., Wilson, D.E., Edwards, C.W., Cannon, W.N., Carmelli, D., et al., The genetics of quantitative plasma Lp(a): Analysis of a large pedigree, Am. J. Med. Genet. 16:179-188 (1983a).

[31] Hasstedt, S.J., Meyers, D.A., and Marsh, D.G., Inheritance of immunoglobulin E. Genetic model fitting, Am J. Med. Genet. **14**:61-66 (1983b).

[32] Iselius, L., A major locus for hyper-β-lipoproteinemia with xanthomatosis, Clin. Genet. **15**:530-533 (1979).

[33] Iselius, L., Complex segregation analysis of hypertriglyceridemia, Hum. Hered. **31**:222-226 (1981).

[34] Kidd, K.K. and Spence, M.A., Genetic analysis of pyloric stenosis suggesting a specific maternal effect, J. Med. Genet. **13**:290-294 (1976).

[35] Kidd, K.K., Bernoco, D., Carbonara, A.O., Daneo, V., Steiger, U., et al., Genetic analysis of HLA associated disease: The illness susceptible gene frequency and sex ratio in ankylosis spondylitis. In: HLA and Disease, Dausset, J. and Svejgaard, A. (eds). Munksgaard, Copenhagen, 72-80 (1977).

[36] Kravitz, K., Skolnick, M., Cannings, C., Carmelli, D., Baty, B., et al., Genetic linkage between hereditary hemochromatosis and HLA, Am. J. Hum. Genet. **31**:601-619 (1979).

[37] Lalouel, J.M., Morton, N.E., and Jackson, J., Neural tube malformation: Complex segregation analysis and calculation of recurrence risks, J. Med. Genet. **16**:8-13 (1979).

[38] Lalouel, J.M. and Morton, N.E., Complex segregation analysis with pointers, Hum. Hered. **31**:312-321 (1981).

[39] Lalouel, J.M., Rao, D.C., Morton, N.E., and Elston, R.C., A unified model for complex segregation analysis, Am. J. Hum. Genet. **35**:816-826 (1983a).

[40] Lalouel, J.M., Darlu, P., Henrotte, J.G., and Rao, D.C., Genetic regulation of plasma and red blood cell magnesium concentrations in man II. Segregation analysis, Am. J. Hum. Genet. **35**:938-950 (1983b).

[41] MacCluer, J.W., Falk, C.T., Spielman, R.S., and Wagener, D.K., Genetic Analysis Workshop II: Analysis of disease-marker associations, Genet. Epidemiol 1: (in press, 1984).

[42] MacLean, C.J., Morton, N.E., and Lew, R., Analysis of family resemblance. IV. Operational characteristics of segregation analysis, Am. J. Hum. Genet. **27**:365-384 (1975).

[43] McKusick, V.A., Mendelian Inheritance in Man. Sixth edition, Johns Hopkins University Press, Baltimore (1983).

[44] Mendell, N.R., Blumenthal, M., Amos, D.B., Yunis, E.J., and Elston, R.C., Ragweed sensitivity: Segregation analysis and linkage to HLA-B, Cytogenet. Cell Genet. **22**:330-334 (1978).

[45] Moll, PP, Berry, T.D., Weidman, W.H., Ellefson, R., Gordon, H., et al., Detection of genetic heterogeneity among pedigrees through complex segregation analysis: An application to hypercholesterolemia, Am. J. Hum. Genet. **36**:197-211 (1984).

[46] Morton, N.E., Segregation Analysis. In: Computer Applications in Genetics, Morton, N.E. (ed). University of Hawaii Press, Honolulu, 129-139 (1969).

[47] Morton, N.E. and MacClean, C.J., Analysis of family resemblance III. Complex segregation of quantitative traits, Am. J. Hum. Genet. 26:489-503 (1974).

[48] Morton, N.E., Gulbrandsen, C.L., Rhoads, G.G., Kagan, A., and Lew, R., Major loci for lipoprotein concentrations, Am. J. Hum. Genet. 30:583-589 (1978).

[49] Morton, N.E., Genetics of hyperuricemia in families with gout, Am. J. Med. Genet. 4:103-106 (1979).

[50] Namboodiri, K.K., Elston, R.C., and Hames, C., Segregation and linkage analyses of a large pedigree with hypertriglyceridemia, Am. J. Med. Genet. 1:157-171 (1977).

[51] Nino, H.E., Noreen, H.J., Dubey, D.P., Resch, J.A., Namboodiri, K.K., et al., A family with hereditary ataxia and HLA type, Neurology 30:12-20 (1980).

[52] O'Rourke, D.H., McGuffin, P., and Reich, T., Genetic analysis of manic depressive illness, Am. J. Phy. Anthro. 62:51-59 (1983).

[53] Ott, J., Estimation of the recombination fraction in human pedigrees. Efficient computation of the likelihood for human linkage studies, Am. J. Hum. Genet. 26:588-597 (1974).

[54] Pauls, D.L., Bucher, K.D., Crowe, R.R., and Noyes, R., A genetic study of panic disorder pedigrees, Am. J. Hum. Genet. 32:639-644 (1980).

[55] Rao, D.C., Lalouel, J.M., Morton, N.E., and Gerrard, J.W., Immunoglobulin E revisited, Am. J. Hum. Genet. 32:620-625 (1980).

[56] Reich, T., Rice, J. and Cloninger, C.R., The detection of a major locus in the presence of multifactorial variation. In: Genetic Research Strategies for Psychobiology and Psychiatry, Gershon. E.S., Matthysse, S., Breakfield, X.O., and Ciaranello, R.D. (eds). Boxwood Press, California, 353-367 (1981).

[57] Rice, J., McGuffin, P., Goldin, L.R., Shaskan, E.G., and Gershon, E.S., Platelet monoamine oxidase (MAO) activity: Evidence for a single major locus, Am. J. Hum. Genet. 36:36-43 (1984).

[58] Siervogel, R.M., Elston, R.C., Lester, R.H., and Graham, J.B., Major gene analysis of quantitative variation in blood clotting factor X levels. Am. J. Hum. Genet. 31:199-213 (1979).

[59] Simpson, S.P. and Morton, N.E., Complex segregation analysis of the locus for beta amino isobutyric acid excretion (BAIB), Hum. Genet. 59:64-67 (1981).

[60] Spence, M.A., Elston, R.C., and Cederbaum, S.D., Pedigree analysis to determine the mode of inheritance in a family with retinitis pigmentosa, Clin. Genet. 5:338-343 (1974).

[61] Vance, J.M., Pericak-Vance, M.A., Elston, R.C., Connealy, P.M., Namboodiri, K.K., et al., Evidence of genetic variation for α-N-acetyl D glucosaminidase in black and white populations, Am. J. Med. Genet. 7:131-140 (1980).

[62] Williams, W.R. and Lalouel, J.M., Complex segregation analysis of hyperlipidemia in a Seattle sample, Hum. Hered. 32:24-36 (1982).

[63] Zavala, C., Morton, N.E., Rao, D.C., Lalouel, J.M. et al., Complex segregation analysis of diabetes mellitus. Hum. Hered. 29, 325-333 (1979).

FIGURE 1: Graphic presentation of the estimated genotypic probability distribution of simulated pedigree data, grouped by their "true" genotypes, columnwise, X1X1, X1X2, X2X2 respectively, using triangular coordinates. With no misclassification, types X1X1 should be clustered in the lower left corner, X1X2 in the right corner, and X2X2 at the apex of the triangle (see text). Rows a-d depict the relative genotypic resolution with increasing levels of genetic information: a) segregation analysis with phenotype held unknown, b) using pedigree and phenotype information, c) using segregation and linkage information with marker G, and d) using all of the above and information on linkage disequilibrium.

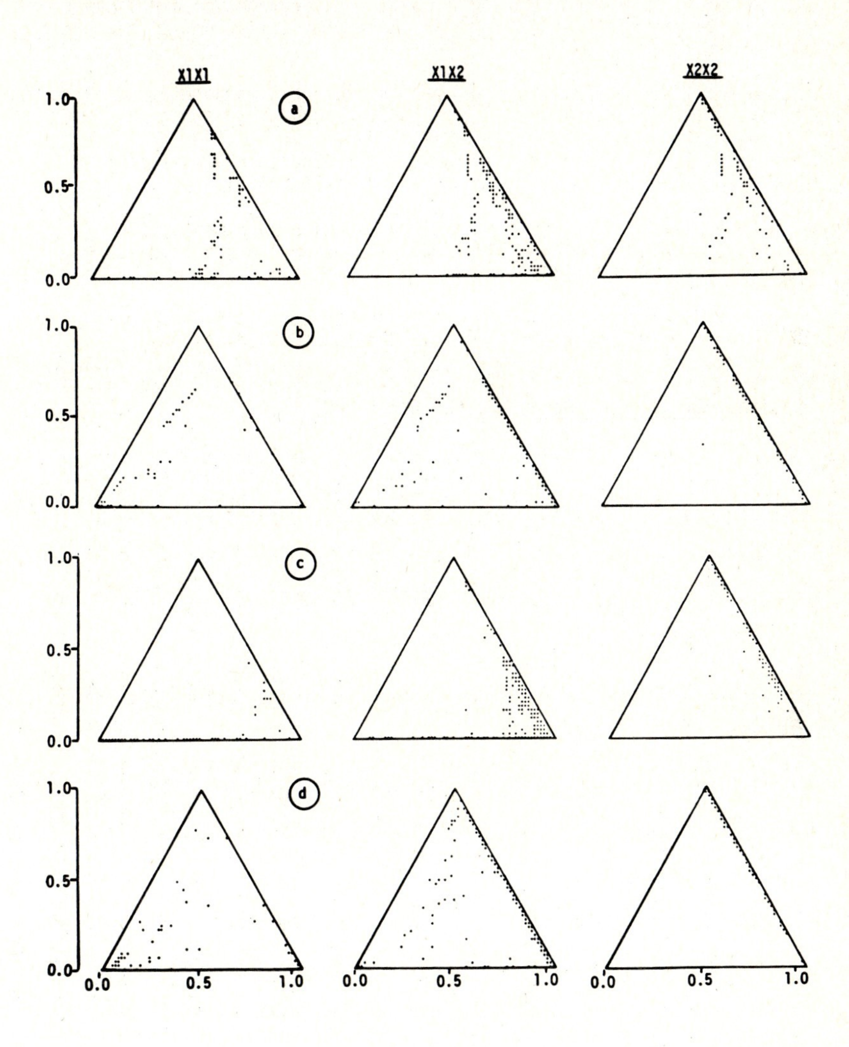

Table 1. Summary of the Component Probabilities in the Likelihood Formulations for Three Widely Used Approaches of Segregation Analyses of Quantitative Traits[1]

I. Component Probabilities	MODELS: Generalized Major Locus (GML)	Multifactorial (MF)	Mixed Model (MM)
(i) Genotypic probabilities: $f(a)$, $f(A)$, or $f(aA)$	ψ_a	$\phi(A,\sigma_g^2)$	$\psi_a\int_{-\infty}^{\infty}\phi(A,\sigma_g^2)$
(ii) Transmission: $t(c\|a,b)$, $t(C\|A,B)$ or $t(cC\|aA,bB)$	P_{abc}	$\phi(C-(A+B)/2,\sigma_g^2/2)$	$P_{abc}\int_{-\infty}^{\infty}\phi(C-(A+B)/2,\sigma_g^2/2)$
(iii) Penetrance functions $g_a(z)$, $g_A(z)$ or $g_{aA}(z)$	$\phi(z-\mu_a,\sigma_e^2)$	$\phi(A+\mu-z,\ \sigma_e^2)$	$\phi(A+\mu_a-z,\ \sigma_e^2)$
II. Parameters[2]	$\psi_i,\tau_i,\mu_i,(i=1,\ldots,k),\sigma_e^2$	$\mu,\ h^2$	$\psi_i,\tau_i,\mu_i,(i=1,\ldots,k)h^2$

[1]Likelihood (L) for a nuclear family with phenotypes z_f, z_m, and z_i, $i=1,2,\ldots,s$, major genotypes indexed by a, b, c for the father, mother and ith child respectively is proportional to $L=\sum f(a)g_a(z_f)\sum f(b)g_b(z_m)\prod_{i=1}^{s}\sum t(c|a,b)g_c(z_i)$ for the GML; for MF, the major genotypes are replaced by polygenotypes A,B,C and summations by integrations (equation (2) in the text); derivation for MM is analogous by combining GML and MF (equation (3)), ϕ denotes the normal density function, σ_g^2, σ_e^2 are the variances due to genes and environment respectively, μ or μ_i are the overall or genotype-specific mean, V the overall variance and h^2 is the heritability (i.e. $\sigma_g^2/(\sigma_g^2+\sigma_e^2)$). The terms P_{abc} represent the appropriate elements in the genetic transition matrix, parameterized in terms of the transmission probabilities (τ_i's).

[2]For alternate parameterization of MM, see text.

Table II. Examples of Complex Traits Analyzed by Generalized Major Gene Model*

Trait	Likely Mode of Inheritance†	Comments	Source
A. Quantitative			
Cholesterol	Dominant	Linkage to C3	Elston et al. (1975b)
Triglycerides	Dominant	Linkage not found	Namboodiri et al. (1977)
Dopamine-β-hydroxylase	Recessive Codominant	Linkage not found; Possible linkage to ABO	Elston et al. (1979b) Goldin et al. (1982)
Ragweed antigen E sensitivity	Dominant	Linked to HLA	Mendell et al. (1978)
Immunoglobulin E	Inconclusive	-	Blumenthal et al. (1981)
Clotting factor X	Recessive	-	Siervogel et al. (1979)
Clotting: von Willebrand disease	Dominant	Analysis multivariate, no conclusive linkage	Goldin et al. (1980)
Hemochromatosis	Recessive	Linked to HLA	Kravitz et al. (1979)
Catechol-O-methyl transferase	Codominant	Linkage not found	Goldin et al.(1982)
Monoamine oxidase	Inconclusive	Linkage not found	Goldin et al. (1982)
N-acetyl-D-glucosaminidase	Codominant	Racial difference in polymorphism	Vance et al. (1980)
B. Qualitative			
Retinitis pigmentosa	X-linked	-	Spence et al. (1974)
Schizophrenia	Inconclusive	Age of onset used	Elston et al. (1978)
Affective disorders	Inconclusive Inconclusive Inconclusive	Age of onset used Age of onset used Age of onset used	Bucher (1977) Goldin et al. (1983) Crowe et al. (1981)
Panic disorder	Dominant	Age of onset used	Pauls et al. (1980)
Breast cancer	Inconclusive	Age of onset used; Tentative linkage with GPT	Go et al. (1983)
Ataxia tangelica	Dominant	Linked to HLA	Nino et al. (1980)
Ankylosing spondylitis	Dominant	Linked to HLA	Kidd et al. (1977)

*Data consisted mostly of large multigenerational pedigrees.

†All autosomal except for retinitis pigmentosa; the dominance/recessives applicable to the high valued genotype.

Table III. Examples of Application of Mixed Model Segregation Analyses to Complex Quantitative Traits*

Trait or Suspected Disease	Likely Mechanism†	Paremeter Estimates				Study Location/ Features	Source
		q	d	t	h		
Cholesterol	MM	.005	.5	4.55	.43	Honolulu Study	Morton et al. (1978)
		.004	.5	5.55	.53	Seattle Study	Williams et al. (1982)
		.001	.5	6.29	.39	Norwegian Study	Iselius (1979)
		.001	1.0	3.42	.49	Rochester Study	Moll et al. (1984)
LDL Cholesterol	MM	.002	.5	6.17	.39	Honolulu Study	Morton et al. (1978)
Triglycerides	I	.001	.5	3.42	--	Honolulu Study‡	Morton et al. (1978)
		.0	.0	.00	.37		
		.140	.0	1.62	.33	Seattle Study	Williams et al. (1982)
		.220	.0	1.94	.29	Cincinnati Study‡	Iselius (1981)
		.0	.0	.00	.29		
		.430	.5	1.84	-		
Lp[a]	M	.100	.5	1.60	.05	Utah Study	Hasstedt et al. (1983a)
Immunoglobulin E	MM	.460	.0	1.60	.23	Canadian Study	Rao et al. (1980)
	MF	.0	.0	.00	.61	Cancer families	Hasstedt et al. (1983b)
Immune response to streptococcus antigen	M	.40	1.0	1.87	.00	Minneapolis Study	Greenberg et al. (1980)
RBC magnesium level	MM	.23	.31	2.62	.56	Paris Study	Lalouel et al. (1983b)
Monoamine oxidase	MM	.28	.00	2.19	.23	NIH Study‡	Rice et al. (1984)
	M	.25	.31	2.67	.00		
β-amino iso-butryic acid excretion	M	.60	.19	2.09	.08	Japanese Study	Simpson et al. (1981)
Uric acid	MM	.001	.5	4.42	.26	Data from literature	Morton (1979)
Vocabulary test	MM	.68	.31	1.68	.17	Hawaii Study on cognition	Borecki et al. (1984)

*Data mostly nuclear families.
†M = Major gene; MF = Multifactorial; MM = Mixed model; I = Inconclusive
‡Similar likelihoods with different sets of estimates.

Table IV. Examples of Application of Mixed Model Segregation Analysis to Complex Qualitative Traits*

Trait/Disease	Prevalence+	Parameter Estimates				Comments	Source
		q	t	d	h		
Cleft lip/palate†	.0013 M .0007 F	.036 0	2.5 0	0 0	- .72	--	Chung et al. (1980)
Multiple sclerosis†	.0002	.10 0	3.82 0	0 0	- .73	--	Haile et al. (1981)
Diabetes mellitus	.02-.10						
a. Insulin dependent†		.09 0	1.70 0	0 0	.50 .50	--	Zavala et al. (1979)
b. Insulin independent		.20	1.79	0	0	Significant heterogeneity	Zavala et al. (1979)
Glaucoma	.001-.005	.006	4.59	0	0	Recessive with sporadics	Demenais (1983)
Pyloric stenosis	.005 M .001 F	.002	0	0.5	0	Codominant; Sex difference in penetrance	Kidd et al. (1976)
Spina bifida	.002-.029	.004	2.20	1.0	0	Sex differences	Lalouel et al. (1979)
Schizophrenia	.01	.017	2.0	0.0	.62	Wide range of parameters feasible	Carter et al. (1980)
Affective disorder	.03 M .04 F	.03	--	--	--	Major gene with sex-specific penetrances gave better fit than multifactorial model	O'Rourke et al. (1983)

*Mode of inheritance unclear with a range of parameter estimates giving similar likelihoods in most and the parsimonius multifactorial model suggested in such instances.
†Similar likelihoods with different sets of estimates
+M = male, F = female

Table V. Segregation Analysis of Workshop Data: Parameter Estimation and Tests of Hypotheses Using Different Pedigree Structures*

	GENPED										POINTER		
	A Intact Pedigrees				B Trimmed Pedigrees			C Nuclear Families			D Nuclear Families†		E
	Transmission‡				Transmission‡			Transmission‡			Transmission‡		"True"
Parameters§	1	2**	3	4	1	2††	3	1	2‡‡	3	3	1	Estimates
τ_1	0.871	0.499	1.0	1.0	0.323	0.547	1.0	0.931	0.665	1.0	1.0	1.0	--
τ_2	0.328	0.499	0.5	0.5	0.000	0.547	0.5	0.510	0.665	0.5	0.5	0.0	--
τ_3	0.094	0.499	0.0	0.0	0.783	0.547	0.0	0.112	0.665	0.0	0.0	0.0	--
ψ_1	0.815	0.815	0.707	0.722	0.996	0.997	0.718	0.680	0.878	0.592	0.575	0.476	0.6999
ψ_2	0.185	0.181	0.268	0.278	0.004	0.003	0.258	0.289	0.118	0.355	0.367	0.428	0.2743
ψ_3	0.000	0.004	0.025	0.000	0.000	0.000	0.023	0.031	0.004	0.053	0.058	0.096	0.0258
g_1	0.002	0.0	0.031	0.0	0.035	0.035	0.015	0.005	0.031	0.003	0.000	0.009	0.0000
g_2	0.002	0.0	0.031	0.0	0.035	0.035	0.015	0.005	0.031	0.003	0.000	0.009	0.0537
g_3	1.000	0.876	1.000	0.925	0.663	1.000	0.903	1.000	0.995	0.617	0.601	0.279	0.7945
P	.002	.0035	.0552	.0005	.0352	.0352	.0352	.0352	.0352	.0352	.0352	.0352	.0352
χ^2	--	2.14	58.66	8.96	--	27.49	63.56	--	85.76	9.38	2.90	--	--
Df	--	2	3	4	--	2	3	--	2	3	1	--	--
Sporadics¶(%)	--	--	39.70	0.0	--	--	30.59	--	--	6.99	0.0	12.17	0.000

*Log likelihoods for the general hypotheses for the different panels are: A = -216.67, B=-251.58, C=-513.43, and D=-121.25.
†Nuclear families with pointers.
‡1=General, 2=None, 3=Mendelian with HWE, and 4=Mendelian without HWE.
§τ_i, ψ_i, and g_i (i=1-3) are transmission probabilities, genotypic proportions, and penetrances of genotypes respectively; P=Prevalence of the trait.
**Offspring genotype frequencies based on τ are: .249, .500, and .251 respectively.
††Offspring genotype frequencies based on τ are: .299, .474, and .227 respectively.
‡‡Offspring genotype frequencies based on τ are: .442, .446, and .112 respectively.
¶Percent of "normal" homozygotes liable to manifest the trait.

NEUROPHYSIOLOGY

BIOSTATISTICS: Statistics in Biomedical, Public Health
and Environmental Sciences, edited by P.K. Sen

NON-STATIONARY STOCHASTIC POINT-PROCESS MODELS IN NEUROPHYSIOLOGY WITH APPLICATIONS TO LEARNING

Muhammad K. Habib and Pranab K. Sen

Department of Biostatistics
University of North Carolina,
Chapel Hill, N.C. 27514, U.S.A.

The basic role and appropriateness of certain non-stationary stochastic point-process models are appraised for the study of discharge activities of neurons. Measures of association (cross-correlations) of spike trains for two or more neurons are also considered. By incorporating random intensity functions, some doubly stochastic Poisson process representations for the counting processes (and histogram processes) arising in the context of spike trains are studied under appropriate regularity assumptions (quite plausible in experimental situations). Applications to synaptic plasticity as a model for neural learning are also discussed.

KEY WORDS AND PHRASES: Counting process, doubly stochastic Poisson process, martingale estimator, neural learning, neuronal modelling, neurons, point process, stochastic intensity, synaptic plasticity.

I. INTRODUCTION

The main objective of this study is to present a systematic account of some non-stationary stochastic point-process models for the discharge activities of nerve cells (neurons) in the central nervous system (CNS). These discharge activities are in the form of brief impulse-like electrical events which are called action potentials (or spikes). The transmission of trains of action potentials between neurons is one of the primary means of communications in the CNS. In this study spike trains are modeled as realizations of stochastic point processes with random intensities. Non-stationary stochastic models may quite appropriately be incorporated to describe this mode of communication and to draw valid statistical conclusions in an experimental situation. Though different regions of the cortex have been shown to subserve different functional roles (e.g. sensory, motor, association), there are hundreds of thousands of neurons in each of these regions, where the adjacent neurons may not act independently of each other. For example, one neuron might determine (or drive) the activity of another neuron. It is therefore of interest to study the association pattern of spike trains recorded simultaneously for two or more neurons, in an attempt to reconstruct the underlying patterns of connectivity. Such measures of association are studied here. These methods may shed light on important aspects of the functional connectivity of the neurons under study as well as the general properties of the neural network involving the studied neurons. The point processes incorporated in this study are not necessarily stationary. The measures of cross-correlations of such processes are functions of two time-parameters (instead of one time-parameter arising in the case of weakly (second order) stationary processes). For this reason, we work with such cross-correlation surfaces (instead of functions).

It should be noted that the incorporation of non-stationary processes is crucial for the study of the discharge activities of neurons since, in the presence of external stimuli, many classes of neurons (such as the ones in the sensory areas in the cortex) fire in a non-stationary fashion [e.g. Johnson and Swami (1983)]. The stochastic models along with the measures of correlation and cross-correlations may be used to study neural synaptic plasticity (i.e., the change of efficacy of connections between neurons as a result of experience) in neural networks which constitutes an important aspect of the study of the neural basis of learning.

Some basic notions of neurophysiology along with some existing statistical studies of spike discharge activities in the nervous system are briefly discussed in Section 2. Section 3 deals with a specific stochastic point process model for the counting process that represents a spike train. Under the condition that the (random) intensity of the counting process depends solely on initial events (and hence, is independent of the process itself), the counting process is modelled as a doubly stochastic Poisson process. This model extends previously proposed spike train models, most of which rest on the basic assumption of weak stationarity of the spike trains. This stringent assumption is not likely to hold in reality in the presence of external stimuli. Thus, the proposed models appear to have both the mathematical generality and also experimental validity. In Section 4, under a multiplicative structure on the random intensity function, we extend the model in Section 3 to the case where the random intensity of the counting process may depend on the process itself. In Section 5 we discuss some applications of cross-correlation surfaces to the study of neuronal plasticity as a model for learning.

2. POINT-PROCESSES AND NEURAL ACTIVITY

The CNS, which comprises the brain and the spinal cord, may be viewed as a highly complex communication system which receives, codes and processes a staggering amount of information. Some degree of uncertainty is inherent in the behavior of the system because of (among other things) its anatomical complexity, the variety of information it receives and the non-deterministic nature of the response of its components to identical experiences or stimuli. This uncertainity (stochastic variability) is reflected in the characteristics of human behavior, neural networks, and single cell responses.

It is widely accepted that higher brain functions are performed by large neural networks. Analytical as well as experimental models of the operational dynamics of neural networks in the CNS that perform sophisticated tasks (such as perception, discrimination, learning and memory) have been proposed [viz., Sokoloff, 1980]. These tasks are fundamental features of any biological system exceeding a certain level of complexity. To date, there are relatively few studies that genuinely integrate theory and experiment in this area of neurophysiology. The current study is an effort to advance some aspects of the theory of stochastic modelling and inference which is applicable to experimental studies of the fundamental mechanisms crucial to the behavior of neural networks. We introduce, briefly, some of these neurophysiological notations, before presenting a review of the existing literature on quantitative and statistical studies of neural behavior.

2.1 Neurons and Action Potentials. A basic unit of the nervous system which receives and transmits information is the nerve cell or neuron. The neuron [see Figure 1] has three morphological regions: The cell body (or soma), the dentrites and the axon (and axon terminals). The soma contains the nucleus and many of the organelles involved in metabolic processes. The dentrites form a series of short highly branched protoplasmic outgrowths from the cell body. The dentrites and the soma are the sites of most specialized junctions where signals are received from other neurons.

The axon is typically a protoplasmic extension which exits the soma at the initial segment (or axon hillock) and which extends from the soma over a short or a long distance (sometimes more than a meter). Near its end, the axon branches into numerous axon terminals, which are responsible for transmitting signals from the neuron to the cells connected to the axon terminals. The junction between two neurons is called the synapse (see Figure 1). A synapse is an anatomically specialized junction between two neurons where the electrical activity in one neuron influences the activity of the other. Most synapses occur between the axon terminals of one neuron (called the presynaptic neuron) and the cell body or dentritic tree of a second one (called the postsynaptic neuron). A single motor neuron in the spinal cord receives, probably, some 15,000 synaptic junctions while neurons in the brain may have more than 100,000 synapses.

Information is relayed between neurons in several forms. An important form is the action potential or spike. In describing action potentials, we notice that the electrical activity in the nervous system is due to the presence of organic

FIGURE 1. A SCHEMATIC PICTURE OF TWO NEURONS A AND B WITH A CIRCLE AROUND A SYNAPSE BETWEEN AND B , MAGNIFIED ON THE LEFT.(Upper Picture)

Pictures of a Post-Synaptic Potential and of two Action Potentials are given in the lower left and right corners respectively.

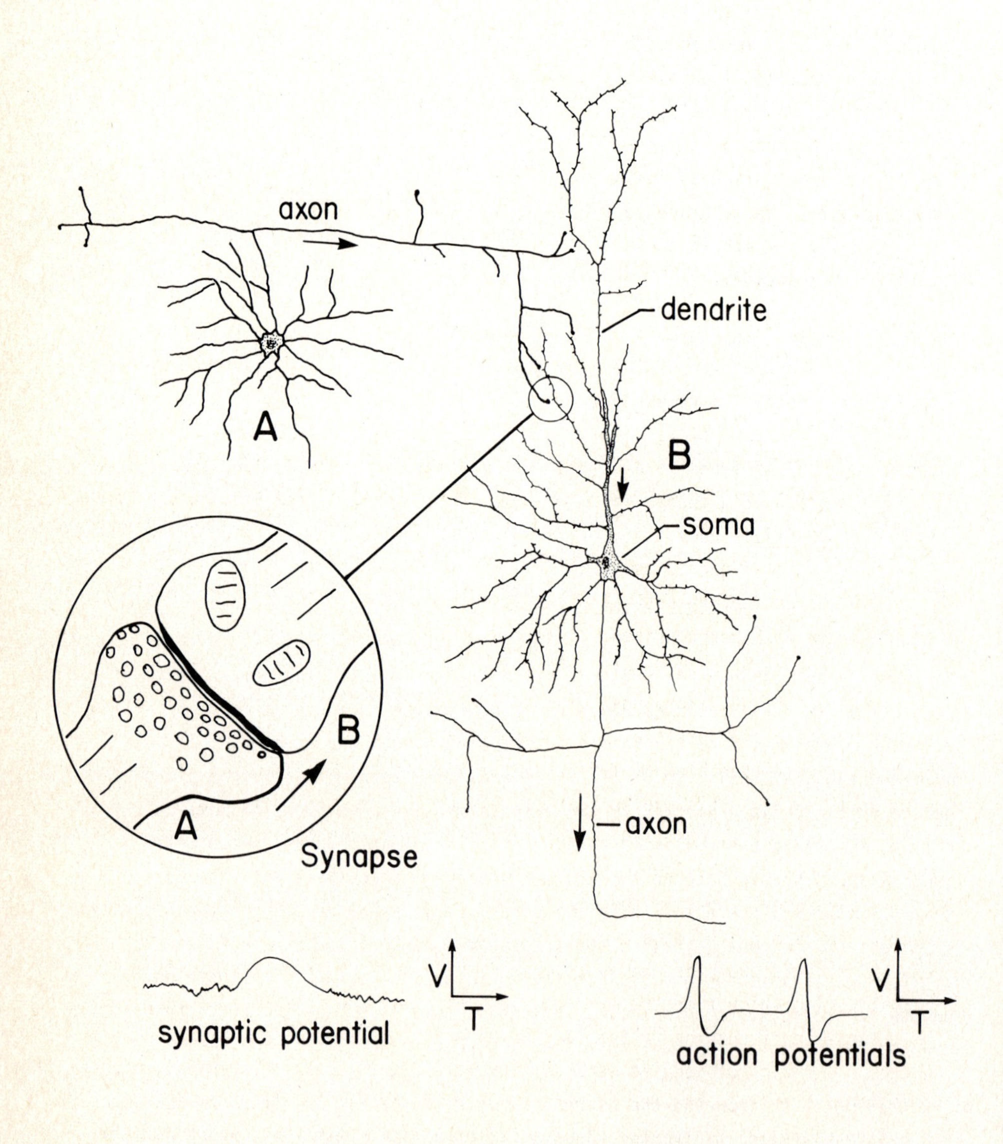

as well as inorganic electrically charged ions. Among the important inorganic ions are sodium (Na^+), potassium (K^+) and chloride (Cl^-). These ions are present outside (i.e. in the extracellular fluid) as well as inside the cell. The cell's membrane is selectively permeable to different ions (permeability is the ease by which ions are transported across a membrane). This leads to a difference in concentration of ions on both sides of the membrane which in turn leads to a difference in potential across the membrane. The transmembrane potential (membrane potential for short) is regulated by, among other things, active as well as passive membrane transport mechanisms (see e.g. Kuffler and Nicholls, 1976), and is defined by convention as inside minus outside potential. In the absence of synaptic events, the membrane potential is kept at a certain level called the resting potential (which is about -60 to -70 m.V., the inside is negative with respect to the outside).

When a (chemical) synapse is activated due to the arrival of action potentials (spikes) along the axon of the presynaptic neuron a chemical substance called neural transmitter is released into the synaptic cleft. The transmitter then crosses the synaptic cleft and combines with the recptor sites of the postsynaptic cell. This leads to a change in the permeability of the postsynaptic membrane and produces a change in potential. This potential change is called postsynaptic potential (PSP). If the PSP results in reducing the potential difference in the post-synaptic membrane (i.e. if the membrane is depolarized) the PSP is called an excitatory post-synaptic potential (EPSP). On the other hand, if the membrane is hyperpolarized, (i.e. the difference in potential across the membrane is increased) as a result of the arrival of the post-synaptic potential, it is called an inhibitory post-synaptic potential (IPSP). Between synaptic events the membrane potential decays exponentially to a resting potential. At a spatially restricted area of the neuronal soma (where the sodium conductance, per unit area, is high relative to that of the remaining somal membrane), the excitatory and inhibitory post-synaptic potentials are integrated. According to this model of neuronal behaviour, the neuronal membrane is sometimes referred to as a leaky integrator. Now, if the membrane potential reaches a level called the neurons threshold, the membrane undergoes a very rapid, transient, and stereotyped change which is known as the action potential. Action potentials may last only 1 m.s. during which time the membrane potential changes from -60 to about +30 m.V. and then it returns to the resting potential. After each action potential there is a period of reduced excitability called the refractory period. At the beginning of this period the neuron cannot fire. This is called the absolute refractory period. Following that period there is an interval during which a second action potential can be produced only when the stimulus is considerably stronger than usual strength. This period is called the <u>relative refractory period</u>.

Action potentials are nearly identical in shape and hence they contain a limited amount of information in their shape of wave form. It is believed then that the temporal patterns of spike trains are the information carriers in many areas in the CNS. The purpose of many studies is to investigate the temporal behavior of spike trains in many areas in the CNS under different controlled conditions; that is, in the presence and absence of external stimuli, and under normal as well as experimentally modified conditions. Next we review some quantitative studies of spike train activity.

2.2 <u>A review of relevant statistical and quantitative neurophysiological studies</u>. Quantitative neurophysiological studies of two or more simultaneously recorded spike trains using measures of cross-correlation and related statistical techniques have proven to be effective in indicating the existence and type of synaptic connections and other sources of functional interaction among the observed neurons. See for example Bryant, Ruiz Marcos, and Segundo (1973), Toyama, Kimura, and Tanaka (1981), Madden and Remmers (1982), and Tanaka (1983).

Spike train data consist of the times of occurrence of the spikes $\{\tau_1, \tau_2, \ldots\}$. Associated with this data is a counting process $N(t)$, where $N(t)$ is the number of spikes in the time interval $(0, t]$. In many quantitative studies of spike train data, a histogram process is used instead of the counting process. The histogram process is formed by partitioning the period of observation $(0, T]$, $T > 0$, into sub-intervals $(t_{k-1}, t_k]$, $k = 1,2,\ldots,q$ of, of equal lengths, i.e. <u>bins</u> of equal lengths $t_k - t_{k-1} = b$. The histogram process is defined by

$$N_k = N(t_k) - N(t_{k-1}) \quad , k = 1,2,\ldots,q.$$

Another important quantitative tool is the <u>post-stimulus time histogram</u> (PSTH). This is formed by repeating the experimental trials m times, say, and adding the number of spikes in corresponding bins with the origin synchronized to the stimulus in each trial. The PSTH, when properly normalized, can be considered as an estimate of the probability of firing during the period of observation.

One of the early studies which discussed the application of the methods of cross-correlation to neurophysiology is by Perkel, Gerstein, and Moore (1967). They discussed several measures of correlation for weakly-stationary spike trains such as the cross-interval histogram, cross-correlation histogram, and the cross-intensity function. For instance, the cross-intensity function of two jointly weakly-stationary spike trains recorded simultaneously from two cells A and B is defined by

$$I_{AB}(u) = \lim_{h \to o} P\{B \text{ spike in } (t, t+h) | A \text{ spike at } t-u\}/h \tag{2.1}$$

The cross-correlation histogram, on the other hand, is defined by

$$\rho_{AB}(k) = E\{[N_\ell^A - \mu_A] [N_{\ell+k}^B - \mu_B]\}/\sigma_A\sigma_B, \tag{2.2}$$

where μ_A is the mean of N^A and $\sigma_A^2 = E(N^A)^2 - \mu_A^2$. μ_B and σ_B^2 are defined similarly. The cross-correlation (3.2) is estimated by

$$\hat{\rho}_{AB}(k) = \frac{n}{(n-1)\ (n-k)} \frac{1}{\hat{\sigma}_A\ \hat{\sigma}_B} \sum_{i=1}^{n-k} (N_i^A - \hat{\mu}_A)\ (N_{i+k}^B - \hat{\mu}_B)$$

$(n-k)^{\frac{1}{2}}\ \rho_{AB}(k)$ is asymptotically normally distributed with mean zero and standard deviation 1 under the null hypothesis. The authors observed that dependencies between two spike trains can arise from one (or both) of two sources: functional interaction or common input. Functional interaction is any mechanism by which the firing of one neuron influences the probability of firing by the other. Such a mechanism could be synaptic (direct or indirect), or due to "field" effects. Common input is any mechanism that simultaneously modulates the firing patterns of both neurons. In addition the authors discussed some of the difficulties of applying the above mentioned statistical techniques in the comparison of two or more simultaneously recorded spike trains. Among them is the difficulty involved with the interpretation of statistical dependency. For instance, it is not always clear whether a given peak (or trough) in the cross-correlation of two spike trains is better explained by synaptic connection between the two observed neurons, mediated perhaps by one or more interneurons, or alternatively by a shared source of input to the two cells. They also discussed briefly complications arising from non-stationarities, focusing only though, on weakly stationary spike trains.

The above mentioned difficulties were addressed in detail by Moore et al. (1970) who provided a theoretical foundation for the interpretation of the cross-correlation histogram in terms of meaningful physiological mechanisms. They discussed how certain elementary prototypical synaptic connections are reflected in the structure of the cross-correlation function and the extent to which an understanding of this relationship can be exploited in the investigation of the structures of small neural networks. They also warned that great care should be taken, though, in drawing physiological conclusions concerning the observation of peaks or troughs in the cross-correlation function, since peaks can arise from monosynaptic excitatory connections and/or from shared excitatory input. Also relatively broad peaks can reflect shared inhibition or shared changes in firing rates (non-stationarity of the spike trains). Similar concerns are valid for the observation of troughs. Several sources of peaks (or troughs) in the cross-correlation histogram were discussed. For instance, direct (or indirect) excitatory (or inhibitory) coupling may account for the appearance of a pronounced peak (or trough), and was termed primary peak (or trough). Secondary peaks, on the other hand, may be due to rhythmicity in the firing patterns of the pre- and/or post-synaptic neurons. For this reason the authors recommend the careful study of the auto-correlation of spike trains of the pre- and post-synaptic cells. This advice has not always been followed. A notable exception is Bryant, Ruiz Marcos, and Segundo (1973) who have successfully applied the theoretical studies mentioned above to their investigation of the firing behavior of small neural networks in *Aplysia Californica*.

Measures of cross-correlation have become quite popular recently in the study of functional connectivity of neurons in many areas of the nervous system. For example, Toyama, Kimura, and Tanaka (1981) employed cross-correlation techniques to study interneuronal connectivity in the primary visual cortex (striate cortex) between cells of several types (i.e. simple, complex, hypercomplex, etc.). They identified three basic types of interneuronal functional connectivity, namely, common excitation, intracortical excitation, and intracortical inhibition. These types of interactions were related to types of cell response, locations of neurons in cortical layers, and location in "ocular dominance and orientation columns." Other interesting studies which employed cross-correlations are Tanaka (1983), and Mickalski et al. (1983). For a review see chapters 6 and 7 of Glaser and Ruchkin (1976).

In an effort to provide a measure of theoretical understanding of the use of the cross-correlation function in studying interaction and synaptic connectivity in small neural networks, Knox (1974) discussed the problem of primary synaptic effect using a simplified model of the nerve cell. In this neuronal model, the somal membrane of the post-synaptic (driven) cell was modeled as a linear time-invariant filter. The neuron is also characterized by: a constant threshold, resetting to resting level (i.e., the refractory period is not taken into consideration), and excitatory post-synaptic potentials only. The pre-synaptic spikes were assumed to arrive at random over a single excitatory channel and are idealized as Dirac delta functions, $X(t) = \sum_k \delta(t-\tau_k)$, where $\{\tau_k\}$ are the instances of arrivals of presynaptic action potentials. Following a resetting to zero at time t_o, the membrane potential, Y, is modeled by

$$Y(t) = \int_{t_0}^{t} X(\tau)h(t-\tau)\, d\tau$$

$$= \sum_k h(t-\tau_k) \qquad , t_0 < \tau_k$$

as long as Y(t) remains less than the constant threshold Θ, where h(t) is the impulse response function of the filter. The author considered both the case of the perfect integrator (i.e., the membrane adds up the synaptic input linearly and without decay in the absence of synaptic events) as well as the case of the leaky integrator (where the membrance potential decays exponentially between synaptic inputs). The cross-correlation function of pre- and post-synaptic spike trains was derived from a linear system as a convolution of the input auto-correlation function with the impulse response function of the system. The cross-correlation was also derived when the input X(t) was a Poisson process. For the case of the leaky integrator the formula of the cross-correlation was slightly more complicated. Knox concluded that the cross-correlation for short lags is related to the derivative h'(t), while for large lags it is related to the auto-correlation function of the input process. See also Knox and Poppele (1977, a and b).

Brillinger, Bryant, and Segundo (1976) applied an identification method of point processes (Brillinger 1974, 1975) to a large data set of spike trains recorded from a neural network in *Aplysia*. The neurons were connected by monosynaptic inhibitory or excitatory post-synaptic potentials and either discharged spontaneously or were driven by intracellular current pulses. Brillinger et al. (1976) developed models for the cross-intensity functions defined by (2.1) above. The cross-intensity function of the spike trains of two cells A and B, $I_{AB}(u)$, is expressed as a functional series, using "kernels," which is a point process analog of the Volterra-Wiener expansion of ordinary time series (Marmarelis and Naka, 1973 a,b). The kernels are expressed as functions of time arguments, and are meant to be invariant of the system, i.e., they retain the same essential characteristics even when the pre-synaptic discharges vary and the kernels are estimated from the corresponding pre- and post-synaptic spike trains. The first-order kernel a(u) describes the effect of a single pre-synaptic spike (or PSP). It is defined as the best linear predictor of the average change in the instantaneous rate at time u in the spike train of the cell B, when a single spike occurs at time 0 in a spike train of the cell A. It is useful, therefore, in predicting whether there will be a B spike u units of time away from an A spike. Brillinger et al. derived the following representation of the cross-intensity function I_{AB}

$$I_{AB}(u) = m_B + a(u) + \int a(u-v)[I_{AA}(v)-m_A]dv \tag{2.3}$$

where $-\infty < u < \infty$, $I_{AA}(u)$ is the auto-intensity function of A, and the constants m_A and m_B are the overall mean rates. The model (2.3) was then extended to include higher order kernels.

The first step of the identification procedure is to estimate certain conditional rate functions each of which is physiologically relevant to synaptic transmission. The one of zero-order μ, i.e., a constant, simply measures the mean rate. The one of first-order (a function of a single time argument), relates the average effect of a pre-synaptic spike and the PSP it elicits. The one of second-order (a function of two time arguments) relates the interactions between pairs of spikes or PSP's, and so on for higher order ones. The second step is to recursively construct the successive models, i.e., that based on the zero-order kernel, that based on the zero- and first-order, that based on the zero-, first- and second-order, and so on. The coherence provides some measure of predictability. A practical difficulty is that all lower order kernels must be recomputed each time a higher order term is added.

As we have previously stated, the above studies are based on the assumptions that the point processes representing the spike trains are weakly-stationary. In addition, the processes were also assumed to be jointly weakly-stationary. It is well documented though that many spike trains are neither weakly-stationary nor jointly so. For example, Corria and Landolt (1977) in a study of the spontaneous activity of anterior semicircular canal afferents that spike activity of 57.5% of the studied units were non-stationary as determined by the Wald-Wolfowitz runs test for trends. It is reasonable, then, to expect that lack of stationarity will be even more pronounced in stimulus driven activity. Furthermore, the assumption of joint weak-stationarity implies that the magnitude of the correlation between the number of spikes (of two simultaneously recorded histogram processes) in two bins with lag k, say, stays constant over the observation period. This assumption then does not allow for the consideration of well established synaptic properties such as synaptic potentiation. This roughly means that the synaptic efficacy of a synapse is enhanced in response to the effect of the action potentials arriving along the pre-synaptic terminals. Synaptic potentiation, in particular long-term potentiation, is widely accepted as a model for neural learning. Therefore, non-stationary models are particularly suited for studies which address alterations in synaptic functioning or efficacy between neurons (see Gerstein, 1970). For a recent review of long-term potentiation see Teyler and Discenna (1984).

Recently, Johnson and Swami (1983) proposed a non-stationary point-process model to describe the spike discharges of single auditory nerve fibers, and discussed the theoretical limitations of the proposed model. The counting process in their model possesses a random intensity of the multiplicative type. The authors exploited this model to study the effect of the refractory period on the PST histogram. This study, though, is concerned with single unit behavior. For other interesting single unit studies, see Corria and Landolt (1977), Yang and Chen (1978), and De Kwaadsteniet (1982). In the following section we propose a non-stationary point-process model for spike-train data.

3. NON-STATIONARY HISTOGRAM-PROCESS MODEL

For spike trains for which the associated counting process may not be stationary (even weakly), a point process model is developed. The cross-correlation function for the associated histogram process may then be expressed as a function of two parameters (instead of the usual one parameter), and the estimates of this correlation function may reveal important information on the dynamics of the synaptic connection between the two observed neurons.

First, we introduce the basic notions. All the random variables (r.v.) and processes to be discussed are defined on some probability space (Ω, F, P). Let τ_i, $i \geq 1$, be the occurrence times of action potentials (spikes). Statistically, we have a realization of a point process on $R^+ = [0,\infty)$. The distribution function (d.f.) $F_k(t) = P\{\tau_k \leq t\}$ assumed to have a continuous probability density function (p.d.f.) $f_k(t)$, $t \in R^+$, $k \geq 1$. A realization of this point process on R^+ can be described by the sequence $\{\tau_k, k \geq 0\}$ (where $\tau_0 = 0$) i.e.,

$$0 = \tau_0 < \tau_1 < \tau_2 < \ldots < \tau_n < \ldots\ldots, \tag{3.1}$$

so that $\tau_k < \infty \Rightarrow \tau_k < \tau_{k+1}$, for every $k \geq 0$. We assume that the realizations of the point process are non-explosive, so that

$$\tau_\infty = \lim_{k\uparrow\infty} \tau_k = +\infty \quad \text{a.s.} \tag{3.2}$$

A complete probabilistic description of $\{\tau_k, k \geq 1\}$ is given by the sequence of conditional probabilities:

$$\tilde{F}_k(t) = P\{\tau_k \leq t \mid \tau_j, j < k\}, k \geq 1. \tag{3.3}$$

Also, (3.1) can equivalently be described by a counting process

$$N(t) = \begin{cases} k, & \text{if } t \,\varepsilon[t_k, \tau_{k+1}) \\ \infty, & \text{if } t > \tau\infty \end{cases}, t\varepsilon R^+ \qquad (3.4)$$

where $N(t) = \sum_{k\geq 1} I(\tau_k \leq t)$, I(a) being the indcator function of the set A. Thus, the sample paths of N(T) are right-continuous step-functions, jumps are upwards of unit magnitude, and $N(0) = 0$ a.s. $\{N(t), t\varepsilon R^+\}$ is called the counting process associated with point process representing the spike train. In practice, we may need to consider a finite domain [0,T] ($T<\infty$) and the contraction $N([0,T]) = \{N(t), t\varepsilon[0,T]\}$.

Consider the probability space (Ω, F, P) and let $\{H_t, t\varepsilon[0,T]\}$ be a non-decreasing family of sub σ-fields of F (i.e., $H_s \subset H_t$, for every s<t). H_t represents the information collected during [0,t], and the family $\{H_t\}$ is called a history. Let $\{X(t), t\varepsilon[0,T]\}$ be a stochastic process defined on (Ω, F, P). The family of sub-σ $\{H_t^X, t\varepsilon[0,T]\}$ generated by the sample paths X(s), s ε[0,t] (i.e., $H_t^X = \sigma\{X(s)$, $0 \leq s \leq t$) is called the internal history of the process $X([0,T]) = \{X(t), 0\leq t\leq T\}$. A family $\{H_t, t\varepsilon[0,T]\}$ is called a history of X([0,T]) if $H_t^X \subset H_t$, $t\varepsilon[0,T]$ and, in this case, X(t) is said to be adapted to H_t. For example, if S is a r.v. representing information concerning the stimulus, and N(t), $\{\tau_k\}$ are defined as in (3.4) and (3.1), then a history H_t of N(t) is

$$H_t = \sigma\{N(s), 0<s<t; \tau_1, \ldots, \tau_{N(t)}; S\}, \; t\varepsilon[0,T] \qquad (3.5)$$

Let $\{M_t, t\varepsilon R^+\}$ be a stochastic process, and let $\{F_t \; t\varepsilon R^+\}$ be an increasing family of σ-fields. Then $\{M_t\}$ is said to be an F_t-martingale if

$$\text{(i) } E|M_t| <\infty, \text{ for all } t\varepsilon R^+ \text{ and (ii) } E(M_t|F_s) = M_s \text{ a.s., for all } s \leq t \qquad (3.6)$$

If the = sign in (ii) be replaced by $\geq$ (or $\leq$), then $\{M_t\}$ is called a sub-(or super-) martingale.

Note that for an integrable r.v. M, whenever $\{F_t\}$ is $\uparrow$, on letting $M_t = E\{M|F_t\}$, $t\varepsilon R^+$, we have $\{M_t\}$ a martingale closed on the right by M. Similarly, for any convex function g(.), whenever g(M) is integrable $M_t^* = E\{g(M)|F_t\}$ is a sub-martingale, while for a concave g(.), this is a super-martingale.

Now consider the counting process $\{N(t), H_t, t\varepsilon[0,T]\}$, $\{H_t\}$ being the history. If $EN(t) <\infty$, then we have a sub-martingale process where by the Doob-Meyer decomposition, we have

$$N(t) = \Lambda(t) + M(t), \text{ for all } t\varepsilon[0,T] \qquad (3.7)$$

where $\Lambda = \{\Lambda(t), t\{[0,T]\}$ is a right-continuous, H_t -predictable, increasing stochastic process and $M = \{M(t), t\varepsilon[0,T]\}$ is a zero-mean H_t-martingale. Intuitively, H_t-predictable means that knowing the family of σ-fields $\{H_s, s<t\}$ one can predict the value at time t, exactly. Also, notice that though N(t) is not H_t-predictable, $N(t-) = \ell im_{s\uparrow t} N(t)$ is so. We assume that Λ(t) is absolutely continuous, so that there exists an H_t-predictable, positive, stochastic process $\lambda = \{\lambda(t), t\varepsilon[0,T]\}$ such that for every $t\varepsilon[0,T]$,

$$\Lambda(t) = \int_0^t \lambda(u)du \text{ a.s.,} \tag{3.8}$$

and $\Lambda(T)<\infty$, λ is called a random intensity of the counting process N([0,T]). The reason for this terminology is that under some mild regularity conditions (e.g. λ is dominated by an integrable r.v.),

$$\lambda(t+) = \ell im_{h\downarrow 0} \{h^{-1}E[\{N(t+h)-N(t)\}|H_t]\} \text{ a.s.,} \tag{3.9}$$

for every $t\varepsilon[0,T]$; see Aalen (1978). Λ(t) is called the compensator of N(t). If N(t) is a homogeneous Poisson process with parameter λ, then (3.7) reduces to: $N(t) = \lambda t + \{N(t) - \lambda t\}$, $t \geq 0$, so that the compensator $\Lambda(t) = \lambda t = \int_0^t \lambda \, du$. Note that the probability law for the process N([0,T]) (and consequently of $\{\tau_k, k \geq 1\}$) is defined (uniquely) by Λ. Note also that by (3.3) and (3.7), for every $k \geq 1$,

$$\Lambda(t) = \int_{\tau_{k-1}}^t \{1-\tilde{F}_k(s-1)\}^{-1} d\tilde{F}_k(s), \quad \tau_{k-1} \leq t < \tau_k; \tag{3.10}$$

see for example, Shirayayev (1981) and Gill (1983), among others. The following definition characterizes doubly stochastic Poisson processes.

Let $\{N(t), t\varepsilon R^+\}$ be a point (counting) process adapted to $\{H_t\}$, and let $\lambda = \{\lambda(t), t\varepsilon R^+\}$ be a non-negative measurable process, such that

$$\lambda(t) \text{ is } H_0\text{-measures, } t \geq 0, \tag{3.11}$$

$$\int_0^t \lambda(s)\, ds < \infty \quad \text{a.s. (P), } t \geq 0 \tag{3.12}$$

If, for all $0 \leq s \leq t$ $(<\infty)$ and all real u,

$$E\{e^{iu[N(t)-N(s)]}|H_s\} = \exp\{(e^{iu} - 1) \int_0^t \lambda(s)ds\}, \tag{3.13}$$

then N(t) is called a (P, H_t)-doubly stochastic Poisson process. Notice that (3.11) and (3.13) imply that for all $0 \leq s \leq t$, $N(t)-N(s)$ is independent of H_s, given H_0, and further, for every $k \geq 0$,

$$P\{N(t) - N(S) = k|H_s\} = \frac{1}{k!}(\int_s^t \lambda(u)du)^k \exp(-\int_s^t \lambda(u)du) \tag{3.14}$$

When $\lambda(t)$ is a deterministic function, $N(t)$ is just a non-homogeneous Poisson process. If $\lambda(t) = \lambda$ is random, but independent of the time parameter t, $N(t)$ is a conditional Poisson process, so that when λ is just a positive constant, $N(t)$ is the familiar homogeneous Poisson process. In a general setup, however, $\lambda(t)$ (and hence, $\Lambda(t)$) may depend on t in a rather involved manner.

Now, recall the definitions in (3.1) through (3.4). Let us partition $(0,T]$ as $\bigcup_{j=1}^{q} (t_{j-1}, t_j]$, where $t_j = jb$, $j \geq 1$, $T=qb$ and $b=t_{j+1}-t_j$ is the common bin width.
Let $N_0 = 0$, $H_0 = H_{t_0}$,

$$N_k = N(t_k) - N(t_{k-1}) \text{ and } H_k = H_{t_k},\ k=1,\dots,q. \tag{3.15}$$

Then, $\{N_k, 0 \leq k \leq q\}$ is the histogram (counting) process. Several copies of the process $\{N_k^{(i)}, 0 \leq k \leq q\}$, $i=1,\dots,m$ are obtained in order to estimate the parameters reliably as well as to form the post-stimulus time PST histogram. In Section 3.1 we consider the case of independent and identically distributed copies. In Section 3.2 we treat the case of independent but not necessarily identically distributed copies.

3.1 Identically distributed copies. In this section we assume that we have $m(\geq 1)$ independent and identically distribute (i.i.d.) copies $\{N_k^{(i)}, 0 \leq k \leq q\}$, $i=1,\dots,m$ of the histogram process defined with respect to a common Λ (in (3.8)), and we let

$$\Lambda_k = \int_{t_{k-1}}^{t_k} \lambda(u)du, \text{ for } k=1,\dots,q. \tag{3.16}$$

Then, the likelihood function is given by

$$\begin{aligned} L_m(\Lambda_1,\dots,\Lambda_q) &= \prod_{i=1}^{m} \prod_{k=1}^{q} P\{N_k^{(i)} = n_{ik}|H_{k-1}^{(i)}\} \\ &= \prod_{i=1}^{m} \prod_{k=1}^{q} \{\frac{1}{n_{ik}!} \Lambda_k^{n_{ik}} e^{-\Lambda_k}\}, \end{aligned} \tag{3.17}$$

so that

$$\frac{\partial}{\partial \Lambda_k} \ell og\ L_m = \Lambda_k^{-1} \sum_{i=1}^{m} N_k^{(i)} - m, \ (1 \leq k \leq q) \tag{3.18}$$

and this leads to the maximum likelihood estimator (MLE)

$$\hat{\Lambda}_k = m^{-1} \sum_{i=1}^{m} N_k^{(i)} = \bar{N}_k, \ 1 \leq k \leq q. \tag{3.19}$$

Note that (3.11), implicit in the above derivation, amounts to assuming that the intensity process $\lambda(t)$ of the counting process $N(t)$ depends only on events at time t_0 (=0), along with the stimulus, but not on events after the beginning of the trial, i.e., $\lambda(t)$ is H_0-measurable and is independent of $N([0,T])$. In many neurophysiological studies, such a model seems appropriate, and there may be some genuine reasons to regard $\lambda(t)$ as a positive r.v. rather than a deterministic function. In this sense, the estimators $\bar{N}_k = \hat{\Lambda}_k$, $k \geq 1$, in (3.12), actually estimate the parameters $\Lambda_k^o = E\Lambda_k$, $k \geq 1$, without assuming a homogeneous model.

Let us proceed on to the study of the association pattern of spike trains recorded simultaneously for two or more neurons. For two cells, say, A and B, let $\{(N_{K,A}^{(i)}, N_{K,B}^{(i)}), 0 \leq k \leq q\}$, $i=1,\ldots,m$ be i.i.d. copies of the simultaneously recorded spike trains. Then, one may consider a cross-covariance function

$$\sigma_{AB}(\ell,k) = E\{N_{\ell,A}\, N_{\ell+k,B}\} - E\{N_{\ell,A}\}E\{N_{\ell+k,B}\} \tag{3.20}$$

for all permissible (ℓ,k), such that $0 \leq \ell \leq q$, $0 \leq \ell+k \leq q$ (k may be positive or negative). We may define similarly $\sigma_{AA}(\ell)$ and $\sigma_{BB}(\ell+k)$, and consider the cross-correlation function

$$\rho_{AB}(\ell,k) = \sigma_{AB}(\ell,k)/\{\sigma_{AA}(\ell)\ \sigma_{BB}(\ell+k)\}^{\frac{1}{2}} \tag{3.21}$$

In a homogeneous model, $\rho_{AB}(\ell,k) = \rho_{AB}(k)$, for all ℓ, but, in a non-homogeneous model, (3.21) is defined in terms of the two-dimensional time-parameters (ℓ,k).

Note that for every permissible combination of (ℓ,k), a symmetric, unbiased and optimal estimator of $\sigma_{AB}(\ell,k)$ is

$$\hat{\sigma}_{AB}(\ell,k) = \frac{1}{m-1} \sum_{i=1}^{m} (N_{\ell,A}^{(i)} - \bar{N}_{\ell,A})(N_{\ell+k,B}^{(i)} - \bar{N}_{\ell+k,B}) \tag{3.22}$$

It is possible to define (3.22) in terms of a U-statistics [viz. Hoeffding (1948)] by considering a kernel of degree 2, defined by

$$\phi_1((a,b),\ (c,d)) = \tfrac{1}{2}\ (a-c)\ (b-d), \tag{3.23}$$

so that

$$\hat{\sigma}_{AB}(\ell,k) = \binom{m}{2}^{-1} \sum_{1\le i<j\le m} \phi_1((N^{(i)}_{\ell,A},\ N^{(i)}_{\ell+k,A}),\ (N^{(j)}_{\ell,A},\ N^{(j)}_{\ell+k,B}). \tag{3.24}$$

Note that the kernel ϕ_1 is an unbiased estimator of σ_{AB} (ℓ,k), so that $\hat{\sigma}_{AB}$ (ℓ,k) unbiasedly estimates σ_{AB} (ℓ,k), for every (ℓ,k) and $m\geq 2$. Similarly

$$\hat{\sigma}_{AA}(\ell) = (m-1)^{-1} \sum_{i=1}^{m} (N^{(i)}_{\ell,A} - \bar{N}_{\ell,A})^2 \tag{3.25}$$

$$\hat{\sigma}_{BB}(\ell+k) = (m-1^{-1} \sum_{i=1}^{m} (N^{(i)}_{\ell+k,B} - \bar{N}_{\ell+k,B})^2 \tag{3.26}$$

are both U-statistics and therefore are unbiased estimates of their population counterparts. We consider then the estimator

$$\hat{\rho}_{AB}(\ell,k) = \{\hat{\sigma}_{AB}(\ell,k)\}/\hat{\sigma}_{AA}(\ell)\ \hat{\sigma}_{BB}(\ell+k)\}^{\frac{1}{2}} \tag{3.27}$$

For various plausible combinations of (ℓ,k), these estimators can be incorporated to generate cross-correlation surfaces such as in Figure 2.

A careful study of such a correlation surface may cast light on the dynamic aspects of the synaptic interactions of the observed neurons. For example, in the case of a direct excitatory synaptic connection between the two neurons (A,B), it may be noticed that the correlation ρ_{AB} (ℓ,k) is increasing in ℓ (for a given k), indicating that the synaptic efficacy is increasing during the presentation of the stimulus. This may reflect, for instance, synaptic potentiation as a result of the firing of the pre-synaptic cell. Important aspects of synaptic and neuronal connectivity may thus be studied using cross-correlation surfaces.

For the study of the statistical properties of the estimators in (3.27), we may exploit the basic results of Hoeffding (1948) on U-statistics. As has been noted earlier, for every plausible (ℓ,k), $\hat{\rho}_{AB}$ (ℓ,k) is a function of three U-statistics $\hat{\sigma}_{AB}$ (ℓ,k), $\hat{\sigma}_{AA}$ (ℓ) and $\hat{\sigma}_{BB}$ $(\ell+k)$. The exact distribution theory of these $\hat{\rho}_{AB}$ (ℓ,k) may be quite involved (in a general non-homogeneous model). Nevertheless, for large m, this can be considerably simplified by incorporating the general results of Hoeffding (1948), and, in neurological investigations, usually m can be taken quite large, so that these asymptotic results remain very much applicable.

By virtue of Theorem 7.5 of Hoeffding (1948), we may conclude that for every

FIGURE 2 . A SIMULATION OF A CROSS-CORRELATION SURFACE SHOWING DELAYED EXCITATION.

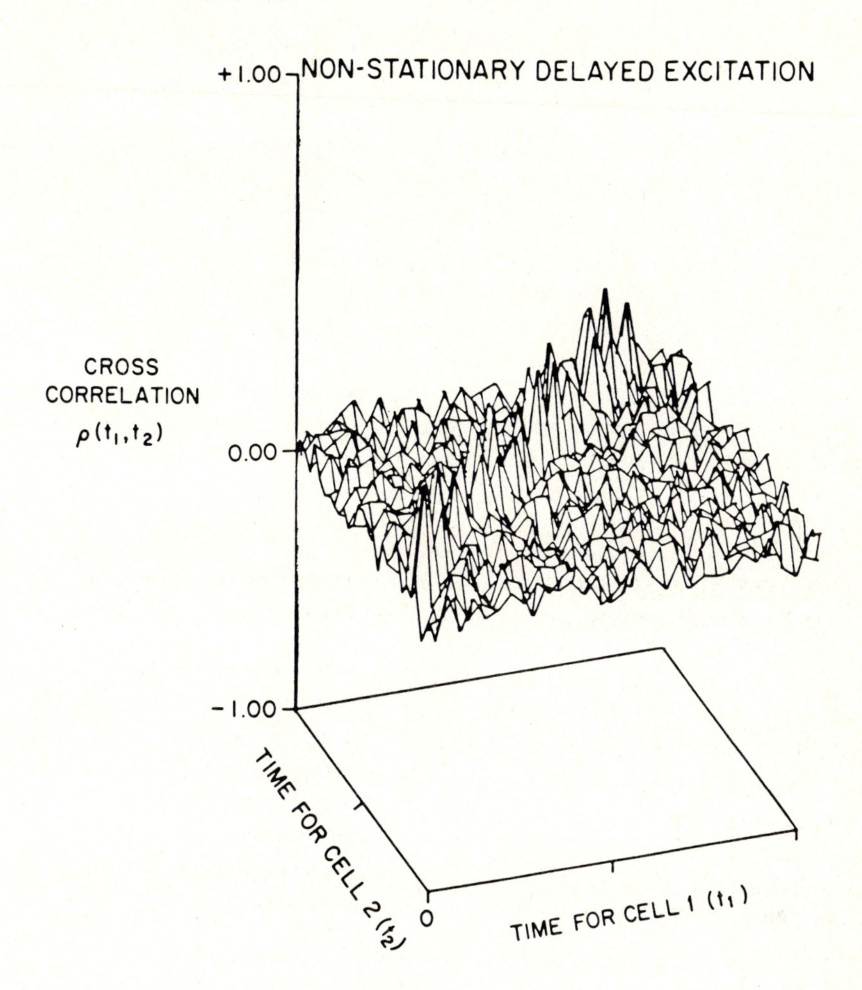

Notice that the magnitude of the cross-correlation changes with time indicating that the spike trains are not jointly stationary. The cross-correlation surface suggests that cell 1 may be driving cell 2 with a certain time delay.

plausible (ℓ,k), as $m\to\infty$,

$$m^{\frac{1}{2}} (\hat{\rho}_{AB}(\ell,k) - \rho_{AB}(\ell,k)) \sim N(0, \sigma^2_{\ell k}) \tag{3.28}$$

where $\sigma^2_{\ell k}$ depends on the underlying probability law and (ℓ,k). In fact, (3.28) naturally extends to the joint asymptotic normality of $(1+q)^2$ elements: $m^{\frac{1}{2}} (\hat{\rho}_{AB}(\ell,k) - \rho_{AB}(\ell,k))$, $0\le\ell\le q$, $0\le\ell+k\le q$. The fact that the dispersion matrix of such a multi-normal distribution is generally unknown, does not raise any serious alarm. For U-statistics, one may use suitable jackknifed estimators [viz., Sen (1960, 1977)] which are (strongly) consistent under quite general regularity conditions. Thus, having obtained such an estimator $\underset{\sim}{V}_m$ of the dispersion matrix, one may readily proceed on to construct suitable (simultaneous) confidence intervals for the $\rho_{AB}(\ell,k)$ or to test suitable hypotheses on these correlations, without necessarily assuming that one has a homogeneous model. For the homogeneous model, one may further simplify the theory by combining the estimates $\{\hat{\rho}_{AB}(\ell,k), \ell=1,....\}$, for a given k, and then studying the pattern for variation on $k(=0, \pm1, \pm2,...)$. We intend to present the technical details of these statistical analyses in a forthcoming article.

3.2 Non-identically distributed copies. In Section 3.1, we considered independent and identically distributed copies of spike trains (i.e., i.i.d. realizations of a point process). It should be noted though that, for neurons with relatively low firing rates (even in the presence of effective or preferred stimuli) it is necessary to have a large number of copies (e.g. in excess of 50 copies) in order to obtain reliable and stable estimates as well as to apply the large sample results discussed above. In this case the response of the neuron under study decreases as the number of presentations of the stimulus increases. This is known as habituation. Formally, habituation is known as a decrease in the "strength" of a behavioral or neural response that occurs when an initially novel effective stimulus is repeatedly presented to the subject. In response to a novel preferred stimulus neurons respond with a relatively high rate of firing. After repeated presentations of the same stimulus the response of the neuron decreases, i.e., it produces fewer spikes per presentation. See e.g. Kandel and Schwartz (1981).

In the light of the above discussion it is clear that, in certain cases, the copies of spike trains obtained by a large number of presentations of the same stimulus may not be identically distributed. It is important then to extend the statistical methods discussed in Section 3.1 to the case of independent but not necessarily identically distributed copies of realizations of the counting process.

In order to accommodate the habituation effect, for the ith copy, in (3.16), we denote the corresponding (random) intensity functions by $\Lambda_k^{(i)}$, for k=1,...,q and i=1,...,m. In this case, $\Lambda_k^{(1)},\ldots,\Lambda_k^{(m)}$ are not generally the same. From what has been discussed earlier, we will expect that in a habituation model,

$$\Lambda_k^{(1)} \geq \Lambda_k^{(2)} \geq \ldots . \geq \Lambda_k^{(m)}, \text{ for every } k=1,\ldots,q \tag{3.29}$$

Thus, one way of extending the MLE in (3.29) to this model would be to use the "isotonic" MLE, whereby we use the MLE of the individual $\Lambda_k^{(i)}$, subject to the restraint in (3.29). However, this would invoke a large class, and it may be difficult to study neatly the large sample properties with so many parameters involved. A more simplifying approach would be to consider a product intensity model:

$$\Lambda_k^{(i)} = \alpha_i \; \Lambda_k^{(0)}, \; i=1,\ldots,m, \; k=1,\ldots,q \tag{3.30}$$

where the α_i are suitable positive constants, and by virtue of (3.29), one may take $\alpha_1 \geq \ldots \geq \alpha_m > 0$. We may consider the model

$$\alpha_i = \exp\{\beta_0+\beta_1 T_i+\ldots+ \beta_r T^r{}_i\}, \quad i\geq 1, \tag{3.31}$$

for some r (≥ 1), where $T_1,\ldots,T_m$ are the time points for the initiation of the stimulus presentations and one may, without any loss of generality, set $\alpha_1=1$, so that $\beta_0+\beta_1 T_1+\ldots+\beta_r T_1^r = 0$. In such a case, one may have natural extension of estimation procedure in (3.19), where we would have r+1 equations (for each k) and, we need to combine the m sets of estimates of the β_j into a combined one, and reiterate to compute the final MLE of Λ_k^o, $1\leq k\leq q$ and the β_j. The cross-correlation study of section 3.1 may then be applied to the estimates Λ_k^o and the α_1 is for the various cells. We may use these estimates to derive estimates of the B_i's, and we shall consider their statistical properties in a forthcoming paper.

4. COUNTING PROCESSES WITH STOCHASTIC INTENSITY

In Section 3, we considered a stochastic model for the counting process N(t) associated with a spike train which was based on the assumption that the intensity process $\lambda(t)$ of N(t) is H_0-measurable. Hence $\lambda(t)$ was assumed to be independent of N(t) and dependent only on initial events including possibly information concerning the stimulus which evoked the discharge activity. These assumptions make N(t) into a doubly stochastic Poisson process. It should be noted that this model is inadequate for modelling systems in which the arrival of an event blanks out the system for a fixed or random period of time during which no arrivals can be observed. That is the model does not take the refractory period into account.

Finally, we notice that in Section 3 we considered only the histogram process which results in loss of information, since only the number of spikes in each bin is considered regardless of the times of arrivals of the spikes within the bin.

In this section we consider more general models for counting processes which avoid the above discussed limitations. In Section 4.1, we consider a doubly stochastic Poisson counting process (rather than the histogram process considered in Section 3). In Section 4.2 we treat a more general model where for a counting process, the (random) intensity may depend on the process itself.

4.1 A Doubly Stochastic Poisson Model. As in Section 3, let $\{N(t), t \in [0,T]\}$ be a counting process adapted to some history $\{H_t, t \in [0,T]\}$, and assume that $N(t)$ possesses a random intensity $\lambda(t)$, $t \in [0,T]$, where for every $t \in [0,T]$, $\lambda(t)$ is H_0-measurable. This implies that $N(t)$ is a doubly stochastic Poisson process. Now, given m i.i.d. copies $\{N^{(i)}(t), t \in [0,T]\}$, $i=1,\dots,m$, we intend to derive the MLE of the cumulative intensity

$$\Lambda(s,t) = \int_s^t \lambda(u)du, \quad 0 \le s < t \le T \tag{4.1}$$

For every (fixed) (s,t), by (3.14), we have the likelihood function

$$L_m[\Lambda(s,t)] = \prod_{i=1}^{m} \{e^{-\Lambda(s,t)}[\Lambda(s,t)]^{N^{(i)}(s,t]}/(N^{(i)}(s,t])!\}, \tag{4.2}$$

where $N^{(i)}(s,t] = N^{(i)}(t) - N^{(i)}(s)$, for all $t \ge s$, $i \ge 1$. Thus,

$$\begin{aligned}(\partial/\partial\Lambda(s,t)) \, \log L_m[\Lambda(s,t)] \\ = -m + (\Sigma_{i=1}^{m} N^{(i)}(s,t])/\Lambda(s,t),\end{aligned} \tag{4.3}$$

so that the MLE $\hat{\Lambda}(s,t)$ is given by

$$\hat{\Lambda}(s,t) = m^{-1} \Sigma_{i=1}^{m} N^{(i)}(s,t) = \bar{N}(s,t], \text{ say.} \tag{4.4}$$

This estimator is very similar to the one in (3.19), though here we do not restrict ourselves to histogram processes.

As in (3.20) - (3.26), we now proceed to study the association pattern of two simultaneously recorded spike trains. Let $\{(N_A^{(i)}(t), N_B^{(i)}(t)), 0 \le t \le T\}$, $i=1,\dots,m$ be m i.i.d. copies of the counting processes associated with simultaneously recorded spike trains of two cells A and B (say), so that as in (3.21), we may now define a cross-correlation surface

$$\rho_{AB}(s,t) = \sigma_{AB}(s,t)/\{\sigma_{AA}(s)\ \sigma_{BB}(t)\}^{\frac{1}{2}} \tag{4.5}$$

for $0 \le s \le t \le T$, where

$$\sigma_{AB}(s,t) = E[N_A(s)\ N_B(t)] - E[N_A(s)]\ E[N_B(t)] \tag{4.6}$$

$$\sigma_{AA}(t) = E[N_A(t)]^2 - E[N_A(t)])^2 \tag{4.7}$$

$$\sigma_{BB}(t) = E[N_B(t)]^2 = E[N_B(t)])^2. \tag{4.8}$$

Parallel to (3.22) - (3.25), we have the estimators

$$\hat{\sigma}_{AB}(s,t) = (m-1)^{-1} \sum_{i=1}^{m} [N_A^{(i)}(s) - \bar{N}_A(s)]\ [N_B^{(i)}(t) - \bar{N}_B(t)], \tag{4.9}$$

$$\hat{\sigma}_{AA}(s) = (m-1)^{-1} \sum_{i=1}^{m} [N_A^{(i)}(s) - \bar{N}_A(s)]^2 \tag{4.10}$$

$$\hat{\sigma}_{BB}(t) = (m-1)^{-1} \sum_{i=1}^{m} [N_B^{(i)}(t) - \bar{N}_B(t)]^2, \tag{4.11}$$

where

$$\bar{N}_A(s) = \frac{1}{m} \sum_{i=1}^{m} N_A^{(i)}(s),\ \bar{N}_B(t) = \frac{1}{m} \sum_{i=1}^{m} N_B^{(i)}(t). \tag{4.12}$$

Parallel to (3.26), we then consider the estimate

$$\hat{\rho}_{AB}(s,t) = \{\hat{\sigma}_{AB}(s,t)\}/\{\hat{\sigma}_{AA}(s)\ \hat{\sigma}_{BB}(t)\}^{\frac{1}{2}} \tag{4.13}$$

The discussions following (3.26) also pertain here. From the statistical point of view, the situation here is more involved than in Section 3. Here we have a two-dimensional (continuous) time-parameter stochastic process $\{\hat{\rho}_{AB}(s,t), 0 \le s, t \le T\}$, and besides, the convergence of finite dimensional distributions (as studied in Section 3), we also need to study the "tightness" of such processes. We intend to take up these technical details in a subsequent paper.

4.2 A Counting-Process Model with Multiplicative Intensity. In this section we consider a more general model for the counting process where the random intensity may depend on the process itself, i.e. $\lambda(t)$ is assumed to be H_t-measurable. We first assume that the intensity is of the multiplicative form and discuss the martingale estimator of Aalen (1978). Then we give an estimator of the cross-correlation surface of the counting processes of two spike trains.

Let $N(t)$, $0 \le t \le T$, be a counting process such that $E[N(t)] < \infty$ for all $t \in [0,T]$, which is adapted to some history H_t, $0 \le t \le T$. As it was stated in Section 3. $N(t)$ is an H_t-sub-martingale. Recall the Doob-Meyer decomposition of $N(t)$:

$$N(t) = \Lambda(t) + M(t) \quad , 0 \le t \le T, \tag{4.14}$$

where $\Lambda(t)$ is a right-continuous, H_t-predictable increasing stochastic process, and $M(t)$ is a (zero-mean) H_t-martingale. Notice that if we only assume that $N(t) < \infty$, a.s., we will still have the decomposition (4.14) with the modification that $M(t)$ is now a local martingale (see e.g. Jacod, 1975). Assume further that $\Lambda(t)$ is absolutely continuous, i.e.

$$\Lambda(t) = \int_0^t \lambda(u)\, du \quad , 0 \le t \le T, \tag{4.15}$$

for some non-negative H_t-predictable process $\lambda(t)$. At this stage, we consider the so-called multiplicative intensity model, where we assume that

$$\lambda(t) = \alpha(t)\, \gamma(t) \quad , 0 \le t \le T, \tag{4.16}$$

where $\alpha(t)$ is a deterministic function and $\gamma(t)$ is a non-negative H_t-predictable process which is assumed to be observable. It should be noted that great care should be taken in modelling $\lambda(t)$ and relating it to important neurobiological properties of the neuron of interest (e.g. anatomical properties, physiological ones such as receptive field properties, etc.).

For the multiplicative model (3.16), Aalen (1978) suggested an estimator of the integrated intensity

$$A(t) = \int_0^t \alpha(u)\, du \tag{4.17}$$

instead of estimating α itself. To motivate for this estimator which is called the martingale (or the Aalen-) estimator, let us rewrite (4.14) as

$$N(y) = \int_0^t \alpha(u)\, \gamma(u)\, du + M(t), \tag{4.18}$$

where (4.15) and (4.16) were used. Now, we formally write (4.18) in the differentiated form

$$dN(u) = \alpha(u)\, \gamma(u)\, du + dM(u) \tag{4.19}$$

In (4.19), $M(u)$ is viewed as "noise", and hence a natural estimator of $A(u)$ is given by

$$\int_0^t \frac{1}{\gamma(u)} \, dN(u)$$

However, no information is gained about α on the (random) set where γ is zero (since no points of N occur there). Therefore, one may estimate the quantity

$$A^*(t) = \int_0^t \alpha(u) \, I \, (\gamma(u) > 0) \, du \tag{4.20}$$

instead of A. Aalen (1978) suggested estimating A^* by the martingale estimator

$$\hat{A}(t) = \int_0^t \frac{I(\gamma(u) > 0}{\gamma(u)} \, dN(u) \tag{4.21}$$

Now to estimate α itself, Ramlau-Hansen (1983) used methods similar to the classic kernel smoothing methods of probability density estimation to recover α from the martingale estimator (3.21). Let K be a bounded function with integral one (K is called the kernel) and b be a positive parameter (the window). The corresponding kernel estimator of α suggested by Ramlau-Hansen is

$$\hat{\alpha}(t) = \frac{1}{b} \int_0^t K(\frac{t-u}{b}) \, d\hat{A}(u) \tag{4.22}$$

If the jump times of N (i.e. times of occurrence of the action potentials) are $\tau_1, \tau_2, \ldots$, then

$$\hat{\alpha}(t) = \frac{1}{b} \sum_{\tau_i} \frac{K((t-\tau_i)/b)}{\gamma(\tau_i)} \, ,$$

which may be used for computational purposes.

Now if we assume that we have m i.i.d. copies of a bivariate counting process $[(N_A^{(i)}(t), N_B^{(i)}(t)), t \, \varepsilon[0,T]\}$, i=1,...,m, we may define the cross-correlations as in (4.13). Asymptotic properties of these cross-correlation surfaces will be studied in a forthcoming paper. On the other hand, let us assume that we have m independent but not necessarily identically distributed copies. This is similar to the situation considered in Section 3.2 where the habituation effect was discussed, In this case, we denote the multiplicative intensity of the ith copy by

$$\lambda^{(i)}(t) = \alpha_i(t) \, \lambda^{(0)}(t)$$

where α_i reflects the effect of habituation on the ith discharge response and $\lambda^{(0)}$ represents an underlying intensity which is a function of certain properties of the neuron of interest as well as information concerning the stimulus evoking the activity. As in Section 3.2, we may assume that α_i, i=1,2,...,m, is a known function which contains a finite number of parameters. The task now would be to estimate those parameters. The discussion then follows the same lines as in Section 3.2.

Before concluding this section, it should be mentioned that Karr (1985) used the method of maximum likelihood to estimate the unknown part of the stochastic intensity for the case of point processes comprising i.i.d. copies of a multiplicative intensity process. It was shown that consistent estimators of the unknown function in the stochastic intensity can be reconstructed using Grenander's method of sieves (Grenander, 1981). For discussion of the martingale method of estimation for point processes see also Jacobsen (1982).

5. APPLICATIONS TO NEURONAL PLASTICITY

In this section we briefly discuss some applications of the analysis of cross-correlation surfaces to studies of synaptic plasticity which is considered as a model of learning in the nervous system (see Kandel, 1976). These applications should lead to the identification of some of the fundamental mechanisms underlying neuronal organization in response to experience. We confine our discussion to two main areas of applications: studies of synaptic connectivity in cortical regions such as the visual cortex (Michalski et al., 1984; Tanaka, 1983), and studies of the relative contribution of excitatory and inhibitory mechanisms in determining the spike discharge behavior of neurons in response to stimuli (Sillito, 1979).

For the study of neuronal plasticity, however, we emphasize that the analysis of cross-correlation surfaces of simultaneously recorded spike trains should be applied to the results of experiments which are conducted systematically and repeatedly during the critical period under normal and experimentally modified conditions. This may lead to identifying the manner in which functional neuronal connectivity changes in response to experience. Similar recommendations concerning the application of cross-correlation techniques in studies of neuronal plasticity has been made by Gerstein (1970, p. 660). For a discussion of applications of these methods to studies of neuronal plasticity in the visual cortex see Habib et al. (1984).

It is well established that the responsiveness of certain types of central and especially cortical neurons to stimuli is highly modifiable by experience during early infancy in animals (e.g. Blakemore and van Sluyers, 1975). The time course of this post-natal period is referred to as the critical period or the period of plasticity. The changes which take place during the critical period persist for varying periods of time according to the type of the cell and the impact of experience as well as its timing and duration. Some changes result in permanent alterations of cellular and synaptic functions.

As mentioned earlier, a systematic analysis of the shape of cross-correlation surfaces of spike trains recorded simultaneously from two or more neurons may reveal some fundamental aspects of the dynamics of changes of functional connectivity in neural networks. These connections may be direct monosynaptic connections, or polysynaptic connections through interneurons, or may reflect the influence of shared synaptic input. This systematic analysis may lead to identifying physical models of functional neuronal connectivity in terms of the types of cells with correlated firing patterns, the types of connections (i.e., excitatory or inhibitory), and magnitudes of time (or synaptic) delays. For example, in a study of cross-correlation of neuronal firing in the cat visual cortex, Michalksi et al. (1983) reported that 61% of neuronal pairs found within an orientation column shared the same input, either excitatory or inhibitory. This type of correlation of firing was found between all combinations of simple and complex cells. Direct connections were found almost exclusively within columns. Tanaka (1983) studied the organization of inputs from the lateral geniculate nucleus to cells in the striate cortex of the cat using cross-correlations between simultaneous extracellular recordings of photically-driven activity. Of 243 pairs of geniculate and striate cells with overlapping receptive fields, 82 showed positive correlations with short delay times (0.9 - 2.7 ms). The delays in 65 of the 82 pairs were short enough to infer that the geniculate cell exerted monosynaptic excitatory influence on the striate cell. It was also found that monosynaptic excitatory input to simple cells originated mostly from X geniculate cells, input to special-complex cells originated from Y geniculate cells, and input to standard-complex cells originated from X and Y cells. The convergence number from geniculate cells to simple striate cells was estimated as more than 10. A larger convergence number (more than 30) was obtained from complex cells. Under stimulation with stationary slits the center fields but not the surrounding fields of geniculate cells were found to contribute to the receptive fields of the simple striate cells, etc. It should be noted here that these two studies were based on the assumption that the simultaneously recorded spike trains were jointly weakly stationary and were performed on adult animals. Thus, the models these studies present are static in nature. The point to be stressed here then is that repeating this study at intervals during the period of plasticity under normal as well as experimentally altered conditions should lead to identifying the way in which physical models of neuronal connectivity change in response to experience rather than to identifying a static picture of functional connectivity of the studied areas.

Another area of application of the analysis of cross-correlation surfaces may be to studies of the relative contribution of excitatory and inhibitory synaptic inputs to the determination of functional properties of neurons. For example, Hubel and Wiesel (1963), Pettigrew (1974), and Blakemore and van Sluyters (1975) examined the development of connections in the visual system of newborn kittens. Hubel and Wiesel reported receptive field organization and binocular interaction characteristic of the adult cat, to some degree, in newborn animals prior to exposure to patterned light. More specifically, Hubel and Wiesel (1963) found that many cortical cells in newborn animals respond best to a stimulus having a specified axis of orientation. During the critical period, however, their properties are developed further; for example, orientation selectivity is enhanced. It is plausible then to assume that, during the period of plasticity, the potency of excitatory synapses connecting the post-synaptic neuron under study to the presynaptic neurons which are most responsive to the optimal orientation is enhanced. This has led to the hypothesis that orientation selectivity is determined primarily by the nature of the excitatory input, although modified by an inhibitory input broadly orientation tuned to the same optimal stimulus. In an effort to test this hypothesis, Sillito (1979) investigated the contribution of GABA-mediated inhibitory processes to the orientation tuning of complex cells in the cat's striate cortex. The GABA antagonist bicuculline was applied to individual complex cells and the modifications produced in their orientation tuning documented. The author concluded that the results of the study support the view that complex cell orientation selectivity is dependent on the action of lateral inhibitory interactions in the orientation domain mediated by inhibitory interneurons releasing GABA.

This indicates the importance of intracortical inhibitory circuits and that excitatory afferent organization is not sufficient to account for orientation in selectivity. The methods we are proposing here may then be used to assess the role played by changes in the efficacy or potency of excitatory and/or inhibitory synapses in neural plasticity.

ACKNOWLEDGEMENTS: We are grateful to Dr. Charles E. Smith for reviewing the manuscript and offering many useful suggestions along with bringing several of the cited references to our attention. We are also grateful to Dr. Thomas M. McKenna for providing us with Figure 1 and for bringing to our attention several references. In addition, we are indebted to Drs. Paul G. Shinkman and Michael R. Isley for reading the manuscript and offering several useful remarks.

REFERENCES:

[1] Aalen, O.O., Non-parametric inference for a family of counting processes, Ann. Statistic. 6:701-726 (1978).

[2] Blakemore, C. and Van Sluytres, R.C., Innate and evnironmental factors in the development of the kitten's visual cortex, J. Physiology, London, 248:663-716 (1975).

[3] Brillinger, D.R., Cross-spectral analysis of processes with stationary increments including the stationary G/G/queue, Ann. Probab. 2:815-827 (1974).

[4] Brillinger, D.R., The identification of point process systems, Ann. Probab. 3:909-924 (1975).

[5] Brillinger, D.R., Bryant Jr., H.L., and Segundo, J.P., Identification of synaptic interactions, Biol. Cybernetics 22:213-228 (1976).

[6] Bryant, H.L., Ruiz Marcos, A., Segundo, J.P., Correlations of neuronal spike discharges produced by monosymaptic connections and of common inputs. J. Neurophysiol. 36:205-225 (1973).

[7] Corriea, M.J. and Landolt, J.P., A point process analysis of spontaneous activity of anterior semicircular canal units in the anesthetized pigeon, Biol. Cybernetics 27:199-213 (1977).

[8] Cox, D.R. and Lewis, P.A.W., The Statistical Analysis of Series of Events, John Wiley & Sons, Inc., New York (1966).

[9] DeKwaadsteniet, J.W., Statistical analysis and stochastic modelling of neuronal spike-train activity, Math. Biosciences 60:17-71 (1982).

[10] Gerstein, G.L., Functional association of neurons: detection and interpretation, pp. 648-661 in The Neurosciences. Second Study Program. Schmitt, F.O. (ed). Rockefeller Univ. Press, New York (1970).

[11] Gill, R.D., Censoring and Stochastic Integrals. Mathematical Centre Tracts. 124: Mathematisch Centrum, Amsterdam (1983).

[12] Glaser, E.M. and Ruchkin, D.S., Principles of Neurobiological Signal Analysis, Academic Press, New York (1976).

[13] Grenander, U., Abstract Inference, John Wiley & Sons, New York (1981).

[14] Habib, M.K., Isley, M.R., Sen, P.K., and Shinkman, P.G., Non-stationary stochastic point-process models and neuronal plasticity in the visual cortex. In preparation (1984).

[15] Hoeffding, W., A class of statistics with asymptotically normal distributions, Ann. Math. Statis. 19:293-325 (1948).

[16] Hubel, D.H. and Wiesel, T.N., Receptive fields of cells in striate cortex of very young, visually inexperienced kittens, J. Neurophysiology 26:994-1002 (1963).

[17] Jacobsen, M., Statistical Analysis of Counting Processes. Lecture Notes in in Statistics. No. 12. Springer-Verlag, New York (1982).

[18] Jacod, J., Multivariate point-processes: predictable projection, Radon-Nikodym derivatives, representation of martingales, Z. Wahrsch. verw. Geb. 31:235-253 (1975).

[19] Johnson, D.H. and Swami, A., The transmission of signals by auditory-nerve fiber discharge patterns, J. Accoust. Soc. Am. 74:493-501 (1983).

[20] Kandel, E.R., Cellular Basis of Behavior: An Introduction to Behavioral Neurobiology, W.H. Freeman and Company, San Francisco (1976).

[21] Kandel, E.R. and Schwartz, J.H., Principles of Neural Science, Elsevier/North-Holland, New York (1981).

[22] Karr, A.F., Maximum likelihood estimation in the multiplicative intensity model, Ann. Statist. Forthcoming (1985).

[23] Knox, C.K., Cross correlation functions for a neuronal model, Biophys. J. 14:567-582 (1974).

[24] Knox, C.K. and Poppele, R.E., Correlation analysis of stimulus-evoked changes in excitability of spontaneously firing neurons, J. Neurophysiol. 31:616-625 (1977a).

[25] Knox, C.K. and Poppele, R.E., A determination of excitability changes in dorsal spinocerebellar tract neurons from spike-train analysis, J. Neurophysiol. 31:626-646 (1977b).

[26] Kuffler, S.W. and Nicholls, J.G., From Neuron to Brain, Sinauer Associates, Inc., Massachusetts (1976).

[27] Madden, K.P. and Remmers, J.E., Short time scale correlations between spike activity of neighboring respiratory neurons of nucleus tractus solitarius, J. Neurophysiol. 48:749-760 (1982).

[28] Marmarelis, P. and Naka, K.I., Non-linear analysis and synthesis of receptive field responses in the catfish retina. I. Horizontal cellganglion cell chains, J. Neurophysiol. 36:605-618 (1973).

[29] Marmarelis, P.Z. and Naki, K.I., Non-linear analysis and synthesis of receptive field responses in the catfish retina. II. One-input white-noise analysis, J. Neurophysiol. 36:619-633 (1973).

[30] Michalski, A., Gerstein, G.L., Czarowska, J., and Tarnecki, R., Interactions between cat striate cortex neurons, Exp. Brain. Res. 31:97-107 (1983).

[31] Moore G.P., Segundo, J.P., Perkel, D.H., and Levitan, H., Statistical signs of synaptic interaction in neurons, Biophys. J. 10:876-900 (1970).

[32] Perkel, J.H., Gerstein, G.L., and Moore, G.P., Neural spike-trains and stochastic point-processes: II. Simultaneous spike-trains, Biophys. J. 7:419-440 (1967).

[33] Pettigrew, J.D., The effect of visual experience on the development of stimulus specificity by kitten cortical neurons: effect of developing cortical neurons, J. Physiology 237:49-74 (1974).

[34] Ramlau-Hansen, H., Smoothing counting process intensities by means of kernel functions, Ann. Statist. 11:453-466 (1983).

[35] Sen, P.K., On some convergence properties of U-statistics, Calcutta Statist. Assoc. Bull. 10:1-18 (1960).

[36] Sen, P.K., Some invariance principles relating to jackknifing and their role in sequential analysis, Ann. Statist. 5:315-329 (1977).

[37] Shiryayev, A.N., Martingales: Recent developments, results and applications, Internation. Stat. Rev. 49:199-233 (1981).

[38] Sillito, A.N., Inhibitory mechanisms influencing complex cell orientation selectivity and their modification at high resting discharge levels, J. Physiology 289:33-53 (1979).

[39] Sokoloff, L., The deoxyglucose method for the measurement of local glucose utilization and the mapping of local functional activity in the central nervous system, Int. Rev. Neurobiol. 19:159-210 (1980).

[40] Tanaka, K., Cross-correlation analysis of geniculostriate neuronal relationships in cats, J. Neurophysiol. 49:1303-1318 (1983).

[41] Teyler, T.J. and Discenna, P., Long-term potentiation as a candidate mnemonic device, Brain Resear. Rev. 7:15-28 (1984).

[42] Toyama, K., Kimura, and Tanaka, K., Cross-correlation analysis of inter neuronal connectivity in cat visual cortex, J. Neurophysiol. 40:191-201 (1981).

[43] Yang, G.L. and Chen, T.C., On statistical methods in neuronal spike-train analysis, Math. Biosciences 39:1-34 (1978).

PUBLICATIONS OF B.G. GREENBERG

1. (with A. Hughes Bryan). "Methods for Studying the Influence of Socio-Economic Factors on the Growth of School Children - Body Measurements," Journal of Elisha Mitchell Scientific Society, Vol. 65, No. 2, December 1949, 311-314.

2. (with John J. Wright and Cecil G. Sheps). "A Technique for Analyzing Some Factors Affecting the Incidence of Syphilis," Journal of American Statistical Association, Vol. 45, September 1950, 373-399.

3. (with Frederick A. Thompson and Harold J. Magnuson). "The Relationship Between Immobilizing and Spirocheticidal Antibodies Against Treponema Pallidum," Journal of Bacteriology, Vol. 60, No. 4, October 1950, 473-480.

4. (with Harold J. Magnuson and Barbara J. Rosenau). "The Effects of Sex, Castration, and Testosterone Upon the Susceptibility of Rabbits to Experimental Syphilis," American Journal of Syphilis, Gonorrhea and Veneral Disease, Vol. 35, No. 2, March 1951, 146-163.

5. (with A. Hughes Bryan), "Methodology in the Study of Physical Measurements of School Children, Part I" , Human Biology, Vol. 23, No. 2, May 1951, 1-20.

6. (with A. Hughes Bryan). "Methodology in the Study of Physical Measurements of School Children, Part II", Human Biology, Vol. 24, No. 2, May 1952, 117-144.

7. "Why Randomize?" Biometrics, Vol. 7, No. 4, 1951, 309-322.

8. (with John E. Larsh and Hazel B. Gilchrist). "A Study of the Distribution and Logevity of Adult Trichinella Spiralis in Immunized and Non-Immunized Mice," Journal of Elisha Mitchell Scientific Society, Vol. 68, No. 1, June 1952, 1-11.

9. "The Use of Analysis of Covariance and Balancing in Analytical Surveys," American Journal of Public Health, Vol. 43, No. 5, June 1953, 692-699.

10. (with Mary Ellen Harris, C. Frances MacKinnon and Sidney S. Chipman). "A Method for Evaluating the Effectiveness of Health Education Literature," American Journal of Public Health, Vol. 43, No. 9, September 1953, 1147-1155.

11. (with Sidney S. Chipman). "Research Problems in the Area of Maternal and Child Health and Child Development," American Journal of Public Health, Vol. 45, No. 4, April 1955, 506-509.

12. (with B.F. Mattison). "The Whys and Wherefores of Program Evaluation," Canadian Journal of Public Health, Vol. 46, No. 7, July 1955, 293-299.

13. (with Charles M. Cameron, Jr.). "The Probable Influence of Salk Poliomyelitis Vaccine on Reported Poliomyelitis in North Carolina", North Carolina Medical Journal, Vol. 16, No. 9, September 1955, 391-395.

14. (with T.T. Mackie, J.W. Mackie, C.M. Vaughn, and N.N. Gleason). "Intestinal Parasitic Infections in Forsyth County, North Carolina, Part I. Diodoquin for the Mass Treatment of Amebiasis," Southern Medical Journal, Vol. 48, No. 7, July 1955, 737-744.

15. "Part II. Amebiasis, A Familial Disease," Annals of Internal Medicine, Vol. 43, No. 3, September 1955, 491-503.

16. "Part III, Amebiasis in School Children, An Index of Prevalence," American Journal of Tropical Medicine and Hygiene, Vol. 4, No. 6, November 1955, 980-988.

17. "Part IV. Domestic Environmental Sanitation and the Prevalence of Entamoeba Histolytica," American Journal of Tropical Medicine and Hygiene, Vol. 5, No. 1, January 1956, 29-39.

18. "Part V. Prevalences of Individual Parasites," American Journal of Tropical Medicine and Hygiene, Vol. 5, No. 1, January 1956, 40-52.

19. (with A.E. Sarhan). "Estimation of Location and Scale Parameters by Order Statistics from Singly and Doubly Censored Samples, Part I," Annals of Mathematical Statistics, Vol. 27, No. 2, June 1956, 427-451.

20. (with Osler L. Peterson, Leon P. Andrews and Robert S. Spain). "An Analytical Study of North Carolina General Practice," Journal of Medical Education, Vol. 31, No. 12, December 1956, 1-165.

21. (with A.E. Sarhan). "Tables for Best Linear Estimates by Order Statistics of the Parameters of Single Exponential Distributions from Singly and Doubly Censored Samples," Journal of American Statistical Association, Vol. 52, March 1957, 58-87.

22. (with James F. Donnelly, Charles E. Flowers, R.N. Creadick and H. Bradley Wells). "Parental, Fetal and Environmental Factors in Perinatal Mortality," American Journal of Obstetrics and Gynecology, Vol. 74, No. 5, December 1957, 1245-1254.

23. (with A.E. Sarhan). "Applications of Order Statistics in Medical Experiments," Proceedings of the Second Conference on the Design of Experiments in Army Research Development and Testing, Office of Ordnance Research, OORR 59-1, 39-49.

24. (with A.E. Sarhan). "Estimation of Location and Scale Parameters by Order Statistics from Singly and Doubly Censored Samples," Part II, Annals of Mathematical Statistics, Vo. 29, No. 1, March 1958, 79-105.

25. (with A.E. Sarhan). "Estimation Problems in the Exponential Distributions Using Order Statistics," Proceedings of the Statistical Techniques in Missile Evaluation Symposium, Office of Ordnance Research, August 1958, 123-175.

26. (with A.E. Sarhan). "Some Applications of Order Statistics," Bulletin de l'Institute International de Statistique, Stockholm, Vol. 36, No. 3, 1958, 172-183.

27. (with Charles E. Flowers, Jr., James F. Donnelly, Robert N. Creadick and H. Bradley Wells). "Spontaneous Premature Rupture of the Membranes," American Journal of Obstetrics and Gynecology, Vol. 76, No. 4, October 1958, 761-722.

28. (with Ahmed E. Sarhan). "Applications of Order Statistics to Health Data," American Journal of Public Health, Vol. 48, No. 10, October 1958, 1388-1394.

29. (with J.F. Donnelly and H. Bradley Wells). "North Carolina Fetal and Neonatal Death Study, I - Study Design and Some Preliminary Results," American Journal of Public Health, Vol. 48, No. 12, December 1958, 1583-1595.

30. (with Andrews, White, Williams, Diamond, Hamrick, and Hunter). "A Study of Patterns of Patient Referral to a Medical Center in a Rural State: Methodology," American Journal of Public Health, Vol. 49, No. 5, May 1959, 634-643.

31. "Conduct of Cooperative Field and Clinical Trials," The American Statistician, Vol. 13, No. 3, June 1959, 13-17, 28.

32. (with James E. Grizzle and R.W. Rundles). "Comparison of Chlorambucil and Myleran in Chronic Lymphocytic and Granulocytic Leukemia," American Journal of Medicine, Vol. 27, No. 3, September 1959, 424-432.

33. (with Frank R. Lock, James F. Donnelly, H. Bradley Wells, Charles Flowers, and Robert Creadick). "Perinatal Mortality in the Primigravida Over 30 Years of Age," American Journal of Obstetrics and Gynecology, Vol. 78, No. 4, October 1959, 755-768.

34. (with A.E. Sarhan). "Matrix Inversion, Its Interest and Application in Analysis of Data," Journal of the American Statistical Association, Vol. 54, December 1959, 755-766. Also published in 31st Session of Bulletin of International Statistical Institute, Brussels, Vol. 2, 1960.

35. (with A.E. Sarhan). "Estimation of Location and Scale Parameters for the Rectangular Population from Censored Samples," Journal of the Royal Statistical Society, Series B., Vol. 21, No. 2, December 1959, 356-363.

36. (with John A. Kirkland and Charles E. Flowers,Jr.), "Suppression of Lactation: A Double-Blind Hormone Study," Journal of Obstetrics and Gynecology, Vol. 15, No. 3, March 1960, 292-298.

37. (with Robert W. Winters and John B. Graham). "The Normal Range of Serum Inorganic Phosphorous and Its Utility as a Discriminant in the Diagnosis of Congenital Hypophosphatemia," Journal of Clinical Endocrinology and Metabolism, Vol. 20, No. 3, March 1960, 364-379.

38. (with Erle E. Peacock, Jr. and Bob W. Brawley). "The Effect of Snuff and Tobacco on the Production or Oral Carcinoma: An Experimental and Epidemiological Study," Annals of Surgery, Vol, 515, No. 4, April 1960, 542-550.

39. (with A.E. Sarhan). "Generalization of Some Results for Inversion of Partitioned Matrices," Contributions to Probability and Statistics: Essays In Honor of Harold Hotelling, (Editor: I. Olkin et al.), Stanford University Press, Stanford, California, 1960, 216-223.

40. "The Present Status of Polio Vaccines," Presnted before the Section on Preventive Medicine and Public Health at the 102th annual meeting of the ISMS in Chicago, May 26, 1960, Illinois Medical Journal, Vol. 118, No. 2, August 1960, 86-89.

41. (with James F. Donnelly, et al.). "Fetal, Parental, and Environmental Factors Associated with Perinatal Mortality in Mothers Under 20 Years of Age," American Journal of Obstetrics and Gynecology, Vol. 80, No. 4, October 1960, 663-671

42. (with T. Franklin Williams, et al.). "Patient Referral to a University Clinic: Patterns in a Rural State," American Journal of Public Health, Vol. 50, No. 10, October 1960, 1493-1507.

43. (with S.N. Roy and A.E. Sarhan). "Evaluation of Determinants, Characteristic Equations and Their Roots for a Class of Patterned Matrices," Journal of the Royal Statistical Society, Series B (Methodological), Vol. 22, No. 2, 1960, 348-359.

44. (with David G. Welton). "Trends in Office Practice of Dermatology," Part I, Archives of Dermatology, Vol. 83, March 1961, 355-378.

45. (with Curtis G. Hames). "A Comparative Study of Serum Cholesterol Levels in School Children and Their Possible Relation to Atherogenesis," American Journal of Public Health, Vol. 51, No. 3, March 1961, 374-385.

46. "The Problems of Definition if Field Investigations of Diseases Other Than Mental," Field Studies in the Mental Disorders, (edited by Joseph Zubin), Grune and Stratton, Inc., 1961, 126-137.

47. (with David J. Newell, T. Franklin Williams and P. Burt Veazey). "Use of Cohort Life Tables in Family Studies of Disease," Journal of Chronic Diseases, Vol. 13, No. 5, May 1961, 439-452.

48. (with T. Franklin Williams, Kerr L. White, and William L. Fleming). "The Referral Process in Medical Care and the University Clinic's Role," The Journal of Medical Education, Vol. 36, No. 8, August 1961, 899-907.

49. (with David G. Welton). "Trends in Office Practice of Dermatology: Part II," Archives of Dermatology, Vol. 84, September 1961, 419-428.

50. (with Kerr L. White and T. Franklin Williams). "The Ecology of Medical Care," New England Journal of Medicine, 265, November 2, 1961, 885-892.

51. (with Joseph M. Hitch). "Adolescent Acne and Dietary Iodine," Archives of Dermatology, Vol. 84, December 1961, 898-911.

52. "Maintaining Interest Amoung Members of Cooperative Clinical Study Groups," Cancer Chemotherapy Reports, 14, October 1961, 189-190.

53. (with A.E. Sarhan and Eleanor Roberts). "Modified Square Root Method of Matrix Inversion," Technometrics, Vol. 4, No. 2, May 1962, 282-297.

54. (with John R. McDonough, Curtis G. Hames, Louis H. Griffin, Jr. and Andrew J. Edwards, Jr.). "Observations on Serum Cholesterol Levels in the Twin Population of Evans County, Georgia," Circulation, Vol. 25, June 1962, 962-969.

55. (with A.E. Sarhan). Contributions to Order Statistics, John Wiley and Sons, New York, New York, 1962.

56. Report of the Medical Exchange Mission to the USSR, Maternal and Child Care, U.S. Department of Health, Education, and Welfare, Public Health Service Publication, No. 954, 1962, 140.

57. (with Charles M. Hughley, et al.). "Comparison of 6-Mercaptopurine and Busulfan in Chronic Granulocytic Leukemia," Blood, Vol. 21, No. 1, January 1963, 89-101.

58. (with David G. Welton). "Acne: Additional Data from a National Survey," Archives of Dermatology, Vol. 87, February 1963, 223-229.

59. (with A.E. Sarhan and Junjiro Ogawa). "Simplified Estimates for the Exponential Distribution, " Annals of Mathematical Statistics, Vol. 34, No. 2, March 1963, 102-116.

60. (with H. Bradley Wells). "Linear Discriminant Analysis in Perinatal Mortality," American Journal of Public Health, Vol. 53, No. 4, April 1963, 594-602.

61. "Field Training for Biostatisticians," The American Statistician, Vol. 18, No. 1. February 1964, 19-22.

62. (with James F. Donnelly, et al.). "Maternal, Fetal, and Environmental Factors in Prematurity," American Journal of Obstetrics and Gynecology, Vol. 88, No. 7, April 1964, 918-931.

63. (with David G. Welton). "Trends in Office Practice of Dermatology, Part III," Archives of Dermatology, Vol. 90, September 1964, 296-304.

64. (with Sarah C. Stulb, John R. McDonough, Curtis G. Hames). "The Relationship of Nutrient Intake and Exercise to Serum Cholesterol Levels in White Males in Evans County, Georgia," American Journal of Clinical Nutrition, Vol. 16, February 1965, 238-242.

65. (with Karl L. Barkley, Dan A. Martin, Robert R. Huntley, T. Franklin Williams, and Kerr L. White). "Hemoglobin and Packed Cell Volume in Negro and Caucasian Subjects: A Comparative Study," Southern Medical Journal, Vol. 58, No. 8, August 1965, 1012-1017.

66. "Biostatistics," Chapter 4 in Preventive Medicine for the Doctor in His Community, Third Edition, by Hugh R. Leavell, and E. Gurney Clark, McGraw-Hill Book Company, New York, 1965, 95-123.

67. (with Robert E. Coker, Jr., and John Kosa). "Authoritarianism and Machiavellianism Among Medical Students," The Journal of Medical Education, Vol. 40, No. 11, November 1965, Part I, 1074-1084.

68. (with Sidney S. Chipman, Abraham M. Lilienfeld, and James F. Donnelly, Editors). Research Methodology and Needs in Perinatal Studies, Charles C. Thomas, Publisher, Springfield, Illinois, 1966, 309.

69. (with James F. Donnelly and H. Bradley Wells). "A Review of Methodology in North Carolina Study of Fetal and Neonatal Deaths," Chapter 1 in Research Methodology and Needs in Perinatal Studies, Edited by S.S. Chipman, A.M. Lilienfeld, B.G. Greenberg, and J.F. Donnelly, Charles C. Thomas, Publishers, Springfield, Illinois, 1966, 3-38.

70. (with James R. Abernathy, H. Bradley Wells, and Todd M. Frazier). "Smoking as an Independent Variable in a Multiple Regression Analysis upon Birth Weight and Gestation," American Journal of Public Health, Vol. 56, No. 4, April 1966, 626-633.

71. (with Robert E. Coker, Jr., Kurt W. Back, Thomas G. Donnelly, Frances S. McConnell, Norman Miller and John Kosa). "The University of North Carolina Study of Public Health Physicians," Milbank Memorial Fund Quarterly, Vol. XLIV, No. 2, April 1966, Part I, 149-154.

72. (with Norman Miller, Robert E. Coker, Jr., Frances S. McConnell and Kurt W. Back). "A Comparison of Career Patterns of Public Health Physicians and Other Medical Specialists," Milbank Memorial Fund Quarterly, Vol. XLIV, No. 2, April 1966, Part I, 181-199.

73. (with Norman Miller, Robert E. Coker, Jr., and Frances S. McConnell). "Toward a Typology of Public Health Careers," Milbank Memorial Fund Quarterly, Vol. XLIV, No. 2, April 1966, Part I, 200-213.

74. (with John Kosa and Robert E. Coker, Jr.). "The Novice Physician in the Local Public Health Service, A Study of the Continuities in the Patterns of Recruitment," Milbank Memorial Fund Quarterly, Vol. XLIV, No. 2, April 1966, Part I, 229-258.

75. (with John Kosa, Robert E. Coker, Jr., and Thomas G. Donnelly). "The Transiency of Physicians in Public Health, Study of a Cost Factor in Institutional Work," Milbank Memorial Fund Quarterly, Vol. XLIV, No. 2, April 1966, Part I, 229-258.

76. (with John Stuart Gaul, Jr.,). "Calcaneous Fractures Involving the Subtolar Joints: A Clinical and Statistical Survey of 98 Cases," Southern Medical Journal, Vol. 59, No. 5, May 1966, 605-613.

77. (with Frances S. McConnell, John Kosa and Robert E. Coker, Jr.). "The Selection of the Field of Public Health by Workers in Local Health Departments," American Journal of Public Health, Vol. 56, No. 5, May 1966, 764-775.

78. (with James R. Abernathy and James F. Donnelly). "Application of Discriminant Functions in Perinatal Death and Survival," American Journal of Obstetrics and Gynecology, Vol. 95, No. 6, July 15, 1966, 860-867.

79. (with James R. Abernathy, James E. Grizzle, and James F. Donnelly). "Birth Weight, Gestation, and Crown-Heel Length as Response Variables in Multivariate Analysis," American Journal of Public Health, Vol. 56, No. 8, August 1966, 1281-1286.

80. "Biostatistics: An Old Tool for the Medical Practitioner," Roche Medical Image, Vol. 8, No. 6, December 1966, 4.

81. (with Abdel-Latif, A. Abul-Ela and Daniel G. Horvitz). "A Multi-Proportions Randomized Response Model," Journal of the American Statistical Association, Vol. 62, September 1967, 990-1008.

82. "Order Statistics," International Encyclopedia of the Social Sciences, (David L. Stills, Editor), Vol. 11, 182-190. The Macmillan Company and the Free Press, New York, 1967.

83. "Goal Setting and Evaluation: Some Basic Principles," Bulletin of the New York Academy of Medicine, Vol. 44, No. 2, February 1968, 131-139.

84. "Evaluation of Social Programs," Review of the International Statistical Institute, Vol. 36, No. 3, 1968, 260-278.

85. (with James E. Grizzle). "Effective Statistical Consultation for Clinical Study Groups," Cancer Chemotherapy Reports (Part I), Vol. 53, No. 1, February 1969, 1-2.

86. (with Richard M. Peters and Edward McG. Hedgpeth, Jr.). "The Effect of Alterations in Acid-Base Balance on Pulmonary Mechanics," Journal of Thoracic and Cardiovascular Surgery, Vol. 57, No. 3, March 1969, 303-311.

87. (with A.E. Sarhan). "Linear Estimates for Doubly Censored Samples from the Exponential Distribution with Observations Also Missing from the Middle," Bulletin of the International Statistical Institute, 36th Session, Sydney, Australia, Vol. 42, Book 2, 1969, 1195-1204.

88. (with Abdel-Latif A. Abul-Ela, Walt R. Simmons and Daniel G. Horvitz). "The Unrelated Question Randomized Response Model: Theoretical Framework," Journal American Statistical Association, Vol. 64, June 1969, 520-539.

89. "Problems of Statistical Inference in Health with Special Reference to the Cigarette Smoking and Lung Cancer Controversy," Journal American Statistical Association, Vol. 64, Septemeber 1969, 739-758.

90. (with James R. Abernathy and Daniel G. Horvitz). "Application of the Randomized Response Technique in Obtaining Quantitative Data," Proceedings of the Social Statistics Section, American Statistical Association, Washington, D.C., 1969, 40-43.

91. (with James R. Abernathy and Daniel G. Horvitz). "A Method for Estimating the Incidence of Abortion in an Open Population," 37th Session of the International Statistical Institute Contributed Papers, 1969, 216-218.

92. (with James R. Abernathy and Daniel G. Horvitz). "Estimates of Induced Abortion in Urban North Carolina," Demography, Vol. 7, February 1970, 19-29.

93. (with James R. Abernathy and Daniel G. Horvitz). "A New Survey Technique and Its Application in the Field of Public Health," Milbank Memorial Fund Quarterly, Vol. 48, October 1970, 39-55.

94. (with A.E. Sarhan). "The Effect of the Extremes in Estimating the Best Linear Unbiased Estimates (BLUE) of the Parameters of the Exponential Distribution," Technometrics, Vol. 13, February 1971, 113-125.

95. (with Roy R. Kuebler, Jr., James R. Abernathy, and Daniel G. Horvitz). "Application of the Randomized Response Technique in Obtaining Quantitative Data," Journal American Statistical Association, Vol. 66, June 1971, 243-250.

96. (with D.R. Brogan). "An Educational Program in Mental Health Statistics," Community Mental Health Journal, Vol. 9(1), 1973.

97. (with Ralph E. Folsom, Daniel G. Horvitz, and James R. Abernathy). "The Two Alternate Questions Randomized Response Model for Human Surveys," Journal American Statistical Association, Vol. 68, September 1973, 525-530.

98. "The Changing Scene in Public Health, Delta Omega Lecture," American Journal of Public Health, Vol. 64, No. 6, June 1974, 534-537.

99. (with D.G. Horvitz and J.R. Abernathy). "A Comparison of Randomized Response Designs," Proceedings of the Conference on Reliability and Biometry at Florida State University, Tallahassee, Florida, Reliability and Biometry: Statistical Analysis of Lifelength, Society for Industrial and Applied Mathematics, 1974, 787-815.

100. "An Overview of Some of the Statistical Problems in Studying the Delivery of Health Services," Medikon International, No. 6/7, 30-9-1974, 32-35.

101. "Commentary," II Public Health and Evaluation," The Challenge of Facts, selected public health papers of Edgar Sydenstricker, (Richard V. Kasius, editor), Prodist, New York, 1974, 21-38.

102. (with D.G. Horvitz and J.R. Abernathy). "Recent Developments in Randomized Response Designs," A Survey of Statistical Design and Linear Models, (J.N. Srivastava, editor), North Holland-Publishing Company, Amsterdam, 1975, 271-285.

103. (with Daniel G. Horvitz and James R. Abernathy). "The Randomized Response Technique," published in Perspectives on Attitude Assessment: Surveys and Their Alternatives - Proceedings of a Confernece. (H. Wallace Sinaiko and Laurie A. Broedling, editors). Smithsonian Institution, Washington, D.C., August 1975, 199-216.

104. (with Gilbert S. Gordan). "Exgeneous Estrogens and Endometrial Cancer: An Invited Review," Postgraduate Medicine, Vol. 59, June 1976, 66-77.

105. (with D.G. Horvitz and J.R. Abernathy). "Randomized Response: A Data-Gathering Device for Sensitive Questions," International Statistical Review, Vol. 44, No. 2, August 1976, 181-196.

106. (with R.R. Kuebler, J.R. Abernathy, and D.G. Horvitz). "Respondent Hazards in the Unrelated Question Randomized Response Model," Journal of Statistical Planning and Inference, Vol. 1, No. 1, 1977, 53-60.

107. (with Barbara S. Hulka and Carol J.R. Hogue). "Methodologic Issues in Epidemiologic Studies of Endometrial Cancer and Exogeneous Estrogen," American Journal of Epidemiology, Vol. 107, No. 4, April 1978, 267-276.

108. (with E.S. Nourse, David D. Mason, James E. Grizzle, and Norman L. Johnson). "Statistical Training and Research: The Univeristy of North Carolina System," International Statistical Review, Vol. 46, No. 2, 1978, 171-207.

109. "Nonparametric Statistics: Order Statistics," International Encyclopedia of Statistics, edited by William H. Kruskal and Judith M. Tanur, 1, 1978, The Free Press, New York, 644-655.

110. "Wither Public Health in North Carolina?" North Carolina Medical Journal, Vol. 40, No. 7, July 1979, 423-426.

111. (with Barbara S. Hulka, Wesley C. Fowler, Jr., David G. Kaufman, Roger Grimson, Carol J.R. Hogue, Gary S. Berger, and Charles C. Pulliam). "Estrogen and Endometrial Cancer: Cases and Two Control Groups from North Carolina," American Journal of Obstetrics and Gynecology, Vol. 137, No. 1, May 1980, 92-101.

112. "Obituary of Chester I. Bliss, 1899-1979", International Statistical Review, Vol. 48, No. 1, April 1980, 135-136.

113. (with Barbara S. Hulka, Roger C. Grimson, David G. Kaufman, Wesley C. Fowler, Jr., Carol J.R. Hogue, Gary S. Berger, and Charles C. Pulliam). "Alternative' Controls in a Case-Control Study of Endometrial Cancer and Exogeneous Estrogen", American Journal of Epidemiology, Vol. 112, No. 3, September 1980, 376-387.

114. (with Barbara S. Hulka, David G. Kaufman, Wesley C. Fowler, Jr., and Roger C. Grimson). "Predominance of Early Endometrial Cancers After Long-term Estrogen Use," Journal of American Medical Association, Vol. 244, No. 21, November 28, 1980, 2419-2422.

115. (with Frank A. Loda and Earl Siegel). "Schools of Public Health and Academic Pediatrics," The Current Status and Future of Academic Pediatrics, (Elizabeth F. Purcell, Editor), 160-177, Josiah Macy, Jr. Foundation, New York, 1981.

116. (with members of Technical Consultant Panel). "The National Health Interview Survey - Recommendations by a Technical Consultant Panel," NCHSR Research Proceedings Series, Health Survey Research Methods, Third Biennial Conference, Reston, Virginia, May 1979, U.S. Department of Health and Human Services, DHHS Publication No. (PHS) 81-3268, 1981, 18-23.

117. (with Barbara S. Hulka, Lloyd E. Chambless, David G. Kaufman, and Wesley C. Fowler, Jr.). "Protection Against Endometrial Carcinoma by Combination Product Oral Contraceptives." Journal of American Medical Association, Vol. 246, No. 4, January 22-29, 1982, 475-477.

118. "Biostatistics", Encyclopedia of Statistical Sciences, (Samuel Kotz and Norman L. Johnson, Editors, Campbell B. Read, Associate Editor), Volume 1, 251-263, John Wiley and Sons, New York, 1982.

119. (with Dennis B. Gillings). "Regional Health Information and Evaluation Systems: Some Concepts and Thoughts", Israel Journal of Medical Sciences, Vol. 18, No. 3, March 1982, 393-409.

120. (with Lloyd E. Chambless). "Statistical Methods in the Study of Toxic Shock Syndrome," Annals of Internal Medicine, Vol. 96 (Part 2), 1982, 912-917.

121. "The Future of Epidemiology," Journal of Chornic Diseases, Vol. 36, No. 4, 1983, 353-359.

122. "An Appreciation of Norman L. Johnson's Contributions to Statistics," Contributions to Statistics: Essays in Honour of Norman L. Johnson, P.K. Sen (Editor), North-Holland Publishing Company, Amsterdam-New York-Oxford, 1983, 185-193.

123. "Joseph Oscar Irwin, 1898-1982, An Obituary Appreciation," Biometrics, Vol. 39. June 1983, 527-528.

124. (with Lawrence L. Kupper, Joseph M. Janis, Ibrahim A. Salama, and Carl N. Yoshizawa). "Age-Period-Cohort Analysis: An Illustration of the Problems in Assessing Interaction in One Observation per Cell Data," Communications in Statistics, Theory and Methods, Vol. 12(23), 1983, 2779-2807.

125. "Discussion, 'Biostatistical science as a discipline: A look into the future'," by M. Zelen, Biometrics, Vol. 39, No. 4, 1983, 831-833.

126. "Interim Clinical Evaluations in Caries Clinical Trials: Discussion of Dr. Poulsen's Presentation," Journal of Dental Research, Vol. 63, Special Issue, May 1984, 739-740.